The Complement FactsBook

The Complement FactsBook

Second Edition

Edited by

Scott Barnum
Theresa Schein

ACADEMIC PRESS
An imprint of Elsevier

Academic Press is an imprint of Elsevier
125 London Wall, London EC2Y 5AS, United Kingdom
525 B Street, Suite 1800, San Diego, CA 92101-4495, United States
50 Hampshire Street, 5th Floor, Cambridge, MA 02139, United States
The Boulevard, Langford Lane, Kidlington, Oxford OX5 1GB, United Kingdom

Notices
Knowledge and best practice in this field are constantly changing. As new research and experience
broaden our understanding, changes in research methods, professional practices, or medical
treatment may become necessary.

Practitioners and researchers must always rely on their own experience and knowledge in
evaluating and using any information, methods, compounds, or experiments described herein. In
using such information or methods they should be mindful of their own safety and the safety of
others, including parties for whom they have a professional responsibility.

To the fullest extent of the law, neither the Publisher nor the authors, contributors, or editors,
assume any liability for any injury and/or damage to persons or property as a matter of products
liability, negligence or otherwise, or from any use or operation of any methods, products,
instructions, or ideas contained in the material herein.

Library of Congress Cataloging-in-Publication Data
A catalog record for this book is available from the Library of Congress

British Library Cataloguing-in-Publication Data
A catalogue record for this book is available from the British Library

ISBN: 978-0-12-810420-0

For information on all Academic Press publications visit our website at
https://www.elsevier.com/books-and-journals

Working together
to grow libraries in
developing countries

www.elsevier.com • www.bookaid.org

Publisher: Sara Tenney
Acquisitions Editor: Linda Versteeg-Buschman
Editorial Project Manager: Tracy Tufaga
Production Project Manager: Mohanambal Natarajan
Designer: Matthew Limbert

Typeset by TNQ Books and Journals

Contents

Part I
Collectins

3. **C1q**

Berhane Ghebrehiwet

4. **Mannose-Binding Lectin**

Maciej Cedzyński, David C. Kilpatrick and Anna S. Świerzko

Part II
Serine Proteases

7.   MASP-1

Ramus Pihl, Jens C. Jensenius and Steffen Thiel

8.   MASP-2

Ramus Pihl, Jens C. Jensenius and Steffen Thiel

9.   MASP-3

Ramus Pihl, Jens C. Jensenius and Steffen Thiel

10.   C1r

Nicole M. Thielens and Christine Gaboriaud

11.   C1s

Nicole M. Thielens, Christine Gaboriaud and Véronique Rossi

12.   Factor D

Steven D. Podos, Atul Agarwal and Mingjun Huang

13.   C2

Kartik Manne and Sthanam V.L. Narayana

Part III
C3 Family

16. C3

Scott R. Barnum

17. C4

David E. Isenman

Part IV
Terminal Pathway Components

20. C7

Richard G. DiScipio

21. C8

Richard G. DiScipio

22. C9

Paul Morgan

Part V
Regulatory Proteins

23. C1 Inhibitor

Christian Drouet, Denise Ponard and Arije Ghannam

24. C4b-Binding Protein

Marcin Okrój and Anna M. Blom

25. Decay-Accelerating Factor

*Joseph M. Christy, Christopher B. Toomey, David
M. Cauvi and Kenneth M. Pollard*

29. CRIg

Menno van Lookeren Campagne and Luz D. Orozco

30. Factor H and Factor H-like Protein 1

Paul N. Barlow

31. Factor H-Related Proteins 1–5

Scott R. Barnum and Paul N. Barlow

32. Clusterin

Valeria Naponelli and Saverio Bettuzzi

33. Vitronectin

Yu-Ching Su and Kristian Riesbeck

34. CD59

Paul Morgan

Part VI
Anaphylatoxin and Leucocyte Receptors

38. C5aR2

*Liam G. Coulthard, Owen A. Hawksworth and
Trent M. Woodruff*

39. C1q Receptors

Suzanne Bohlson

41. CR3

Daniel C. Bullard

42. CR4

Daniel C. Bullard

List of Contributors

Atul Agarwal Achillion Pharmaceuticals, Inc., New Haven, CT, United States

John P. Atkinson Washington University School of Medicine, St. Louis, MO, United States

Paul N. Barlow University of Edinburgh, Edinburgh, United Kingdom

Scott R. Barnum University of Alabama at Birmingham, Birmingham, AL, United States

Saverio Bettuzzi University of Parma, Parma, Italy

Anna M. Blom Lund University, Malmö, Sweden

Susan A. Boackle University of Colorado School of Medicine, Aurora, CO, United States

Suzanne Bohlson Des Moines University, Des Moines, IA, United States

Daniel C. Bullard University of Alabama at Birmingham, Birmingham, AL, United States

David M. Cauvi University of California, San Diego, La Jolla, CA, United States

Maciej Cedzyński Polish Academy of Sciences, Lodz, Poland

Joseph M. Christy The Scripps Research Institute, La Jolla, CA, United States

Liam G. Coulthard Royal Brisbane and Women's Hospital, Herston, QLD, Australia; University of Queensland, Herston, QLD, Australia

Richard G. DiScipio Torrey Pines Institute for Molecular Studies, San Diego, CA, United States; Sanford Burnham Prebys Medical Discovery Institute, La Jolla, CA, United States

Christian Drouet Université Grenoble Alpes, Grenoble, France

Viviana P. Ferreira University of Toledo, Toledo, OH, United States

Zvi Fishelson Tel Aviv University, Tel Aviv, Israel

Christine Gaboriaud Institut de Biologie Structurale (IBS), Univ. Grenoble Alpes, CEA, CNRS, Grenoble, France

Arije Ghannam Université Grenoble Alpes, Grenoble, France

Berhane Ghebrehiwet Stony Brook University School of Medicine, Stony Brook, NY, Uniter States

Ionita Ghiran Beth Israel Deaconess Medical Center, Boston, MA, United States

Owen A. Hawksworth University of Queensland, Herston, QLD, Australia; University of Queensland, St. Lucia, QLD, Australia

Mingjun Huang Achillion Pharmaceuticals, Inc., New Haven, CT, United States

David E. Isenman University of Toronto, Toronto, ON, Canada

Jens C. Jensenius Aarhus University, Aarhus, Denmark

Claudia Kemper King's College London, London, United Kingdom

David C. Kilpatrick Scottish National Blood Transfusion Service, Edinburgh, Scotland, United Kingdom

Jennifer Laskowski University of Colorado School of Medicine, Aurora, CO, United States

M. Kathryn Liszewski Washington University School of Medicine, St. Louis, MO, United States

Kartik Manne University of Alabama at Birmingham, Birmingham, AL, United States

Misao Matsushita Tokai University, Hiratsuka, Japan

Paul Morgan Cardiff University, Cardiff, United Kingdom

Valeria Naponelli University of Parma, Parma, Italy

Sthanam V.L. Narayana University of Alabama at Birmingham, Birmingham, AL, United States

Anne Nicholson-Weller Beth Israel Deaconess Medical Center, Boston, MA, United States

Katsuki Ohtani Asahikawa Medical University, Asahikawa, Japan

Marcin Okrój Medical University of Gdańsk, Gdańsk, Poland

Luz D. Orozco Genentech Inc., South San Francisco, CA, United States

Michael K. Pangburn The University of Texas Health Science Center at Tyler, Tyler, TX, United States

Ramus Pihl Aarhus University, Aarhus, Denmark

Steven D. Podos Achillion Pharmaceuticals, Inc., New Haven, CT, United States

Kenneth M. Pollard The Scripps Research Institute, La Jolla, CA, United States

Denise Ponard Université Grenoble Alpes, Grenoble, France

Kristian Riesbeck Lund University, Malmö, Sweden

Véronique Rossi Institut de Biologie Structurale (IBS), Univ. Grenoble Alpes, CEA, CNRS, Grenoble, France

Theresa N. Schein University of Alabama at Birmingham, Birmingham, AL, United States

Yu-Ching Su Lund University, Malmö, Sweden

Anna S. Świerzko Polish Academy of Sciences, Lodz, Poland

Nicole M. Thielens Institut de Biologie Structurale (IBS), Univ. Grenoble Alpes, CEA, CNRS, Grenoble, France

Steffen Thiel Aarhus University, Aarhus, Denmark

Joshua M. Thurman University of Colorado School of Medicine, Aurora, CO, United States

Christopher B. Toomey University of California, San Diego, La Jolla, CA, United States

Menno van Lookeren Campagne Genentech Inc., South San Francisco, CA, United States

Nobutaka Wakamiya Asahikawa Medical University, Asahikawa, Japan

Rick A. Wetsel The University of Texas Health Science Center at Houston, Houston, TX, United States

Trent M. Woodruff University of Queensland, Herston, QLD, Australia; University of Queensland, St. Lucia, QLD, Australia

Preface

The editors wish to thank all those who contributed entries for the second edition of *The Complement FactsBook*. We are especially appreciative of those authors who contributed two or three chapters including Drs. Paul Barlow, Dan Bullard, Richard DiScipio, Paul Morgan, Nicole Thielens, Christine Gaboriaud, Rasmus Phil, Jens Christian Jensenius, Steffen Thielens, Anna Blom, Liam Coulthard, Owen Hawksworth and Trent Woodruff. A special acknowledgement goes to Dr. Narayana Sthanam and Kartik Manne (UAB, Departments of Optometry and Biochemistry and Molecular Genetics) who generated the crystal structures used throughout the book. We also thank Drs. Doryen Bubeck and Marina Serna (Imperial College London, Department of Life Sciences) for providing the spectacular cover art images of the membrane attack complex. A special thanks goes to Donny Bliss who animated the activation of the complement pathways. This animation brings to life a complex set of pathways and makes complement more accessible to those in and out of the field.

The editors would also like to thank the Elsevier publishing team, especially Linda Versteeg-Buschman, Halima Williams and Tracy Tufuga for their support, guidance and patience throughout the preparation of this text.

Chapter 1

Introduction

Scott R. Barnum, Theresa N. Schein
University of Alabama at Birmingham, Birmingham, AL, United States

AIMS AND SCOPE OF THE BOOK

The aim of this book is to present biochemical, functional and relevant biological information about the proteins of the complement system. This edition of *The Complement FactsBook* contains updated information for all sections of most chapters in the first edition, particularly with respect to the structural domains and, where available, the crystal structure of the proteins. In addition, this edition covers information on complement mutant mice including their generation and disease phenotypes. Since the publication of the first edition of this book, many new complement proteins have been characterised and these proteins are now included in this edition.

As with the first edition, the focus of the book is on the human system, although accession numbers have been included for other species, particularly mouse and rat. In contrast to the first edition, the book is organised around functional families, namely the collectins, serine proteases, the C3 family, regulatory proteins and the anaphylatoxin and leukocyte receptors. The various protein modules that compose complement proteins are also highlighted, but their overlapping use between families makes them less useful as an organisational tool.

ORGANISATION OF THE DATA

Entries are subdivided into the following sections described below.

Other Names

Entries are identified by the recently established complement nomenclature.[1] This section also lists additional names used over the years for many of the proteins that may have been classified as members of other protein families (e.g., CR3, Mac-1; now known as CD18/CD11b) and for those characterised many decades ago when protein nomenclature was frequently based on electrophoretic mobility and other physicochemical features.

The Complement FactsBook. http://dx.doi.org/10.1016/B978-0-12-810420-0.00001-8

Physicochemical Properties

This section includes information on the number of amino acids in the mature protein and the leader peptide (if present), the pI, the molecular weight (predicted based on amino acid sequence and observed under reducing and non-reducing conditions where the data are available), the number and location of putative N-glycosylation sites, and if known whether the sites are occupied, the number and location of phosphorylation and tyrosination sites and the number and location of interchain disulfide bonds.

Structure

Details of the three-dimensional structure along with other unique structural features for the protein are described. For each protein a figure depicting the protein motifs is shown and, where available, a partial or complete crystal structure is provided. A key for the protein motifs is provided in Table 1.1 along with the complete names and abbreviations used throughout the book. Proteins such as CD59, which do not have a well-defined modular structure, are not included in the table.

Function

The mechanism of activation, the role of the protein and its biologically active fragments in the complement system are described in this section. Additional functions of the protein outside the complement system may also be described.

Degradation Pathway

Where appropriate a description and/or diagram of the degradation of the protein during the course of complement activation is provided. This is especially relevant for C3 and C4, which undergo extensive proteolytic degradation on complement activation.

Tissue Distribution

For secreted proteins, the known serum concentration is provided. In some cases, the plasma concentration is listed if there are known differences from that of the serum level. The levels in other biological fluids (cerebrospinal fluid, urine, tears, etc.) are also provided if known. The primary and secondary sites of synthesis are listed, but may not reflect the full repertoire of cells capable of synthesising a given protein. For membrane-bound proteins, the cell types clearly shown to express the protein are listed.

TABLE 1.1 Definitions of Protein Domain Diagrams Used Throughout This Book

◇	Complement control protein domain
⬤	Serine protease domain
▭	CUB domain
⊙	Ca²⁺-binding epidermal growth factor-like repeat
○	Epidermal growth factor-like repeat
⬤	Von Willebrand factor type A
⊙	Low density lipoprotein receptor class A repeat
■	Scavenger receptor cysteine-rich domain
⬭	Factor I / membrane attack complex C6/7 module
▮	Transmembrane domain
▯	Cytoplasmic domain
▬	C4BP oligomerization domain
■	Serine, threonine, proline-rich mucin-like domain
⬤	Glycosylphosphatidylinositol anchor
◨	Thrombospondin type 1 repeat
▭	Membrane attack complex/perforin domain
⬭	Immunoglobulin superfamily domain
○	α-macroglobulin domain
◢	C345C domain
○	Thioester domain
▲	Anaphylatoxin domain
▯	Linker domain (LNK)
▭	Carbohydrate-rich domain
⬭	Serpin domain
◇	WD domain (Propellor subunits)
▦	I domain
▦	I-like domain
▢	Thigh domain (C2-set Ig fold)
▢	Calf domain
◉	Integrin EGF domain
◁	Plexin semaphorin integrin domain
▭	Hybrid domain
⬭	β tail domain
⬤	Hemopexin domain
○	Somatomedin B-like domain
—	Collagen-like domain
⬤	CHO binding domain
⬤	C1q globular domain
⬤	Cysteine-rich region
∿∿∿	G protein-coupled 7-transmembrane receptor

Regulation of Expression

This section details the mechanisms that alter baseline synthesis of the proteins, whether or not the protein is an acute phase reactant and, disease-specific changes in expression.

Human Protein Sequence

Sequence is shown in single-letter amino acid code. Numbering starts with the initiator methionine residue. The leader sequence is underlined, as are cleavage sites between chains. Putative and known N-linked glycosylation sites are denoted by **N**. Additional specialised notations, such as unique posttranslation modifications, are described on a protein-by-protein basis.

Protein Modules

For the protein modules given in Table 1.1, the leader sequence, amino acid boundaries and exons are listed. Special structures such as the thioester bonds in C3 and C4 and the catalytic triad residues in the serine proteases are also listed.

Chromosomal Location

The chromosomal location in human, mouse and other species is provided. Closely linked genes are indicated.

cDNA Sequence

The cDNA sequence is shown. Where known, the sequence starts with the 5' end of the message, otherwise the most 5' sequence is given. Stop codons and poly-adenylation signals where known are shown. All possible exons are included and exon–intron boundaries are denoted by underlining the first five nucleotides in each exon. Alternative splicing of exons is noted. No intron sequences are included. Discrepancies between published sequences are noted.

Genomic Structure

For each protein the structure of the gene is provided drawn to scale. The gene is represented as a single horizontal line and vertical lines indicate the positions of exons. In some cases unique genetic elements and their position within the gene are denoted.

Accession Numbers

Accession numbers from various databases are provided for human, mouse and other available species (where space permitted).

Deficiency

The mode of inheritance of deficiency in humans, as well as the phenotypic outcome of the deficiency, is described. Clinical syndromes associated with deficiencies are listed. The molecular basis of the deficiency is denoted as follows:

C123 to G, S480 to L; two chromosomes/patients/families, where C is the normal nucleotide; 123 is the position in the presented nucleotide cDNA sequence; G is the mutant nucleotide; S is the normal amino acid; 480 is the position of the presented amino acid sequence; L is the mutant, non- or aberrantly functional amino acid; and two chromosomes/patients/families represents the number of times this mutation has been described.

Polymorphic Variants

Polymorphic variants at the protein or single nucleotide level are listed. In some cases the functional or structural consequences of the variation are described.

Mutant Animals

This is a new feature of *The Complement FactsBook*. Where mutant or transgenic mice have been generated, the following information is included: a description of the targeting construct, the targeting approach, genotype of injected eggs and crosses into different genetic backgrounds (if performed), phenotype in routine husbandry, phenotype in infectious disease and autoimmune models (if performed) and commercial availability (JAX labs and others).

REFERENCE

1. Kemper C, Pangburn MK, Fishelson Z. Complement nomenclature 2014. *Mol Immunol* 2014;**61**:56–8.

Chapter 2

The Complement System

Scott R. Barnum, Theresa N. Schein
University of Alabama at Birmingham, Birmingham, AL, United States

HISTORICAL PERSPECTIVE

Complement has been known by many names over the decades including alexin, cytase, addiment and opsonin of normal serum and its activity was divided into midpiece and endpiece.[1] Since the studies from the late 1880s and early 1900s, we now appreciate that complement comprises several activation pathways and mediates a plethora of biological activities including host defence, synaptic pruning and modulation of metabolism (Fig. 2.1).[2] The history of complement is complex. It is intertwined with the earliest studies of innate and adaptive immunity, infectious disease and the first vaccines. Space limitations preclude a detailed presentation of complement history in this text, so a series of bullet-point highlights are shown below. For a more detailed history the reader is referred to an excellent review and the references therein[1] and earlier perspectives by Pillemer[3] and Lachmann.[4,5]

1790–1900	Recognition of as yet uncharacterised bactericidal activity in serum
1890s	Demonstrations of the lytic activity of complement; recognition that complement is a heat-sensitive system
1900s	Development of haemolytic assays to monitor complement activity in vitro
Early 1990s	Opsonic nature of complement first described
1907	Initial biochemical characterisation of complement into 'midpiece' and 'endpiece' (C′1 and C′2 – essentially C1 and the remaining complement proteins)
1910–40	Biochemical fractionation and characterisation of some the classical pathway components; identification of various complement inactivators (zymosan, cobra venom factor, ammonia sensitivity)
1950	Initial characterisation of properdin and the alternative pathway
1950–70	Characterisation of the components and order of reaction of the classical pathway proteins with the advent of improved chromatographic and electrophoretic techniques; purification of alternative pathway proteins; complement deficiencies and disease phenotypes described
1970–80	Identification of many regulatory components and their complement-derived ligands; addition of protein and cDNA sequencing and monoclonal antibodies for biochemical and functional studies of the pathways; characterisation of the thioester bond in C3 and C4

Continued

The Complement FactsBook. http://dx.doi.org/10.1016/B978-0-12-810420-0.00002-X

1990 Characterisation of complement genes and protein domain-based identification of families of complement proteins; structural characterisation of many complement proteins via NMR and X-ray crystallography

1990–present Characterisation of the lectin binding pathway and additional complement proteins; interrelationship between complement and the coagulation system established; therapeutic targeting of complement for multiple disease indications

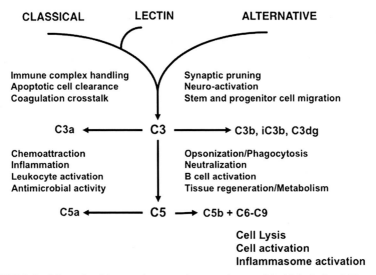

FIGURE 2.1 Schematic of the complement pathways and some of the biological activities mediated by the complement.

MODULAR STRUCTURE OF THE COMPONENTS

Sequencing of the complement proteins started in the early 1980s, and it quickly became apparent the proteins were comprised of multiple structural domains.[6] In the first edition of this book, complement proteins were classified into functional groups based on these motifs. Since then several additional complement proteins have been identified and characterised, adding more motifs and complexity. The significant sharing of protein motifs across the functional families makes structural motif classification less valuable. Therefore, we have divided the proteins into five groups based on their *overall functions* within the complement system to allow for inclusion of all proteins rather than the older classification style based on structural motifs.

The Collectins

For the purposes of this edition, the collectins group will be limited to C1q, mannose-binding protein (MBL) and the ficolins/collectins. Although there are other structurally related family members, such as the surfactant proteins A and

D, conglutinin and the C1q domain family members,[7,8] their contribution to complement activity is minimal. The members of this family are readily identified by their dominant structural features: collagen stalks and C-terminal globular domains (Fig. 2.2). The globular domains interact with the Fc region of IgM and IgG antibodies in the case of C1q and carbohydrate moieties in the case of MBL and ficolins/collectins. All these proteins are composed of multiple polypeptide chains that are disulfide-linked in the collagen stalk region (the N-terminal portion), while the rest of the polypeptide chains form the globular domains.[9–13]

Serine Proteases

All the enzymes of the complement system are serine proteases of the trypsin subfamily and chymotrypsin family. Each enzyme has a serine protease domain with the classic catalytic triad of histidine, aspartic acid and serine (Fig. 2.3). Unlike most other serine proteases, the complement enzymes have a more restricted substrate specificity and generally slower enzyme kinetics.[14–16] The remaining domains in these proteins are important for interaction with the collectins, each other (C1r/C1s and MASP1/2) or other components in the convertases (C2, Factors B and I).

C3 Family

The C3 family of proteins, C3, C4 and C5, is thought to have evolved from a common ancestral gene. The members share a modest homology within a given species (~25%), but share higher homology for the same protein between different species.[17] All three proteins are synthesised as a single-chain molecule, but undergo posttranslational modification including

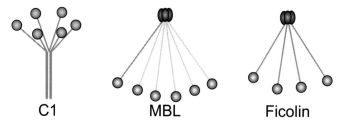

FIGURE 2.2   Modular structure of the collectins. See Table 1.1 for the key to the domains.

MASP-1/2/3
C1r/s
Factor D
Factor B
C2
Factor I

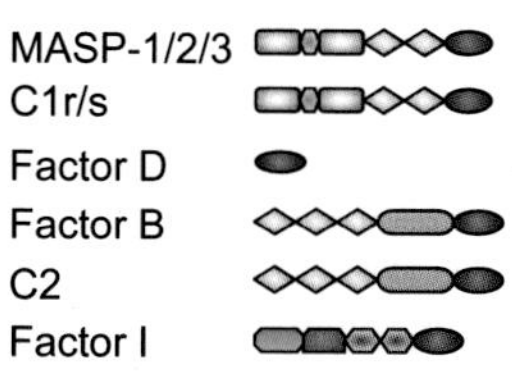

FIGURE 2.3   Modular structure of the serine proteases. See Table 1.1 for the key to the domains.

glycosylation, sulphation (C4), and cleavage to two (C3 and C5) or three (C4) polypeptide chains at conserved linker regions[18–20] (Fig. 2.4). The most important feature of these molecules is the presence of an internal thioester bond in C3 and C4.[21,22] This chemical moiety is formed between cysteine and glutamic acid residues found in the α-chain of both proteins (Fig. 2.5). When C3 and C4 are cleaved on activation of the complement, the thioester bond becomes reactive allowing C3b and C4b to covalently attach to proteins or carbohydrates on the surface of pathogens, or host tissues, in the immediate vicinity.

Terminal Pathway

The terminal pathway is composed of the proteins C6 through C9, all of which share significant modular similarity (Fig. 2.6). The most important domain is the so-called membrane attack complex/perforin-like domain (MAC/PF) required for the formation of the MAC. These proteins are members of the MACPF/cholesterol-dependent cytolysins family of pore-forming toxins that undergo conformational changes and insert into lipid bilayers.[23,24] These proteins are structurally similar to perforin, a molecule released by cytotoxic T cells to lyse target cells. The MAC forms by the association of C5b, C6, C7, C8 and multiple C9 molecules to form a pore of varying size based on the number of C9 molecules that insert into the complex. C9 can also polymerise under certain conditions in the absence of the C6–C8 complex.[25]

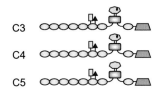

FIGURE 2.4 Modular structure of the C3 family of proteins. See Table 1.1 for the key to the domains.

Intact Thioester   Activated Thioester   R Ester or Amide Bond

$$\overset{O}{\underset{\parallel}{C}} - S \longrightarrow \overset{O}{\underset{\parallel}{\overset{\oplus}{C}}} \quad \overset{\ominus}{S} \longrightarrow O = C \quad SH$$

FIGURE 2.5 Schematic of the intact, metastable and transacylated thioester bond.

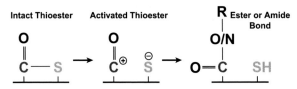

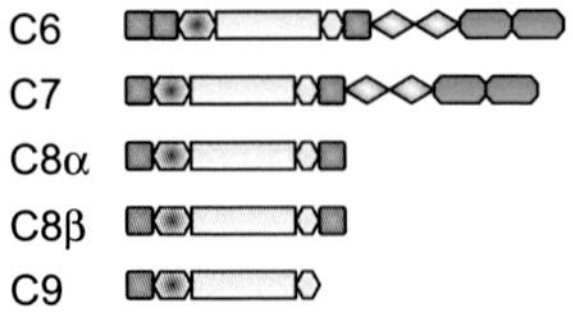

FIGURE 2.6 Modular structure of the terminal pathway proteins. See Table 1.1 for the key to the domains.

Regulatory Proteins

Proteins that regulate the activation and functions of complement comprise a diverse group, but many are members of the so-called regulators of complement activation or RCA.[26] The RCA proteins are composed exclusively or almost exclusively of the complement control protein domain (CCP), also known as a short consensus repeat, or a sushi domain (Fig. 2.7A). This domain is approximately 60 amino acids in size with four conserved cysteine residues in which the first and third and second and fourth residues are cross-linked. The number of CCPs ranges from two in C2 and Factor B to as many as 37 in one of the CR1 alleles. A new addition to the RCA family since that last edition is the Factor H-related family of proteins.[27] A structural exception to the RCA proteins is CRIg, the only complement regulatory protein composed of immunoglobulin domains.[28,29] All the RCA proteins bind proteolytic fragments of C3 and C4 and either serve as cofactors in their degradation or inhibit their biological activities.[30] The majority of these regulators are membrane-bound but some are found in plasma.

A small subset of the regulators (Fig. 2.7B) serve to inhibit the formation of the MAC either on cell surfaces or in the fluid phase and share no structural homology (CD59, clusterin, vitronectin).[31] C1-inhibitor (C1-INH) blocks the activity of the classical and lectin pathways by binding to the serine proteases that initiate activation of these two pathways.[32] The last member of this group, properdin, serves to stabilise the C3 convertase of the alternative pathway, thus perpetuating C3b and

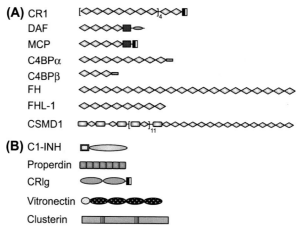

FIGURE 2.7   Modular structure of the complement regulatory proteins. See Table 1.1 for the key to the domains. (A) The RCA family members are composed almost exclusively of complement control protein domains with the exception of CSMD1, which has a large number of CUB (complement C1r/C1s, Ugf, Bmp1) domains. These proteins exert their regulatory activity primarily on C3 and C4. (B) The remaining regulatory proteins, comprised of a mix of protein domains, control complement activity at multiple levels in the pathways.

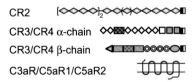

CR2

CR3/CR4 α-chain

CR3/CR4 β-chain

C3aR/C5aR1/C5aR2

FIGURE 2.8 Modular structure of the leucocyte receptors. See Table 1.1 for the key to the domains.

C3 convertase generation through this pathway.[33] Properdin is important in the so-called amplification loop of the alternative complement pathway.

The Leucocyte Receptors

The proteins in this group are expressed predominantly on leucocytes including T and B cells, dendritic cells and myeloid cells (Fig. 2.8). The exception is the expression of the anaphylatoxin receptors, which have been found on nearly every cell type examined with the exception of RBCs.[34,35] As a group these receptors mediate many of the host defence functions of the complement that includes activation and chemoattraction of lymphocytes and myeloid cells, phagocytosis and induction of inflammation and the acute phase response.[35–37] CR2 is traditionally grouped with the RCA proteins based on its chromosomal location in the RCA locus and its CCP domain composition. However, CR2 does not inhibit events early in complement activation or at the level of the convertases, but rather modulates complement activities related to activation/enhancement of humoral immune responses.[38,39]

PATHWAYS

The complement system is composed of four main pathways, but now it has become apparent that the complement and coagulation pathways are inter-twined.[40] This interrelationship, seen for decades in isolated reports, has matured into what is now termed the extrinsic protease pathway.[40–42] In this pathway, several activated coagulation proteins, including factors IXa, Xa, XIa; plasmin; and thrombin can directly cleave C3 and C5, bypassing the canonical activation pathways. Furthermore, there is evidence that complement can activate the coagulation system and potentially focus activation of the coagulation system to sites of complement deposition.[43–46] In addition to these fluid-phase and membrane-bound pathways, there is evidence of intracellular complement activation that serves at least as a feedback loop, if not an entirely new pathway. In several studies it has been demonstrated that locally produced C3a and C5a bind to their respective receptors on T cells providing tonic signals to T cells. Intracellular production of C3a, derived from cathepsin L-mediated cleavage of C3 occurs during T cell activation and contributes to T cell survival and effector responses.[47–51] Undoubtedly there is additional complement biology to be uncovered in all these areas.

Classical Pathway

The classical pathway is activated when C1q, the pattern-recognition portion of the C1 complex binds to one of several types of activators including antigen–antibody complexes (Ag–Ab), C-reactive protein, apoptotic cells and other structures (Fig. 2.9). Antibodies must be bound to their antigenic targets to activate this pathway; it is not activated by simple interaction with fluid-phase antibodies in blood or other body fluids. The classical pathway is activated by Ag–Ab complexes containing IgM and IgG (IgG3>IgG1>IgG2>IgG4) by binding to the Fc region of the antibodies. One pentameric IgM complex with antigen will activate the pathway, while multiple IgG molecules (ideally IgG hexamers) are required.[52,53] Activation is Ca^{2+} dependent and inhibited by divalent cation chelators. When C1q engages an activator, C1r is autoactivated, which leads to activation of the other C1r in the complex, followed by C1r-mediated activation of the two C1s molecules.[16] C1-inhibitor (C1-INH) modulates the activation of the C1 complex prior to its binding to an activator and can bind to an activated C1 complex to inactivate and disassociate the complex.

Activated C1s cleaves C4 to C4a and C4b. C4a was initially thought to be an anaphylatoxin similar to C3a and C5a; however, this is no longer believed to

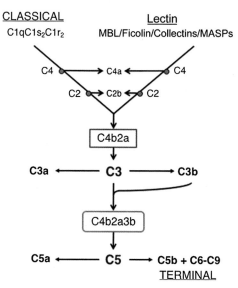

FIGURE 2.9 Classical and lectin pathway activation. Shown is a schematic representation of the activation steps for the classical and lectin pathways. Both pathways use unique pattern recognition receptors (C1 for the classical pathway and MBL or ficolin/collectin for the lectin pathway). On engagement with their ligands, they undergo a conformationally induced autoactivation step leading to successive cleavage of C4 and C2 and the formation of the C3 convertase (C4b2a, *red box*) responsible for cleaving C3 to C3b and C3a. Some of the C3b generated will bind to the C3 convertase resulting in the formation of the C5 convertase (C4b2a3b, *orange box*) responsible for cleaving C5 to C5b and C5a.

be the case.[35,54] C4a does appear to have antimicrobial activity.[55,56] C4b has a highly reactive thioester bond that will form an ester linkage with –OH groups on carbohydrates or an amide linkage with –NH$_2$ groups on proteins[57] at the site of activation. The thioester bond in the majority of the nascent C4b molecules will be hydrolysed and these C4b proteins will remain in the fluid phase. Covalently attached C4b has a binding site for C2, which when complexed with C4b is cleaved by C1s in the active C1 complex to C2a and C2b. The smaller C2b fragment is released, while C2a remains bound to C4b forming the classical pathway C3 convertase, C4b2a. The C2a portion of the convertase has a serine protease domain and cleaves C3 near the amino terminus of the alpha chain into C3a and C3b. C3a is an anaphylatoxin and mediates both pro- and antiinflammatory functions and, like C4a, has antimicrobial activity.[35,58] C3b, similar to C4b, has a reactive thioester bond allowing it to covalently attach to the activating surface and thus serve as an opsonin, targeting the activator for removal by phagocytic cells. C3b that associates with the C3 convertase generates C4b2a3b, the classical pathway C5 convertase. Similar to C4b, the thioester bond in the majority of the C3b molecules will be hydrolysed; however, C3b surface deposition appears to be more efficient than that of C4b.[59]

Both fluid-phase and membrane-bound complement regulatory molecules tightly control activation of the classical pathway. C1-inhibitor is key for modulating the initial activation steps. Mutations or deficiency of this protein result in dysregulation of the pathway and is associated with the well-described clinical syndrome hereditary angioedema.[60] C4b and C3b are degraded to their inactive forms by Factor I (iC4b and iC3b, respectively) when bound to cofactors including CR1, DAF, MCP, C4BP (C4b in the fluid phase) and CSMD1. Factor H also serves as a cofactor for C3b degradation, but not C4b.

Lectin Pathway

The lectin pathway is very similar to the classical pathway (Fig. 2.9). Activation of this pathway starts with macromolecular protein complexes structurally and functionally similar to the C1 complex of the classical pathway. MBL, ficolin and collectin resemble C1q with collagen stalks and carbohydrate-binding domains. These proteins pair with homodimers of MBL-associated serine proteases, MASP-1 and MASP-2, analogous to C1r and C1s, in a Ca^{2+}-dependent complex. The majority of MBL in plasma is associated with either MASP-1 or MASP-2.[61] The ficolins associate with the MASP proteins to activate the lectin pathway.[62] The heterodimer of the collectins CL-L1 and CL-K1 (termed CL-LK) have been found in complex with MASP-1/3 and MASP-2 and this complex activates the lectin pathway.[13] Once MBL, ficolin or collectin bind to an activator, a conformational change leads to autoactivation of MASP-1, which then activates MASP-2. MASP-1 activation of MASP-2 can occur in *cis*, if both proteins are in the same complex, or in *trans* if they are located on adjacent complexes.[13,63] Once activated by carbohydrates such as N-acetyl-D-glucosamine and D-mannose found on bacteria, fungi and other pathogens,[64] MASP-2 cleaves both C4 and C2 analogous to C1s in the classical pathway. MASP-1 can also cleave C2, enhancing activation of the pathway. Once both C4 and C2 have been cleaved, a C3 convertase identical to that of the classical pathway (C4b2a) is formed and subsequent activation steps in this pathway are identical to that of the classical pathway. Regulation of the lectin pathway overlaps with that of the classical pathway; however, MASP-3, Map44 and Map19, which are homologous to MASP-1 and 2, can bind to MBL, ficolins and collectins and inhibit activation of the pathway.[62,65,66]

Alternative Pathway

This pathway is unique among the complement pathways in that it is always active at a very low level and, as such, is primed to respond rapidly to infection. It is activated by pathogen-associated molecular pattern molecules, such as lipopolysaccharide in the cell wall of Gram-negative bacteria, bacterial cell wall teichoic acid, carbohydrates in the cell wall of fungi and other structures. The low-level constitutive activation of the alternative pathway is known as 'tickover' (Fig. 2.10). In this process, native C3 undergoes a reversible conformational change exposing the thioester-containing domain (TED). The thioester is hydrolysed and the resulting $C3(H_2O)$ molecule undergoes a significant conformational change exposing a binding site for Factor B.[67] In this complex, Factor B is cleaved by Factor D to its active form and the resulting complex, $C3(H_2O)Bb$, serves as an unstable fluid-phase C3 convertase (sometimes known as the 'initiation' convertase) that cleaves C3 to C3a and C3b. The thioester bond in the newly generated C3b molecule will be hydrolysed, and the protein is degraded in the absence of a pathogen or activating surface. In the presence of a pathogen, C3b will covalently attach to the surface and bind Factor B, which will be cleaved by Factor D to Bb and Ba forming the alternative pathway C3 convertase (C3bBb). The alternative pathway C3 convertase is stabilised by the binding of properdin, which extends the half-life of the convertase 5- to 10-fold.[68] The C3 convertase cleaves C3 to C3a and C3b, and each newly generated C3b molecule can start the formation of a new C3 convertase. Through this mechanism, activation of the alternative pathway can be rapidly amplified (the so-called amplification loop; Fig. 2.9) generating thousands of C3b molecules to opsonise and neutralise pathogens. As with the classical pathway, some of the C3b generated will bind to the C3 convertase forming the alternative pathway C5 convertase, C3bBb3b. Interestingly, C3b generated by activation of the classical pathway can serve as the nidus for an alternative pathway convertase, triggering cross-pathway activation. In fact, alternative pathway activation may account for most of the complement activity (up to 80%) seen when the classical pathway was the first pathway activated.[69]

The alternative pathway has to be tightly regulated because of its potential to rapidly amplify and target host tissues for C3b deposition and subsequent

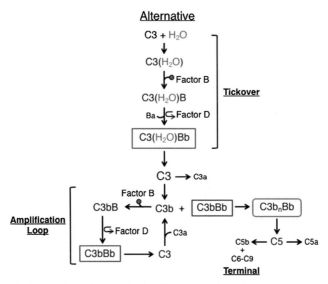

FIGURE 2.10 Alternative pathway activation. Shown is a schematic representation of the activation steps for the alternative pathway. This pathway is continuously active due to the hydrolysis of the thioester bond in C3. In this so-called tickover pathway, the addition of water to C3 generates $C3(H_2O)$, which has a binding site for Factor B. Factor B binds to $C3(H_2O)$ and is then cleaved to Bb, its active form, by Factor D. This generates the C3 initiation convertase $C3(H_2O)Bb$ (*blue box*), which cleaves C3 to C3b and C3a. C3b generated by this convertase binds Factor B, which is again cleaved by Factor D leading to the formation of alternative pathway C3 convertase C3Bb (*red box*) that cleaves C3 to C3b and C3a. C3b generated by this convertase can participate in the formation of additional C3 convertases very rapidly generating large amounts of C3b (the amplification loop). Some of the C3b generated will bind to the C3 convertase forming the alternative pathway C5 convertase ($C3b_nBb$, *orange box*).

complement-mediated destruction. Factor H is a critical regulator of this pathway serving as a cofactor for the Factor I-mediated cleavage of $C3(H_2O)$ and C3b. The membrane-bound regulatory proteins, DAF, MCP, CSMD1, CRIg and CR1, serve as cofactors in the Factor I-mediated cleavage of C3b.[70–73]

Terminal Pathway

The terminal or lytic pathway starts when C5 is cleaved to C5a and C5b (Figs. 2.9 and 2.10). Similar to C3 family members, C5b undergoes a conformational change that allows it to associate with C6 and C7 initiating the formation of an irreversible complex (C5b-7) that associates with the membrane of a pathogen or host cell.[74,75] C8 then associates with the complex and undergoes a conformational change allowing the β-hairpins of the MACPF domains to extend long enough to insert through the bacterial membrane. At this point, multiple C9 molecules associate with the C5b-8 complex and begin pore formation. Theoretically, only one C9 is required to begin pore formation; however, a MAC complex may have as many as 22 C9 molecules generating a structure over 300 Å 'tall' and 240 Å in diameter.[25] Although initially thought to form a complete circle, there is evidence to indicate that the fully formed MAC is an asymmetric pore with incomplete closure with respect to the C9 subunits.[25] Beyond its ability to lyse cells, the MAC has been shown to induce cell cycling and proliferation, protein synthesis, apoptosis and other effects.[76] Recently, the MAC has been demonstrated to activate the NLRP3 inflammasome complex through a Ca^{2+}-dependent mechanism.[77]

The terminal pathway is regulated primarily at the levels of the MAC complex. In the case of insertion into eukaryotic cells, the MAC can be endocytosed or ectocytosed.[78,79] The membrane-bound inhibitor CD59 blocks formation of the MAC by interacting with the C5b-8 complex inhibiting the addition of C9 molecules and also binds to forming MAC complexes to limit the number of C9 molecules in the complex. In the fluid phase the forming MAC complex is inhibited by clusterin and vitronectin, which bind to the complex, limiting the addition of C9 molecules, and prevent interaction with the host cell.

REFERENCES

1. Sim RB, Schwaeble W, Fujita T. Complement research in the 18th–21st centuries: progress comes with new technology. *Immunobiology* 2016. http://dx.doi.org/10.1016/j.imbio.2016.06.011.
2. Hess C, Kemper C. Complement-mediated regulation of metabolism and basic cellular process. *Immunity* 2016;**45**:240–54.
3. Pillemer L. Recent advances in the chemistry of complement. *Chem Rev* 1943;**33**:1–26.
4. Lachmann P. Complement before molecular biology. *Mol Immunol* 2006;**43**:496–508.
5. Nesargikar PN, Spiller B, Chavez R. The complement system: history, pathways, cascade and inhibitors. *Eur J Microbiol Immunol* 2012;**2**:103–11.
6. Reid KB, Day AJ. Structure-function relationships of the complement components. *Immunol Today* 1989;**10**:177–80.
7. Haagsman HP, Hogenkamp A, van Eijk M, Veldhuizen EJ. Surfactant collectins and innate immunity. *Neonatology* 2008;**93**:288–94.
8. Ghai R, Waters P, Roumenina LT, et al. C1q and its growing family. *Immunobiology* 2007;**212**:253–66.
9. Reid KB, Day AJ. Ig-binding domains of C1q. *Immunol Today* 1990;**11**:387–8.
10. Carland TM, Gerwick L. The C1q domain containing proteins: where do they come from and what do they do? *Dev Comp Immunol* 2010;**34**:785–90.
11. Endo Y, Matsushita M, Fujita T. The role of ficolins in the lectin pathway of innate immunity. *Int J Biochem Cell Biol* 2011;**43**:705–12.
12. Kjaer TR, Thiel S, Andersen GR. Toward a structure-based comprehension of the lectin pathway of complement. *Mol Immunol* 2013;**56**:222–31.
13. Hansen SW, Ohtani K, Roy N, Wakamiya N. The collectins CL-L1, CL-K1 and CL-P1, and their roles in complement and innate immunity. *Immunobiology* 2016;**221**:1058–67.
14. Kam CM, McRae BJ, Harper JW, Niemann MA, Volanakis JE, Powers JC. Human complement proteins D, C2, and B. Active site mapping with peptide thioester substrates. *J Biol Chem* 1987;**262**:3444–51.
15. Kam CM, Oglesby TJ, Pangburn MK, Volanakis JE, Powers JC. Substituted isocoumarins as inhibitors of complement serine proteases. *J Immunol* 1992;**149**:163–8.

16. Gal P, Dobo J, Zavodszky P, Sim RB. Early complement proteases: C1r, C1s and MASPs. A structural insight into activation and functions. *Mol Immunol* 2009;**46**:2745–52.

17. Campbell RD, Law SK, Reid KB, Sim RB. Structure, organization, and regulation of the complement genes. *Annu Rev Immunol* 1988;**6**:161–95.

18. Belt KT, Carroll MC, Porter RR. The structural basis of the multiple forms of human complement component C4. *Cell* 1984;**36**:907–14.

19. Wetsel RA, Lemons RS, Le Beau MM, Barnum SR, Noack D, Tack BF. Molecular analysis of human complement component C5: localization of the structural gene to chromosome 9. *Biochemistry* 1988;**27**:1474–82.

20. Barnum SR, Fey G, Tack BF. Biosynthesis and genetics of C3. *Curr Top Microbiol Immunol* 1990;**153**:23–43.

21. Tack BF. The beta-Cys-gamma-Glu thiolester bond in human C3, C4, and alpha 2-macroglobulin. *Springer Semin Immunopathol* 1983;**6**:259–82.

22. Law SK, Dodds AW. The internal thioester and the covalent binding properties of the complement proteins C3 and C4. *Protein Sci* 1997;**6**:263–74.

23. Rosado CJ, Kondos S, Bull TE, et al. The MACPF/CDC family of pore-forming toxins. *Cell Microbiol* 2008;**10**:1765–1774.

24. Dunstone MA, Tweten RK. Packing a punch: the mechanism of pore formation by cholesterol dependent cytolysins and membrane attack complex/perforin-like proteins. *Curr Opin Struct Biol* 2012;**22**:342–9.

25. Serna M, Giles JL, Morgan BP, Bubeck D. Structural basis of complement membrane attack complex formation. *Nat Commun* 2016;**7**:10587.

26. Hourcade D, Holers VM, Atkinson JP. The regulators of complement activation (RCA) gene cluster. *Adv Immunol* 1989;**45**:381–416.

27. Skerka C, Chen Q, Fremeaux-Bacchi V, Roumenina LT. Complement factor H related proteins (CFHRs). *Mol Immunol* 2013;**56**:170–80.

28. Langnaese K, Colleaux L, Kloos DU, Fontes M, Wieacker P. Cloning of Z39Ig, a novel gene with immunoglobulin-like domains located on human chromosome X. *Biochim Biophys Acta* 2000;**1492**:522–5.

29. Helmy KY, Katschke Jr KJ, Gorgani NN, et al. CRIg: a macrophage complement receptor required for phagocytosis of circulating pathogens. *Cell* 2006;**124**:915–27.

30. Alcorlo M, Lopez-Perrote A, Delgado S, et al. Structural insights on complement activation. *FEBS J* 2015;**282**:3883–91.

31. Zipfel PF, Skerka C. Complement regulators and inhibitory proteins. *Nat Rev Immunol* 2009;**9**:729–40.

32. Cicardi M, Zingale L, Zanichelli A, Pappalardo E, Cicardi B. C1 inhibitor: molecular and clinical aspects. *Springer Semin Immunopathol* 2005;**27**:286–98.

33. Lesher AM, Nilsson B, Song WC. Properdin in complement activation and tissue injury. *Mol Immunol* 2013;**56**:191–8.

34. Klos A, Tenner AJ, Johswich KO, Ager RR, Reis ES, Kohl J. The role of the anaphylatoxins in health and disease. *Mol Immunol* 2009. http://dx.doi.org/10.1016/j.molimm.2009.04.027. pii:S0161-5890(09)00196-5.

35. Klos A, Wende E, Wareham KJ, Monk PN. International union of basic and clinical pharmacology. [corrected]. LXXXVII. Complement peptide C5a, C4a, and C3a receptors. *Pharmacol Rev* 2013;**65**:500–43.

36. Staunton DE, Lupher ML, Liddington R, Gallatin WM. Targeting integrin structure and function in disease. *Adv Immunol* 2006;**91**:111–57.

37. Rosetti F, Mayadas TN. The many faces of Mac-1 in autoimmune disease. *Immunol Rev* 2016;**269**:175–93.

38. Carroll MC. Complement and humoral immunity. *Vaccine* 2008;**26**(Suppl. 8):I28–33.
39. Carroll MC, Isenman DE. Regulation of humoral immunity by complement. *Immunity* 2012;**37**:199–207.
40. Amara U, Flierl MA, Rittirsch D, et al. Molecular intercommunication between the complement and coagulation systems. *J Immunol* 2010;**185**:5628–36.
41. Markiewski MM, Nilsson B, Ekdahl KN, Mollnes TE, Lambris JD. Complement and coagulation: strangers or partners in crime? *Trends Immunol* 2007;**28**:184–92.
42. Huber-Lang M, Sarma JV, Zetoune FS, et al. Generation of C5a in the absence of C3: a new complement activation pathway. *Nat Med* 2006;**12**:682–7.
43. Reinartz J, Hansch GM, Kramer MD. Complement component C7 is a plasminogen-binding protein. *J Immunol* 1995;**154**:844–50.
44. Krarup A, Wallis R, Presanis JS, Gal P, Sim RB. Simultaneous activation of complement and coagulation by MBL-associated serine protease 2. *PLoS One* 2007;**2**:e623.
45. Gulla KC, Gupta K, Krarup A, et al. Activation of mannan-binding lectin-associated serine proteases leads to generation of a fibrin clot. *Immunology* 2010;**129**:482–95.
46. Oikonomopoulou K, Ricklin D, Ward PA, Lambris JD. Interactions between coagulation and complement–their role in inflammation. *Semin Immunopathol* 2012;**34**:151–65.
47. Lalli PN, Strainic MG, Yang M, Lin F, Medof ME, Heeger PS. Locally produced C5a binds to T cell-expressed C5aR to enhance effector T-cell expansion by limiting antigen-induced apoptosis. *Blood* 2008;**112**:1759–66.
48. Strainic MG, Liu J, Huang D, et al. Locally produced complement fragments C5a and C3a provide both costimulatory and survival signals to naive CD4+ T cells. *Immunity* 2008;**28**:425–35.
49. Cravedi P, Leventhal J, Lakhani P, Ward SC, Donovan MJ, Heeger PS. Immune cell-derived C3a and C5a costimulate human T cell alloimmunity. *Am J Transplant* 2013;**13**:2530–9.
50. Liszewski MK, Kolev M, Le Friec G, et al. Intracellular complement activation sustains T cell homeostasis and mediates effector differentiation. *Immunity* 2013;**39**:1143–57.
51. Le Friec G, Kohl J, Kemper C. A complement a day keeps the Fox(p3) away. *Nat Immunol* 2013;**14**:110–2.
52. Wang G, de Jong RN, van den Bremer ET, et al. Molecular basis of assembly and activation of complement component C1 in complex with immunoglobulin G1 and antigen. *Mol Cell* 2016;**63**:135–45.
53. Cook EM, Lindorfer MA, van der Horst H, et al. Antibodies that efficiently form hexamers upon antigen binding can induce complement-dependent cytotoxicity under complement limiting conditions. *J Immunol* 2016;**197**:1762–75.
54. Barnum SR. C4a: an anaphylatoxin in name only. *J Innate Immun* 2015;**7**:333–9.
55. Malmsten M, Davoudi M, Walse B, et al. Antimicrobial peptides derived from growth factors. *Growth Factors* 2007;**25**:60–70.
56. Nordahl EA, Rydengard V, Nyberg P, et al. Activation of the complement system generates antibacterial peptides. *Proc Natl Acad Sci USA* 2004;**101**:16879–84.
57. Nilsson B, Ekdahl KN. The tick-over theory revisited: is C3 a contact-activated protein? *Immunobiology* 2012;**217**:1106–10.
58. Pasupuleti M, Walse B, Nordahl EA, Morgelin M, Malmsten M, Schmidtchen A. Preservation of antimicrobial properties of complement peptide C3a, from invertebrates to humans. *J Biol Chem* 2007;**282**:2520–8.
59. Ollert MW, Kadlec JV, David K, Petrella EC, Bredehorst R, Vogel CW. Antibody-mediated complement activation on nucleated cells. A quantitative analysis of the individual reaction steps. *J Immunol* 1994;**153**:2213–21.
60. Davis 3rd AE, Mejia P, Lu F. Biological activities of C1 inhibitor. *Mol Immunol* 2008;**45**:4057–63.

61. Heja D, Kocsis A, Dobo J, et al. Revised mechanism of complement lectin-pathway activation revealing the role of serine protease MASP-1 as the exclusive activator of MASP-2. *Proc Natl Acad Sci USA* 2012;**109**:10498–503.

62. Endo Y, Matsushita M, Fujita T. New insights into the role of ficolins in the lectin pathway of innate immunity. *Int Rev Cell Mol Biol* 2015;**316**:49–110.

63. Degn SE, Kjaer TR, Kidmose RT, et al. Complement activation by ligand-driven juxtaposition of discrete pattern recognition complexes. *Proc Natl Acad Sci USA* 2014;**111**:13445–50.

64. Bajic G, Degn SE, Thiel S, Andersen GR. Complement activation, regulation, and molecular basis for complement-related diseases. *EMBO J* 2015;**34**:2735–57.

65. Degn SE, Hansen AG, Steffensen R, Jacobsen C, Jensenius JC, Thiel S. MAp44, a human protein associated with pattern recognition molecules of the complement system and regulating the lectin pathway of complement activation. *J Immunol* 2009;**183**:7371–8.

66. Gaboriaud C, Gupta RK, Martin L, et al. The serine protease domain of MASP-3: enzymatic properties and crystal structure in complex with ecotin. *PLoS One* 2013;**8**:e67962.

67. Janssen BJ, Christodoulidou A, McCarthy A, Lambris JD, Gros P. Structure of C3b reveals conformational changes that underlie complement activity. *Nature* 2006;**444**:213–6.

68. Fearon DT, Austen KF. Properdin: binding to C3b and stabilization of the C3b-dependent C3 convertase. *J Exp Med* 1975;**142**:856–63.

69. Harboe M, Ulvund G, Vien L, Fung M, Mollnes TE. The quantitative role of alternative pathway amplification in classical pathway induced terminal complement activation. *Clin Exp Immunol* 2004;**138**:439–46.

70. Kim DD, Song WC. Membrane complement regulatory proteins. *Clin Immunol* 2006;**118**:127–36.

71. Schmidt CQ, Herbert AP, Hocking HG, Uhrin D, Barlow PN. Translational mini-review series on complement factor H: structural and functional correlations for factor H. *Clin Exp Immunol* 2008;**151**:14–24.

72. Nilsson SC, Sim RB, Lea SM, Fremeaux-Bacchi V, Blom AM. Complement factor I in health and disease. *Mol Immunol* 2011;**48**:1611–20.

73. Liszewski MK, Atkinson JP. Complement regulator CD46: genetic variants and disease associations. *Hum Genom* 2015;**9**:7.

74. Aleshin AE, Discipio RG, Stec B, Liddington RC. Crystal structure of c5b-6 suggests structural basis for priming assembly of the membrane attack complex. *J Biol Chem* 2012;**287**:19642–52.

75. Aleshin AE, Schraufstatter IU, Stec B, Bankston LA, Liddington RC, DiScipio RG. Structure of complement C6 suggests a mechanism for initiation and unidirectional, sequential assembly of membrane attack complex (MAC). *J Biol Chem* 2012;**287**:10210–22.

76. Morgan BP. The membrane attack complex as an inflammatory trigger. *Immunobiology* 2015. http://dx.doi.org/10.1016/j.imbio.2015.04.006.

77. Triantafilou K, Hughes TR, Triantafilou M, Morgan BP. The complement membrane attack complex triggers intracellular Ca^{2+} fluxes leading to NLRP3 inflammasome activation. *J Cell Sci* 2013;**126**:2903–13.

78. Morgan BP, Dankert JR, Esser AF. Recovery of human neutrophils from complement attack: removal of the membrane attack complex by endocytosis and exocytosis. *J Immunol* 1987;**138**:246–53.

79. Scolding NJ, Morgan BP, Houston WA, Linington C, Campbell AK, Compston DA. Vesicular removal by oligodendrocytes of membrane attack complexes formed by activated complement. *Nature* 1989;**339**:620–2.

Part I

Collectins

Chapter 3

C1q

Berhane Ghebrehiwet

Stony Brook University School of Medicine, Stony Brook, NY, Uniter States

PHYSICOCHEMICAL PROPERTIES

C1q is composed of three polypeptide chains, A, B and C, which are synthesised as precursors with leader sequences of 22, 27 and 28 amino acids in length, respectively.

pI	9.3
M_r (K)	459.3

	A-Chain	B-Chain	C-Chain
M_r (K)	27.5	25.2	23.8
N-linked glycosylation site	1(146)	–	–
Interchain disulphide bonds	26	29	32

Approximately 8% (w/w) of C1q is carbohydrate in the form of glucosyl-galactosyl disaccharide units linked to hydroxylysine residues in the collagen region (cC1q) and six-asparagine-linked sugar chains located in the globular head region (gC1q).[1–4] The two types of sugars account for 69% and 31% respectively of the total carbohydrate.[1–3] C1q also contains 4.42% hydroxyproline, 1.81% hydroxylysine and 18.7% glycine (guinea pig).[2,4]

STRUCTURE

Human C1q is made up of 18 polypeptide chains (6A, 6B and 6C each of approximately A = 28 kDa, B = 26 kDa and C = 24 kDa, respectively). Each chain contains an N-terminal collagen region of ~81 amino acid residues and a C-terminal region of ~136 residues.[1–3] The three polypeptide chains A, B and C are arranged to form six triple helical strands with three peptide chains – A, B and C – forming a single strand. In each strand, the A-chain is disulphide linked to the B-chain, whereas the C-chain is disulphide linked to the C-chain of a neighbouring strand to form a doublet and three doublets are associated by strong noncovalent forces to give intact C1q its characteristic 'bouquet-of-flowers-like' hexameric structure (Fig. 3.1). The triple helix in each strand is

The Complement FactsBook. http://dx.doi.org/10.1016/B978-0-12-810420-0.00003-1

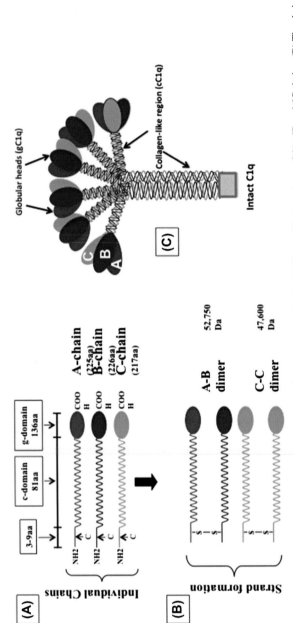

FIGURE 3.1 Diagram of protein domains for C1q. Structural organisation of the C1q molecule. (A) Intact C1q is made up of 6A, 6B and 6C chains. (B) The chains are arranged to form six individual strands and each strand is made up of a disulphide bonded A-B dimer, noncovalently associated with a C-chain. (C) Two strands are disulphide-bonded through adjacent C-chains to form a doublet, and three doublets are noncovalently linked to form the intact hexameric C1q.

formed between the N-terminal collagen-like sequences of the three chains while the globular '*head*' (gC1q) is formed from the C-terminal portion of each of these chains. The crystal structure of the globular heads of C1q[5] shows striking similarity to the TNF-α family of proteins indicating that both arose by divergence from the same ancestral gene (Fig. 3.2).[6]

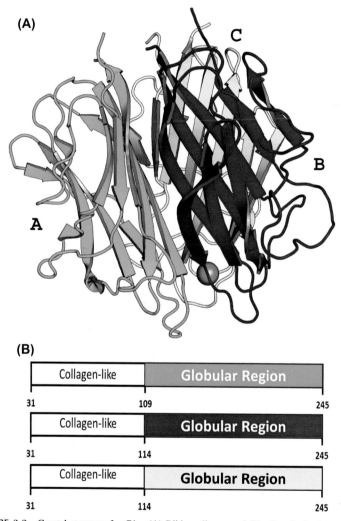

FIGURE 3.2   Crystal structure for C1q. (A) Ribbon diagram of C1q 'head', the heterotrimeric globular domain crystal structure (PDB ID: 5KC5). Modules A, B and C are shown in *green, red and yellow coloured ribbons*, respectively, and the Ca²⁺ ion is represented as an *orange sphere*. (B) Schematic representation of one of the six chains of C1q, and each chain is made of three collagen-like helices that associate to form an N-terminal 'stalk' and a C-terminal globular domain that associates to form the 'head'.

FUNCTION

Human C1q, which is a member of the collectin/ficolin family of proteins, is the recognition unit of the classical pathway of complement and circulates in plasma with the Ca^{2+}-dependent $C1r_2C1s_2$ tetramer to form the pentameric C1. The role of C1q within this complex is to sense and bind to immune complexes, as well as other activators of the classical pathway, such as pathogen-associated or danger-associated danger molecular patterns, via its versatile globular heads.[5,7] This in turn leads to an intramolecular rearrangement within the C1 molecule thereby facilitating autoactivation of the proenzyme C1r into an active enzyme. Activated C1r then converts its only substrate – the C1s proenzyme – into an active C1s enzyme and the latter then sequentially cleaves its two natural substrates – first C4 and then C2 – thereby triggering activation of downstream events that eventually lead to the formation of the lytic molecule – the membrane attack complex (C5b-9_n). In addition to being the recognition unit of the classical pathway of complement, C1q has a plethora of other functions. These include:

- Recognises immune complexes and other antigens' initiation of the classical pathway of complement activation
- Enhances phagocytosis by binding to pathogen associated molecular patterns or danger associated molecular patterns including danger signals on apoptotic cells
- Binds to cells through two membrane receptors: cC1qR/CR (calreticulin) and gC1qR/p33[8,9]
- Serves to promote angiogenesis, enhance chemotaxis and cell adhesion and modulate adhesion of trophoblasts to decidual endothelial cells[10]
- Some of C1q's functions reflects its genetic ancestry and therefore mirrors those of TNF-α
- C1q induces the production of cytokines such as IL-6, IL-8, IL-10 and MCP-1[11]
- Induces expression of adhesion molecules on endothelial cells: P-selectin, E-selectin, ICAM-1 and VCAM-1[12]
- C1q is also involved in a wide range of pathological processes including: autoimmune diseases like systemic lupus erythematosus (SLE), recurrent infections, Alzheimer's disease, cancer and preeclampsia

DEGRADATION PATHWAY

C1q does not undergo degradation during activation. However, bacterial enzymes such as pseudomonas protease are capable of degrading C1q.

TISSUE DISTRIBUTION

The C1q molecule is primarily synthesised in cells of myeloid origin including monocytes, macrophages, dendritic cells, but also by other cells such as Kupffer cells, glial cells, mast cells, trophoblasts, endothelial cells, epithelial cells and fibroblasts. Interestingly a number of cell types and in particular activated or proliferating cells, such as cancer cells, appear to express C1q on their surface. In addition, C1q is also expressed on and around cancer cells, atherosclerotic plaques and in aging brain tissues,[13] where it is assumed to contribute to the inflammatory processes associated with these conditions.

REGULATION OF EXPRESSION

The expression of C1q is modulated by cytokines such as IFN-γ or IL-6, and powerful steroid antiinflammatory agents such as dexamethasone and prednisone can also upregulate C1q synthesis and expression.[14] However, LPS appears to be an inhibitor of C1q expression.

HUMAN PROTEIN SEQUENCE

The protein sequence of all the genes encoding the C1q chains, A–, B– and C–chains, have been described in detail.[1–4]

A chain

```
MEGPRGWLVL CVLAISLASM VTEDLCRAPD GKKGEAGRPG RRGRPGLKGE  50
QGEPGAPGIR TGIQGLKGDQ GEPGPSGNPG KVGYPGPSGP LGARGIPGIK 100
GTKGSPGNIK DQPRPAFSAI RRNPPMGGNV VIFDTVITNQ EEPYQNHSGR 150
FVCTVPGYYY FTFQVLSQWE ICLSIVSSSR GQVRRSLGFC DTTNKGLFQV 200
VSGGMVLQLQ QGDQVWVEKD PKKGHIYQGS EADSVFSGFL IFPSA
```

B chain

```
MMMKIPWGSI PVLMLLLLLG LIDISQAQLS CTGPPAIPGI PGIPGTPGPD  50
GQPGTPGIKG EKGLPGLAGD HGEFGEKGDP GIPGNPGKVG PKGPMGPKGG 100
PGAPGAPGPK GESGDYKATQ KIAFSATRTI NVPLRRDQTI RFDHVITNMN 150
NNYEPRSGKF TCKVPGLYYF TYHASSRGNL CVNLMRGRER AQKVVTFCDY 200
AYNTFQVTTG GMVLKLEQGE NVFLQATDKN SLLGMEGANS IFSGFLLFPD 250
MEA
```

C chain

```
MDVGPSSLPH LGLKLLLLLL LLPLRGQANT GCYGIPGMPG LPGAPGKDGY  50
DGLPGPKGEP GIPAIPGIRG PKGQKGEPGL PGHPGKNGPM GPPGMPGVPG 100
PMGIPGEPGE EGRYKQKFQS VFTVTRQTHQ PPAPNSLIRF NAVLTNPQGD 150
YDTSTGKFTC KVPGLYYFVY HASHTANLCV LLYRSGVKVV TFCGHTSKTN 200
QVNSGGVLLR LQVGEEVWLA VNDYYDMVGI QGSDSVFSGF LLFPD
```

The leader sequences are underlined and the putative *N*-linked glycosylation site in the A-chain is denoted by **N**.

PROTEIN MODULES

A-Chain

1–22	Leader peptide	exon 1
31–109	Collagen-like region	exon 1/2
110–245	Globular region	exon 2

B-Chain

1–25	Leader peptide	exon 1
31–114	Collagen-like region	exon 1/2
115–251	Globular region	exon 2

C-Chain

1–28	Leader peptide	exon 1
31–114	Collagen-like region	exon 1/2
15–245	Globular region	exon 2

Human C1q has weak affinity for monomeric IgG in plasma or in solution and does not lead to complement activation. Its affinity increases when two of its globular heads are bound to two adjacent Fc2 domains of IgG as in immune complexes or when IgG is aggregated. Because the flexibility of C1q globular heads is limited, the clustering of IgG molecules is requisite to facilitate the binding of C1q via at least two of its gC1q modules to either two adjacent IgG molecules or one IgM molecule. Not all subclasses of IgG activate complement with the hierarchy being: IgG3 > IgG1 > IgG2 > IgG4. The C1q binding site on IgG is Cγ2 domain, whereas the binding site on IgM is Cμ3 domain. The three charged residues Glu-318, Lys-320 and Lys-322 in the Cγ2 domain have been implicated in the binding of C1q to IgG,[15] whereas Asp-417, Glu-418 and His-420 in the Cμ3 region of IgM are considered to be predominantly involved in binding to the C1q globular head regions. There may also be a preferential binding of the gC1q modules when it comes to IgG binding. The gC1qA (or ghA) module binds aggregated IgG and IgM to the same extent, while gC1qB (ghB) binds aggregated IgG in preference to IgM.[15]

CHROMOSOMAL LOCATION

The genes for C1q polypeptide chains are located on the short arm of chromosome at 1p34-1p36.3.

HUMAN cDNA SEQUENCE

A chain

```
GGCAGAGGCA TCATGGAGGG TCCCCGGGGA TGGCTGGTGC TCTGTGTGCT GGCCATATCG    60
CTGGCCTCTA TGGTGACCGA GGACTTGTGC CGAGCACCAG ACGGGAAGAA GGGGAGGCA    120
GGAAGACCTG GCAGACGGGG GCGGCCAGGC CTCAAGGGGG AGCAAGGGGA GCCGGGGGCC   180
CCTGGCATCC GGACAGGCAT CCAAGGCCTT AAAGGAGACC AGGGGGAACC TGGGCCCTCT   240
GGAAACCCCG GCAAGGTGGG CTACCCAGGG CCCAGCGGCC CCCTCGGGGC CCGTGGCATC   300
CCGGGAATTA AAGGCACCAA GGGCAGCCCA GGAAACATCA AGGACCAGCC GAGGCCAGCC   360
TTCTCCGCCA TTCGGCGGAA CCCCCCAATG GGGGGCAACG TGGTCATCTT CGACACGGTC   420
ATCACCAACC AGGAAGAACC GTACCAGAAC CACTCCGGCC GATTCGTCTG CACTGTACCC   480
GGCTACTACT ACTTCACCTT CCAGGTGCTG TCCCAGTGGG AAATCTGCCT GTCCATCGTC   540
TCCTCCTCAA GGGGCCAGGT CCGACGCTCC CTGGGCTTCT GTGACACCAC CAACAAGGGG   600
CTCTTCCAGG TGGTGTCAGG GGGCATGGTG CTTCAGCTGC AGCAGGGTGA CCAGGTCTGG   660
GTTGAAAAAG ACCCCAAAAA GGGTCACATT TACCAGGGCT CTGAGGCCGA CAGCGTCTTC   720
AGCGGCTTCC TCATCTTCCC ATCTGCCTGA GCCAGGGAAG GACCCCCTCC CCCACCCACC   780
TCTCTGGCTT CCATGCTCCG CCTGTAAAAT GGGGGCGCTA TTGCTTCAGC TGCTGAAGGG   840
AGGGGGCTGG CTCTGAGAGC CCCAGGACTG GCTGCCCCGT GACACATGCT CTAAGAAGCT   900
CGTTTCTTAG ACCTCTTCCT GGAATAAA
```

B chain

```
CTCACAGGACACCAGCTTCCCAGGAGGCGTCTGACACAGTATGATGATGAAGATCCCATG        60
GGGCAGCATCCCAGTACTGATGTTGCTCCTGCTCCTGGGCCTAATCGATATCTCCCAGGC       120
CCAGCTCAGCTGCACCGGGCCCCCAGCCATCCCTGGCATCCCGGGTATCCCTGGGACACC       180
TGGCCCCGATGGCCAACCTGGGACCCCAGGGATAAAAGGAGAGAAAGGGCTTCCAGGGCT       240
GGCTGGAGACCATGGTGAGTTCGGAGAGAAGGGAGACCCAGGGATTCCTGGGAATCCAGG       300
AAAAGTCGGCCCCAAGGGCCCCATGGGCCCTAAAGGTGGCCCAGGGGCCCCTGGAGCCCC       360
AGGCCCCAAAGGTGAATCGGGAGACTACAAGGCCACCCAGAAAATCGCCTTCTCTGCCAC       420
AAGAACCATCAACGTCCCCCTGCGCCGGGACCAGACCATCCGCTTCGACCACGTGATCAC       480
CAACATGAACAACAATTATGAGCCCCGCAGTGGCAAGTTCACCTGCAAGGTGCCCGGTCT       540
CTACTACTTCACCTACCACGCCAGCTCTCGAGGGAACCTGTGCGTGAACCTCATGCGTGG       600
CCGGGAGCGTGCACAGAAGGTGGTCACCTTCTGTGACTATGCCTACAACACCTTCCAGGT       660
CACCACCGGTGGCATGGTCCTCAAGCTGGAGCAGGGGGAGAACGTCTTCCTGCAGGCCAC       720
CGACAAGAACTCACTACTGGGCATGGAGGGTGCCAACAGCATCTTTTCCGGGTTCCTGCT       780
CTTTCCAGATATGGAGGCCTGACCTGTGGGCTGCTTCACATCCCACCCCGGCTCCCCCTGC       840
CAGCAACGCTCACTCTACCCCCAACACCACCCCTTGCCCAACCAATGCACACAGTAGGGC       900
TTGGTGAATGCTGCTGAGTGAATGAGTAAATAAACTCTTCAAGGCCAAGGGACA
```

C chain

```
GCCAGAAACCGCCCACCTGCAGGTGAGGCCCGGACCCCTGCCCAGTTCCTTCTCCGGGAT    60
GGACGTGGGGCCCAGCTCCCTGCCCCACCTTGGGCTGAAGCTGCTGCTGCTCCTGCTGCT   120
GCTGCCCCTCAGGGGCCAAGCCAACACAGGCTGCTACGGGATCCCAGGGATGCCCGGCCT   180
GCCCGGGGCACCAGGGAAGGATGGGTACGACGGACTGCCGGGGCCCAAGGGGGAGCCAGG   240
AATCCCAGCCATTCCCGGGATCCGAGGACCCAAAGGGCAGAAGGGAGAACCCGGCTTACC   300
CGGCCATCCTGGGAAAAATGGCCCCATGGGACCCCCTGGGATGCCAGGGGTGCCCGGCCC   360
CATGGGCATCCCTGGAGAGCCAGGTGAGGAGGGCAGATACAAGCAGAAATTCCAGTCAGT   420
GTTCACGGTCACTCGGCAGACCCACCAGCCCCCTGCACCCAACAGCCTGATCAGATTCAA   480
CGCGGTCCTCACCAACCCGCAGGGAGATTATGACACGAGCACTGGCAAGTTCACCTGCAA   540
AGTCCCCGGCCTCTACTACTTTGTCTACCACGCGTCGCATACAGCCAACCTGTGCGTGCT   600
GCTGTACCGCAGCGGCGTCAAAGTGGTCACCTTCTGTGGCCACACGTCCAAAACCAATCA   660
GGTCAACTCGGGCGGTGTGCTGCTGAGGTTGCAGGTGGGCGAGGAGGTGTGGCTGGCTGT   720
CAATGACTACTACGACATGGTGGGCATCCAGGGCTCTGACAGCGTCTTCTCCGGCTTCCT   780
GCTCTTCCCCGACTAGGGCGGGCAGATGCGCTCGAGCCCCACGGGCCTTCCACCTCCCTC   840
AGCTTCCTGCATGGACCCACCTTACTGGCCAGTCTGCATCCTTGCCTAGACCATTCTCCC   900
CACCAGATGGACTTCTCCTCCAGGGAGCCCACCCTGACCCACCCCCACTGCACCCCCTCC   960
CCATGGGTTCTCTCCTTCCTCTGAACTTCTTTAGGAGTCACTGCTTGTGTGGTTCCTGGG  1020
ACACTTAACCAATGCCTTCTGGTACTGCCATTCTTTTTTTTTTTTTTTTCAAGTATTGGA  1080
AGGGGTGGGGAGATATATAAATAAATCATGAAATCAATACATA
```

The first five nucleotides in each exon are underlined to indicate the intron/exon boundaries. The methionine initiation codon (<u>ATG</u>), the termination codon (<u>TGA</u>) and the polyadenylation signal (<u>ATTAAA</u>) are double-underlined.

GENOMIC STRUCTURE

The chains of C1q are aligned 5'⇒3', in the same orientation, in the order A-C-B on a 24-kb stretch of DNA on chromosome 1[3]. The A and C are separated by 4 kb and B and C are separated by 11 kb (Fig. 3.3).

ACCESSION NUMBERS

Human: P02745-C1QA, P02746-C1QB, P02747-C1QC

DEFICIENCY

There are two forms of C1q deficiency: a complete absence or the presence of a dysfunctional molecule. C1q deficiency is a rare autosomal recessive disorder

FIGURE 3.3 Diagram of genomic structure.

and is associated with autoimmune disorders such as SLE, lupus nephritis, skin lesions, proliferative nephritis, rheumatoid arthritis (RA), glomerulonephritis and chronic and recurrent infections.[16] The most common deficiency in C1q is a homozygous mutation in the A-chain of the molecule in which the C-to-T transition in codon 186 of exon 2 results in a Gln-to-stop (Q186X) substitution. This mutation has been shown in patients with C1q deficiency in Slovakian and Turkish families.[16,17]

POLYMORPHIC VARIANTS

There is a possible implication of the G allele in rs172378 as a risk factor for lupus nephritis in a homozygous status, at least for the Bulgarian population.[18] Single nucleotide polymorphisms in and around the C1q genes, *C1qA*, *C1qB* and *C1qC*, in a Dutch set of 845 RA cases and 1046 controls show a risk for the development of RA.[19]

MUTANT ANIMALS

C1q-deficient mice have been developed,[20] by targeting the A-chain of C1q (C1qa$^{-/-}$). The C1qa$^{-/-}$ mice spontaneously develop high titers of autoantibodies with glomerulonephritis and apoptotic cells. These observations appear to be consistent with the hypothesis that C1q deficiency leads to the development of SLE in part through impairment of apoptotic cell clearance. The mouse C1q genes are clustered on chromosome 4 and show conservation of gene organisation.[21]

REFERENCES

1. Reid KBM. Molecular cloning and characterization of the complementary DNA and gene coding for the B chain of the subcomponent C1q of the human complement system. *Biochem J* 1985;**231**:729–35.
2. Reid KBM. Chemistry and molecular genetics of C1q. *Behring Inst Mitt* 1989;**84**:8–19.
3. Sellar GC, Blake DJ, Reid KBM. Characterization and organization of the genes encoding the A-, B- and C-chains of human complement subcomponent C1q. The complete derived amino acid sequence of human C1q. *Biochem J* 1991;**274**:481–90.
4. Yonemasu K, Stroud RM, Niedermeier W, Butler W. Chemical studies on C1q: a modulator of immunoglobulinbiology. *Biochem Biophys Res Commun* 1971;**43**:1388–94.
5. Gaboriaud C, Juanhuix J, Gruez A, Lacroix M, Darnault C, Pignol D, Verger D, Fontecilla-Camps JC, Arlaud GJ. The crystal structure of the globular head of complement protein C1q provides a basis for its versatile recognition properties. *J Biol Chem* 2003;**278**:46974–82.
6. Shapiro L, Scherer PE. The crystal structure of a complement-1q family suggests an evolutionary link to tumor necrosis factor. *Curr Biol* 1998;**12**:335–8.
7. Kishore U, Reid KBM. Modular organization of proteins containing C1q-like globular domain. *Immunopharmacology* 1999;**42**:15–21.
8. Ghebrehiwet B, Hosszu KK, Valentino A, Peerschke EIB. The C1q family of proteins: insights into the emerging non-traditional functions. *Front Immunol* 2012;**3**:1–9.

9. Peerschke EI, Ghebrehiwet B. cC1qR/CR and gC1qR/p33: observations in cancer. *Mol Immunol* 2014;**61**:100–9.

10. Agostinis A, Bulla R, Tripodo C, Gismondi A, Stabile H, Bossi F, Guamotta C, Garlanda C, De Stata F, Spessotto P, Santonno A, Ghebrehiwet B, Girardi G, Tedesco F. An alternative role of C1q in cell migration and tissue remodeling: contribution to trophoblasts invasion and placental development. *J Immunol* 2010;**185**:4420–9.

11. van den Berg RH, Faber-Krol MC, Sim RB, Daha MR. The first subcomponent of complement C1q triggers the production of IL-6, IL-8, and monocyte chemoattractant protein-1 by human umbilical vein endothelial cells. *J Immunol Lett* 1998;**161**:6924–30.

12. Lozada C, Levin RI, Huie M, Hirschhorn R, Naime D, Whitlow M, Recht PA, Golden B, Cronstein B. Identification of C1q as the heat-labile serum cofactor required for immune complexes to stimulate endothelial expression of the adhesion molecules E-selectin and intracellular and vascular cell adhesion molecules 1. *Proc Natl Acad Sci USA* 1995;**92**:8378–82.

13. Stephan AH, Madison DV, Meteos JM, Fraser DA, Lovelette EA, Coutellier L, Kim L, Tsai H-H, Huang EJ, Rowitch DH, Berns DS, Tenner AJ, Shamloo M, Barres BA. A dramatic increase of C1q protein in the CNS during normal aging. *J Neurosci* 2013;**33**:13460–74.

14. Walker D. Expression and regulation of complement C1q by human THP-1-derived macrophages. *Mol Chem Neuropathol* 1998;**34**:197–218.

15. Duncan AR, Winter G. The binding site for C1q on IgG. *Nature* 1988;**332**:738–40.

16. Petry F, Izzet Berkel A, Loos M. Multiple identification of a particular type of hereditary C1q deficiency in the Turkish population: review of the cases and additional genetic and functional analysis. *Hum Genet* 1997;**100**:51–6.

17. Topaloglu R, Bakkaloglu A, Slingsby JH, Mihatsch MJ, Pascual M, Norsworthy P, Morley BJ, Saatci U, Schifferli JA, Walport MJ. Molecular basis of hereditary C1q deficiency associated with SLE and IgA nephropathy in a Turkish family. *Kidney Int* 1996;**50**:635–42.

18. Radanova M, Vasilev V, Dimitrov T, Deliyska B, Ikonomov V, Ivanova D. Association of rs172378 C1q gene cluster polymorphism with lupus nephritis in Bulgarian patients. *Lupus* 2015;**24**:280–9.

19. Trouw LA, Daha N, Kurreeman FA, Bohringer S, Goulielmos GN, Westra HJ, Zhrnakova A, Franlke L, Stahl EA, Stoeken-Rijsbergen G, Verdujin W, Roos A, Li Y, Houwing-Dusitermaat JJ, Huizinga TW, Toes RE. Genetic variants in the region of C1q genes are associated with rheumatoid arthritis. *Clin Exp Immunol* 2013;**173**:76–83.

20. Botto M, Dell Agnola C, Bygrave AE, Thompson EM, Cook HT, Petry F, Loos M, Pandolfi PP, Walport MJ. Homozygous C1q deficiency causes glomerulonephritis associated with multiple apoptotic bodies. *Nat Genet* 1998;**19**:56–9.

21. Petry F, McClive PJ, Botto M, Morley BJ, Morahan G, Loos M. The mouse C1q genes are clustered on chromosome 4 and show conservation of gene organisation. *Immunogenetics* 1996;**43**:370–6.

FURTHER READING

1. Perkins SJ, Nealis AS, Sutton BJ, Feinstein A. Solution structure of human and mouse immunoglobulin M by synchrotron X-ray scattering and molecular modelling: a possible mechanism for complement activation. *J Mol Biol* 1991;**221**:1345–66.

Chapter 4

Mannose-Binding Lectin

Maciej Cedzyński[1], David C. Kilpatrick[2], Anna S. Świerzko[1]

[1]*Polish Academy of Sciences, Lodz, Poland;* [2]*Scottish National Blood Transfusion Service, Edinburgh, Scotland, United Kingdom*

OTHER NAMES OF MANNOSE-BINDING LECTIN

Mannan-binding lectin; MBL; mannose-binding protein; mannan-binding protein; *MBL2*.

PHYSICOCHEMICAL PROPERTIES

Human Mannose-binding lectin (MBL) consists of *O*-glycosylated (~4% carbohydrate) polypeptide chains. The gene product consists of 248 amino acids, of which 228 are found in the mature polypeptide (20 comprise the leader sequence). The basic subunit is a trimer that can form a series of higher oligomers.

pI	5.39[1]
M_r	24 kDa (predicted)
	25.3–25.5 kDa (mass spectrometry)[2,3]
	26–32 kDa (SDS-PAGE)
Hydroxylation sites	Eight[3,4]
O-glycosylation sites	Four[3,4]
N-glycosylation sites	None
Interchain disulphide bonds	Three sites[5]

STRUCTURE

MBL belongs to the collectin subfamily of C-type lectins. Multimers of the basic triplet subunit possess a 'bunch of tulips', or a three-dimensional structure resembling that of C1q. Each polypeptide has four regions: (1) a short N-terminal cysteine-rich domain (21 AA, including 3 Cys); (2) the collagen-like region (59 AA, including 19 Gly-Xaa-Yaa triplets, where Xaa and Yaa correspond to any amino acid); (3) an α-helical neck or hinge region (30 AA) and (4) the C-terminal carbohydrate-recognition domain (CRD, 118 AA) (reviewed in Refs. 6,7) (Fig. 4.1).

The cysteine-rich domain is responsible for the disulphide bond-dependent arrangement of subunits. The collagen-like domain interacts with MBL-associated

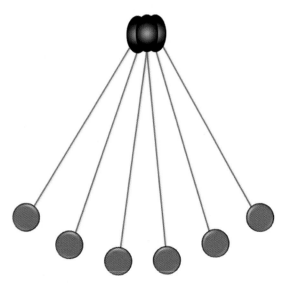

FIGURE 4.1   Diagram of protein domains for mannose-binding lectin.

serine proteases (MASPs). It is also the region implicated in the attachment of putative collectin receptors like cC1qR (calreticulin), CR1 (CD35)[1,8,9] and C1qRp (CD93)[1,8,10] on phagocytes. The collagen-like domain is stabilised by hydroxylation of proline residues and hydroxylation and O-glycosylation of lysines. After the seventh Gly-Xaa-Yaa repeat, it is interrupted by a Gly–Gln (GQ) sequence, giving rise to a flexible joint, which may lead to formation of a kink. The CRD is responsible for pattern recognition. It has a single binding site. The EPN amino acid motif, in association with Ca^{2+} cations, forms coordination bonds with hydroxyl groups at the three and four carbon atoms of hexose rings in the equatorial plane (e.g., D-mannose or N-acetyl-D-glucosamine). Although the affinity of a single CRD for carbohydrate is weak (K_d: $1–1.5 \times 10^{-3}$ M), high-avidity (K_d: 10^{-9} M) interaction is possible due to involvement of multiple CRDs.[11,12] The CRDs in the basic triplet subunit are separated by 4.5 nm.[11,13] The most common angles between neighbouring subunits are 35–45 degrees. Approximately 30% of subunits have kinks (the commonest angles: 140–150 degrees). The mean length of the subunit was determined to be 19 ± 4 nm.[14] Crystal structure of a trimer of human MBL CRDs with the neck domain was shown by Sheriff et al.[13] (Fig. 4.2).

FUNCTION

MBL can bind to saccharides, phospholipids, nucleic acids and other proteins. It is a pattern-recognition molecule that distinguishes self from nonself or altered self. Its ability to bind to putative collectin receptors and its well-established capacity to activate the lectin pathway of complement (in association with MASPs) enable MBL to act as an opsonin, promoting phagocytosis (reviewed

(A)

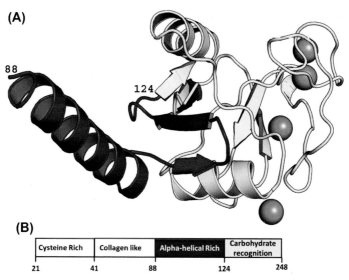

88

124

(B)

Cysteine Rich	Collagen like	Alpha-helical Rich	Carbohydrate recognition
21	41	88	124 248

FIGURE 4.2 Crystal structure for mannose-binding lectin (MBL) of human complement. (A) Ribbon diagram of human MBL crystal structure (PDB ID: 1HUP). Calcium atoms are represented by *orange spheres*. (B) Schematic representation of MBL domains.

in Refs. 1,6,7). Its interaction with dendritic cells suggests that MBL influences cellular immunity.[15,16] MBL can bind to and presumably cooperate with other innate immunity molecules like pentraxins.[17]

Additionally, activation of MASPs may lead to networking of complement components with the coagulation and kallikrein-kinin systems (reviewed in Ref. 7). MBL can bind to late apoptotic and necrotic cells,[18] senescent fibroblasts[19] and various cancer cells displaying aberrant glycosylation. This may lead to complement activation or MBL-dependent cell-mediated cytotoxicity and neoplastic disease.[20,21] Finally, MBL may also function in metabolism, as MBL concentration can influence insulin resistance and obesity.[22,23]

TISSUE DISTRIBUTION

MBL is primarily expressed in liver. The mean plasma concentration is approximately 1 µg/mL (ranging from undetectable to >10 µg/mL). Trace amounts of MBL have been reported in human milk,[24] amniotic fluid,[25] cervico-vaginal lavage[26] and cerebrospinal fluid.[27] MBL has also been detected in foetal lung[28] and first-trimester placenta.[29] Under pathological conditions, low concentrations have been detected in multiple body fluids including synovium,[30,31] small intestine,[32] peritoneum,[33] ovarian cysts,[33] bronchoalveolar lavage,[34] as well as ear effusions and nasopharyngeal secretions.[35] Glomerular deposition of MBL–MASP complexes was found in patients with Henoch-Schönlein purpura nephritis.[36]

MBL-specific mRNA expression has been found in small intestine,[37] testis,[37] ovary,[38] vagina,[38] bone marrow,[37] macrophages[37] and human corneal epithelial cells.[39] Furthermore, elevated *MBL2* mRNA expression was observed in ovarian cancer.[38] It was also detected in intestinal biopsies of coeliac patients[40] and in gastric biopsies from children with chronic gastritis.[41]

REGULATION OF EXPRESSION

MBL is an acute phase protein whose plasma concentration can increase up to threefold in response to infection or surgery. Its expression is upregulated by IL-6, thyroid hormones T3 and T4, growth hormone, peroxisome proliferator-activated receptors (PPAR) α and γ and possibly progesterone.[26,35,42–44] Heat shock and drugs such as dexamethasone[45] or fenofibrate[46] also induce expression.[42,43] In contrast, it is downregulated by IL-1 and microRNA-27a (miR-27a).[47,48] Several studies have reported a correlation between MBL concentration and gestational age.[49–51]

PROTEIN SEQUENCE

```
  1    MSLFPSLPLL LLSMVAASYS ETVTCEDAQK TCPAVIACSS PGINGFPGKD
 51    GRDGTKGEKG EPGQQGLRGLQ GPPGKLGPPG NPGPSGSPGP KGQKGDPGKS
101    PDGDSSLAAS ERKALQTEMA RIKKWLTFSL GKQVGNKFFL TNGEIMTFEK
151    VKALCVKFQA SVATPRNAAE NGAIQNLIKE EAFLGITDEK TEGQFVDLTG
201    NRLTYTNWNE GEPNNAGSDE DCVLLLKNGQ WNDVPCSTSH LAVCEFPI
```

The MBL protein sequence according to Dinasarapu et al.[5]: The leader sequence (amino acids 1–20) is shown in italics. It is encoded by exon 1, as is the cysteine-rich domain and part of the collagen-like region. The remaining part of the latter is encoded by exon 2. The 'neck' region is encoded by exon 3 and the CRD by exon 4.[6,7] Residues 63–64 ([43–44 in mature protein]; GQ, underlined) form the interruption among collagen-like repeats previously mentioned. *O*-glycosylation sites (lysines 56 [36], 59 [39], 91 [71] and 94 [74])[3] are shown in bold while cysteines (residues 25 [5], 32 [12], 38 [18])[1] involved in disulphide bonds are shown in bold italics.

PROTEIN MODULES

1–20	Leader	exon 1
21–41	N-terminal cysteine-rich domain	exon 1
42–100	Collagen-like region	exons 1/2
101–124	Alpha-helical coiled-coil neck region	exon 3
125–248	Carbohydrate-recognition domain	exon 4

CHROMOSOMAL LOCATION

In humans, *MBL1* is a pseudogene while *MBL2* codes for a functional protein. The latter is located on the long arm of chromosome 10 (10q11.2-q21),[52] between the regions 52765380 and 52772841 bp making the gene 7461 bp long.[52]

cDNA SEQUENCE

```
1    CCCTGAGTTT TCTCACACCA AGGTGAGGAC CATGTCCCTG TTTCCATCAC TCCCTCTCCT
61   TCTCCTGAGT ATGGTGGCAG CGTCTTACTC AGAAACTGTG ACCTGTGAGG ATGCCCAAAA
121  GACCTGCCCT GCAGTGATTG CCTGTAGCTC TCCAGGCATC AACGGCTTCC CAGGCAAAGA
181  TGGGCGTGAT GGCACCAAGG GAGAAAAGGG GGAACCAGGC CAAGGGCTCA GAGGCTTACA
241  GGGCCCCCCT GGAAAGTTGG GGCCTCCAGG AAATCCAGGG CCTTCTGGGT CACCAGGACC
301  AAAGGGCCAA AAAGGAGACC CTGGAAAAAG TCCGGATGGT GATAGTAGCC TGGCTGCCTC
361  AGAAAGAAAA GCTCTGCAAA CAGAAATGGC ACGTATCAAA AAGTGGCTCA CCTTCTCTCT
421  GGGCAAACAA GTTGGGAACA AGTTCTTCCT GACCAATGGT GAAATAATGA CCTTTGAAAA
481  AGTGAAGGCC TTGTGTGTCA AGTTCCAGGC CTCTGTGGCC ACCCCCAGGA ATGCTGCAGA
541  GAATGGAGCC ATTCAGAATC TCATCAAGGA GGAAGCCTTC CTGGGCATCA CTGATGAGAA
601  GACAGAAGGG CAGTTTGTGG ATCTGACAGG AAATAGACTG ACCTACACAA ACTGGAACGA
661  GGGTGAACCC AACAATGCTG GTTCTGATGA AGATTGTGTA TTGCTACTGA AAAATGGCCA
721  GTGGAATGAC GTCCCCTGCT CCACCTCCCA TCTGGCCGTC TGTGAGTTCC CTATCTGAAG
781  GGTCATATCA CTCAGGCCCT CCTTGTCTTT TTACTGCAAC CCACAGGCCC ACAGTA
```

The *MBL2* gene cDNA (836 bp) sequence provided by GeneBank (accession BC096180; version BC096180.3 GI:109730114). Initiation (ATG) and termination (TGA) codons are shown in bold. First five nucleotides in exons 2, 3 and 4 are underlined.

GENOMIC STRUCTURE

The *MBL2* gene consists of four exons and three introns. Exon 1 (251 bp) encodes for the 5′UTR, signal peptide, Cys-rich domain and seven Gly-Xaa-Yaa repeats of the collagen region. Exon 2 (117 bp) encodes the remainder of the collagen-like region. Exon 3 (69 bp) encodes the neck region, while exon 4 (3.1 kbp) encodes the CRD and 3′UTR. The size of introns was determined as 600 bp, 1350 bp and 800 bp.[52,53] There are multiple promoters for the MBL gene[37,54] (Fig. 4.3).

ACCESSION NUMBERS

Human (*Homo sapiens*) MBL:

XP_011538118.1 (predicted: mannose-binding protein C isoform X1)
XM_011539816.2 (predicted: mannose-binding lectin 2 (*MBL2*), transcript variant X1, mRNA)

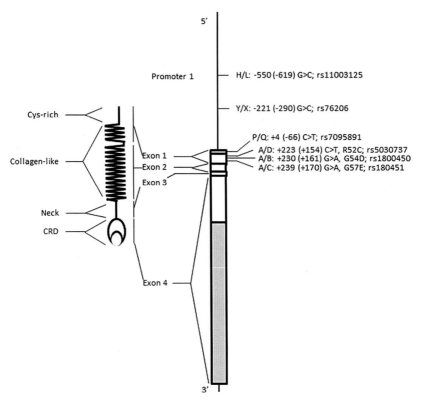

FIGURE 4.3 Scheme of *MBL2* gene and mannose-binding lectin (MBL) protein. Promoter 0, exon 0 and introns have been omitted. The 5' (exon 1) and 3' (exon 4) untranslated regions are shown in *pale grey*. The common promoter 1 and exon 1 polymorphisms are marked; positions corresponding to the distance from the transcriptional start site are given in parentheses. Amino acid substitutions R52C, G54D, G57E may be referred as R32C, G34D, G37E, respectively (in mature protein).

XP_006717924.1 (predicted: mannose-binding protein C isoform X1)
XM_006717861.3 (predicted: mannose-binding lectin 2 (*MBL2*), transcript variant X2, mRNA)
NP_000233.1 (mannose-binding protein C precursor)
NM_000242.2 (mannose-binding lectin 2 (*MBL2*), mRNA)
DQ217939.1 (mannose-binding protein C (*MBL2*) gene, complete cds)
EU596574.1 (mannan-binding lectin (*MBL2*) gene, complete cds)

Mouse (*Mus musculus*)

ENSMUSG00000037780.2 (*Mbl1* gene; chromosome 14)
ENSMUST00000047095.2 (*Mbl1* transcript)
ENSMUSG00000024863.5 (*Mbl2* gene; chromosome 19)
ENSMUST00000025797.5 (*Mbl2* transcript)

DEFICIENCY

MBL deficiency is a common congenital immunodeficiency and individuals with levels below 100 ng/mL are considered to have an MBL-associated opsonic defect.[45] There are low serum MBL-conferring genotype groups with *MBL2* O/O and XA/O genotypes considered to be associated with MBL deficiency.[7,55] The relationship between MBL genotype and circulating protein is, however, not predictable. Serum MBL can differ by more than 90-fold in individuals within the same genotype. Most (O/O or LXA/O) individuals have low (<200 ng/mL) serum MBL, although most individuals with serum MBL <200 ng/mL will not be (O/O or LXA/O). Most, but not all, A/A subjects will have relatively high serum MBL, but (YA/O) heterozygotes have values ranging from virtually undetectable to above-average.[55]

POLYMORPHIC VARIANTS

Eighty-seven polymorphic sites (comprising 85 SNPs and 2 insertion/deletion polymorphisms) have been identified within the human *MBL2* gene.[46,56] The six most extensively investigated are two SNPs in the promoter region (H/L, X/Y), three nonsynonymous SNPs in exon 1 (A/D, A/B, A/C) and another located in its 5′ untranslated end (P/Q). For the structural gene, the minority or variant alleles D, B and C are often collectively designated 'O'.[1,11,46,55,56] For details (positions, AA changes, reference numbers, etc.) see Fig. 4.1.

The A/O polymorphisms are associated with disruption to the collagen domain structure and reduced complement activation.[7,46,55,56] The shortened biological half-life of the protein is due, in part, to increased sensitivity to serum metalloproteases.[57] The promoter SNPs (H/L and Y/X) and the exon 1 untranslated region polymorphism, P/Q, all influence gene expression and MBL serum concentration.[46,55,56]

Very strong linkage disequilibria exist between the structural polymorphisms and the promoter polymorphisms. Seven haplotypes (HYPA, LYPA, LYQA, LXPA, HYPD, LYPB, LYQC) are considered to be relatively common. These haplotypes combine to form 28 genotypes. Although there are major differences between ethnic groups in the prevalence of certain SNPs (e.g., the B allele in Quechuan Indians, the C allele in Kenyans[46]), the overwhelming majority of the global population possesses one of only 28 genotypes. Rare haplotypes include LYPD, LYQB, HXPA and HYPB.[55,56,58]

Several other promoter region polymorphisms segregate between LYP and LYQ haplotypes.[56,59,60] A common SNP at position 3130 from ATG (3130G>C, exon 4, L126L) may affect MBL protein function.[58] SNPs of suspected clinical importance, located within the 3′-untranslated region of the exon 4 include Ex4-1483T>C (rs10082466), Ex4-1067G>A (rs10824792), Ex4-1047T>G (rs12254577), Ex4-901A>G (rs2120132), Ex4-710A>G (rs2099902) and 3238 bp 3′ STP (rs10450310)[48,61,62]. Six other polymorphisms (396A>C, rs1103124; 474A>G, rs7084554; 487A>G, rs36014597; 753C>T, rs11003123; 5411C>T, rs 2165813 and 5537G>T, rs2099903) have been reported to influence *MBL2* expression, structure and/or function and may be clinically relevant.[63]

MUTANT ANIMALS

MBL knockout mice appear healthy, fertile and have no obvious developmental disorders.[64–66] MBL-A-deficient animals were the first to be generated. The *Mbl1* gene was disrupted by introducing the neomycin resistance gene into its exon 5 (thereby disrupting the CRD of the protein). The embryonic stem cell clones were injected into C57B1/6J blastocysts and the chimeric animals were crossed with C57B1/6J mice.[64] The MBL-C-deficient mice were generated in a similar way.[65] Finally, by crossing MBL-A KO and MBL-C KO animals, MBL-null mice were obtained.[65]

MBL-null mice have been used in numerous disease-association studies (reviewed in detail by Genster et al.[66]). MBL deficiency increases susceptibility to a variety of infections. Reconstitution with human recombinant MBL restores the MBL-dependent immune response.[66] However, harmful effects including an excessive inflammatory response during sepsis or to ischaemia/reperfusion (I/R) injury were noted. Mice lacking MBL-A and MBL-C but expressing human *MBL* (under the control of *Mbl1* promoter) were protected from myocardial I/R injury on treatment with antihuman MBL antibody.[67]

REFERENCES

1. Teillet F, Dublet B, Andrieu J-P, et al. The two major oligomeric forms of human mannan-binding lectin: chemical characterization, carbohydrate-binding properties, and interaction with MBL-associated serine proteases. *J Immunol* 2005;**174**:2870–7.
2. Jensen PH, Laursen I, Matthiesen F, Hojrup P. Posttranslational modifications in human plasma MBL and human recombinant MBL. *Biochim Biophys Acta* 2007;**1774**:335–44.
3. Jensen PH, Weilguny D, Matthiesen P, et al. Characterization of the oligomer structure of recombinant human mannan-binding lectin. *J Biol Chem* 2005;**280**:11043–51.
4. Dinasarapu AR, Chandrasekhar A, Fujita T, Subramaniam S. Mannose/mannan-binding lectin. *Signal Gatew – UCSD Mol Pages* 2013. http://dx.doi.org/10.6072/H0.MP.A004276.01.
5. Larsen F, Madsen HO, Sim RB, et al. Disease-associated mutations in human mannose-binding lectin compromise oligomerization and activity of final protein. *J Biol Chem* 2004;**279**:21302–11.
6. Thiel S. Complement activating soluble pattern recognition molecules with collagen-like regions, mannan binding lectin, ficolins and associated proteins. *Mol Immunol* 2007;**44**:3875–88.
7. Pągowska-Klimek I, Cedzyński M. Mannan-binding lectin in cardiovascular disease. *Biomed Res Int* 2014;**2014**:616817. http://dx.doi.org/10.1155/2014/616817.
8. Pagh R, Duus K, Laursen I, et al. The chaperome and potential mannan-binding lectin (MBL) co-receptor calreticulin interacts with MBL through the binding site for MBL-associated serine proteases. *FEBS J* 2008;**275**:515–26.

9. Ghiran I, Barbashov SF, Klickstein LB, et al. Complement receptor 1/CD35 is a receptor for mannan-binding lectin. *J Exp Med* 2000;**192**:1797–808.

10. Tenner AJ, Robinson SL, Ezekowitz RAB. Mannose binding protein (MBP) enhances mononuclear phagocyte function via a receptor that contains the 126,000 M(r) component of the C1q receptor. *Immunity* 1995;**3**:485–93.

11. Petersen SV, Thiel S, Jensenius JC. The mannan-binding lectin pathway of complement activation: biology and disease association. *Mol Immunol* 2001;**38**:133–49.

12. Thiel S, Gadjeva M. Humoral pattern recognition molecules: mannan-binding lectin and ficolins. In: Kishore U, editor. *Target pattern recognition in innate immunity*. New York: Springer-Verlag; 2009. p. 58–73.

13. Sheriff S, Chang CY, Ezekowitz RA. Human mannose-binding protein carbohydrate recognition domain trimerizes through a triple alpha-helical coiled-coil. *Nat Struct Biol* 1994;**1**:789–94.

14. Jensenius H, Klein DCG, van Hecke M, et al. Mannan-binding lectin: structure, oligomerization, and flexibility studied by atomic force microscopy. *J Mol Biol* 2009;**391**:246–59.

15. MacDonald SL, Downing I, Atkinson AP, et al. Dendritic cells previously exposed to mannan-binding lectin enhance cytokine production in allogeneic mononuclear cell cultures. *Hum Immunol* 2010;**71**:1077–83.

16. Dean MM, Flower RL, Eisen DP, et al. Mannose-binding lectin deficiency influences innate and antigen-presenting functions of blood myeloid dendritic cells. *Immunology* 2011;**132**:296–305.

17. Ma YJ, Doni A, Skjoedt M-O, et al. Heterocomplexes of mannose-binding lectin and the pentraxins PTX3 or serum amyloid P component trigger cross-activation of the complement system. *J Biol Chem* 2011;**286**:3405–17.

18. Nauta AJ, Raashou-Jensen N, Roos A, et al. Mannose-binding lectin engagement with late apoptotic and necrotic cells. *Eur J Immunol* 2003;**33**:2853–63.

19. Tomaiuolo R, Ruocco A, Salapete C, et al. Activity of mannose-binding lectin (MBL) in centenarians. *Aging Cell* 2012;**11**:394–400.

20. Nakagawa T, Kawasaki N, Ma Y, et al. Antitumor activity of mannan-binding protein. *Methods Enzymol* 2003;**363**:26–33.

21. Swierzko AS, Kilpatrick DC, Cedzynski M. Mannan-binding lectin in malignancy. *Mol Immunol* 2013;**55**:16–21.

22. Fernandez-Real JM, Straczkowski M, Vendrell J, et al. Protection from inflammatory disease in insulin resistance: the role of mannan-binding lectin. *Diabetologia* 2006;**49**:2402–11.

23. Kowalska I, Fernandez-Real JM, Straczkowski M, et al. Insulin resistance is associated with decreased circulating mannan-binding lectin concentrations in women with polycystic ovary syndrome. *Diabetes Care* 2008;**31**:e20. http://dx.doi.org/10.2337/dc07-1872.

24. Tregoat V, Montagne P, Bene MC, Faure G. Changes in the mannan binding lectin (MBL) concentration in human milk during lactation. *J Clin Lab Anal* 2002;**16**:304–7.

25. Malhotra R, Willis AC, Lopez Bernal A, et al. Mannan-binding protein levels in human amniotic fluid during gestation and its interaction with collectin receptor from amnion cells. *Immunology* 1994;**82**:439–44.

26. Bulla R, De Seta F, Radillo O, et al. Mannose-binding lectin is produced by vaginal epithelial cell and its level in the vaginal fluid is influenced by progesterone. *Mol Immunol* 2010;**48**:281–6.

27. Reiber H, Padilla-Docal B, Jensenius JC, Dorta-Contreras AJ. Mannan-binding lectin in cerebrospinal fluid: a leptomeningeal protein. *Fluids Barriers CNS* 2012;**9**:17. http://dx.doi.org/10.1186/2045-8118-9-17.

28. Garred P, Honore C, Ma YJ, et al. *MBL2, FCN1, FCN2* and *FCN3* – The genes behind the initiation of the lectin pathway of complement. *Mol Immunol* 2009;**46**:2737–44.

29. Kilpatrick DC, Bevan BH, Liston WA. Association between mannan binding protein deficiency and recurrent miscarriage. *Hum Reprod* 1995;**10**:2501–5.
30. Malhotra R, Wormald MR, Rudd PM, et al. Glycosylation changes of IgG associated with rheumatoid arthritis can activate complement *via* the mannose-binding protein. *Nat Med* 1995;**1**:237–43.
31. Petri C, Thiel S, Jensenius JC, Herlin T. Investigation of complement-activating pattern recognition molecules and associated enzymes as possible inflammatory markers in oligoarticular and systemic juvenile idiopathic arthritis. *J Rheumatol* 2015;**42**:1252–8.
32. Kelly P, Jack DL, Naeem A, et al. Mannose-binding lectin is a component of innate mucosal defense against *Cryptosporidium parvum* in AIDS. *Gastroenterology* 2000;**119**:1236–46.
33. Świerzko AS, Florczak K, Cedzyński M, et al. Mannan-binding lectin (MBL) in women with tumours of the reproductive system. *Cancer Immunol Immunother* 2007;**56**:959–71.
34. Fidler KJ, Hillard TN, Bush A, et al. Mannose-binding lectin is present in the infected airway: a possible pulmonary defence mechanism. *Thorax* 2009;**64**:150–5.
35. Sorensen CM, Hansen TK, Steffensen R, et al. Hormonal regulation of mannan-binding lectin synthesis in hepatocytes. *Clin Exp Immunol* 2006;**145**:173–82.
36. Hisano S, Matsushita M, Fujita T, Iwasaki H. Activation of the lectin complement pathway in Henoch-Schonlein purpura nephritis. *Am J Kidney Dis* 2005;**45**:295–302.
37. Seyfarth J, Garred P, Madsen HO. Extrahepatic transcription of the human mannose-binding gene (*mbl2*) and the MBL-associated serine protease 1-3 genes. *Mol Immunol* 2006;**43**:962–71.
38. Świerzko AS, Szala A, Sawicki S, et al. Mannose-binding lectin (MBL) and MBL-associated serine protease-2 (MASP-2) in women with malignant and benign ovarian tumours. *Cancer Immunol Immunother* 2014;**63**:1129–40.
39. Che CY, Zhang JF, Lee JE, et al. Early expression of mannose-binding lectin 2 during *Aspergillus fumigatus* infection in human corneal epithelial cells. *Int J Ophtalmol* 2015;**18**:35–8.
40. Boniotto M, Radillo O, Braida L, et al. Detection of *MBL-2* gene expression in intestinal biopsies of celiac patients by in situ reverse transcription polymerase chain reaction. *Eur J Histochem* 2003;**47**:177–80.
41. Bąk-Romaniszyn L, Cedzyński M, Szemraj J, Świerzko A, et al. Mannan-binding lectin in children with chronic gastritis. *Scand J Immunol* 2006;**63**:131–5.
42. Arai T, Tabona P, Summerfield JA. Human mannose-binding protein gene is regulated by interleukins, dexamethasone and heat shock. *Q J Med* 1993;**86**:575–82.
43. Rakhshandehroo M, Stienstra R, de Wit NJ, et al. Plasma mannose-binding lectin is stimulated by PPARα in humans. *Am J Physiol Endocrinol Metab* 2012;**302**:E595–602.
44. Tachibana K, Takeuchi K, Inada H, et al. Human mannose-binding lectin 2 is directly regulated by peroxisome proliferator-activated receptors via a peroxisome proliferator responsive element. *J Biochem* 2013;**154**:265–73.
45. Richardson VF, Larcher VF, Price JF. A common congenital immunodeficiency predisposing to infection and atopy in infancy. *Arch Dis Child* 1983;**58**:799–802.
46. Bernig T, Taylor JG, Foster CB, et al. Sequence analysis of the mannose-binding lectin (*MBL2*) gene reveals a high degree of heterozygosity with evidence of selection. *Genes Immun* 2004;**5**:461–76.
47. Thiel S, Holmskov U, Hviid L, et al. The concentration of the C-type lectin, mannose-binding protein, in human plasma increases during an acute phase response. *Clin Exp Immunol* 1992;**90**:31–5.
48. Zanetti KA, Haznadar M, Welsh JA, et al. 3′UTR and functional secretor haplotypes in mannose-binding lectin 2 are associated with increased colon cancer risk in African Americans. *Cancer Res* 2012;**72**:1467–77.

49. Sallenbach S, Thiel S, Aebi C, et al. Serum concentrations of lectin pathway components in healthy neonates, children and adults: mannan-binding lectin (MBL), M-, L-, and H-ficolin, and MBL-associated serine protease-2 (MASP-2). *Pediatr Allergy Immunol* 2011;**22**:424–30.

50. Swierzko AS, Szala A, Cedzynski M, et al. Mannan-binding lectin genotypes and genotype-phenotype relationships in a large cohort of Polish neonates. *Hum Immunol* 2009;**70**:68–72.

51. Świerzko AS, Szala-Poździej A, Kilpatrick DC, et al. Components of the lectin pathway of complement activation in paediatric patients of intensive care units. *Immunobiology* 2016;**221**:657–69.

52. Sastry K, Herman GA, Day L, et al. The human mannose-binding protein gene exon structure reveals its evolutionary relationship to a human pulmonary surfactant gene and localization to chromosome 10. *J Exp Med* 1989;**170**:1175–89.

53. Taylor ME, Brickell PM, Craig RK, Summerfield JA. Structure and evolutionary origin of the gene encoding a human serum mannose-binding protein. *Biochem J* 1989;**262**:763–71.

54. Naito H, Ikeda A, Hasegawa K, et al. Characterization of human serum mannan-binding protein promoter. *J Biochem* 1999;**126**:1004–12.

55. Kilpatrick DC. Clinical significance of mannan-binding lectin and L-ficolin. In: Kilpatrick D, editor. *Collagen-related lectins in innate immunity.* Trivandrum (Kerala): Research Signpost; 2007. p. 57–84.

56. Garred P, Larsen F, Seyfarth J, et al. Mannose-binding lectin and its genetic variants. *Genes Immun* 2006;**7**:85–94.

57. Butler GS, Sim D, Tam E, et al. Mannose-binding lectin (MBL) mutants are susceptible to matrix metalloproteinase proteolysis. Potential role in human MBL deficiency. *J Biol Chem* 2002;**277**:17511–9.

58. Heitzender S, Seidel M, Forster-Waldl E, Heitger A. Mannan-binding lectin deficiency – good news, bad news, doesn't matter? *Clin Immunol* 2012;**143**:23–38.

59. Boldt ABW, Messias-Reason IJ, Meyer D, et al. Phylogenetic nomenclature and evolution of mannose-binding lectin (*MBL2*) haplotypes. *BMC Genet* 2010;**11**:38. http://dx.doi.org/10.1186/1471-2156-11-38.

60. Adamek M, Heyder J, Heinold A, et al. Characterization of mannose-binding lectin (MBL) variants by allele-specific sequencing of *MBL2* and determination of serum MBL protein levels. *Tissue Antigens* 2013;**82**:410–5.

61. Bernig T, Breunis W, Brouwer N, et al. An analysis of genetic variation across the *MBL2* locus in Dutch Caucasians indicates that 3′ haplotypes could modify circulating levels of mannose-binding lectin. *Hum Genet* 2005;**118**:404–15.

62. Bernig T, Boersma BJ, Howe TM, et al. The mannose-binding lectin (*MBL2*) haplotype and breast cancer: an association study in African-American and caucasian women. *Carcinogenesis* 2007;**28**:828–36.

63. Kalia N, Sharma A, Kaur M, et al. A comprehensive in silico analysis of non-synonymous and regulatory SNPs of human *MBL2* gene. *SpringerPlus* 2016;**5**:811. http://dx.doi.org/10.1186/s40064-016-2543-4.

64. Takahashi K, Gordon J, Liu H, et al. Lack of mannose-binding lectin-A enhances survival in a mouse model of acute septic peritonitis. *Microb Infect* 2002;**4**:773–84.

65. Shi L, Takahashi K, Dundee J, et al. Mannose-binding lectin-deficient mice are susceptible to infection with *Staphylococcus aureus. J Exp Med* 2004;**199**:1379–90.

66. Genster N, Takahashi M, Sekine H, et al. Lessons learned from mice deficient in lectin complement pathway molecules. *Mol Immunol* 2014;**61**:59–68.

67. Pavlov VI, Tan YS, McClure EE, et al. Human mannose-binding lectin inhibitor prevents myocardial injury and arterial thrombogenesis in a novel animal model. *Am J Pathol* 2015;**185**:347–55.

Chapter 5

Ficolins

Misao Matsushita
Tokai University, Hiratsuka, Japan

OTHER NAMES FOR FICOLINS

Ficolin-1: FCN1, M-ficolin, P35-related protein, Ficolin-2: FCN2, L-ficolin, P35, Hucolin, Elastin-binding protein (EBP)-37, Ficolin-3: FCN3, H-ficolin, Hakata antigen, thermolabile β-2 macroglycoprotein, thermolabile substance.

PHYSICOCHEMICAL PROPERTIES

The ficolins are a group of lectins defined by a collagen-like domain combined with a fibrinogen-like domain.[1] Ficolin-1,[2,3] ficolin-2[4] and ficolin-3[5] are human ficolins, each of which is composed of identical chains of around 32 kDa. Each protein chain is linked by disulphide bonds to form homooligomers of trimeric subunits.

FICOLIN-1

Ficolin-1 is a 326 amino acid molecule including a 29 amino acid leader sequence.

Mature protein:		
pI	Predicted	6.1
M_r (K)	Predicted	32
	Observed	34 (reduced)
		300~400 (nonreduced)
N-linked glycosylation sites	None	
Interchain disulphide bonds	Cys33, Cys53 (arrangement unknown)	

FICOLIN-2

Ficolin-2 is a 313 amino acid molecule including a 25 amino acid leader sequence.

Mature protein:

pI	Predicted	6.7
M_r (K)	Predicted	31
	Observed	35 (reduced)
		400 (nonreduced)

N-linked glycosylation sites
Potential 2 (Asn240, Asn300)
Interchain disulphide bonds are formed between Cys32-Cys32, Cys32-Cys52 and Cys52-Cys52 to form homooligomers.[6]

FICOLIN-3

Ficolin-3 is a 299 amino acid molecule including a 23 amino acid leader sequence.

Mature protein:

pI	Predicted	6.2
M_r (K)	Predicted	30
	Observed	34 (reduced)
		400~600 (nonreduced)

N-linked glycosylation sites
Potential 1 (Asn189)
Interchain disulphide bonds Cys29, Cys46 (arrangement unknown)

STRUCTURE

The human ficolins are homooligomers. Each subunit is composed of an N-terminal region including two cysteine residues, a collagen-like domain, a neck region and a fibrinogen-like domain (Fig. 5.1). The collagen-like domain forms a triple helix, leading to the formation of homotrimers. The assembly of four to six homotrimer units finally forms native, oligomeric ficolin, which is held together by interchain disulphide bonds formed by the cysteine residues in the N-terminal region. The fibrinogen-like domain of ficolin shows a globular, ellipsoidal structure. Electron microscopy of the whole proteins of human ficolin-2 and ficolin-3 show bouquet-like images, similar to those first described for C1q. The carbohydrate-binding site is located in the fibrinogen-like domain. Percent identities between ficolin-1 and ficolin-2, ficolin-1 and ficolin-3 and ficolin-2 and ficolin-3 are 79.9, 48.3 and 48.6, respectively.

FUNCTION

Lectin and Binding Activities

All human ficolins recognise N-acetylated carbohydrates and compounds such as N-acetylglucosamine (GlcNAc) in common, although the spectrum

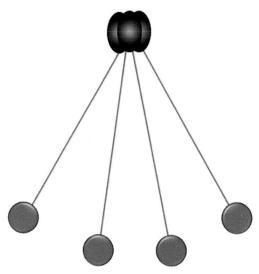

FIGURE 5.1   Diagram of protein domains for ficolin (for domain definitions, see key in front of book).

of ligands for each ficolin is slightly different.[1] Human ficolin-1 binds sialic acid, and human ficolin-3 recognises D-fucose and galactose. Ficolin-2 exhibits broad binding specificities due to it having four distinct binding sites in the fibrinogen-like domain that allow it to bind various carbohydrates, including β-(1,3)-D-glucan and peptideglycan, and sulphated carbohydrates, including heparin, while ficolin-1 and ficolin-3 have only one binding site.[7,8] In addition to carbohydrates, ficolin-1 binds to tectonin,[9] and ficolin-2 binds elastin,[10] corticosteroid[11] and DNA.[12]

Binding of Ficolins to Microorganisms

Each ficolin is able to bind to a spectrum of bacteria.[1] Ficolin-1 binds to *Staphylococcus aureus* derived from clinical specimens and *Salmonella typhimurium* LT2. Human ficolin-2 binds to many types of bacteria including *S. typhimurium* TV119, *Escherichia coli*, capsulated *S. aureus*, capsulated *Staphylococcus pneumoniae* serotypes, *Mycobacterium tuberculosis* and *Pseudomonas aeruginosa*. It recognises bacterial components such as lipopolysaccharides and lipoteichoic acid. Ficolin-3 binds *Aerococcus viridans* and its polysaccharide component[13] and certain lipopolysaccharides of *Hafnia alvei*.[14] All human ficolins bind to Influenza A virus. Ficolin-2 and ficolin-3 bind to Hepatitis viruses and *Trypanosoma cruzi*. Ficolin-2 exhibits an opsonic activity towards *S. typhimurium* TV119.[4]

Lectin Pathway Activation

Human ficolins activate the lectin complement pathway. In human serum, fico-lin-2 and ficolin-3 form complexes with all three types of mannose-binding lectin (MBL)-associated serine proteases (MASP) and their truncated proteins, sMAP (MAp19) and MAp44 (MAP-1). Upon binding to ligands for these fico-lins, the MASPs in the complexes become activated, resulting in the cleavage of C4 and C2 to generate the C3 convertase, C4bC2a.[15,16] Ficolin-3 was dem-onstrated to exhibit in vitro bactericidal activity against *Aerococcus viridans* in the presence of human serum.[17] MASP-2 is reported to be complexed with ficolin-1 in serum.[18]

Cross Talk and Collaboration of Ficolins With Noncomplement Proteins

Ficolin-1 and ficolin-2 interact with C-reactive protein (CRP), boosting both the lectin and classical pathway.[19,20] Ficolin-1 on monocytes regulates IL-8 production by modulating the GPR43-dependent signalling.[21] The binding of ficolin-1 released from neutrophils to leukosialin (CD43) on neutrophils induces neutrophil adhesion in an autocrine manner, suggesting its role in neutrophil recruitment.[22] Ficolin-2 interacts with pentraxin 3 (PTX3), boosting the recognition and subsequent comple-ment activation on *Aspergillus fumigatus*.[23] At acidic pH, ficolin-3 forms complexes with natural antibodies. Upon recognising microbes, the complex binds to FcγR1 on monocytes, inducing a killing signal.[24]

Roles in Homeostasis

Ficolin-1 interacts with PTX3 to form a complex on dying self-cells.[25] The interaction of ficolin-1 with PTX3 enhances phagocytosis of apoptotic and necrotic cells by macrophages. Ficolin-2 binds to late apoptotic cells, apoptotic bodies, necrotic cells and DNA.[12] Ficolin-3 binds to late apoptotic and necrotic cells.[26] These facts suggest that ficolins are involved in homeostasis by clear-ance of dying host cells.

TISSUE DISTRIBUTION

Ficolin-1 is produced by peripheral blood monocytes and neutrophils, bone marrow cells and type II alveolar epithelial cells in the lungs.[27] Ficolin-1 is present on the surface of monocytes and neutrophils through association with sialic acid, GPR43 or leukosialin on these cells. A low concentration of ficolin-1 is also present in the serum with the mean level of 1 μg/mL. Ficolin-2 and ficolin-3 are expressed in the liver and are secreted into serum. In the liver, ficolin-3 is produced by bile duct epithelial cells and hepatocytes and is secreted into bile and serum.[28] Liver is the only expression site for ficolin-2, while ficolin-3 is also expressed in the lung and ovary.[29] Ficolin-3 produced in the lungs is secreted into the bronchi and alveoli.[28] The mean serum levels of ficolin-2 and ficolin-3 are 3.3–5 μg/mL and 18.4–32.6 μg/mL, respectively.

PROTEIN SEQUENCE

Ficolin-1

```
MELSGATMAR GLAVLLVLFL HIKNLPAQAA DTCPEVKVVG LEGSDKLTIL   50
RGCPGLPGAP GPKGEAGVIG ERGERGLPGA PGKAGPVGPK GDRGEKGMRG  100
EKGDAGQSQS CATGPRNCKD LLDRGYFLSG WHTIYLPDCR PLTVLCDMDT  150
DGGGWTVFQR RMDGSVDFYR DWAAYKQGFG SQLGEFWLGN DNIHALTAQG  200
SSELRVDLVD FEGNHQFAKY KSFKVADEAE KYKLVLGAFV GGSAGNSLTG  250
HNNNFFSTKD QDNDVSSSNC AEKFQGAWWY ADCHASNLNG LYLMGPHESY  300
ANGINWSAAK GYKYSYKVSE MKVRPA
```

Ficolin-2

```
MELDRAVGVL GAATLLLSFL GMAWALQAAD TCPEVKMVGL EGSDKLTILR   50
GCPGLPGAPG PKGEAGTNGK RGERGPPGPP GKAGPPGPNG APGEPQPCLT  100
GPRTCKDLLD RGHFLSGWHT IYLPDCRPLT VLCDMDTDGG GWTVFQRRVD  150
GSVDFYRDWA TYKQGFGSRL GEFWLGNDNI HALTAQGTSE LRVDLVDFED  200
NYQFAKYRSF KVADEAEKYN LVLGAFVEGS AGDSLTFHN**N** QSFSTKDQDN  250
DLNTGNCAVM FQGAWWYKNC HVSNLNGRYL RGTHGSFANG INWKSGKGY**N**  300
YSYKVSEMKV RPA
```

Ficolin-3

```
MDLLWILPSL WLLLLGGPAC LKTQEHPSCP GPRELEASKV VLLPSCPGAP   50
GSPGEKGAPG PQGPPGPPGK MGPKGEPGDP VNLLRCQEGP RNCRELLSQG  100
ATLSGWYHLC LPEGRALPVF CDMDTEGGGW LVFQRRQDGS VDFFRSWSSY  150
RAGFGNQESE FWLGNENLHQ LTLQGNWELR VELEDFNG**NR** TFAHYATFRL  200
LGEVDHYQLA LGKFSEGTAG DSLSLHSGRP FTTYDADHDS SNSNCAVIVH  250
GAWWYASCYR SNLNGRYAVS EAAAHKYGID WASGRGVGHP YRRVRMMLR
```

The leader peptides are underlined. Potential *N*-linked glycosylation sites are indicated [**N**].

PROTEIN MODULES

Ficolin-1

1–29	Leader peptide	exon 1
30–51	N-terminal cysteine rich region	exon 1/2
52–108	Collagen-like domain	exon 2/3/4/5
109–114	Neck region	exon 5
115–326	Fibrinogen-like domain	exon 6/7/8/9

Ficolin-2

1–25	Leader peptide	exon 1
26–50	N-terminal cysteine rich region	exon 1/2
51–95	Collagen-like domain	exon 2/3/4
96–101	Neck region	exon 4
102–313	Fibrinogen-like domain	exon 5/6/7/8

Ficolin-3

1–23	Leader peptide	exon 1
24–47	N-terminal cysteine rich region	exon 1/2
48–80	Collagen-like domain	exon 2/3/4
81–89	Neck region	exon 4
90–299	Fibrinogen-like domain	exon 5/6/7/8

Splicing Variants of Ficolins

Ficolin-2: Three different splicing variants have been identified.[2] The first variant lacks the entire second exon sequence and has an additional 76-bp sequence extending from 3′UT. The second variant is generated by the insertion of the fifth intron and also has an additional 3′UT sequence of 1060 bp. The third variant is generated by the insertion of both the fourth and fifth introns.

Ficolin-3: There is a splicing variant lacking the fourth exon.[30]

CHROMOSOMAL LOCATION

Ficolin-1	9q34
Ficolin-2	9q34
Ficolin-3	1p35.3

cDNA SEQUENCE
Ficolin-1

```
GGAGTTGAGA AACTGTGGCA CAAGGCGAGA GCTGGTTTCC TCTGCCCTGT TAGAGCTGGG     60
GGACTCTTCA GAGTCAAAGG CCAGAGAGCA TGGAGCTGAG TGGAGCCACC ATGGCCCGGG    120
GGCTCGCTGT CCTGCTAGTC TTGTTCCTGC ATATCAAGAA CCTGCCTGCC CAGGCTGCGG    180
ACACATGTCC AGAGGTGAAG GTGGTGGGCC TGGAGGGCTC TGACAAGCTC ACCATTCTCC    240
GAGGCTGCCC GGGGCTGCCC GGGGCCCCAG GGCCAAAGGG AGAGGCAGGT GTCATTGGAG    300
AGAGAGGAGA ACGCGGTCTC CCTGGAGCCC CTGGAAAGGC AGGACCAGTG GGGCCCAAAG    360
GAGACCGAGG AGAGAAGGGG ATGCGTGGAG AGAAAGGAGA CGCTGGGCAG TCTCAGTCGT    420
GTGCGACAGG CCCACGCAAC TGCAAGGACC TGCTAGACCG GGGGTATTTC CTGAGCGGCT    480
GGCACACCAT CTACCTGCCC GACTGCCGGC CCCTGACTGT GCTCTGTGAC ATGGACACGG    540
ACGGAGGGGG CTGGACCGTT TTCCAGCGGA GGATGGATGG CTCTGTGGAC TTCTATCGGG    600
ACTGGGCCGC ATACAAGCAG GGCTTCGGCA GTCAGCTGGG GGAGTTCTGG CTGGGGAATG    660
ACAACATCCA CGCCCTGACT GCCCAGGGAA GCAGCGAGCT CCGTGTAGAC CTGGTGGACT    720
TTGAGGGCAA CCACCAGTTT GCTAAGTACA AATCATTCAA GGTGGCTGAC GAGGCAGAGA    780
AGTACAAGCT GGTACTGGGA GCCTTTGTCG GGGGCAGTGC GGGTAATTCT CTAACGGGCC    840
ACAACAACAA CTTCTTCTCC ACCAAAGACC AAGACAATGA TGTGAGTTCT TCGAATTGTG    900
CTGAGAAGTT CCAAGGAGCC TGGTGGTACG CCGACTGTCA TGCTTCAAAC CTCAATGGTC    960
TCTACCTCAT GGGACCCCAT GAGAGCTATG CCAATGGTAT CAACTGGAGT GCGGCGAAGG   1020
GGTACAAATA TAGCTACAAG GTGTCAGAGA TGAAGGTGCG GCCCGCCTAG ACGGGCCAGG   1080
ACCCCTCCAC ATGCACCTGC TAGTGGGGAG GCCACACCCA CAAGCGCTGC
```

Ficolin-2

```
ACCAGAAGAG ATGGAGCTGG ACAGAGCTGT GGGGGTCCTG GGCGCTGCCA CCCTGCTGCT     60
CTCTTTCCTG GGCATGGCCT GGGCTCTCCA GGCGGCAGAC ACCTGTCCAG AGGTGAAGAT    120
GGTGGGCCTG GAGGGCTCTG ACAAGCTCAC CATTCTCCGA GGCTGTCCGG GGCTGCCTGG    180
GGCCCCTGGG CCCAAGGGAG AGGCAGGCAC CAATGGAAAG AGAGGAGAAC GTGGCCCCCC    240
TGGACCTCCT GGGAAGGCAG GACCACCTGG GCCCAACGGA GCACCTGGGG AGCCCCAGCC    300
GTGCCTGACA GGCCCGCGTA CCTCCAACCA CCTGCTACAC CCAGGCACTT CCTGAGCGG    360
CTGGCACACC ATCTACCTGC CCGACTGCCG GCCCCTGACT GTGCTCTGTG ACATGGACAC    420
GGACGGAGGG GGCTGGACCG TTTTCCAGCG GAGGGTGGAT GGCTCTGTGG ACTTCTACCG    480
GGACTGGGCC ACGTACAAGC AGGGCTTCGG CAGTCGGCTG GGGGAGTTCT GGCTGGGGAA    540
TGACAACATC CACGCCCTGA CCGCCCAGGG AACCAGCGAG CTCCGTGTAG ACCTGGTGGA    600
CTTTGAGGAC AACTACCAGT TTGCTAAGTA CAGATCATTC AAGGTGGCCG ACGAGGCGGA    660
GAAGTACAAT CTGGTCCTGG GGGCCTTCGT GGAGGGCAGT GCGGGGAGATT CCCTGACGTT    720
CCACAACAAC AGTCCTTCT CCACCAAAGA CCAGGACAAT GATCTTAACA CCGGAAATTG    780
TGCTGTGATG TTTCAGGGAG CTTGGTGGTA CAAAAACTGC CATGTGTCAA ACCTGAATGG    840
TCGCTACCTC AGGGGGACTC ATGGCAGCTT TGCAAATGGC ATCAACTGGA AGTCGGGGAA    900
AGGATACAAT TATAGCTACA AGGTGTCAGA GATGAAGGTG CGACCTGCCT AGCCCAGGCC    960
GGCCTCAGGG TCAGGACGCC TCCACACATA GTTGGTTGGG GGGTAGGGTT GGGAGCTTGG   1020
CCCTACGGTT TGTAAAAGAA ACACATGTCG TGATTCT
```

Ficolin-3

```
AATCTGGCAT CCTTCCCCTG GTGGGCCCAG CAAGATGGAT CTACTGTGGA TCCTGCCCTC   60
CCTGTGGCTT CTCCTGCTTG GGGGGCCTGC CTGCCTGAAG ACCCAGGAAC ACCCCAGCTG  120
CCCAGGACCC AGGGAACTGG AAGCCAGCAA AGTTGTCCTC CTGCCCAGTT GTCCCGGAGC  180
TCCAGGAAGT CCTGGGGAGA AGGGAGCCCC AGGTCCTCAA GGGCCACCTG GACCACCAGG  240
CAAGATGGGC CCCAAGGGTG AGCCAGGAGA TCCAGTGAAC CTGCTCCGGT GCCAGGAAGG  300
CCCCAGAAAC TGCCGGGAGC TGTTGAGCCA GGGCGCCACC TTGAGCGGCT GGTACCATCT  360
GTGCCTACCT GAGGGCAGGG CCCTCCCAGT CTTTTGTGAC ATGGACACCG AGGGGGGCGG  420
CTGGCTGGTG TTTCAGAGGC GCCAGGATGG TTCTGTGGAT TTCTTCCGCT CTTGGTCCTC  480
CTACAGAGCA GGTTTTGGGA ACCAAGAGTC TGAATTCTGG CTGGGAAATG AGAATTTGCA  540
CCAGCTTACT CTCCAGGGTA ACTGGGAGCT GCGGGTAGAG CTGGAAGACT TTAATGGTAA  600
CCGTACTTTC GCCCACTATG CGACCTTCCG CCTCCTCGGT GAGGTAGACC ACTACCAGCT  660
GGCACTGGGC AAGTTCTCAG AGGGCACTGC AGGGGATTCC CTGAGCCTCC ACAGTGGGAG  720
GCCCTTTACC ACCTATGACG CTGACCACGA TTCAAGCAAC AGCAACTGTG CAGTGATTGT  780
CCACGGTGCC TGGTGGTATG CATCCTGTTA CCGATCAAAT CTCAATGGTC GCTATGCAGT  840
GTCTGAGGCT GCCGCCCACA AATATGGCAT TGACTGGGCC TCAGGCCGTG GTGTGGGCCA  900
CCCCTACCGC AGGGTTCGGA TGATGCTTCG ATAGGGCACT CTGGCAGCCA GTGCCCTTAT  960
CTCTCCTGTA CAGCTTCCGG ATCGTCAGCC ACCTTGCCTT TGCCAACCAC CTCTGCTTGC 1020
CTGTCCACAT TTAAAAATAA AATCATTTTA GCCCTTTCAA AAAAAAAAAA AAAAAAAA
```

GENOMIC STRUCTURE

The first five nucleotides in each exon are underlined to indicate intron–exon boundaries. The methionine initiation codon (ATG) and the termination codon (TAG) are double underlined (Fig. 5.2).

Ficolin-1

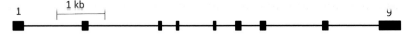

Ficolin-2

Ficolin-3

FIGURE 5.2 Genomic structure of human ficolins.

ACCESSION NUMBERS

	cDNA	Genome
Ficolin-1	NM_002003.4	NC_000009.12
Ficolin-2	NM_004108.2	NC_000009.12
	D49353.1	
Ficolin-3	NM_003665.3	NC_000001.11

DEFICIENCY

Two cases of ficolin-3 deficiency due to homozygosity for a *FCN3* frame-shift mutation (*c.349delC*, rs28357092) in exon 5 leading to the abnormal ficolin-3 lacking the entire fibrinogen-like domain have been reported. One patient with ficolin-3 deficiency had several episodes of pneumonia caused by *Haemophilus influenzae* and *P. aeruginosa*, and showed severe bronchiectasis and pulmonary fibrosis.[31] The other patient was a premature infant suffering from severe necrotising enterocolitis.[32] To date, deficiencies of ficolin-1 or ficolin-2 have not been reported. In the case of ficolin-2, low serum levels are associated with premature, low birthweight neonates, perinatal infections and respiratory infections.[33]

POLYMORPHIC VARIANTS

Many SNPs have been identified in the promoter and exons for human ficolins as shown in Table 5.1.

FICOLIN-DEFICIENT MOUSE

In mice, there are two ficolin isoforms named ficolin A and ficolin B. Three lineages of ficolin-deficient mice, i.e., ficolin A-deficient (FcnAKO), ficolin B-deficient (FcnBKO) and ficolins A- and B-double deficient mice (FcnABKO), have been generated.[34] No abnormality was observed in these knockout lineages in their appearance, body weight, reproductive fitness and tissues including the liver, spleen, lung and bone marrow. The lectin pathway activity of mouse serum was significantly lower in FcnAKO and FcnABKO than in wild-type serum, while that of serum from FcnBKO mouse was normal. The infection with *S. pneumoniae* D39 strain resulted in significantly reduced survival rates in all of the ficolin-knockout lineages. The injection of ficolin A-encoding vector into the mice improved survival rate in FcnAKO but not FcnABKO mice.

TABLE 5.1 SNPs in the Human Ficolin Genes

	Region	Position	Amino Acid Substitution	rs Number
FCN1	Promoter	c.-1981G>A		rs2989727
	Promoter	c.-1524T>C		rs7857015
	Promoter	c.-791A>G		rs28909068
	Promoter	c.-542G>A		rs10120023
	Promoter	c.-144C>A		rs10117466
	Exon 1	c.33G>T	p.Gly11Gly	rs10858293
	Exon 6	c.371G>A	p.Arg124Gln	rs147309328
	Exon 6	c.570T>C	p.Asn190Asn	rs2274845
	Exon 8	c.652G>A	p.Ala218Thr	rs148649884
	Exon 9	c.802T>C	p.Ser268Pro	rs150625869
	Exon 9	c.825A>G	p.Gln275Gln	rs1071583
	Exon 9	c.866A>G	p.Asn289Ser	rs138055828
FCN2	Promoter	c.-986G>A		rs3124952
	Promoter	c.-602A>G		rs3124953
	Promoter	c.-557A>G		rs3811140
	Promoter	c.-4A>G		rs17514136
	Promoter	c.-64A>C		rs7865453
	Exon 5	c.307C>T	p.Arg103Cys	rs55895215
	Exon 8	c.707C>T	p.Thr236Met	rs17549193
	Exon 8	c.772G>T	p.Ala258Ser	rs7851696
FCN3	Exon 5	c.349delC	p.Leu117Serfs	rs28357092

REFERENCES

1. Endo Y, Matsushita M, Fujita T. New insights into the role of ficolins in the lectin pathway of innate immunity. *Int Rev Cell Mol Biol* 2015;**316**:49–110.
2. Endo Y, Sato Y, Matsushita M, Fujita T. Cloning and characterization of the human lectin P35 gene and its related gene. *Genomics* 1996;**36**:515–21.
3. Lu J, Tay PN, Kon OL, Reid KB. Human ficolin: cDNA cloning, demonstration of peripheral blood leucocytes as the major site of synthesis and assignment of the gene to chromosome 9. *Biochem J* 1996;**313**:473–8.

4. Matsushita M, Endo Y, Taira S, Sato Y, Fujita T, Ichikawa N, et al. A novel human serum lectin with collagen- and fibrinogen-like domains that functions as an opsonin. *J Biol Chem* 1996;**271**:2448–54.

5. Sugimoto R, Yae Y, Akaiwa M, Kitajima S, Shibata Y, Sato H, et al. Cloning and characterization of the Hakata antigen, a member of the ficolin/opsonin p35 lectin family. *J Biol Chem* 1998;**273**:20721–7.

6. Hummelshoj T, Thielens NM, Madsen HO, Arlaud GJ, Sim RB, Garred P. Molecular organization of human ficolin-2. *Mol Immunol* 2007;**44**:401–11.

7. Tanio M, Kondo S, Sugio S, Kohno T. Trivalent recognition unit of innate immunity system: crystal structure of trimeric human M-ficolin fibrinogen-like domain. *J Biol Chem* 2007;**282**:3889–95.

8. Garlatti V, Belloy N, Martin L, Lacroix M, Matsushita M, Endo Y, et al. Structural insights into the innate immune recognition specificities of L- and H-ficolins. *EMBO J* 2007;**26**:623–33.

9. Low DH, Ang Z, Yuan Q, Frecer V, Ho B, Chen J, et al. A novel human tectonin protein with multivalent beta-propeller folds interacts with ficolin and binds bacterial LPS. *PLoS One* 2009;**4**:e6260.

10. Harumiya S, Omori A, Sugiura T, Fukumoto Y, Tachikawa H, Fujimoto D. EBP-37, a new elastin-binding protein in human plasma: structural similarity to ficolins, transforming growth factor-β 1J. *Biochem J* 1995;**117**:1029–35.

11. Edgar PF. Hucolin, a new corticosteroid-binding protein from human plasma with structural similarities to ficolins, transforming growth factor-beta 1-binding proteins. *FEBS Lett* 1995;**375**:159–61.

12. Jensen ML, Honore C, Hummelshoj T, Hansen BE, Madsen HO, Garred P. Ficolin-2 recognizes DNA and participates in the clearance of dying host cells. *Mol Immunol* 2007;**44**:856–65.

13. Tsujimura M, Ishida C, Sagara Y, Miyazaki T, Murakami K, Shiraki H, et al. Detection of serum thermolabile β-2 macroglycoprotein (Hakata antigen) by enzyme-linked immunosorbent assay using polysaccharide produced by *Aerococcus viridans*. *Clin Diagn Lab Immunol* 2001;**8**:454–9.

14. Swierzko A, Lukasiewicz J, Cedzynski M, Maciejewska A, Jachymek W, Niedziela T, et al. New functional ligands for ficolin-3 among lipopolysaccharides of *Hafnia alvei*. *Glycobiology* 2012;**22**:267–80.

15. Matsushita M, Endo Y, Fujita T. Cutting edge: complement-activating complex of ficolin and mannose-binding lectin-associated serine protease. *J Immunol* 2000;**164**:2281–4.

16. Matsushita M, Kuraya M, Hamasaki N, Tsujimura M, Shiraki H, Fujita T. Activation of the lectin complement pathway by H-ficolin (Hakata antigen). *J Immunol* 2002;**168**:3502–6.

17. Tsujimura M, Miyazaki T, Kojima E, Sagara Y, Shiraki H, Okochi K, et al. Serum concentration of Hakata antigen, a member of ficolins, is linked with inhibition of *Aerococcus viridans* growth. *Clin Chim Acta* 2002;**325**:139–46.

18. Kjaer TR, Hansen AG, Sørensen UBS, Nielsen O, Thiel S, Jensenius JC. Investigations on the pattern recognition molecule M-ficolin: quantitative aspects of bacterial binding and leukocyte association. *J Leukoc Biol* 2011;**90**:425–37.

19. Tanio M, Wakamatsu K, Kohno T. Binding site of C-reactive protein on M-ficolin. *Mol Immunol* 2009;**47**:215–21.

20. Zhang J, Koh J, Lu J, Thiel S, Leong BS, Sethi S, et al. Local inflammation induces complement crosstalk which amplifies the antimicrobial response. *PLoS Pathog* 2009;**5**:e1000282.

21. Zhang J, Yang L, Ang Z, Yoong SL, Tran TT, Anand GS, et al. Secreted M-ficolin anchors onto monocyte transmembrane G protein-coupled receptor 43 and cross talks with plasma C-reactive protein to mediate immune signaling and regulate host defense. *J Immunol* 2010;**185**:6899–910.

22. Moreno-Amaral AN, Gout E, Danella-Polli C, Tabarin F, Lesavre P, Pereira-da-Silva G, et al. M-ficolin and leukosialin (CD43): new partners in neutrophil adhesion. *J Leukoc Biol* 2012;**91**:469–74.

23. Ma YJ, Doni A, Hummelshoj T, Honore C, Bastone A, Mantovani A, et al. Synergy between ficolin-2 and pentraxin 3 boosts innate immune recognition and complement deposition. *J Biol Chem* 2009;**284**:28263–75.

24. Panda S, Zhang J, Yang L, Anand GS, Ding JL. Molecular interaction between natural IgG and ficolin–mechanistic insights on adaptive-innate immune crosstalk. *Sci Rep* 2014;**4**:3675.

25. Ma YJ, Doni A, Romani L, Jurgensen HJ, Behrendt N, Mantovani A, et al. Ficolin-1-PTX3 complex formation promotes clearance of altered self-cells and modulates IL-8 production. *J Immunol* 2013;**191**:1324–33.

26. Honore C, Hummelshoj T, Hansen BE, Madsen HO, Eggleton P, Garred P. The innate immune component ficolin 3 (Hakata antigen) mediates the clearance of late apoptotic cells. *Arthritis Rheum* 2007;**56**:1598–607.

27. Liu Y, Endo Y, Iwaki D, Nakata M, Matsushita M, Wada I, et al. Human M-ficolin is a secretory protein that activates the lectin complement pathway. *J Immunol* 2005;**175**:3150–6.

28. Akaiwa M, Yae Y, Sugimoto R, Suzuki SO, Iwaki T, Izuhara K, et al. Hakata antigen, a new member of the ficolin/opsonin p35 family, is a novel human lectin secreted into bronchus/alveolus and bile. *J Histochem Cytochem* 1999;**47**:777–86.

29. Szala A, Sawicki S, Swierzko AS, Szemraj J, Sniadecki M, Michalski M, et al. Ficolin-2 and ficolin-3 in women with malignant and benign ovarian tumours. *Cancer Immunol Immunother* 2013;**62**:1411–9.

30. Garred P, Honoré C, Ma YJ, Munthe-Fog L, Hummelshøj T. MBL2, FCN1, FCN2 and FCN3-The genes behind the initiation of the lectin pathway of complement. *Mol Immunol* 2009;**46**:2737–44.

31. Munthe-Fog L, Hummelshoj T, Honore C, Madsen HO, Permin H, Garred P. Immunodeficiency associated with FCN3 mutation and ficolin-3 deficiency. *N Engl J Med* 2009;**360**:2637–44.

32. Schlapbach LJ, Thiel S, Kessler U, Ammann RA, Aebi C, Jensenius JC. Congenital H-ficolin deficiency in premature infants with severe necrotising enterocolitis. *Gut* 2011;**60**:1438–9.

33. Kilpatrick DC, Chalmers JD. Human L-ficolin (ficolin-2) and its clinical significance. *J Biomed Biotech* 2012:138797. doi:10.1155/2012/138797.

34. Endo Y, Takahashi M, Iwaki D, Ishida Y, Nakazawa N, Kodama T, et al. Mice deficient in ficolin, a lectin complement pathway recognition molecule, are susceptible to *Streptococcus pneumoniae* infection. *J Immunol* 2012;**189**:5860–6.

Chapter 6

The Collectins

Katsuki Ohtani, Nobutaka Wakamiya
Asahikawa Medical University, Asahikawa, Japan

OTHER NAMES

CL-L1: collectin subfamily member 10, collectin 10, COLEC10, collectin liver 1, collectin liver protein 1, collectin 34.

CL-K1: collectin subfamily member 11, collectin 11, COLEC11, collectin kidney 1, collectin kidney protein 1.

PHYSICOCHEMICAL PROPERTIES

CL-L1 and CL-K1 belong to the collectin family that has at least two characteristic structures, a collagen-like domain and a carbohydrate recognition domain (CRD).[1,2]

CL-L1

pl: predicted	7.2588
Predicted mol. wt.:	30,705.03 g/mol
Observed mol. wt.:	34 kDa (SDS-PAGE in reduced state)
Number of amino acids:	277
Location of N-linked glycosylation sites:	Arg 252
Interchain disulphide bonds:	39

CL-K1 (CL-K1-I, isoform a, transcript variant 1)

pl: predicted	5.0547
Predicted mol. wt.:	28,665.37 g/mol
Observed mol. wt.:	34 kDa (SDS-PAGE in reduced state)
Number of amino acids:	271
Location of N-linked glycosylation sites:	Occupied
Interchain disulphide bonds:	33

The Complement FactsBook. http://dx.doi.org/10.1016/B978-0-12-810420-0.00006-7

STRUCTURE

The N-terminal domain in both CL-L1 and CL-K1 has one cysteine residue participating in cross-linking of polypeptide chains into subunits and further into oligomers of subunits. The N-terminal domain is followed by a collagen-like domain, an alpha-helical coiled neck domain and a CRD. There is a striking primary structural homology between CL-L1 and CL-K1, with identical numbers of amino acid residues in every domain. The sugar frame motif of EPS (CL-L1) and EPN (CL-K1) in CRD preferentially binds mannose-containing carbohydrates. Both CL-L1 and CL-K1 include a C-X-C motif in the neck domain.[1,2] CL-L1 forms a heteromeric complex with CL-K1 in human blood. CL-L1 and CL-K1 are present in blood and they are secreted mainly by the liver.[3] The heteromeric complex of CL-L1 and CL-K1 was named CL-LK (Fig. 6.1).[4]

CRYSTAL STRUCTURE

A recombinant fragment of CL-K1 (representing the neck) and CRD was expressed in *E. coli*, solubilised from inclusion bodies, and crystallised to a resolution of 2.45 Å, with and without carbohydrate ligands. In parallel with the structures of other collectins, the structure was trimeric and stabilised by noncovalent interpolypeptide chain interactions in the alpha-helical coiled-coil region.[5]

FUNCTION

CL-LK activates the lectin pathway of the complement system via mannose-binding lectin-associated serine proteases, MASP-2 and MASP-1, similar to mannose-binding lectin.[4,6] CL-K1 regulation and function have been associated with the coagulation system. The incidence of elevated plasma levels of CL-K1 was significantly higher in patients with disseminated intravascular coagulation.[7] High serum levels of CL-K1 among Nigerian populations are associated with decreased appearance of urogenital infection with *Schistosoma haematobium*.[8] CL-LK has also been shown to bind to *Mycobacterium tuberculosis* via interaction with the lipoglycan

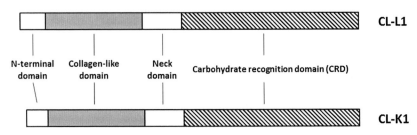

FIGURE 6.1   Diagram of protein domains of CL-L1 and CL-K1.

lipoarabinomannan, which is a major constituent of their cell envelope. The serum level of CL-K1 was further reduced in patients with tuberculosis, suggesting either a consumption of CL-K1 or that low levels are associated with increased risk of infection or colonisation.[9] In a cross-sectional cohort study of systemic lupus erythematosus (SLE), decreased serum levels of CL-L1 were associated with SLE.[10] CL-K1 has shown to interact calcium independently with negatively charged ligands, e.g., DNA, RNA and sulphated compounds, i.e., heparin.[4] CL-K1 served as a chemotactic attractant to guide migration of neural crest cells (in zebra fish model).[11] Mutations in both CL-K1 and MASP-3 genes are involved in the autosome recessive hereditary disease 3MC (Mingarelli, Malpuech, Michels and Carnevale) syndrome.[11] This fact indicates that CL-K1 might play an important role in foetal development. See the details in 'Deficiency' section.

Sugar Specificity

CL-L1: D-Man, L-Fuc, D-Fuc, Gal, GlcNAc[1,12]
CL-K1: D-Man, Mannose (1–2)-Mannose,[5] L-Fuc, ManNAc[1,13]

Ligand Specificity

CL-L1: None known to date
CL-K1: DNA, RNA, sulphated compounds, apoptotic cells, *E. coli, Klebsiella pneumoniae, Pseudomonas aeruginosa, M. tuberculosis, Saccharomyces cerevisiae, Candida albicans* and Influenza A virus, *S. haematobium*

TISSUE DISTRIBUTION

CL-L1

Liver, stomach, small intestine, colon, prostate, placenta, lung, heart, spleen, brain, kidney, bone marrow, mammary gland, uterus, pancreas, adrenal gland and thyroid.

The average serum concentration of CL-L1 was 306 ng/mL in a healthy Danish population.[10,12]

CL-K1

Adrenals, liver, kidney, pancreas, central nervous system , stomach, small intestine, colon, testis, ovaries, placenta, lung, heart, thymus, large intestine, skeletal muscle and white adipose tissue.

In the adrenals, all three layers, zona fasciculate, glomerulosa and reticularis, were associated with CL-K1 immunoreactivity. In the liver, hepatocytes appear as the cellular source of CL-K1 and in the kidneys, CL-K1 immunoreactivity was associated particularly with the distal tubules but also with the glomerulus and proximal tubules.[3,13] The average serum concentration of CL-K1 in healthy populations in Japan and Denmark was estimated to be 340 and 284 ng/mL, respectively.[14,15]

REGULATION OF EXPRESSION

To date little work has been performed examining the regulation of expression of CL-L1 and CL-K1.

HUMAN PROTEIN SEQUENCE

CL-L1

```
MNGFASLLRR NQFILLVLFL LQIQSLGLDI DSRPTAEVCA THTISPGPKG   50
DDGEKGDPGE EGKHGKVGRM GPKGIKGELG DMGDQGNIGK TGPIGKKGDK  100
GEKGLLGIPG EKGKAGTVCD CGRYRKFVGQ LDISIARLKT SMKFVKNVIA  150
GIRETEEKFY YIVQEEKNYR ESLTHCRIRG GMLAMPKDEA ANTLIADYVA  200
KSGFFRVFIG VNDLEREGQY MFTDNTPLQN YSNWNEGEPS DPYGHEDCVE  250
MLSSGRW**N**DT ECHLTMYFVC EFIKKKK
```

CL-K1

CL-K1-I, isoform a (transcript variant 1)

```
MRGNLALVGV LISLAFLSLL PSGHPQPAGD DACSVQILVP GLKGDAGEKG   50
DKGAPGRPGR VGPTGEKGDM GDKGQKGSVG RHGKIGPIGS KGEKGDSGDI  100
GPPGPNGEPG LPCECSQLRK AIGEMDNQVS QLTSELKFIK NAVAGVRETE  150
SKIYLLVKEE KRYADAQLSC QGRGGTLSMP KDEAANGLMA AYLAQAGLAR  200
VFIGINDLEK EGAFVYSDHS PMRTFNKWRS GEPNNAYDEE DCVEMVASGG  250
WNDVACHTTM YFMCEFDKEN M
```

The leader sequences are underlined. Glycosylation sites are denoted as **N**.

PROTEIN MODULES

CL-L1

1–27	Leader peptide	exon 1
28–46	N-terminal domain	exon 1
47–118	Collagen-like domain	exons 1–5
119–147	Neck domain	exon 5
148–277	Carbohydrate recognition domain (CRD)	exon 6

CL-K1
CL-K1-I, isoform a (transcript variant 1)

	(untranslated region)	(exon 1)
1–25	Leader peptide	exon 4
25–40	N-terminal domain	exon 4
41–112	Collagen-like domain	exons 4, 6, 8, 9, 10
113–141	Neck domain	exon 10
142–271	Carbohydrate recognition domain (CRD)	exon 11

CHROMOSOMAL LOCATION

CL-L1:

Human	8q24.12
Mouse	15 D1; 15 21.15 cM
Rat	7q31

CL-K1:

Human	2p25.3
Mouse	12; 12 B1
Rat	6q16

HUMAN cDNA SEQUENCE

CL-L1

```
CAGCAATGAA TGGCTTTGCA TCCTTGCTTC GAAGAAACCA ATTTATCCTC CTGGTACTAT   60
TTCTTTTGCA AATTCAGAGT CTGGGTCTGG ATATTGATAG CCGTCCTACC GCTGAAGTCT  120
GTGCCACACA CACAATTTCA CCAGGACCCA AAGGAGATGA TGGTGAAAAA GGAGATCCAG  180
GAGAAGAGGG AAAGCATGGC AAAGTGGGAC GCATGGGGCC GAAAGGAATT AAAGGAGAAC  240
TGGGTGATAT GGGAGATCGG GGCAATATTG GCAAGACTGG GCCCATTGGG AAGAAGGGTG  300
ACAAAGGGGA AAAAGGTTTG CTTGGAATAC CTGGAGAAAA AGGCAAAGCA GGTACTGTCT  360
GTGATTGTGG AAGATACCGG AAATTTGTTG GACAACTGGA TATTAGTATT GCCCGGCTCA  420
AGACATCTAT GAAGTTTGTC AAGAATGTGA TAGCAGGGAT TAGGGAAACT GAAGAGAAAT  480
TCTACTACAT CGTGCAGGAA GAGAAGAACT ACAGGGAATC CCTAACCCAC TGCAGGATTC  540
GGGGTGGAAT GCTAGCCATG CCCAAGGATG AAGCTGCCAA CACACTCATC GCTGACTATG  600
TTGCCAAGAG TGGCTTCTTT CGGGTGTTCA TTGGCGTGAA TGACCTTGAA AGGGAGGGAC  660
AGTACATGTT CACAGACAAC ACTCCACTGC AGAACTATAG CAACTGGAAT GAGGGGGAAC  720
CCAGCGACCC CTATGGTCAT GAGGACTGTG TGGAGATGCT GAGCTCTGGC AGATGGAATG  780
ACACAGAGTG CCATCTTACC ATGTACTTTG TCTGTGAGTT CATCAAGAAG AAAAAGTAAC  840
TTCCCTCATC CTACGTATTT GCTATTTTCC TGTGACCGTC ATTACAGTTA TTGTTATCCA  900
TCCTTTTTTT CCTGATTGTA CTACATTTGA TCTGAGTCAA CATAGCTAGA AAATGCTAAA  960
CTGAGGTATG GAGCCTCCAT CATCATGCTC TTTTGTGATG ATTTCATAT TTTCACACAT 1020
GGTATGTTAT TGACCCAATA ACTCGCCAGG TTACATGGGT CTTGAGAGAG AATTTTAATT 1080
ACTAATTGTG CACGAGATAG TTGGTTGTCT ATATGTCAAA TGAGTTGTTC TCTTGGTATT 1140
TGCTCTACCA TCTCTCCCTA GAGCACTCTG TGTCTATCCC AGTGGATAAT TTCCCAGTTT 1200
ACTGGTGATG ATTAGGAAGG TTGTTGATGG TTAGGCTAAC CTGCCCTGGC CCAAAGCCAG 1260
ACATGTACAA GGGCTTTCTG TGAGCAATGA TAAGATCTTT GAATCCAAGA TGCCCAGATG 1320
TTTTACCAGT CACACCCTAT GGCCATGGCT ATACTTGGAA GTTCTCCTTG TTGGCACAGA 1380
CATAGAAATG CTTTAACCCC AAGCCTTTAT ATGGGGGACT TCTAGCTTTG TGTCTTGTTT 1440
CAGACCATGT GGAATGATAA ATACTCTTTT TGTGCTTCTG ATCTATCGAT TTCACTAACA 1500
TATACCAAGT AGGTGCTTTG AACCCCTTTC TGTAGGCTCA CACCTTAATC TCAGGCCCCT 1560
ATATAGTCAC ACTTTGATTT AAGAAAAACG GAGC
```

CL-K1

isoform a, c, d, e

```
GGACGGTGGA CGCAGCGCAG ACAGGAAGCT CCCCGAGATA ACGCTGCGGC CGGGCGGCCT   60
GATTTGCTGG GCTGTCTGAT GGCCCGGGCC GAGGCTTCTC CCTGCGCCTG GGACTGCGGC  120
CGCCTCTCTA AATAGCAGCC ATGAGGCGCC TGGGGGCAGT GTCCTCGCGG CCGCGTCGAC  180
CGACGGCCGC AGTCGACGCC CCGTTCGCCT AGCGCGTGCT CAGGAGTTGG TGTCCTGCCT  240
GCGCTCAGGA TGAGGGGGAA TCTGGCCCTG GTGGGCGTTC TAATCAGCCT GGCCTTCCTG  300
TCACTGCTGC CATCTGGACA TCCTCAGCCG GCTGGCGATG ACGCCTGCTC TGTGCAGATC  360
CTCGTCCCTG GCCTCAAAGG GGATGCGGGA GAGAAGGGAG ACAAAGGCGC CCCCGGACGG  420
CCTGGAAGAG TCGGCCCCAC GGGAGAAAAA GGAGACATGG GGGACAAAGG ACAGAAAGGC  480
AGTGTGGGTC GTCATGGAAA AATTGGTCCC ATTGGCTCTA AAGGTGAGAA AGGAGATTCC  540
GGTGACATAG GACCCCCTGG TCCTAATGGA GAACCAGGCC TCCCATGTGA GTGCAGCCAG  600
CTGCGCAAGG CCATCGGGGA GATGGACAAC CAGGTCTCTC AGCTGACCAG CGAGCTCAAG  660
TTCATCAAGA ATGCTGTCGC CGGTGTGCGC GAGACGGAGA GCAAGATCTA CCTGCTGGTG  720
AAGGAGGAGA AGCGCTACGC GGACGCCCAG CTGTCCTGCC AGGGCCGCGG GGGCACGCTG  780
AGCATGCCCA AGGACGAGGC TGCCAATGGC CTGATGGCCG CATACCTGGC GCAAGCCGGC  840
CTGGCCCGTG TCTTCATCGG CATCAACGAC CTGGAGAAGG AGGGCGCCTT CGTGTACTCT  900
GACCACTCCC CCATGCGGAC CTTCAACAAG TGGCGCAGCG GTGAGCCCAA CAATGCCTAC  960
GACGAGGAGG ACTGCGTGGA GATGGTGGCC TCGGGCGGCT GGAACGACGT GGCCTGCCAC 1020
ACCACCATGT ACTTCATGTG TGAGTTTGAC AAGGAGAACA TGTGAGCCTC AGGCTGGGGC 1080
TGCCCATTGG GGGCCCCACA TGTCCCTGCA GGGTTGGCAG GGACAGAGCC CAGACCATGG 1140
TGCCAGCCAG GGAGCTGTCC CTCTGTGAAG GGTGGAGGCT CACTGAGTAG AGGGCTGTTG 1200
TCTAAACTGA GAAAATGGCC TATGCTTAAG AGGAAAATGA AAGTGTTCCT GGGGTGCTGT 1260
CTCTGAAGAA GCAGAGTTTC ATTACCTGTA TTGTAGCCCC AATGTCATTA TGTAATTATT 1320
ACCCAGAATT GCTCTTCCAT AAAGCTTGTG CCTTTGTCCA AGCTATACAA TAAAATCTTT 1380
AAGTAGTGCA GT
```

The methionine initiation codon (<u>ATG</u>), the termination codon (<u>TAA</u>, <u>TGA</u>) and the probable polyadenylation signal (<u>AATAA</u>, <u>AATAAA</u>) are indicated. The first five nucleotides in each exon are underlined.

GENOMIC STRUCTURE

Human CL-L1 is encoded by six exons, 39 kb long. Human CL-K1 is encoded by eight exons 49.9 kb long (Fig. 6.2).

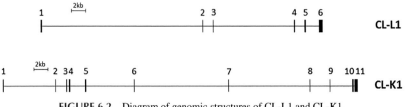

FIGURE 6.2   Diagram of genomic structures of CL-L1 and CL-K1.

ACCESSION NUMBERS

CL-L1

Human: AB002631.1
Gene ID 10584
NM_006438.3

CL-K1

Human: AB119525, AB119650, AB119651, AB119652
Gene ID 78989
NM_024027.4, NM_199235.2, NM_001255982.1, NM_001255983.1,
NM_001255984.1, NM_001255985.1, NM_001255986.1,
NM_001255987.1, NM_001255988.1, NM_001255989.1

DEFICIENCY

CL-K1

Deficiency of either CL-K1 or MASP-3 was shown to cause the rare autosomal recessive syndrome known as 3MC syndrome, characterised by various symptoms including facial, genital, renal, mental and limb abnormalities.[11]

Ser169Pro, rs387907075 (in the CRD)
Gly204Ser, rs387907076 (in the CRD)
Ser217del, c.648-659delCTC (in the CRD)
Phe16SerfsX85, c.56delC (frame shift or lack of transcription)
Gly101ValfsX113, c.300delT (frame shift or lack of transcription)
an exon 1–3 deletion

Homozygosity for any of the six 3MC-associated mutations results in the development of the syndrome and lack of CL-K1 in the circulation, while heterozygotes are healthy with approximately half of the normal CL-K1 level.[11,15] Developmental delay/intellectual disability was associated with hypermethylation of *COLEC11*, but eventual CL-K1 deficiency in the affected individuals was not investigated further, although a single patient had some overlapping symptoms with that of 3MC patients.[16]

POLYMORPHIC VARIANTS

CL-L1

Glu78Asp (rs150828850, collagen domain, minor allele frequency 0.003)[17]
Arg125Trp (rs149331285, neck domain, minor allele frequency 0.007)[17]
The CL-L1 serum concentration was increased by approximately 40% ($P = .0478$).

CL-K1

His219Arg (rs7567833, CRD, minor allele frequency 0.033)[17]
9570C>T promoter region (rs3820897)[17]

Increased serum levels (at least 10%) … three homozygous individuals tested ($P=.044$). A minor but similar (insignificant) trend was observed among 41 heterozygotes.

His219Arg (rs7567833) Nigerian individuals of Yoruba ethnicity[8,18]

Decreased serum levels, with approximately a 40% decrease in the median concentration between homozygotes ($P=.03$, $n=41$).

MUTANT ANIMALS

CL-K1$^{-/-}$ mice have been generated and their phenotype is currently under investigation (Mori et al., manuscript in preparation).

REFERENCES

1. Ohtani K, Suzuki Y, Eda S, Kawai T, Kase T, Yamazaki H, et al. Molecular cloning of a novel human collectin from liver (CL-L1). *J Biol Chem* 1999;**274**(19):13681–9.
2. Keshi H, Sakamoto T, Kawai T, Ohtani K, Katoh T, Jang SJ, et al. Identification and characterization of a novel human collectin CL-K1. *Microbiol Immunol* 2006;**50**(12):1001–13.
3. Motomura W, Yoshizaki T, Ohtani K, Okumura T, Fukuda M, Fukuzawa J, et al. Immunolocalization of a novel collectin CL-K1 in murine tissues. *J Histochem Cytochem* 2008;**56**(3):243–52.
4. Henriksen ML, Brandt J, Andrieu JP, Nielsen C, Jensen PH, Holmskov U, et al. Heteromeric complexes of native collectin kidney 1 and collectin liver 1 are found in the circulation with MASPs and activate the complement system. *J Immunol* 2013;**191**(12):6117–27.
5. Venkatraman Girija U, Furze CM, Gingras AR, Yoshizaki T, Ohtani K, Marshall JE, et al. Molecular basis of sugar recognition by collectin-K1 and the effects of mutations associated with 3MC syndrome. *BMC Biol* 2015;**13**:27.
6. Kjaer T, Jensen L, Hansen A, Dani R, Jensenius JC, Dobó J, et al. Oligomerization of mannan-binding lectin dictates binding properties and complement activation. *Scand J Immunol* 2016;**84**(1):12–9.
7. Takahashi K, Ohtani K, Larvie M, Moyo P, Chigweshe L, Van Cott EM, et al. Elevated plasma CL-K1 level is associated with a risk of developing disseminated intravascular coagulation (DIC). *J Thromb Thrombolysis* 2014;**38**(3):331–8.
8. Antony JS, Ojurongbe O, Meyer CG, Thangaraj K, Mishra A, Kremsner PG, et al. Correlation of interleukin-6 levels and lectins during *Schistosoma haematobium* infection. *Cytokine* 2015;**76**(2):152–5.
9. Troegeler A, Lugo-Villarino G, Hansen S, Rasolofo V, Henriksen ML, Mori K, et al. Collectin CL-LK is a novel soluble pattern recognition receptor for mycobacterium tuberculosis. *PLoS One* 2015;**10**:e0132692.
10. Troldborg A, Thiel S, Jensen L, Hansen S, Laska MJ, Deleuran B, et al. Collectin liver 1 and collectin kidney 1 and other complement-associated pattern recognition molecules in systemic lupus erythematosus. *Clin Exp Immunol* 2015;**182**(2):132–8.
11. Rooryck C, Diaz-Font A, Osborn DP, Chabchoub E, Hernandez-Hernandez V, Shamseldin H, et al. Mutations in lectin complement pathway genes COLEC11 and MASP1 cause 3MC syndrome. *Nat Genet* 2011;**43**(3):197–203.
12. Axelgaard E, Jensen L, Dyrlund TF, Nielsen HJ, Enghild JJ, Thiel S, et al. Investigations on collectin liver 1. *J Biol Chem* 2013;**288**(32):23407–20.

13. Hansen S, Selman L, Palaniyar N, Ziegler K, Brandt J, Kliem A, et al. Collectin 11 (CL-11, CL-K1) is a MASP-1/3-associated plasma collectin with microbial-binding activity. *J Immunol* 2010;**185**(10):6096–104.

14. Yoshizaki T, Ohtani K, Motomura W, Jang SJ, Mori K, Kitamoto N, et al. Comparison of human blood concentrations of collectin kidney 1 and mannan-binding lectin. *J Biochem* 2012;**151**(1):57–64.

15. Selman L, Hansen S. Structure and function of collectin liver 1 (CL-L1) and collectin 11 (CL-11, CL-K1). *Immuonobiology* 2012;**217**(9):851–63.

16. Kolarova J, Tangen I, Bens S, Gillessen-Kaesbach G, Gutwein J, Kautza M, et al. Array-based DNA methylation analysis in individuals with developmental delay/intellectual disability and normal molecular karyotype. *Eur J Med Genet* 2015;**58**(8):419–25.

17. Bayarri-Olmos R, Hansen S, Henriksen ML, Storm L, Thiel S, Garred P, et al. Genetic variation of COLEC10 and COLEC11 and association with serum levels of collectin liver 1 (CL-L1) and collectin kidney 1 (CL-K1). *PLoS One* 2015;**10**(2):e0114883.

18. Antony JS, Ojurongbe O, Kremsner PG, Velavan TP. Lectin complement protein collectin 11 (CL-K1) and susceptibility to urinary schistosomiasis. *PLoS Negl Trop Dis* 2015;**9**(3):e0003647.

Part II

Serine Proteases

Chapter 7

MASP-1

Ramus Pihl, Jens C. Jensenius, Steffen Thiel
Aarhus University, Aarhus, Denmark

PHYSICOCHEMICAL PROPERTIES

Immature Protein

Mannan-binding lectin-associated serine protease 1 (MASP-1) is synthesised as a single-chain polypeptide of 699 residues. The 19 N-terminal residues constitute a leader peptide that is removed during secretion, resulting in a protein with a theoretical molecular weight of 77.0 kDa. However, due to the presence of four *N*-glycosylation sites it migrates as a 93-kDa protein in SDS-PAGE. MASP-1 is generated as a proenzyme that later becomes activated by cleavage between residues R448 and I449, yielding an A-chain and a B-chain linked by a disulphide bond.

Mature Protein		
pI (theoretical)	5.21	
Molecular weight	93 kDa	

	A-Chain	**B-Chain**
Residues	20–448	449–699
Molecular weight	49 kDa (theoretical)	28 kDa (theoretical)
	66 kDa (observed)	31 kDa (observed)
N-Linked glycosylation		
Sites	4 (residues 49, 178, 385, 407)	0
Disulphide bond between		
A-chain and B-chain	C436	C572

The Complement FactsBook. http://dx.doi.org/10.1016/B978-0-12-810420-0.00007-9

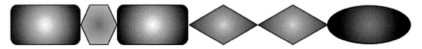

FIGURE 7.1 Protein domains of MASP1.

STRUCTURE

MASP-1 consists of six discrete domains, analogous to C1r, C1s, MASP-2 and MASP-3 (Fig. 7.1), and exists as a homodimer due to Ca^{2+}-dependent dimerisation mediated by the CUB1-EGF domains in an antiparallel orientation. The MASP-1 dimer forms an elongated structure with the two serine protease (SP) domains separated by 200–300 Å.[1] Binding to the collagen-like domains of pattern-recognition molecules of the lectin pathway (LP) (i.e., mannan-binding lectin, H-ficolin, M-ficolin, L-ficolin and collectin LK) is facilitated by the CUB1 and CUB2 domains,[2] while the CCP1 and CCP2 domains likely influence the catalytic properties of MASP-1, as found in other complement proteases.[3]

The SP domain of MASP-1 possesses an accessible and wide substrate binding groove that allows it to cleave a number of substrates (Fig. 7.2).[4]

FUNCTION

The isoforms MASP-1, MASP-3 and MAp44 are generated through alternative splicing from the *MASP1* gene. MASP-1 circulates bound to the ficolins and collectins of the LP, and when these complexes are concentrated on an activating surface, MASP-1 will undergo autoactivation.[5] Upon activation, MASP-1 is able to activate MASP-2, which subsequently cleaves C2 and C4 resulting in C3 convertase formation (C4b2a).[6] These events might similarly lead to activation of MASP-3.[7] Furthermore, MASP-1 also contributes directly to C2 cleavage. Although MASP-1 cleaves C2 significantly less efficiently than MASP-2, the higher abundance of MASP-1 means that approximately half of the C2 that is cleaved during LP activation is possibly cleaved by MASP-1.[8] The LP is not strictly dependent on MASP-1, as evidenced by the fact that chickens, which lack MASP-1, still have an active LP.[9] This activity is due to the slower autoactivation of MASP-2. However, in scenarios where MASP-1 is present, but inhibited, the LP will be nonfunctional, which highlights the role of MASP-1 as the physiologically relevant activator of MASP-2.[8]

MASP-1 may also serve complement-independent functions. Evidence is emerging that MASP-1 plays a role as a procoagulant by cleaving a number of coagulation factors, including the conversion of prothrombin to thrombin.[10,11]

TISSUE DISTRIBUTION

The mean serum concentration of MASP-1 in healthy donors is estimated as 11.0 μg/mL (143 nM of MASP-1 monomer) with concentrations ranging from 4.2 to 29.8 μg/mL. The concentration is similar in citrate plasma, slightly lower in EDTA plasma, and approximately 1.5-fold higher in heparin plasma.[12]

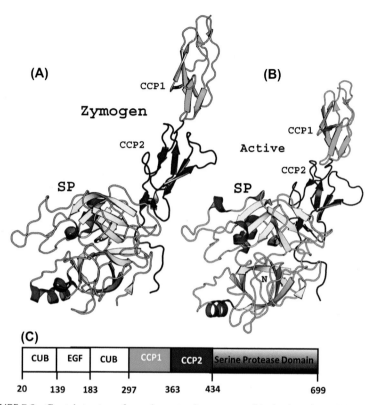

FIGURE 7.2 Crystal structure of complement serine protease of the lectin pathway, MASP1. (A) Ribbon diagram of catalytic region in zymogen form of human mannose-binding lectin (MBL) – associated serine protease 1 (MASP1) crystal structure (PDB ID: 4IGD). (B) Ribbon diagram of activated catalytic region of human MASP1 crystal structure (PDB ID: 3GOV). (C) Schematic representation of MASP1 domains that are coloured similar to the ribbon representation.

MASP-1 has been reported to be undetectable in cerebrospinal fluid and urine.[13]

MASP-1 is predominantly synthesised in the liver. Lower levels of expression are also seen in the small intestine and the kidney.[14]

REGULATION OF EXPRESSION

MASP-1 does not behave as a classical acute-phase protein, as it is quickly downregulated during an acute-phase response, followed by a slower rise in concentration to approximately twice the normal level.[12]

The transcriptional regulation of MASP-1 is relatively unstudied, although there are indications of a modest upregulation by IL-1β and a similarly small downregulation by IL-6 and IFN-γ.[15]

It remains unknown what defines the balance between the alternative splicing products MASP-1, MASP-3 and MAp44 from the *MASP1* gene.

HUMAN PROTEIN SEQUENCE

The sequence below is from uniprot.org and has the identifier P48740-1:

```
MRWLLLYYAL CFSLSKASAH TVELNNMFGQ IQSPGYPDSY PSDSEVTWNI    50
TVPDGFRIKL YFMHFNLESS YLCEYDYVKV ETEDQVLATF CGRETTDTEQ   100
TPGQEVVLSP GSFMSITFRS DFSNEERFTG FDAHYMAVDV DECKEREDEE   150
LSCDHYCHNY IGGYYCSCRF GYILHTDNRT CRVECSDNLF TQRTGVITSP   200
DFPNPYPKSS ECLYTIELEE GFMVNLQFED IFDIEDHPEV PCPYDYIKIK   250
VGPKVLGPFC GEKAPEPIST QSHSVLILFH SDNSGENRGW RLSYRAAGNE   300
CPELQPPVHG KIEPSQAKYF FKDQVLVSCD TGYKVLKDNV EMDTFQIECL   350
KDGTWSNKIP TCKIVDCRAP GELEHGLITF STRNNLTTYK SEIKYSCQEP   400
YYKMLNNNTG IYTCSAQGVW MNKVLGRSLP TCLPVCGLPK FSRKLMARIF   450
NGRPAQKGTT PWIAMLSHLN GQPFCGGSLL GSSWIVTAAH CLHQSLDPED   500
PTLRDSDLLS PSDFKIILGK HWRLRSDENE QHLGVKHTTL HPQYDPNTFE   550
NDVALVELLE SPVLNAFVMP ICLPEGPQQE GAMVIVSGWG KQFLQRFPET   600
LMEIEIPIVD HSTCQKAYAP LKKKVTRDMI CAGEKEGGKD ACAGDSGGPM   650
VTLNRERGQW YLVGTVSWGD DCGKKDRYGV YSYIHHNKDW IQRVTGVRN    699
```

The signal peptide is underlined. The *N*-glycosylation sites are depicted in bold, while the residues that constitute the catalytic triad are shown in italic and coloured grey (residues H490, D552, S646). The cleavage site between the A- and B-chains is written in bold and underlined.

PROTEIN MODULES

1–19	Leader peptide	exon 1+2
20–138	CUB1	exon 2+3
139–182	EGF-like	exon 4
185–297	CUB2	exon 5+6
299–364	CCP1	exon 7+8
365–434	CCP2	exon 10+11
449–696	SP	exons 13–18

CHROMOSOMAL LOCATION[16]

Human:	3q27-q28
Mouse:	16 B2-B3
Rat:	11q23

HUMAN cDNA SEQUENCE

The sequence is the NCBI Reference Sequence: NM_001879.5

```
acacacagag tgatacaaat acctgcttga gcccctcagt tattttctct caagggctga    60
agtcagccac acaggataaa ggagggaagg gaaggagcag atcttttcgg taggaagaca   120
gattttgttg tcaggttcct gggagtgcaa gagcaagtca aaggagagag agaggagaga   180
ggaaaagcca gagggagaga gggggagagg ggatctgttg caggcagggg aaggcgtgac   240
ctgaatggag aatgccagcc aattccagag acacacaggg acctcagaac aaagataagg   300
catcacggac accacaccgg gcacgagctc acaggcaagt caagctggga ggaccaaggc   360
cggcagccg ggagcaccca aggcaggaaa **atg**aggtggc tgcttctcta ttatgctctg   420
tgcttctccc tgtcaaaggc ttcagcccac accgtggagc taaacaatat gtttggccag   480
atccagtcgc ctggttatcc agactcctat cccagtgatt cagaggtgac ttggaatatc   540
actgtcccag atgggtttcg gatcaagctt tacttcatgc acttcaactt ggaatcctcc   600
tacctttgtg aatatgacta tgtgaaggta gaaactgagg accaggtgct ggcaaccttc   660
tgtggcaggg agaccacaga cacagagcag actcccggcc aggaggtggt cctctcccct   720
ggctccttca tgtccatcac tttccggtca gatttctcca atgaggagcg tttcacaggc   780
tttgatgccc actacatggc tgtggatgtg gacgagtgca aggagaggga ggacgaggag   840
ctgtcctgtg accactactg ccacaactac attggcggct actactgctc ctgccgcttc   900
ggctacatcc tccacacaga caacaggacc tgccgagtgg agtgcagtga caacctcttc   960
actcaaagga ctggggtgat caccagccct gacttcccaa acccttaccc caagagctct  1020
gaatgcctgt ataccatcga gctggaggag ggtttcatgg tcaacctgca gtttgaggac  1080
atatttgaca ttgaggacca tcctgaggtg ccctgcccct atgactacat caagatcaaa  1140
gttggtccaa aagttttggg gcctttctgt ggagagaaag ccccagaacc catcagcacc  1200
cagagccaca gtgtcctgat cctgttccat agtgacaact cgggagagaa ccggggctgg  1260
aggctctcat acagggctgc aggaaatgag tgcccagagc tacagcctcc tgtccatggg  1320
aaaatcgagc cctcccaagc caagtatttc ttcaaagacc aagtgctcgt cagctgtgac  1380
acaggctaca aagtgctgaa ggataatgtg gagatggaca cattccagat tgagtgtctg  1440
aaggatggga cgtggagtaa caagattccc acctgtaaaa ttgtagactg tagagcccca  1500
ggagagctgg aacacgggct gatcaccttc tctacaagga acaacctcac cacatacaag  1560
tctgagatca aatactcctg tcaggagccc tattacaaga tgctcaacaa taacacaggt  1620
atatatacct gttctgccca aggagtctgg atgaataaag tattgggag aagcctaccc  1680
acctgccttc cagtgtgtgg gctccccaag ttctcccgga agctgatggc caggatcttc  1740
aatggacgcc cagcccagaa aggcaccact ccctggattg ccatgctgtc acacctgaat  1800
gggcagccct tctgcggagg ctcccttcta ggctccagct ggatcgtgac cgccgcacac  1860
tgcctccacc agtcactcga tccggaagat ccgaccctac gtgattcaga cttgctcagc  1920
ccttctgact tcaaaatcat cctgggcaag cattggaggc tccggtcaga tgaaaatgaa  1980
cagcatctcg gcgtcaaaca caccactctc caccccagt atgatcccaa cacattcgag  2040
aatgacgtgg ctctggtgga gctgttggag agcccagtgc tgaatgcctt cgtgatgccc  2100
atctgtctgc ctgagggacc ccagcaggaa ggagccatgg tcatcgtcag cggctggggg  2160
aagcagttct tgcaaaggtt cccagagacc ctgatggaga ttgaaatccc gattgttgac  2220
cacagcacct gccagaaggc ttatgccccg ctgaagaaga aagtgaccag ggacatgatc  2280
tgtgctgggg agaaggaagg gggaaaggac gcctgtgcgg gtgactctgg aggccccatg  2340
gtgaccctga atagagaaag aggccagtgg tacctggtgg gcactgtgtc ctggggtgat  2400
gactgtggga agaaggaccg ctacggagta tactcttaca tccaccacaa caaggactgg  2460
atccagaggg tcaccgggagt gaggaac**tga** atttggctcc tcagcccag caccaccagc  2520
tgtgggcagt cagtagcaga ggacgatcct ccgatgaaag cagccatttc tcctttcctt  2580
cctcccatcc cccctccttc ggcctatcca ttactgggca atagagcagg tatcttcacc  2640
```

```
cccttttcac tctctttaaa gagatggagc aagagagtgg tcagaacaca ggccgaatcc 2700
aggctctatc acttactagt ttgcagtgct gggcaggtga cttcatctct tcgaacttca 2760
gtttcttcat aagatggaaa tgctatacct tacctacctc gtaaaagtct gatgaggaaa 2820
agattaacta atagatgcat agcacttaac agagtgcata gcatacactg ttttcaataa 2880
atgcacctta gcagaaggtc gatgtgtcta ccaggcagac gaagctctct tacaaacccc 2940
tgcctgggtc ttagcattga tcagtgacac acctctcccc tcaaccttga ccatctccat 3000
ctgcccttaa atgctgtatg ctttttttgcc accgtgcaac ttgcccaaca tcaatcttca 3060
ccctcatccc taaaaaagta aaacagacaa ggttctgagt cctgtggtat gtcccctagc 3120
aaatgtaact aggaacatgc actagatgac agattgcggg agggcctgag agaagcaggg 3180
acaggaggga gcctggggat tgtggtttgg gaaggcagac acctggttct agaactagct 3240
ctgcccttag cccctgtat gaccctatgc aagtcctcct ccctcatctc aaagggtcct 3300
caaagctctg acgatctaag atacaatgaa gccattttcc ccctgataag atgaggtaaa 3360
gccaatgtaa ccaaaaggca aaaattacaa tcggttcaaa ggaactttga tgcagacaaa 3420
atgctgctgc tgctgctcct gaaataccca cccctttcca ctacgggtgg gttcccaagg 3480
acatgggaca ggcaaagtgt gagccaaagg atccttcctt attcctaagc agagcatctg 3540
ctctgggccc tggcctcctt cccttcttgg gaaactgggc tgcatgaggt gggccctggt 3600
agtttgtacc ccaggcccct atactcttcc ttcctatgtc cacagctgac cccaagcagc 3660
cgttccccga ctcctcaccc ctgagcctca ccctgaactc cctcatcttg caaggccata 3720
agtgttttcc aagcaaaatg cctctcccat cctctctcag gaagcttcta gagactttat 3780
gccctccaga gctccaagat ataagccctc caagggatca gaagctccaa gttcctgtct 3840
tctgttttat agaaattgat cttccctggg ggactttaac tcttgacctg tatgcagctg 3900
ttggagtaat tccaggtctc ttgaaaaaaa agaggaagat aatggagaat gagaacatat 3960
atatatatat attaagcccc aggctgaata ctcagggaca gcaattcaca gcctgcctct 4020
ggttctataa acaagtcatt ctacctcttt gtgccctgct gtttattctg taaggggaag 4080
gtggcaatgg gacccagctc catcagacac ttgtcaagct agcagaaact ccattttcaa 4140
tgccaaagaa gaactgtaat gctgtttttgg aatcatccca aggcatccca agacaccata 4200
tcttcccatt tcaagcactg cctgggcaca ccccaacatc ccaggctgtg gtggctcctg 4260
tgggaactac ctagatgaag agagtatcat ttataccttc taggagctcc tattgggaga 4320
catgaaacat atgtaattga ctaccatgta atagaacaaa ccctgccaag tgctgctttg 4380
guaagtcauy yaygtaaaag aaagaccatt ctggtatgaa ggtttttgggg gagaagatat 4440
caatcaagaa ggcttcccag aagaggtgac tggaccagag ccttgtccac aggtaagacg 4500
gaggaggcct tccacatgga gggagaacaa tagtaaatgt ccactcaaga tgtcctttat 4560
tataccagct cctcccacaa aaacacatgt ccagtggact cttttttctgg gatcagaacc 4620
aacaccaaaa agagctttc tccttaaagt tagaattcta aacaggactt gaaatggcct 4680
caaggtttgt gcacaaatac tgacttctgt ctggacccag cttattctgt ttatttctcc 4740
aattgcaatt tcatccttat cctgagaaaa tgtctaaata ggccatggaa cccaggcttc 4800
cccgtgacct acaagcactt attagctgtg ccagctcctg cactgctgct aaggtccaag 4860
aaacccagat ctctcacaga gccatagaag cagagggctg gagtatctgt gaggacaaca 4920
accttgtcta acttcgcgac ctcattcttg agcatttcta ctgatgagaa actcactacc 4980
cccaattgca gctcattcaa ctttaaaatt gctgcttttt gaatcactga ttgtgaatat 5040
taatttaaaa aaataagtaa gaaaatgttt aaatgtgct gcctctttaa aaggtctcct 5100
ctttgtgcaa ccaaaaccca cctctctaga atacagtttg tataactgaa gctataattt 5160
cataccatga gtgctgctgt tagcaataat aatcatgccc ggattttatt aacaacagaa 5220
gctgttgctc gtatgaaaaa acaaacatta gttctaataa acatctgcat tgagtcaaag 5280
```

```
ctccctgttt gtttgtatgt cttttatgca ctgatgatta tagtgagttg ctttcattta   5340
ccaacatttt gttgtattcg tgtaggatca ctgtaccatg aagggagaga gactatgatg   5400
ggaagattgt tgtagataca aaagcatgtc ttaggttttt gggtcagttc tgtttaaata   5460
cctgtcctat tattcctgta aattatcaaa atatcccaga atgtcaatgt ttctgcatcc   5520
acattacaat tattaaatgc cactcattta ttaaatttac tattatcagt ggcatttaat   5580
aaatttgaat catatgttca gtgtttggtt tagaaaatat ggtgccatgt ctatgagtgg   5640
cctgttctgg attggagtac atgccttctt tctgccttga gttaatctta ctcaatggag   5700
aacaagaatc aaagaaacac caccaccaag aagcccttca agctagagtt gggcaagagt   5760
cagggagggg aatgtagacc actcatatga cagaggtgga aaccaatctt ggtctagaat   5820
aagtctcaaa atcaaaagac ttgaattcta gtgcagcgta ggttgactcc cttatttatt   5880
taattttccc atctctacac cgctagaata acctctctcc tgaggctgtt gaatctgatg   5940
aattagcaga taggaaagaa cttagaaaat tataagttca ctcaaatgta aaaggttata   6000
tgggaaataa tcaccactaa catttttgag tacttactat ctgcttgtta tacacattct   6060
ctaatttaat tttcacaaga aaattcatga aaggactata cttatccccc ttttacaggt   6120
gagcaacctg gagtgcagtg aagtgtaaaa tgtgggccga cttaggagca caaatacccc   6180
agcaacaatg agcactctta gtacacaggt cttggtttct aaaatatcat catccgataa   6240
aaggaacaca ggctctttga agaaatgact gattccggga ttggggcaag aaacacacaa   6300
gctgagactg gagcatcttg cagtcccaga aagtgaagaa acgctgagga tatgtcaaag   6360
ggacacagga gccaaatgaa agagcttcca ctggccaaag ctggaatgtt gagcaacaaa   6420
gcaacatagc attggataat aacccaaagt ataa**aataa**a tattcatgag tacata       6476
```

The start (**ATG**) and stop (**TGA**) codons are highlighted in the sequence. Furthermore, the first five nucleotides of each exon is underlined to indicate the intron/exon boundaries. Lastly, the polyadenylation signal is written in bold.

GENOMIC STRUCTURE

The *MASP1* gene covers approximately 83 kbp and has the composition shown in Fig. 7.3.

The different exons encode the domains described in 'protein modules'. Exon 9 is unique to the C-terminal of MAp44, exon 12 encodes the SP domain of MASP-3, while exons 13–18 encode the SP domain of MASP-1.

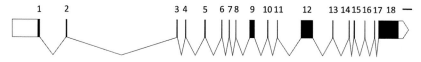

FIGURE 7.3   Genomic structure of *MASP1*: the diagram depicts the structure of the *MASP1* gene. Exons are shown as *black boxes*, while the introns are shown as *lines* interconnecting the exons. The *white boxes* in the beginning and the end of the gene are the 5′- and 3′-UTR, respectively. The scale bar corresponds to 2 kbp.

ACCESSION NUMBERS

Organism	Protein	Gene ID	cDNA
Human	P48740-1	5648	NM_001879.5
	NP_001870		
Mouse	P98064-1	17174	NM_008555.2
	NP_032581.2		
Rat	Q8CHN8-1	64023	NM_022257.1
	NP_071593.1		

DEFICIENCY

MASP-1 deficiency has only been reported in connection with the developmental syndrome 3MC1, where the phenotype is driven by the splice variant MASP-3. None of these mutations reside in the MASP-1-exclusive exons (exons 13–18), but instead lead to deficiency in all three splice variants.[17]

POLYMORPHIC VARIANTS

More than 450 different single nucleotide polymorphisms, which are widely distributed across the MASP-1-encoding exons, have been reported to result in changes at the protein level (reference: GRCh38.p7 at NCBI). However, besides the 3MC1-causing mutations, none of the functional consequences of these have been examined.

It has been shown that SNPs affect the serum levels of MASP-1, and the relative levels between the splice variants MASP-1, MASP-3 and MAp44.[18]

MUTANT ANIMALS

A *MASP1* knockout mouse has been generated by replacing exon 2 in embryonic stem cells with a gene cassette containing a neomycin resistance gene, and chimeric mice were subsequently backcrossed into a C57BL/6J line.[6] This resulted in mice that are deficient in MASP-1, MASP-3 and MAp44 and are therefore termed *MASP1/3*$^{-/-}$. The *MASP1* knockout mice display skeletal abnormalities and are significantly smaller than wild type mice, partly due to a loss of adipose tissue.[19,20] Due to this phenotype, homozygotes are created by mating heterozygotic pairs.

The *MASP-1* knockout mouse has provided knowledge about the interplay between MASP-1 and coagulation. However, the majority of studies in this animal model have focused on the cross talk between the LP and the alternative pathway.

REFERENCES

1. Kjaer TR, Le LTM, Pedersen JS, Sander B, Golas MM, Jensenius JC, et al. Structural insights into the initiating complex of the lectin pathway of complement activation. *Structure* 2015;**23**(2):342–51.

2. Teillet F, Gaboriaud C, Lacroix M, Martin L, Arlaud GJ, Thielens NM. Crystal structure of the CUB1-EGF-CUB2 domain of human MASP-1/3 and identification of its interaction sites with mannan-binding lectin and ficolins. *J Biol Chem* 2008;**283**(37):25715–24.

3. Ambrus G, Gál P, Kojima M, Szilágyi K, Balczer J, Antal J, et al. Natural substrates and inhibitors of mannan-binding lectin-associated serine protease-1 and -2: a study on recombinant catalytic fragments. *J Immunol* 2003;**170**(3):1374–82.

4. Dobó J, Harmat V, Beinrohr L, Sebestyén E, Závodszky P, Gál P. MASP-1, a promiscuous complement protease: structure of its catalytic region reveals the basis of its broad specificity. *J Immunol* 2009;**183**(2):1207–14.

5. Degn SE, Kjaer TR, Kidmose RT, Jensen L, Hansen AG, Tekin M, et al. Complement activation by ligand-driven juxtaposition of discrete pattern recognition complexes. *Proc Natl Acad Sci* 2014;**111**(37):13445–50.

6. Takahashi M, Iwaki D, Kanno K, Ishida Y, Xiong J, Matsushita M, et al. Mannose-binding lectin (MBL)-associated serine protease (MASP)-1 contributes to activation of the lectin complement pathway. *J Immunol* May 01, 2008;**180**(9):6132–8.

7. Iwaki D, Kanno K, Takahashi M, Endo Y, Matsushita M, Fujita T. The role of mannose-binding lectin-associated serine protease-3 in activation of the alternative complement pathway. *J Immunol* 2011;**187**(7):3751–8.

8. Heja D, Kocsis A, Dobo J, Szilagyi K, Szasz R, Zavodszky P, et al. Revised mechanism of complement lectin-pathway activation revealing the role of serine protease MASP-1 as the exclusive activator of MASP-2. *Proc Natl Acad Sci* 2012;**109**(26):10498–503.

9. Lynch NJ, Khan S, Cordula M, Sandrini SM, Marston D, Presanis JS, et al. Composition of the lectin pathway of complement in *Gallus gallus*: absence of mannan-binding lectin-associated serine protease-1 in birds. *J Immunol* 2005;**174**(174):4998–5006.

10. Dobó J, Schroeder V, Jenny L, Cervenak L, Závodszky P, Gál P. Multiple roles of complement MASP-1 at the interface of innate immune response and coagulation. *Mol Immunol* 2014;**61**:69–78.

11. Jenny L, Dobó J, Gál P, Schroeder V. MASP-1 induced clotting – the first model of prothrombin activation by MASP-1. *PLoS One* 2015;**10**(12):1–13.

12. Thiel S, Jensen L, Degn SE, Nielsen HJ, Gál P, Dobó J, et al. Mannan-binding lectin (MBL)-associated serine protease-1 (MASP-1), a serine protease associated with humoral pattern-recognition molecules: normal and acute-phase levels in serum and stoichiometry of lectin pathway components. *Clin Exp Immunol* July 2012;**169**(1):38–48.

13. Terai I, Kobayashi K, Matsushitat M. Human serum mannose-binding lectin (MBL)-associated serine protease-1 (MASP-1): determination of levels in body fluids and identification of two forms in serum. *Clin Exp Immunol* 1997;**110**:317–23.

14. Seyfarth J, Garred P, Madsen HO. Extra-hepatic transcription of the human mannose-binding lectin gene (mbl2) and the MBL-associated serine protease 1-3 genes. *Mol Immunol* 2006;**43**(7):962–71.

15. Endo Y, Takahashi M, Kuraya M, Matsushita M, Stover CM, Schwaeble WJ, et al. Functional characterization of human mannose-binding lectin-associated serine protease (MASP)-1/3 and MASP-2 promoters, and comparison with the C1s promoter. *Int Immunol* 2002;**14**(10):1193–201.

16. https:www.ncbi.nlm.nih.gov/gene.
17. Atik T, Koparir A, Bademci G, Foster J, Altunoglu U, Mutlu GY, et al. Novel MASP1 mutations are associated with an expanded phenotype in 3MC1 syndrome. *Orphanet J Rare Dis* 2015;**10**(1):128.
18. Ammitzbøll CG, Steffensen R, Nielsen HJ, Thiel S, Stengaard-Pedersen K, Bøgsted M, et al. Polymorphisms in the MASP1 gene are associated with serum levels of MASP-1, MASP-3, and MAp44. Ojcius DM, editor. *PLoS One* September 02, 2013;**8**(9):e73317. http://journals.plos.org/plosone/article?id=10.1371/journal.pone.0073317.
19. Takahashi M, Iwaki D, Endo Y, Fujita T. The study of MASPs knockout mice. In: Abdelmohsen K, editor. *Binding protein*. 2012. p. 165–80.
20. Takahashi M, Endo Y, Fujita T. Developmental abnormalities in Masp1/3-deficient mice. *Immunobiology* 2012;**217**:1164.

Chapter 8

MASP-2

Ramus Pihl, Jens C. Jensenius, Steffen Thiel

Aarhus University, Aarhus, Denmark

PHYSICOCHEMICAL PROPERTIES

MASP-2 is synthesised as a single-chain polypeptide of 686 residues. The 15N-terminal residues constitute a leader peptide that is removed during secretion, resulting in a protein with a theoretical molecular weight of 74.2 kDa. Due to the absence of any glycosylation sites on MASP-2, it migrates in accordance with its predicted size on SDS-PAGE.

MASP-2 is generated as a proenzyme that later becomes activated by cleavage between residues R444 and I445, yielding an A-chain and a B-chain linked by a disulphide bond.

pI (theoretical)	5.34	
Molecular weight	74.2 kDa	
	A chain	B chain
Residues	16–444	445–686
Molecular weight	47.7 kDa	26.5 kDa
N-Linked glycosylation sites	0	0
Disulphide bond between A- and B-chains	C434	C552

STRUCTURE

MASP-2 consists of six discrete domains, analogous to C1r, C1s, MASP-1 and MASP-3 (Fig. 8.1), and exists as a homodimer due to Ca^{2+}-dependent dimerisation mediated by the CUB1-EGF domains in an antiparallel orientation. The MASP-2 dimer likely forms an elongated structure with the two serine protease (SP) domains separated by 200–300 Å, as seen for MASP-1.[1] Binding to the collagen-like domains of pattern-recognition molecules of the lectin pathway (LP) (i.e., mannan-binding lectin, H-ficolin, M-ficolin,

FIGURE 8.1 Protein domains of MASP-2.

L-ficolin and collectin LK) is facilitated by the CUB1 and CUB2 domains,[2] while the complement control protein (CCP1 and CCP2) domains contain an important exosite for the substrate C4.[3]

The SP domain is the catalytic domain and is organised around a classic chymotrypsin-like structure (Fig. 8.2).[4]

FUNCTION

The isoforms MASP-2 and MAp19 are generated through alternative splicing from the *MASP2* gene. MASP-2 circulates bound to the ficolins and collectins of the LP, and when these complexes are concentrated on an activating surface MASP-2 becomes activated from its zymogenic form. MASP-2 is capable of autoactivating by a mechanism in which a high degree of flexibility allows it to attain a catalytic conformation despite being a one-chain enzyme.[5] However, MASP-1 seems to be an important activator of MASP-2 in normal physiological settings.[6] On activation, MASP-2 cleaves C2 and C4, thus generating the LP C3 convertase, C4b2a. As MASP-2 is the only protease of the LP that cleaves C4, it is considered essential for LP activity.[7] On the contrary, the in vivo cleavage of C2 seems to be a collaboration between MASP-2 and MASP-1.[6]

Additionally, it has been reported that MASP-2 can lead to C3b deposition by a route that bypasses C4. The mechanism of the C4-bypass activation remains opaque, but it is suggested to involve MASP-1 and C2.[8] MASP-2 likewise seems to be implicated in coagulation as it promotes clotting by activation of prothrombin.[9]

TISSUE DISTRIBUTION

The median serum concentration of MASP-2 in healthy donors is estimated as 0.41 µg/mL (5.5 nM of MASP-2 monomer) with an interquartile range of 0.29–0.53 µg/mL. The levels are somewhat higher in EDTA plasma with a median concentration of 0.49 µg/mL (6.6 nM of MASP-2 monomer) and an interquartile range of 0.39–0.63 µg/mL (data not published). MASP-2 is almost exclusively synthesised in the liver.[10] MASP-2 has been reported to be undetectable in cerebrospinal fluid or present in levels so low that the sensitivity of detection is challenged (20–40 ng/mL).[11] MASP-2 was not detected in urine.[10]

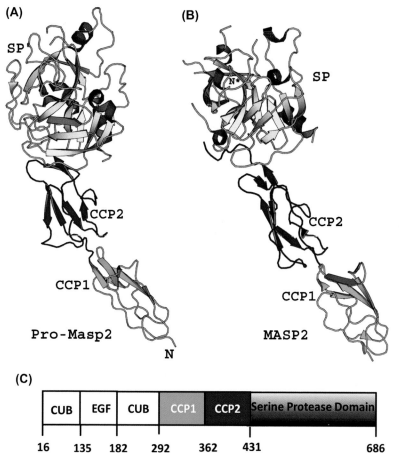

FIGURE 8.2 Crystal structure of complement serine protease of the lectin pathway MASP2. (A) Ribbon diagram of the human Pro-MASP2 (Mannose-binding lectin-associated serine protease 2) crystal structure (PDB ID:1ZJK). (B) Ribbon diagram of the active human MASP2 crystal structure (PDB ID: 1Q3X). (C) Schematic representation of MASP2 domains, coloured similar to the ribbon representation.

REGULATION OF EXPRESSION

MASP-2 does not behave as an acute phase protein.[12] The transcriptional regulation of MASP-2 is relatively unstudied, although there is indication of a modest upregulation by IL-1β, which is counteracted by simultaneous stimulation with IL-6.[13] It remains unknown what defines the balance between the alternative splicing products MASP-2 and MAp19 from the *MASP2* gene.

HUMAN PROTEIN SEQUENCE

The sequence below is from uniprot.org and has the identifier O00187-1:

```
MRLLTLLGLL CGSVATPLGP KWPEPVFGRL ASPGFPGEYA NDQERRWTLT    50
APPGYRLRLY FTHFDLELSH LCEYDFVKLS SGAKVLATLC GQESTDTERA   100
PGKDTFYSLG SSLDITFRSD YSNEKPFTGF EAFYAAEDID ECQVAPGEAP   150
TCDHHCHNHL GGFYCSCRAG YVLHRNKRTC SALCSGQVFT QRSGELSSPE   200
YPRPYPKLSS CTYSISLEEG FSVILDFVES FDVETHPETL CPYDFLKIQT   250
DREEHGPFCG KTLPHRIETK SNTVTITFVT DESGDHTGWK IHYTSTAQPC   300
PYPMAPPNGH VSPVQAKYIL KDSFSIFCET GYELLQGHLP LKSFTAVCQK   350
DGSWDRPMPA CSIVDCGPPD DLPSGRVEYI TGPGVTTYKA VIQYSCEETF   400
YTMKVNDGKY VCEADGFWTS SKGEKSLPVC EPVCGLSART TGG**RI**YGGQK   450
AKPGDFPWQV LILGGTTAAG ALLYDNWVLT AA*H*AVYEQKH DASALDIRMG   500
TLKRLSPHYT QAWSEAVFIH EGYTHDAGFD N*D*IALIKLNN KVVINSNITP   550
ICLPRKEAES FMRTDDIGTA SGWGLTQRGF LARNLMYVDI PIVDHQKCTA   600
AYEKPPYPRG SVTANMLCAG LESGGKDSCR GD*S*GGALVFL DSETERWFVG   650
GIVSWGSMNC GEAGQYGVYT KVINYIPWIE NIISDF                  686
```

The signal peptide is underlined. The residues that constitute the catalytic triad are shown in italic and coloured grey (residues H483, D532, S633). The cleavage site between the A- and B-chains is written in bold.

PROTEIN MODULES

1–15	Leader peptide	exon 1+2
16–137	CUB1	exon 2+3
138–181	EGF-like	exon 4
184–296	CUB2	exon 6+7
298–363	CCP1	exon 8+9
364–432	CCP2	exon 10+11
445–686	SP	exon 12

CHROMOSOMAL LOCATION[14]

Human	1p36.3–p36.2
Mus musculus	4 E1
Rattus norvegicus	5q36

HUMAN cDNA SEQUENCE

The sequence is the NCBI Reference Sequence: NM_006610.3

```
agaccaggcc aggccagctg gacgggcaca cc**atg**aggct gctgaccctc ctgggccttc    60
tgtgtggctc ggtggccacc cccttgggcc cgaagtggcc tgaacctgtg ttcgggcgcc   120
tggcatcccc cggctttcca ggggagtatg ccaatgacca ggagcggcgc tggaccctga   180
ctgcacccccc cggctaccgc ctgcgcctct acttcaccca cttcgacctg gagctctccc   240
acctctgcga gtacgacttc gtcaagctga gctcggggc caaggtgctg gccacgctgt   300
gcgggcagga gagcacagac acggagcggg cccctggcaa ggacactttc tactcgctgg   360
gctccagcct ggacattacc ttccgctccg actactccaa cgagaagccg ttcacggggt   420
tcgaggcctt ctatgcagcc gaggacattg acgagtgcca ggtggccccg ggagaggcgc   480
ccacctgcga ccaccactgc cacaaccacc tgggcggttt ctactgctcc tgccgcgcag   540
gctacgtcct gcaccgtaac aagcgcacct gctcagccct gtgctccggc caggtcttca   600
cccagaggtc tggggagctc agcagccctg aatacccacg gccgtatccc aaactctcca   660
gttgcactta cagcatcagc ctggaggagg ggttcagtgt cattctggac tttgtggagt   720
ccttcgatgt ggagacacac cctgaaaccc tgtgtcccta cgactttctc aagattcaaa   780
cagacagaga agaacatggc ccattctgtg ggaagacatt gccccacagg attgaaacaa   840
aaagcaacac ggtgaccatc acctttgtca cagatgaatc aggagaccac acaggctgga   900
agatccacta cacgagcaca gcgcagcctt gcccttatcc gatggcgcca cctaatggcc   960
acgtttcacc tgtgcaagcc aaatacatcc tgaaagacag cttctccatc ttttgcgaga  1020
ctggctatga gcttctgcaa ggtcacttgc ccctgaaatc ctttactgca gtttgtcaga  1080
aagatggatc ttgggaccgg ccaatgcccg cgtgcagcat tgttgactgt ggccctcctg  1140
atgatctacc cagtggccga gtggagtaca tcacaggtcc tggagtgacc acctacaaag  1200
ctgtgattca gtacagctgt gaagagacct tctacacaat gaaagtgaat gatggtaaat  1260
atgtgtgtga ggctgatgga ttctggacga gctccaaagg agaaaaatca ctcccagtct  1320
gtgagcctgt ttgtggacta tcagcccgca caacaggagg gcgtatatat ggagggcaaa  1380
aggcaaaacc tggtgatttt ccttggcaag tcctgatatt aggtggaacc acagcagcag  1440
gtgcactttt atatgacaac tgggtcctaa cagctgctca tgccgtctat gagcaaaaac  1500
atgatgcatc cgccctggac attcgaatgg gcaccctgaa aagactatca cctcattata  1560
cacaagcctg gtctgaagct gtttttatac atgaaggtta tactcatgat gctggctttg  1620
acaatgacat agcactgatt aaaattgaata acaaagttgt aatcaatagc aacatcacgc  1680
ctatttgtct gccaagaaaa gaagctgaat cctttatgag gacagatgac attggaactg  1740
catctggatg gggattaacc caaaggggtt ttcttgctag aaatctaatg tatgtcgaca  1800
taccgattgt tgaccatcaa aaatgtactg ctgcatatga aaagccaccc tatccaaggg  1860
gaagtgtaac tgctaacatg ctttgtgctg cgttagaaag tggggggcaag gacagctgca  1920
gaggtgacag cggagggca ctggtgtttc tagatagtga aacagagagg tggtttgtgg  1980
gaggaatagt gtcctggggt tccatgaatt gtggggaagc aggtcagtat ggagtctaca  2040
caaaagttat taactatatt ccctggatcg agaacataat tagtgatttt **taa**cttgcgt  2100
gtctgcagtc aaggattctt cattttaga aatgcctgtg aagaccttgg cagcgacgtg  2160
gctcgagaag cattcatcat tactgtggac atggcagttg ttgctccacc caaaaaaaca  2220
gactccaggt gaggctgctg tcatttctcc acttgccagt ttaattccag ccttacccat  2280
tgactcaagg gga**cataa**ac cacgagagtg acagtcatct ttgcccaccc agtgtaatgt  2340
cactgctcaa attacatttc attaccttaa aaagccagtc tcttttcata ctggctgttg  2400
gcatttctgt aaactgcctg tccatgctct ttgtttttaa acttgttctt attg*aaaaaa*  2460
*aaaaaaaaaa a*                                                        2471
```

FIGURE 8.3 Genomic structure of *MASP2*: Exons are shown as *black boxes*, while the introns are shown as *lines* interconnecting the exons. The *white boxes* in the beginning and the end of the gene are the 5'- and 3'-UTR, respectively. The scale bar corresponds to 1 kbp.

The start (**<u>atg</u>**) and stop (**<u>taa</u>**) codons are highlighted in the sequence. Furthermore, the first five nucleotides of each exon are underlined to indicate the intron/exon boundaries. Lastly, the polyadenylation signal is written in bold and the polyadenylation tail is written in italic.

GENOMIC STRUCTURE

The *MASP2* gene covers approximately 27.7 kbp and has the composition shown in Fig. 8.3.

The different exons encode the domains described in 'protein modules'. Exon 12 encodes the SP domain of MASP-2, while exon 5 is unique to the C-terminal of MAp19.

ACCESSION NUMBERS

Organism	Protein	Gene ID	cDNA
Human	O00187-1	10747	NM_006610.3
	NP_006601.2		
Mus musculus	Q91WP0-1	17175	NM_001003893.2
	NP_001003893.1		
Rattus norvegicus	Q9JJS8-1	64459	NM_172043.1
	NP_742040		

DEFICIENCY

A number of studies have reported on individuals deficient in MASP-2, estimated as low MASP-2 levels in the circulation. LP activity in such individuals may or may not be present depending on the nature of the polymorphism causing the deficiency.[15] Interestingly, MASP-2 deficiency is found both in healthy individuals, as well as in patients suffering from recurrent infections. This indicates that MASP-2 deficiency displays a low penetrance and might only manifest itself pathologically when additional elements are compromised.[16]

POLYMORPHIC VARIANTS

More than 300 single nucleotide polymorphisms in *MASP2* have been reported to result in changes at the protein level (reference: GRCh38.p7 at NCBI).

However, few have been characterised on a molecular level. Examples of these are mentioned here.

p.D120G (c.359A>G): This missense mutation yields MASP-2 that is unable to bind to the lectins and collectins of the LP and is found with a gene frequency of 3.9% in Caucasians. This results in low levels of MASP-2 and a defective LP. This variant has been found to clinically manifest itself as recurrent infections and chronic inflammatory disease.[17]

p.156_159dupCHNH (c.466_477dupTGCCACAACCAC): This duplication disrupts MASP-2 binding to the lectins and collectins of the LP. The MASP-2 variant is furthermore synthesised at very low levels by HEK293 cells. The polymorphism is found with a gene frequency of 0.26% in Hong Kong Chinese.[18]

p.R439H (c.1316G4>A): This mutation leads to inactive MASP-2, since it is located in the vicinity of the activation site, and the mutant is thus incapable of autoactivating. The polymorphism is found with a gene frequency of 9% in Zambian Africans.[18]

MUTANT ANIMALS

Two types of mice deficient in MASP-2 have been generated. The first was established when the MAp19-specific exon 5 was replaced by a neomycin resistance gene cassette. This dramatically reduced the levels of MASP-2 in the mouse, most likely due to an effect on splicing, and the mice are therefore termed the MASP2/MAp19[−/−] mice.[19] The mice were backcrossed to create homozygotic mice on a C57BL/6J background. Later, a knockout strain that only lacks MASP-2 was generated by replacing exon 11 and part of exon 12 with a neomycin resistance gene cassette. The resulting homozygotes were backcrossed onto a C57BL/6J background.[8] It should be noted that the level of MAp19 is upregulated in this Masp2[−/−] strain. Both homozygotic strains were fertile and without any apparent abnormalities.

MASP-2 knockout mice have been used to deduce the importance of MASP-2 for LP activation,[19] as it was found that lack of MASP-2 increases the susceptibility for pneumococcal infection.[20] Additionally, it has been shown that MASP2[−/−] mice are protected against myocardial, gastrointestinal and renal ischaemia reperfusion injury due to a C4-independent mechanism.[8,21]

REFERENCES

1. Kjaer TR, Le LTM, Pedersen JS, Sander B, Golas MM, Jensenius JC, et al. Structural insights into the initiating complex of the lectin pathway of complement activation. *Structure* 2015;**23**(2):342–51.
2. Teillet F, Gaboriaud C, Lacroix M, Martin L, Arlaud GJ, Thielens NM. Crystal structure of the CUB1-EGF-CUB2 domain of human MASP-1/3 and identification of its interaction sites with mannan-binding lectin and ficolins. *J Biol Chem* 2008;**283**(37):25715–24.
3. Kidmose RT, Laursen NS, Dobo J, Kjaer TR, Sirotkina S, Yatime L, et al. Structural basis for activation of the complement system by component C4 cleavage. *Proc Natl Acad Sci* 2012;**109**(38):15425–30.

4. Harmat V, Gál P, Kardos J, Szilágyi K, Ambrus G, Végh B, et al. The structure of MBL-associated serine protease-2 reveals that identical substrate specificities of C1s and MASP-2 are realized through different sets of enzyme–substrate interactions. *J Mol Biol* 2004;**342**(5): 1533–46.

5. Gál P, Harmat V, Kocsis A, Bián T, Barna L, Ambrus G, et al. A true autoactivating enzyme: structural insight into mannose-binding lectin-associated serine protease-2 activations. *J Biol Chem* 2005;**280**(39):33435–44.

6. Heja D, Kocsis A, Dobo J, Szilagyi K, Szasz R, Zavodszky P, et al. Revised mechanism of complement lectin-pathway activation revealing the role of serine protease MASP-1 as the exclusive activator of MASP-2. *Proc Natl Acad Sci* 2012;**109**(26):10498–503.

7. Ambrus G, Gál P, Kojima M, Szilágyi K, Balczer J, Antal J, et al. Natural substrates and inhibitors of mannan-binding lectin-associated serine protease-1 and -2: a study on recombinant catalytic fragments. *J Immunol* 2003;**170**(3):1374–82.

8. Schwaeble WJ, Lynch NJ, Clark JE, Marber M, Samani NJ, Ali YM, et al. Targeting of mannan-binding lectin-associated serine protease-2 confers protection from myocardial and gastrointestinal ischemia/reperfusion injury. *Proc Natl Acad Sci USA* 2011;**108**(18):7523–8.

9. Krarup A, Wallis R, Presanis JS, Gál P, Sim RB. Simultaneous activation of complement and coagulation by MBL-associated serine protease 2. Sommer P, editor. PLoS One July 18, 2007;**2**(7):e623.

10. Degn SE, Thiel S, Nielsen O, Hansen AG, Steffensen R, Jensenius JC. MAp19, the alternative splice product of the MASP2 gene. *J Immunol Methods* 2011;**373**(1–2):89–101.

11. Kwok JY, Augst RM, Yu DY, Singh KK. Sensitive CSF ELISAs for the detection of MBL, MASP-2 and functional MBL/MASP-2. *J Neurosci Methods* 2012;**209**(1):255–7.

12. Ytting H, Christensen IJ, Basse L, Lykke J, Thiel S, Jensenius JC, et al. Influence of major surgery on the mannan-binding lectin pathway of innate immunity. *Clin Exp Immunol* 2006;**144**(2):239–46.

13. Endo Y, Takahashi M, Kuraya M, Matsushita M, Stover CM, Schwaeble WJ, et al. Functional characterization of human mannose-binding lectin-associated serine protease (MASP)-1/3 and MASP-2 promoters, and comparison with the C1s promoter. *Int Immunol* 2002;**14**(10):1193–201.

14. https://www.ncbi.nlm.nih.gov/gene/.

15. Thiel S, Steffensen R, Christensen IJ, Ip WK, Lau YL, Reason IJM, et al. Deficiency of mannan-binding lectin associated serine protease-2 due to missense polymorphisms. *Genes Immun* 2007;**8**(2):154–63.

16. Sokolowska A, Szala A, St. Swierzko A, Kozinska M, Niemiec T, Blachnio M, et al. Mannan-binding lectin-associated serine protease-2 (MASP-2) deficiency in two patients with pulmonary tuberculosis and one healthy control. *Cell Mol Immunol* 2015;**12**(1):119–21.

17. Stengaard-Pedersen K, Thiel S, Gadjeva M, Møller-Kristensen M, Sørensen R, Jensen LT, et al. Inherited deficiency of mannan-binding lectin-associated serine protease 2. *N Engl J Med* 2003;**349**(6):554–60.

18. Thiel S, Kolev M, Degn S, Steffensen R, Hansen AG, Ruseva M, et al. Polymorphisms in mannan-binding lectin (MBL)-associated serine protease 2 affect stability, binding to MBL, and enzymatic activity. *J Immunol* 2009;**182**:2939–47. 1550–6606 (Electronic).

19. Iwaki D, Kanno K, Takahashi M, Lynch NJ, Schwaeble WJ, Matsushita M, et al. Small mannose-binding lectin-associated proteins plays a regulatory role in the lectin complement pathway. *J Immunol* December 15, 2006;**177**(12):8626–32.

20. Ali YM, Lynch NJ, Haleem KS, Fujita T, Endo Y, Hansen S, et al. The lectin pathway of complement activation is a critical component of the innate immune response to pneumococcal infection. 2012.
21. Asgari E, Farrar CA, Lynch N, Ali YM, Roscher S, Stover C, et al. Mannan-binding lectin-associated serine protease 2 is critical for the development of renal ischemia reperfusion injury and mediates tissue injury in the absence of complement C4. *FASEB J* September 2014;**28**(9):3996–4003.

Chapter 9

MASP-3

Ramus Pihl, Jens C. Jensenius, Steffen Thiel
Aarhus University, Aarhus, Denmark

PHYSICOCHEMICAL PROPERTIES

Immature Protein

MASP-3 is synthesised as a single-chain polypeptide of 728 residues. The 19 N-terminal residues constitute a leader peptide that is removed during secretion, resulting in a protein with a theoretical molecular weight of 79.6 kDa. However, due to the presence of six N-glycosylation sites it migrates as a 100 kDa protein on SDS-PAGE. MASP-3 is generated as a proenzyme that later becomes activated by cleavage between residues 449 and 450, yielding an A-chain and a B-chain linked by a disulphide bond.

Mature protein		
pI (theoretical)	4.91	
Molecular weight	93 kDa	
	A-chain	B-chain
Residues	20–449	450–728
Molecular weight	49 kDa (theoretical)	30 kDa (theoretical)
	66 kDa (observed)	40 kDa (observed)
N-Linked glycosylation sites	4 (res 49, 178, 385, 407)	2 (res 533 and 599)
Disulphide bond between A- and B-chains	C436	C573

STRUCTURE

MASP-3 consists of six discrete domains, analogous to C1r, C1s, MASP-1 and MASP-2 (Fig. 9.1. Protein structure), and exists as a homodimer due to Ca^{2+}-dependent dimerisation mediated by the CUB1-EGF domains in an antiparallel orientation. The MASP-3 dimer likely forms an elongated structure with the two serine protease (SP) domains separated by 200–300 Å, as seen for MASP-1.[1] Binding to the collagen-like domains of pattern-recognition molecules of the lectin pathway (LP) (i.e., mannan-binding lectin, H-ficolin,

89

FIGURE 9.1   Diagram of protein domains for *MASP3*.

M-ficolin, L-ficolin and collectin LK) is facilitated by the CUB1 and CUB2 domains,[2] while the complement control protein domains, CCP1 and CCP2, potentially influence the catalytic properties of MASP-3, as found in other complement proteases.[3] The SP domain of MASP-3 resembles those of the other proteases of the LP and classical pathways, but noticeably contains a number of dynamic loops enabling a certain degree of conformational flexibility.[4]

FUNCTION

The isoforms MASP-3, MASP-1 and MAp44 are generated through alternative splicing from the *MASP1* gene. The function of MASP-3 has long been unknown, as it does not cleave any of the complement components downstream of LP activation.[5] However, it has become evident that MASP-3 bridges the LP and the alternative pathway by being a main enzyme that cleaves zymogenic profactor D to mature factor D.[6,7] In turn, this enables factor D to cleave factor B in the C3bB complex to generate the alternative pathway convertase C3bBb.

MASP-3 circulates bound to the ficolins and collectins of the LP, and when these complexes are concentrated on an activating surface, MASP-3 becomes activated.[8] The mechanism by which MASP-3 is activated is not fully clear. However, evidence suggests that MASP-1, MASP-2 and MASP-3 itself can contribute to converting the proenzyme into active MASP-3.[9] Furthermore, it has been shown that the proenzyme form of MASP-3 does not convert profactor D to factor D.[10] The mechanism by which MASP-3 becomes activated and continuously converts profactor D to factor D in the fluid phase is not yet resolved.

Lack of MASP-3 activity results in the syndrome 3MC1, which indicates that MASP-3 plays an important role during development.[11] It has been suggested that MASP-3 (together with CL-K1) acts as a guidance cue for neural crest cell migration.[12] However, this role of MASP-3 has not been firmly consolidated. Likewise, MASP-3 can cleave the insulin-like growth factor binding protein 5 in vitro in a process that is yet to be shown to have biological relevance.[13]

TISSUE DISTRIBUTION

The median serum concentration of MASP-3 in healthy donors is estimated as 5.0 μg/mL (54 nM of MASP-3 monomer) with concentrations ranging from 1.8 to 10.6 μg/mL. The concentrations in serum was found to be approximately 1.5-fold higher than in plasma. The concentration is low at birth but reaches adult levels within 6 months.[14]

No reports address the level of MASP-3 in cerebrospinal fluid or urine.

The main site of MASP-3 production is the liver. However, in contrast to MASP-1 and MASP-2 significant production is also observed in a wide range of extra-hepatic tissues including colon, heart and skeletal muscle.[15,16]

REGULATION OF EXPRESSION

MASP-3 does not act as an acute-phase protein.[14]

The transcriptional regulation of MASP-3 is relatively unstudied, although there are indications of a modest upregulation of transcription from the *MASP1* gene by IL-1β and a similarly small downregulation by IL-6 and IFN-γ.[17]

HUMAN PROTEIN SEQUENCE

The sequence below is from uniprot.org and has the identifier P48740-2:

```
MRWLLLYYAL CFSLSKASAH TVELNNMFGQ IQSPGYPDSY PSDSEVTWNI    50
TVPDGFRIKL YFMHFNLESS YLCEYDYVKV ETEDQVLATF CGRETTDTEQ   100
TPGQEVVLSP GSFMSITFRS DFSNEERFTG FDAHYMAVDV DECKEREDEE   150
LSCDHYCHNY IGGYYCSCRF GYILHTDNRT CRVECSDNLF TQRTGVITSP   200
DFPNPYPKSS ECLYTIELEE GFMVNLQFED IFDIEDHPEV PCPYDYIKIK   250
VGPKVLGPFC GEKAPEPIST QSHSVLILFH SDNSGENRGW RLSYRAAGNE   300
CPELQPPVHG KIEPSQAKYF FKDQVLVSCD TGYKVLKDNV EMDTFQIECL   350
KDGTWSNKIP TCKIVDCRAP GELEHGLITF STRNNLTTYK SEIKYSCQEP   400
YYKMLNNNTG IYTCSAQGVW MNKVLGRSLP TCLPECGQPS RSLPSLVKRI   450
IGGRNAEPGL FPWQALIVVE DTSRVPNDKW FGSGALLSAS WILTAAHVLR   500
SQRRDTTVIP VSKEHVTVYL GLHDVRDKSG AVNSSAARVV LHPDFNIQNY   550
NHDIALVQLQ EPVPLGPHVM PVCLPRLEPE GPAPHMLGLV AGWGISNPNV   600
TVDEIISSGT RTLSDVLQYV KLPVVPHAEC KTSYESRSGN YSVTENMFCA   650
GYYEGGKDTC LGDSGGAFVI FDDLSQRWVV QGLVSWGGPE ECGSKQVYGV   700
YTKVSNYVDW VWEQMGLPQS VVEPQVER                          728
```

The signal peptide is underlined. The *N*-glycosylation sites are depicted in bold, while the residues that constitute the catalytic triad are shown in italic and coloured grey (residues H490, D552, S646). The cleavage site between the A- and B-chains is likewise written in bold and underlined.

PROTEIN MODULES

1–19	Leader peptide	exon 1 + 2
20–138	CUB1	exon 2 + 3
139–182	EGF-like	exon 4
185–297	CUB2	exon 5 + 6
299–364	CCP1	exon 7 + 8
365–434	CCP2	exon 10 + 11
450–728	SP	exon 12

CHROMOSOMAL LOCATION[18]

Human: 3q27-q28
Mus musculus: 16 B2-B3
Rattus norvegicus: 11q23

HUMAN cDNA SEQUENCE

The sequence is the NCBI Reference Sequence: NM_139125.3.

```
acacacagag tgatacaaat acctgcttga gcccctcagt tattttctct caagggctga     60

agtcagccac acaggataaa ggagggaagg gaaggagcag atcttttcgg taggaagaca    120

gattttgttg tcaggttcct gggagtgcaa gagcaagtca aaggagagag agaggagaga    180

ggaaaagcca gagggagaga gggggagagg ggatctgttg caggcagggg aaggcgtgac    240

ctgaatggag aatgccagcc aattccagag acacacaggg acctcagaac aaagataagg    300

catcacggac accacaccgg gcacgagctc acaggcaagt caagctggga ggaccaaggc    360

cgggcagccg ggagcaccca aggcaggaaa atgaggtggc tgcttctcta ttatgctctg    420

tgcttctccc tgtcaaaggc ttcagcccac accgtggagc taaacaatat gtttggccag    480

atccagtcgc ctggttatcc agactcctat cccagtgatt cagaggtgac ttggaatatc    540

actgtcccag atgggtttcg gatcaagctt tacttcatgc acttcaactt ggaatcctcc    600

tacctttgtg aatatgacta tgtgaaggta gaaactgagg accaggtgct ggcaaccttc    660

tgtggcaggg agaccacaga cacagagcag actcccggcc aggaggtggt cctctcccct    720

ggctccttca tgtccatcac tttccggtca gatttctcca atgaggagcg tttcacaggc    780

tttgatgccc actacatggc tgtggatgtg gacgagtgca aggagaggga ggacgaggag    840

tggtggtgtg accaataatg ggagaactac attggcggct actactgctc ctgccgcttc    900

ggctacatcc tccacacaga caacaggacc tgccgagtgg agtgcagtga caacctcttc    960

actcaaagga ctggggtgat caccagccct gacttcccaa acccttaccc caagagctct   1020

gaatgcctgt ataccatcga gctggaggag ggtttcatgg tcaacctgca gtttgaggac   1080

atatttgaca ttgaggacca tcctgaggtg ccctgcccct atgactacat caagatcaaa   1140

gttggtccaa aagttttggg gcctttctgt ggagagaaag ccccagaacc catcagcacc   1200

cagagccaca gtgtcctgat cctgttccat agtgacaact cgggagagaa ccggggctgg   1260

aggctctcat acagggctgc aggaaatgag tgcccagagc tacagcctcc tgtccatggg   1320

aaaatcgagc cctcccaagc caagtatttc ttcaaagacc aagtgctcgt cagctgtgac   1380

acaggctaca aagtgctgaa ggataatgtg gagatggaca cattccagat tgagtgtctg   1440

aaggatggga cgtggagtaa caagattccc acctgtaaaa ttgtagactg tagagcccca   1500
```

ggagagctgg aacacgggct gatcaccttc tctacaagga acaacctcac cacatacaag 1560

tctgagatca aatactcctg tcaggagccc tattacaaga tgctcaacaa taacacag<u>gt</u> 1620

<u>at</u>atatacct gttctgccca aggagtctgg atgaataaag tattggggag aagcctaccc 1680

acctgccttc ca<u>gagtgt</u>gg tcagccctcc cgctccctgc caagcctggt caagaggatc 1740

attgggggcc gaaatgctga gcctggcctc ttcccgtggc aggccctgat agtggtggag 1800

gacacttcga gagtgccaaa tgacaagtgg tttgggagtg gggccctgct ctctgcgtcc 1860

tggatcctca cagcagctca tgtgctgcgc tcccagcgta gagacaccac ggtgatacca 1920

gtctccaagg agcatgtcac cgtctacctg ggcttgcatg atgtgcgaga caaatcgggg 1980

gcagtcaaca gctcagctgc ccgagtggtg ctccacccag acttcaacat ccaaaactac 2040

aaccacgata tagctctggt gcagctgcag gagcctgtgc ccctgggacc ccacgttatg 2100

cctgtctgcc tgccaaggct tgagcctgaa ggcccggccc cccacatgct gggcctggtg 2160

gccggctggg gcatctccaa tcccaatgtg acagtggatg agatcatcag cagtggcaca 2220

cggaccttgt cagatgtcct gcagtatgtc aagttacccg tggtgcctca cgctgagtgc 2280

aaaactagct atgagtcccg ctcgggcaat tacagcgtca cggagaacat gttctgtgct 2340

ggctactacg agggcggcaa agacacgtgc cttggagata gcggtggggc ctttgtcatc 2400

tttgatgact tgagccagcg ctgggtggtg caaggcctgg tgtcctgggg gggacctgaa 2460

gaatgcggca gcaagcaggt ctatggagtc tacacaaagg tctccaatta cgtggactgg 2520

gtgtgggagc agatgggctt accacaaagt gttgtggagc cccaggtgga acgg**tga**gct 2580

gacttacttc ctcggggcct gcctcccctg agcgaagcta caccgcactt ccgacagcac 2640

actccacatt acttatcaga ccatatggaa tggaacacac tgacctagcg gtggcttctc 2700

ctaccgagac agcccccagg accctgagag gcagagtgtg gtatagggaa aaggctccag 2760

gcaggagacc tgtgttcctg agcttgtcca agtctctttc cctgtctggg cctcactcta 2820

ccgagtaata caatgcagga gctcaaccaa ggcctctgtg ccaatcccag cactcctttc 2880

caggccatgc ttcttacccc agtggccttt attcactcct gaccacttat caaacccatc 2940

ggtcctactg ttggtataac tgagcttgga cctgactatt agaaaatggt ttctaacatt 3000

gaactgaatg ccgcatctgt atattttcct gctctgcctt ctgggactag ccttggccta 3060

atccttcctc taggagaaga gcattcaggt tttgggagat ggctcatagc caagcccctc 3120

tctcttagtg tgatcccttg gagcaccttc atgcctgggg tttctctccc aaaagcttct 3180

tgcagtctaa gccttatccc ttatgttccc cattaaagga atttcaaaag acatggagaa 3240

```
agttgggaag gtttgtgctg actgctggga gcagaatagc cgtgggaggc ccaccaagcc    3300

cttaaattcc cattgtcaac tcagaacaca tttgggccca tatgccaccc tggaacacca    3360

gctgacacca tgggcgtcca cacctgctgc tccagacaag cacaaagcaa tctttcagcc    3420

ttgaaatgta ttatctgaaa ggctacctga agcccaggcc cgaatatggg gacttagtcg    3480

attacctgga aaaagaaaag acccacactg tgtcctgctg tgcttttggg caggaaaatg    3540

gaagaaagag tggggtgggc acattagaag tcacccaaat cctgccaggc tgcctggcat    3600

ccctggggca tgagctgggc ggagaatcca ccccgcagga tgttcagagg gacccactcc    3660

ttcatttttc agagtcaaag gaatcagagg ctcacccatg gcaggcagtg aaaagagcca    3720

ggagtcctgg gttctagtcc ctgctctgcc cccaactggc tgtataacct ttgaaaaatc    3780

attttctttg tctgagtctc tggttctccg tcagcaacag gctggcataa ggtcccctgc    3840

aggttccttc tagctggagc actcagagct tccctgactg ctagcagcct ctctggccct    3900

cacagggctg attgttctcc ttctccctgg agctctctct cctgaaaatc tccatcagag    3960

caaggcagcc agagaagccc ctgagaggga atgattggga agtgtccact ttctcaaccg    4020

gctcatcaaa cacactcctt tgtctatgaa tggcacatgt aaatgatgtt atattttgta    4080

tcttttatat catatgcttc accattctgt aaagggcctc tgcattgttg ctcccatcag    4140

gggtctcaag tgga**aataaa** ccctcgtgga taaccaac*a aaaa*                    4184
```

The start (<u>**atg**</u>) and stop (<u>**tga**</u>) codons are highlighted in the sequence. Furthermore, the first five nucleotides of each exon is underlined to indicate the intron/exon boundaries. Lastly, the polyadenylation signal is written in bold, while the polyadenylation site is written in italic.

GENOMIC STRUCTURE

The *MASP1* gene covers approximately 83 kbp and has the composition shown in Fig. 9.2. The different exons encode the domains described in 'protein modules'. The SP domain of MASP-3 is encoded by a single exon (exon 12), while the SP domain of MASP-1 is encoded by several exons (exons 13–18). Exon 9 is unique to the C-terminal of MAp44.

ACCESSION NUMBERS

Organism	Protein	Gene ID	cDNA
Human	P48740-2	5648	NM_139125.3
	NP_624302.1		
Mus musculus	P98064-2	17174	XM_006521828.3
	XP_006521891.1		
Rattus norvegicus	Q8CHN8-2	64023	XM_006248526.3
	XP_006248588.1		

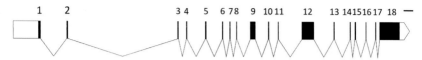

FIGURE 9.2   Genomic structure of *MASP1*: Exons are shown as *black boxes*, while the introns are shown as *lines* interconnecting the exons. The *white boxes* in the beginning and the end of the gene are the 5'- and 3'-UTR, respectively. The scale bar corresponds to 2 kbp.

DEFICIENCY

MASP-3 deficiency is found in the developmental syndrome termed 3MC1, which is a rare autosomal recessive disorder. Some patients are homozygous for mutations affecting the domains shared by MASP-1, MASP-3, MAp44, but the majority of the disease-causing mutations cluster in exon 12, thus highlighting the central role of MASP-3 in the etiology of the disease.[19]

The syndrome is characterised by facial abnormalities including cleft lip/palate, bilateral ptosis, highly arched eyelids and craniosynostosis. Furthermore, phenotypes such as mental retardation, caudal appendage and hearing loss are found in patients.[11]

The gene frequencies of the polymorphisms causing 3MC are currently not known.

POLYMORPHIC VARIANTS

More than 450 different SNPs, which are widely distributed across the *MASP1* gene, have been reported to result in changes at the protein level (reference: GRCh38.p7 at NCBI). However, besides the 3MC1-causing mutations none of the functional consequences of these have been examined.

The polymorphisms that are found in 3MC1 syndrome can be grouped into mutations that abolish the expression of MASP-3 (p.G484E, p.C630R and W290stop) and mutations that result in MASP-3 variants without enzymatic activity (p.H497Y, p.D553N, p.D663Y, p.G665S, p.G666E, p.G687R) (data not published).

MUTANT ANIMALS

A *MASP1* knockout mouse has been generated by replacing exon 2 in embryonic stem cells with a gene cassette containing a neomycin resistance gene and chimeric mice were subsequently backcrossed into a C57BL/6J line.[20] This results in mice that are deficient in MASP-1, MASP-3 and MAp44, and are therefore also termed *MASP1/3*$^{-/-}$. The *MASP1* knockout mice display skeletal abnormalities and are significantly smaller than wild-type mice, partly due to a loss of adipose tissue.[21,22] Due to this phenotype, homozygotes are created by mating heterozygotic pairs.

The *MASP-1* knockout mouse has been the key in deciphering the role of MASP-3 as a link between the LP and the alternative pathway. The first suggestion for this role was thus the observation that *MASP1* knockout mice lack the ability to convert profactor D into factor D, and sera from these animals have a significantly reduced alternative pathway activity.[6]

REFERENCES

1. Kjaer TR, Le LTM, Pedersen JS, Sander B, Golas MM, Jensenius JC, et al. Structural insights into the initiating complex of the lectin pathway of complement activation. *Structure* 2015;**23**(2):342–51.
2. Teillet F, Gaboriaud C, Lacroix M, Martin L, Arlaud GJ, Thielens NM. Crystal structure of the CUB1-EGF-CUB2 domain of human MASP-1/3 and identification of its interaction sites with mannan-binding lectin and ficolins. *J Biol Chem* 2008;**283**(37):25715–24.
3. Ambrus G, Gál P, Kojima M, Szilágyi K, Balczer J, Antal J, et al. Natural substrates and inhibitors of mannan-binding lectin-associated serine protease-1 and -2: a study on recombinant catalytic fragments. *J Immunol* 2003;**170**(3):1374–82.
4. Gaboriaud C, Gupta RK, Martin L, Lacroix M, Serre L, Teillet F, et al. The serine protease domain of MASP-3: enzymatic properties and crystal structure in complex with ecotin. *PLoS One* 2013;**8**(7).
5. Zundel S, Cseh S, Lacroix M, Dahl MR, Matsushita M, Andrieu J-P, et al. Characterization of recombinant mannan-binding lectin-associated serine protease (MASP)-3 suggests an activation mechanism different from that of MASP-1 and MASP-2. *J Immunol* 2004;**172**(7):4342–50.
6. Takahashi M, Ishid1 Y, Iwaki D, Kanno K, Suzuki T, Endo Y, et al. Essential role of mannose-binding lectin-associated serine protease-1 in activation of the complement factor D. *J Exp Med* 2010;**207**(1):29–37.
7. Dobó J, Szakács D, Oroszlán G, Kortvely E, Kiss B, Boros E, et al. MASP-3 is the exclusive pro-factor D activator in resting blood: the lectin and the alternative complement pathways are fundamentally linked. *Sci Rep* August 2016;**6**:31877.
8. Degn SE, Kjaer TR, Kidmose RT, Jensen L, Hansen AG, Tekin M, et al. Complement activation by ligand-driven juxtaposition of discrete pattern recognition complexes. *Proc Natl Acad Sci* 2014;**111**(37):13445–50.
9. Iwaki D, Kanno K, Takahashi M, Endo Y, Matsushita M, Fujita T. The role of mannose-binding lectin-associated serine protease-3 in activation of the alternative complement pathway. *J Immunol* 2011;**187**(7):3751–8.
10. Oroszlan G, Kortvely E, Szakacs D, Kocsis A, Dammeier S, Zeck A, et al. MASP-1 and MASP-2 Do not activate pro-factor D in resting human blood, whereas MASP-3 is a potential activator: kinetic analysis involving specific MASP-1 and MASP-2 inhibitors. *J Immunol* 2016;**196**:857–65.
11. Sirmaci A, Walsh T, Akay H, Spiliopoulos M, Şakalar YB, Hasanefendioğlu-Bayrak A, et al. MASP1 mutations in patients with facial, umbilical, coccygeal, and auditory findings of carnevale, malpuech, OSA, and michels syndromes. *Am J Hum Genet* 2010;**87**(5):679–86.
12. Rooryck C, Diaz-Font A, Osborn DPS, Chabchoub E, Hernandez-Hernandez V, Shamseldin H, et al. Mutations in lectin complement pathway genes COLEC11 and MASP1 cause 3MC syndrome. *Nat Genet* 2011;**43**(3):197–203.
13. Cortesio CL, Jiang W. Mannan-binding lectin-associated serine protease 3 cleaves synthetic peptides and insulin-like growth factor-binding protein 5. *Arch Biochem Biophys* 2006;**449**(1–2):164–70.

14. Degn SE, Jensen L, Gál P, Dobó J, Holmvad SH, Jensenius JC, et al. Biological variations of MASP-3 and MAp44, two splice products of the *MASP1* gene involved in regulation of the complement system. *J Immunol Methods* 2010;**361**(1–2):37–50.

15. Seyfarth J, Garred P, Madsen HO. Extra-hepatic transcription of the human mannose-binding lectin gene (mbl2) and the MBL-associated serine protease 1–3 genes. *Mol Immunol* 2006;**43**(7):962–71.

16. Skjoedt MO, Hummelshoj T, Palarasah Y, Honore C, Koch C, Skjodt K, et al. A novel mannose-binding lectin/ficolin-associated protein is highly expressed in heart and skeletal muscle tissues and inhibits complement activation. *J Biol Chem* 2010;**285**(11):8234–43.

17. Endo Y, Takahashi M, Kuraya M, Matsushita M, Stover CM, Schwaeble WJ, et al. Functional characterization of human mannose-binding lectin-associated serine protease (MASP)-1/3 and MASP-2 promoters, and comparison with the C1s promoter. *Int Immunol* 2002;**14**(10):1193–201.

18. https://www.ncbi.nlm.nih.gov/gene/.

19. Atik T, Koparir A, Bademci G, Foster J, Altunoglu U, Mutlu GY, Tekin M, et al. Novel MASP1 mutations are associated with an expanded phenotype in 3MC1 syndrome. *Orphanet J Rare Dis* 2015;**10**(1):128.

20. Takahashi M, Iwaki D, Kanno K, Ishida Y, Xiong J, Matsushita M, et al. Mannose-binding lectin (MBL)-associated serine protease (MASP)-1 contributes to activation of the lectin complement pathway. *J Immunol* May 1, 2008;**180**(9):6132–8.

21. Takahashi M, Iwaki D, Endo Y, Fujita T. The study of MASPs knockout mice. In: Abdelmohsen K, editor. *Binding protein*. 2012. p. 165–80.

22. Takahashi M, Endo Y, Fujita T. Developmental abnormalities in Masp1/3-deficient mice. *Immunobiology* 2012;**217**:1164.

Chapter 10

C1r

Nicole M. Thielens, Christine Gaboriaud

Institut de Biologie Structurale (IBS), Univ. Grenoble Alpes, CEA, CNRS, Grenoble, France

PHYSICOCHEMICAL PROPERTIES

Human C1r is a noncovalent homodimer of M_r (K) 173. Each monomer is synthesised as a single-chain proenzyme molecule of 705 amino acids including a 17 amino acid leader sequence. Activation occurs through cleavage of a single bond (R463-I464), yielding two disulphide-linked chains A and B.

Mature protein		
pI	4.9	
M_r (K)	172.6	
	A-chain	B-chain
Amino acids	18–463	464–705
M_r (K) (predicted)	55.3	31.2
N-linked glycosylation sites	2 (125, 221)	2 (514, 581)
Interchain disulphide bonds 1 A-B	451	577

The murine gene is duplicated and encodes two variants, C1rA expressed in the liver, and C1rB expressed only in male reproductive tissues.[1] The corresponding proteins have 95.5% amino acid identity and are highly homologous to human C1r (80.9% and 79.9% identity for C1rA and C1rB, respectively).

STRUCTURE

Each monomer comprises an interaction region (CUB1-EGF-CUB2) derived from the N-terminal part of the A-chain, and a catalytic region comprising two complement control protein (CCP) modules and the serine protease (SP) domain (B-chain).[2,3] Assembly of the C1r-C1r dimer occurs through the catalytic regions.[4] The interaction regions are located at the ends of the dimer and mediate calcium-dependent interaction of C1r with C1s within the C1s-C1r-C1r-C1s tetramer and of the tetramer with C1q within the C1 complex (Fig. 10.1).[5,6]

The Complement FactsBook. http://dx.doi.org/10.1016/B978-0-12-810420-0.00010-9

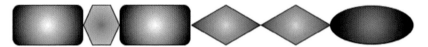

FIGURE 10.1 Stylised protein domains for C1r.

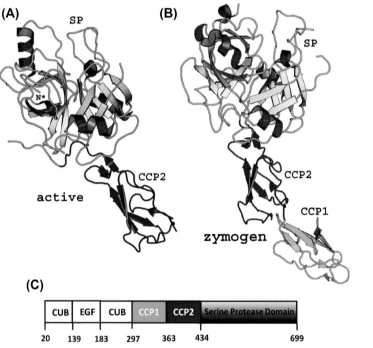

FIGURE 10.2 Complement serine protease C1r. (A) Ribbon diagram of the activated C1 catalytic domain (CCP2-SP) (PDB ID: 2QY0). (B) Zymogen (CCP1-CCP2-SP) form (PDB ID: 1GPZ). Activated N terminus is marked as N*. (C) Schematic representation of the six domain complement serine protease C1r, and the crystal structure is not available for the N-terminal CUB and EGF domains and depicted as colourless.

3D STRUCTURE

Crystal structures are available for the CCP1-CCP2-SP dimer in proenzyme (1GPZ)[7] and active (2QY0)[8] forms and for the CCP2-SP monomer in proenzyme (1MD7)[9] and active (1MD8)[9] forms. The NMR solution structure of the EGF-like module C1r is available (1APQ) (Fig. 10.2).[10]

FUNCTION

C1r is a SP (EC 3.4.21.41) with very narrow trypsin-like specificity that is responsible for activation of the C1 complex, which is triggered when C1 binds to an activating target through its recognition protein C1q. C1 activation is a two-step process involving (1) C1r intramolecular autoactivation and (2) C1s cleavage by activated C1r.[11,12] Both reactions occur through cleavage of a single

Arg–Ile bond that generates two-chain active proteases. The only known protease inhibitor of activated C1r is the serpin C1-inhibitor.

TISSUE DISTRIBUTION

The primary site of synthesis is hepatocytes. Plasma concentration is 34 µg/mL. Secondary sites of synthesis are monocytes, bone marrow-derived macrophages, epithelial and endothelial cells, myoblasts, smooth muscle cells, chondrocytes and cells of the central nervous system.[13–19]

REGULATION OF EXPRESSION

C1r and C1s are coordinately expressed in the liver.

The expression of the C1r gene (*C1R*) is upregulated upon activation of bone marrow-derived macrophages to M1 macrophages (induced by lipopolysaccharides and interferon γ) and M2b macrophages (induced by immune complex and toll-like receptor ligation).[19]

HUMAN PROTEIN SEQUENCE[20–23]

```
MWLLYLLVPA LFCRAGGSIP IPQKLFGEVT SPLFPKPYPN NFETTTVITV  50
PTGYRVKLVF QQFDLEPSEG CFYDYVKISA DKKSLGRFCG QLGSPLGNPP 100
GKKEFMSQGN KMLLTFHTDF SNEENGTIMF YKGFLAYYQA VDLDECASRS 150
KSGEEDPQPQ CQHLCHNYVG GYFCSCRPGY ELQEDTHSCQ AECSSELYTE 200
ASGYISSLEY PRSYPPDLRC NYSIRVERGL TLHLKFLEPF DIDDHQQVHC 250
PYDQLQIYAN GKNIGEFCGK QRPPDLDTSS NAVDLLFFTD ESGDSRGWKL 300
RYTTEIIKCP QPKTLDEFTI IQNLQPQYQF RDYFIATCKQ GYQLIEGNQV 350
LHSFTAVCQD DGTWHRAMPR CKIKDCGQPR NLPNGDFRYT TTMGVNTYKA 400
RIQYYCHEPY YKMQTRAGSR ESEQGVYTCT AQGIWKNEQK GEKIPRCLPV 450
CGKPVNPVEQ RQRIIGGQKA KMGNFPWQVF TNIHGRGGGA LLGDRWILTA 500
AHTLYPKEHE AQSNASLDVF LGHTNVEELM KLGNHPIRRV SVHPDYRQDE 550
SYNFEGDIAL LELENSVTLG PNLLPICLPD NDTFYDLGLM GYVSGFGVME 600
EKIAHDLRFV RLPVANPQAC ENWLRGKNRM DVFSQNMFCA GHPSLKQDAC 650
QGDSGGVFAV RDPNTDRWVA TGIVSWGIGC SRGYGFYTKV LNYVDWIKKE 700
MEEED
```

The leader sequence and the cleavage site (RI) between the A- (N-terminal) and B- (C-terminal) chains are underlined. The *N*-linked glycosylation sites (all occupied) are indicated (**N**). N167 undergoes full posttranslational hydroxylation.

PROTEIN MODULES

1–17	Leader sequence	exon 2/3
18–141	CUB1	exon 3/4
142–192	EGF-Ca^{2+}	exon 5
193–304	CUB2	exon 6/7
305–373	CCP1	exon 8/9
374–447	CCP2	exon 10/11
448–463	Activation peptide	exon 12
464–705	Serine protease domain	exon 12

Catalytic triad: H502, D557, S654.

CHROMOSOMAL LOCATION

Human: 12p13.[24] The human *C1R* and *C1S* genes lie in a tail-to-tail orientation, with a distance of about 9.5 kb between their 3′ ends[25]

Mouse: 6F2

Rat: 4q42

A gene encoding a C1r-like protease (C1r-LP) is located 2 kb upstream of human *C1R* and 3 kb upstream of the mouse *C1ra* gene.[26,27]

HUMAN CDNA SEQUENCE[22]

```
TGCACGAAGA CGCTGTCGGG AGAGCCCAGG ATTCAACACG GGCCTTGAGA AATGTGGCTC   60
TTGTACCTCC TGGTGCCGGC CCTGTTCTGC AGGGCAGGAG GCTCCATTCC CATCCCTCAG  120
AAGTTATTTG GGGAGGTGAC TTCCCCTCTG TTCCCCAAGC CTTACCCCAA CAACTTTGAA  180
ACAACCACTG TGATCACAGT CCCCACGGGA TACAGGGTGA AGCTCGTCTT CCAGCAGTTT  240
GACCTGGAGC CTTCTGAAGG CTGCTTCTAT GATTATGTCA AGATCTCTGC TGATAAGAAA  300
AGCCTGGGGA GGTTCTGTGG GCAACTGGGT TCTCCACTGG GCAACCCCCC GGGAAAGAAG  360
GAATTTATGT CCCAAGGGAA CAAGATGCTG CTGACCTTCC ACACAGACTT CTCCAACGAG  420
GAGAATGGGA CCATCATGTT CTACAAGGGC TTCCTGGCCT ACTACCAAGC TGTGGACCTT  480
GATGAATGTG CTTCCCGGAG CAAATTAGGG GAGGAGGATC CCCAGCCCCA GTGCCAGCAC  540
CTGTGTCACA ACTACGTTGG AGGCTACTTC TGTTCCTGCC GTCCAGGCTA TGAGCTTCAG  600
GAAGACAGGC ATTCCTGCCA GGCTGAGTGC AGCAGCGAGC TGTACACGGA GGCATCAGGC  660
TACATCTCCA GCCTGGAGTA CCCTCGGTCC TACCCCCCTG ACCTGCGCTG CAACTACAGC  720
ATCCGGGTGG AGCGGGGCCT CACCCTGCAC CTCAAGTTCC TGGAGCCTTT TGATATTGAT  780
GACCACCAGC AAGTACACTG CCCCTATGAC CAGCTACAGA TCTATGCCAA CGGGAAGAAC  840
ATTGGCGAGT TCTGTGGGAA GCAAAGGCCC CCCGACCTCG ACACCAGCAG CAATGCTGTG  900
GATCTGCTGT TCTTCACAGA TGAGTCGGGG GACAGCCGGG GCTGGAAGCT GCGCTACACC  960
ACCGAGATCA TCAAGTGCCC CCAGCCCAAG ACCCTAGACG AGTTCACCAT CATCCAGAAC 1020
CTGCAGCCTC AGTACCAGTT CCGTGACTAC TTCATTGCTA CCTGCAAGCA AGGCTACCAG 1080
CTCATAGAGG GGAACCAGGT GCTGCATTCC TTCACAGCTG TCTGCCAGGA TGATGGCACG 1140
TGGCATCGTG CCATGCCCAG ATGCAAGATC AAGGACTGTG GCAGCCCCG AAACCTGCCT 1200
AATGGTGACT TCCGTTACAC CACCACAATG GGAGTGAACA CCTACAAGGC CCGTATCCAG 1260
TACTACTGCC ATGAGCCATA TTACAAGATG CAGACCAGAG CTGGCAGCAG GGAGTCTGAG 1320
CAAGGGGTGT ACACCTGCAC AGCACAGGGC ATTTGGAAGA ATGAACAGAA GGGAGAGAAG 1380
ATTCCTCGGT GCTTGCCAGT GTGTGGGAAG CCCGTGAACC CCGTGGAACA GAGGCAGCGC 1440
ATCATCGGAG GGCAAAAAGC CAAGATGGGC AACTTCCCCT GGCAGGTGTT CACCAACATC 1500
CACGGGCGCG GGGGCGGGGC CCTGCTGGGC GACCGCTGGA TCCTCACAGC TGCCCACACC 1560
CTGTATCCCA AGGAACACGA AGCGCAAAGC AACGCCTCTT TGGATGTGTT CCTGGGCCAC 1620
ACAAATGTGG AAGAGCTCAT GAAGCTAGGA AATCACCCCA TCCGCAGGGT CAGCGTCCAC 1680
CCGGACTACC GTCAGGATGA GTCCTACAAT TTTGAGGGGG ACATCGCCCT GCTGGAGCTG 1740
GAAAATAGTG TCACCCTGGG TCCCAACCTC CTCCCCATCT GCCTCCCTGA CAACGATACC 1800
TTCTACGACC TGGGCTTGAT GGGCTATGTC AGTGGCTTCG GGGTCATGGA GGAGAAGATT 1860
GCTCATGACC TCAGGTTTGT CCGTCTGCCC GTAGCTAATC CACAGGCCTG TGAGAACTGG 1920
CTCCGGGGAA AGAATAGGAT GGATGTGTTC TCTCAAAACA TGTTCTGTGC TGGACACCCA 1980
TCTCTAAAGC AGGACGCCTG CCAGGGGGAT AGTGGGGGCG TTTTTGCAGT AAGGGACCCG 2040
AACACTGATC GCTGGGTGGC CACGGGCATC GTGTCCTGGG GCATCGGGTG CAGCAGGGGC 2100
TATGGCTTCT ACACCAAAGT GCTCAACTAC GTGGACTGGA TCAAGAAAGA GATGGAGGAG 2160
GAGGACTGAG CCCAGAATTC ACTAGGTTCG AATCCAGAGA GCAGTGTGGA AAAAAAAAAA 2220
CAAAAAACAA CTGACCAGTT GTTGATAACC ACTAAGAGTC TCTATTAAAA TTACTGATGC 2280
AGAAAGACCG TGTGTGAAAT TCTCTTTCCT GTAGTCCCAT TGATGTACTT TACCTGAAAC 2340
AACCAAAGGG CCCCTTTCTT TCTTCTGAGG ATTGCAGAGG ATATAG
```

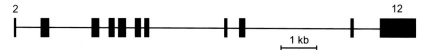

FIGURE 10.3   Diagram indicating the relative position of each exon in the *C1R* gene.

By analogy with the *C1S* gene, it is assumed that the dinucleotide AT of the initiation codon is present in exon 2 (G is present in exon 3). However, it is still unknown whether the *C1R* gene contains exon 1.[28] The first five nucleotides in each exon (starting from exon 3) are underlined to indicate intron–exon boundaries. The initiation codon (ATG), the termination codon (TGA) and the probable polyadenylation signal (AATAAA) are indicated.

GENOMIC STRUCTURE

The human *C1R* gene spans 11 kb from the initiation codon to the stop codon and is assumed to comprise 12 exons.[28] The exon/intron structure of the *C1R* gene is very similar to that of the *C1S* gene. The SP domain and the preceding connecting segment are encoded by a single exon (exon 12) (Fig. 10.3). Four ALU elements have been identified in introns 8, 10 and 11. The murine *C1ra* gene has a similar intron/exon organisation.[1]

ACCESSION NUMBERS

Human	X04701
	M14058
Mouse C1rA	AF14216
Mouse C1rB	AF459018

DEFICIENCY

Cases of C1r deficiencies are rare and usually combined with a C1s deficiency. Twelve cases (in eight families) of C1r deficiencies have been reported and the serum levels of C1s were consistently 30% of the normal level. C1r deficiencies are associated with an increased susceptibility to pyogenic infection and autoimmune disease such as systemic lupus erythematosus.[29] The molecular defect leading to C1r deficiency has been determined in one case, where a homozygous C to T substitution in exon 10, corresponding to the R380X nonsense mutation in CCP2 module, resulted in no detectable protein in the serum.[30]

A study has shown that periodontal Ehlers-Danlos Syndrome, mainly characterised by early onset periodontitis, easy bruising and pretibial hyperpigmentation, is caused by heterozygous missense or in-frame insertion/deletion mutations in *C1R* (14 variants, 15 families) or *C1S* (2 variants, 2 families).[31]

Pathogenic human C1r mutations in periodontal Ehlers-Danlos syndrome

V50D (CUB1)	15 patients (1 family)
D290G (CUB1)	1 patient
G297D (CUB2)	1 patient
L300P (CUB2)	3 patients (1 family)
R301P (CUB2)	13 patients (1 family)
T302C (CUB2)	7 patients (1 family)
I306-C309 deletion - RR insertion (CCP1)	1 patient
C309W (CCP1)	4 patients (2 families)
C338R (CCP1)	4 patients (1 family)
C358F (CCP1)	3 patients (1 family)
W364C (CCP1)	1 patient
C371W (CCP1)	1 patient
R401-Y405 deletion - HVI insertion (CCP2)	10 patients (1 family)
W435R (CCP2)	12 patients (1 family)

POLYMORPHIC VARIANTS

The *C1R* locus is highly polymorphic; six common and rare alleles are characterised by amino acid mutations (Y131H, S152L, H163Y, E184K, T186R and G261R) in exons 4, 5 and 7.[22,23,28] In addition, nine nucleotide substitutions and one length polymorphism occur in introns 2, 3, 4, 8 and 10.[28]

MUTANT ANIMALS

There are no knockout mice for C1r available.

REFERENCES

1. Garnier G, Circolo A, Xu Y, Volanakis JE. Complement C1r and C1s genes are duplicated in the mouse: differential expression generates alternative isomorphs in the liver and in the male reproductive system. *Biochem J* 2003;**371**(2):631–40.
2. Busby TF, Ingham KC. Calcium-sensitive thermal transitions and domain structure of human complement subcomponent C1r. *Biochemistry* 1987;**26**(17):5564–71.
3. Villiers CL, Arlaud GJ, Colomb MG. Domain structure and associated functions of subcomponents C1r and C1s of the first component of human complement. *Proc Natl Acad Sci USA* 1985;**82**(13):4477–81.
4. Lacroix M, Rossi V, Gaboriaud C, Chevallier S, Jaquinod M, Thielens NM, et al. Structure and assembly of the catalytic region of human complement protease C1r: a three-dimensional model based on chemical cross-linking and homology modeling. *Biochemistry* 1997;**36**(21):6270–82.
5. Thielens NM, Aude CA, Lacroix MB, Gagnon J, Arlaud GJ. Ca²⁺ binding properties and Ca²(+)-dependent interactions of the isolated NH²-terminal alpha fragments of human complement proteases C1-r and C1-s. *J Biol Chem* 1990;**265**(24):14469–75.
6. Bally I, Rossi V, Lunardi T, Thielens NM, Gaboriaud C, Arlaud GJ. Identification of the C1q-binding sites of human C1r and C1s: a refined three-dimensional model of the C1 complex of complement. *J Biol Chem* 2009;**284**(29):19340–8.
7. Budayova-Spano M, Lacroix M, Thielens NM, Arlaud GJ, Fontecilla-Camps JC, Gaboriaud C. The crystal structure of the zymogen catalytic domain of complement protease C1r reveals that a disruptive mechanical stress is required to trigger activation of the C1 complex. *EMBO J* 2002;**21**(3):231–9.

8. Kardos J, Harmat V, Pallo A, Barabas O, Szilagyi K, Graf L, et al. Revisiting the mechanism of the autoactivation of the complement protease C1r in the C1 complex: structure of the active catalytic region of C1r. *Mol Immunol* 2008;**45**(6):1752–60.

9. Budayova-Spano M, Grabarse W, Thielens NM, Hillen H, Lacroix M, Schmidt M, et al. Monomeric structures of the zymogen and active catalytic domain of complement protease c1r: further insights into the c1 activation mechanism. *Structure* 2002;**10**(11):1509–19.

10. Bersch B, Hernandez JF, Marion D, Arlaud GJ. Solution structure of the epidermal growth factor (EGF)-like module of human complement protease C1r, an atypical member of the EGF family. *Biochemistry* 1998;**37**(5):1204–14.

11. Arlaud GJ, Thielens NM. Human complement serine proteases C1r and C1s and their proenzymes. *Methods Enzymol* 1993;**223**:61–82.

12. Cooper NR. The classical complement pathway: activation and regulation of the first complement component. *Adv Immunol* 1985;**37**:151–216.

13. Bradley K, North J, Saunders D, Schwaeble W, Jeziorska M, Woolley DE, et al. Synthesis of classical pathway complement components by chondrocytes. *Immunology* 1996;**88**(4):648–56.

14. Drouet C, Reboul A. Biosynthesis of C1r and C1s subcomponents. *Behring Inst Mitt* 1989;**84**:80–8.

15. Gulati P, Lemercier C, Guc D, Lappin D, Whaley K. Regulation of the synthesis of C1 subcomponents and C1-inhibitor. *Behring Inst Mitt* 1993;**93**:196–203.

16. Laufer J, Oren R, Farzam N, Goldberg I, Passwell J. Differential cytokine regulation of complement proteins in human glomerular epithelial cells. *Nephron* 1997;**76**(3):276–83.

17. Legoedec J, Gasque P, Jeanne JF, Scotte M, Fontaine M. Complement classical pathway expression by human skeletal myoblasts in vitro. *Mol Immunol* 1997;**34**(10):735–41.

18. Walker DG, Dalsing-Hernandez JE, Lue LF. Human postmortem brain-derived cerebrovascular smooth muscle cells express all genes of the classical complement pathway: a potential mechanism for vascular damage in cerebral amyloid angiopathy and Alzheimer's disease. *Microvasc Res* 2008;**75**(3):411–9.

19. Luo C, Chen M, Madden A, Xu H. Expression of complement components and regulators by different subtypes of bone marrow-derived macrophages. *Inflammation* 2012;**35**(4):1448–61.

20. Arlaud GJ, Gagnon J. Complete amino acid sequence of the catalytic chain of human complement subcomponent C1-r. *Biochemistry* 1983;**22**(8):1758–64.

21. Arlaud GJ, Willis AC, Gagnon J. Complete amino acid sequence of the A chain of human complement-classical-pathway enzyme C1r. *Biochem J* 1987;**241**(3):711–20.

22. Journet A, Tosi M. Cloning and sequencing of full-length cDNA encoding the precursor of human complement component C1r. *Biochem J* 1986;**240**(3):783–7.

23. Leytus SP, Kurachi K, Sakariassen KS, Davie EW. Nucleotide sequence of the cDNA coding for human complement C1r. *Biochemistry* 1986;**25**(17):4855–63.

24. Nguyen VC, Tosi M, Gross MS, Cohen-Haguenauer O, Jegou-Foubert C, de Tand MF, et al. Assignment of the complement serine protease genes C1r and C1s to chromosome 12 region 12p13. *Hum Genet* 1988;**78**(4):363–8.

25. Kusumoto H, Hirosawa S, Salier JP, Hagen FS, Kurachi K. Human genes for complement components C1r and C1s in a close tail-to-tail arrangement. *Proc Natl Acad Sci USA* 1988;**85**(19):7307–11.

26. Circolo A, Garnier G, Volanakis JE. A novel murine complement-related gene encoding a C1r-like serum protein. *Mol Immunol* 2003;**39**(14):899–906.

27. Ligoudistianou C, Xu Y, Garnier G, Circolo A, Volanakis JE. A novel human complement-related protein, C1r-like protease (C1r-LP), specifically cleaves pro-C1s. *Biochem J* 2005;**387**(1):165–73.

28. Nakagawa M, Yuasa I, Irizawa Y, Umetsu K. The human complement component C1R gene: the exon-intron structure and the molecular basis of allelic diversity. *Ann Hum Genet* 2003;**67**(3):207–15.

29. Lintner KE, Wu YL, Yang Y, Spencer CH, Hauptmann G, Hebert LA, et al. Early components of the complement classical activation pathway in human systemic autoimmune diseases. *Front Immunol* 2016;**7**:36.

30. Wu YL, Brookshire BP, Verani RR, Arnett FC, Yu CY. Clinical presentations and molecular basis of complement C1r deficiency in a male African-American patient with systemic lupus erythematosus. *Lupus* 2011;**20**(11):1126–34.

31. Kapferer-Seebacher I, Pepin M, Werner R, Aitman T, Nordgren A, Stoiber H, et al. Periodontal Ehlers-Danlos syndrome is caused by mutations in *C1R* and *C1S*, which encode subcomponents C1r and C1s of complement. *Am J Hum Genet* 2016, 1005–1014;**99**. http://dx.doi.org/10.1016/j.ajhg.2016.08.019. [in press].

Chapter 11

C1s

Nicole M. Thielens, Christine Gaboriaud, Véronique Rossi

Institut de Biologie Structurale (IBS), Univ. Grenoble Alpes, CEA, CNRS, Grenoble, France

PHYSICOCHEMICAL PROPERTIES

Human C1s is a glycoprotein of M_r (K) 79.8. It is synthesised as a single-chain proenzyme molecule of 688 amino acids including a 15 amino acid leader sequence. Activation occurs through cleavage of a single bond (R437-I438), yielding two disulphide-linked chains A and B.

Mature protein		
pI	4.5	
M_r (K)	79.8	
	A-chain	B-chain
Amino acids	16–437	438–688
M_r (K) (observed)	52.2	27.7
N-linked glycosylation sites	2 (174, 406)	
Interchain disulphide bonds A-B	425	549

The murine gene is duplicated and encodes two variants, C1sA expressed in the liver and C1sB expressed only in the male reproductive tissues.[1] The corresponding proteins have 93.0% amino acid identity and are highly homologous to human C1s (74.0% and 74.5% identity for C1sA and C1sB, respectively).

STRUCTURE

C1s comprises an interaction region (CUB1-EGF-CUB2) derived from the N-terminal part of the A-chain, a catalytic region comprising two complement control protein modules and the serine protease (SP) domain (B-chain).[2–5] The interaction regions mediate interaction with the corresponding regions of C1r within the C1s-C1r-C1r-C1s tetramer and binding of the tetramer to the collagen stalks of C1q within the C1 complex (Fig. 11.1).[4,6]

The Complement FactsBook. http://dx.doi.org/10.1016/B978-0-12-810420-0.00011-0

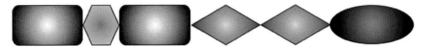

FIGURE 11.1 Stylised protein domains for C1s.

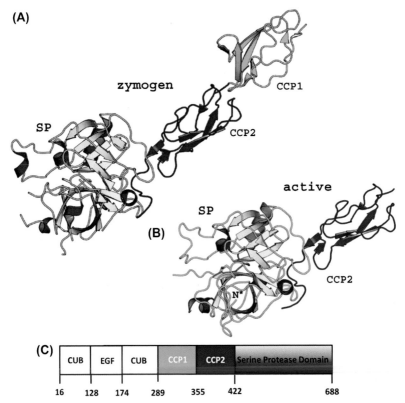

FIGURE 11.2 Crystal structure for complement serine protease C1s. (A) Ribbon diagram of the C1s zymogen crystal structure (PDB ID: 4J1Y). (B) Activated C1s crystal structure (PDB ID: 1ELV) with activated N-terminus marked as N*. (C) Schematic representation of the six domain complement serine protease C1s. Crystal structure is not available for the N-terminal CUB and EGF domains.

3D STRUCTURE

Crystal structures are available for the CUB1(complement C1r/C1s, Uegf, Bmp1)-EGF(epidermal growth factor) fragment (1NZI),[7] the CUB1-EGF-CUB2 fragment[8] [alone (4MLF) and in association with a collagen-like peptide (4LOR)], the CUB2-CCP1 (4LOS) and CUB2-CCP1-CCP2 (4LOT) fragments[8] and for zymogen CCP1-CCP2-SP (4J1Y)[9] and activated CCP2-SP (1ELV)[10] fragments from the catalytic region (Fig. 11.2).

FUNCTION

C1s is a highly specific SP (EC 3.4.21.42) that mediates proteolytic activity of the C1 complex towards its complement substrates C4 and C2. It cleaves a single Arg–Ala bond in C4 to yield C4a and C4b and a single Arg–Lys bond in C2 to yield C2a and C2b.[11,12] Assembly of the C3 convertase C4bC2a requires binding of C2 to C4b before cleavage by C1s. C1s activity is inhibited by the serpin C1-Inhibitor. Noncomplement C1s protein substrates include β2-microglobulin,[13] heavy chain of the major histocompatibility complex class I antigens,[14] insulin-like growth factor-binding protein 5[15] and low density lipoprotein receptor-related protein 6 (LRP6).[16] In addition, C1s has been shown to cleave the nuclear protein high-mobility group box 1, an alarmin released from apoptotic cells.[17]

TISSUE DISTRIBUTION

Primary site of synthesis is hepatocytes. Plasma concentration is 31 μg/mL. Secondary sites of synthesis are monocytes, bone marrow-derived macrophages, epithelial and endothelial cells, myoblasts, smooth muscle cells, chondrocytes and cells of the central nervous system.[18–24]

REGULATION OF EXPRESSION

C1s and C1r are coordinately expressed in the liver. The level of *C1S* is upregulated upon activation of bone marrow-derived macrophages to M1 macrophages [induced by lipopolysaccharides and interferon (IFN)-γ] and M2b macrophages (induced by immune complex and toll-like receptor ligation).[24]Activation of human monocytes by conditioned media from mitogen-, antigen- or allogeneic-stimulated lymphocyte cultures increases C1s secretion.[25] Monocyte C1s secretion is also enhanced by IFN-α and IFN-γ.[19]

HUMAN PROTEIN SEQUENCE[26–28]

```
MWCIVLFSLL AWVYAEPTMY GEILSPNYPQ AYPSEVEKSW DIEVPEGYGI  50
HLYFTHLDIE LSENCAYDSV QIISGDTEEG RLCGQRSSNN PHSPIVEEFQ 100
VPYNKLQVIF KSDFSNEERF TGFAAYYVAT DINECTDFVD VPCSHFCNNF 150
IGGYFCSCPP EYFLHDDMKN CGVNCSGDVF TALIGEIASP NYPKPYPENS 200
RCEYQIRLEK GFQVVVTLRR EDFDVEAADS AGNCLDSLVF VAGDRQFGPY 250
CGHGFPGPLN IETKSNALDI IFQTDLTGQK KGWKLRYHGD PMPCPKEDTP 300
NSVWEPAKAK YVFRDVVQIT CLDGFEVVEG RVGATSFYST CQSNGKWSNS 350
KLKCQPVDCG IPESIENGKV EDPESTLFGS VIRYTCEEPY YYMENGGGGE 400
YHCAGNGSWV NEVLGPELPK CVPVCGVPRE PFEEKQRIIG GSDADIKNFP 450
WQVFFDNPWA GGALINEYWV LTAAHVVEGN REPTMYVGST SVQTSRLAKS 500
KMLTPEHVFI HPGWKLLEVP EGRTNFDNDI ALVRLKDPVK MGPTVSPICL 550
PGTSSDYNLM DGDLGLISGW GRTEKRDRAV RLKAARLPVA PLRKCKEVKV 600
EKPTADAEAY VFTPNMICAG GEKGMDSCKG DSGGAFAVQD PNDKTKFYAA 650
GLVSWGPQCG TYGLYTRVKN YVDWIMKTMQ ENSTPRED
```

The leader sequence and the cleavage site (R̲I̲) between the A- (N-terminal) and B-chains (C-terminal) are underlined. The *N*-linked glycosylation sites (all occupied[29]) are indicated (**N**). N149 undergoes partial posttranslational hydroxylation.[30]

PROTEIN MODULES

1–15	Leader sequence	exon 2/3
16–128	CUB1	exon 3/4
129–174	EGF-Ca^{2+}	exon 5
175–289	CUB2	exon 6/7
290–356	CCP1	exon 8/9
357–422	CCP2	exon 10/11
423–437	Activation peptide	exon 12
438–688	Serine protease domain	exon 12

Catalytic triad: H475, D529, S632.

CHROMOSOMAL LOCATION

Human: 12p13.[31] The human *C1R* and *C1S* genes lie in a tail-to-tail orientation, with a distance of about 9.5 kb between their 3′ ends[26] and are derived from a common ancestral gene through a gene duplication event.[31]

Mouse: 6F2

Rat: 4q42

HUMAN CDNA SEQUENCE[26]

```
GGGCCGGAGT TCCTGCAGAG GGAGCGTCAA GGCCCTGTGC TGCTGTCCCT GGGGGCCAGA   60
GGGGTTGCCC AGCATGCCCA CTGGCAGGAG AGAGGGAACT GACCCACTTG CTCCTACCAG  120
CTTCTGAAGG CTCCAAAGTC CGGAGTGCAG AAAGCCAGGA CCAAGAGACA GGCAGCTCAC  180
CAGGGTGGAC AAATCGCCAG AGATGTGGTG CATTGTCCTG TTTTCACTTT TGGCATGGGT  240
TTATGCTGAG CCTACCATGT ATGGGGAGAT CCTGTCCCCT AACTATCCTC AGGCATATCC  300
CAGTGAGGTA GAGAAATCTT GGGACATAGA AGTTCCTGAA GGGTATGGGA TTCACCTCTA  360
CTTCACCCAT CTGGACATTG AGCTGTCAGA GAACTGTGCG TATGACTCAG TGCAGATAAT  420
CTCAGGAGAC ACTGAAGAAG GGAGGCTCTG TGGACAGAGG AGCAGTAACA ATCCCCACTC  480
TCCAATTGTG GAAGAGTTCC AAGTCCCATA CAACAAACTC CAGGTGATCT TTAAGTCAGA  540
CTTTTCCAAT GAAGAGCGTT TTACGGGGTT TGCTGCATAC TATGTTGCCA CAGACATAAA  600
TGAATGCACA GATTTTGTAG ATGTCCCTTG TAGCCACTTC TGCAACAATT TCATTGGTGG  660
TTACTTCTGC TCCTGCCCCC CGGAATATTT CCTCCATGAT GACATGAAGA ATTGCGGAGT  720
TAATTGCAGT GGGGATGTAT TCACTGCACT GATTGGGGAG ATTGCAAGTC CCAATTATCC  780
CAAACCATAT CCAGAGAACT CAAGGTGTGA ATACCAGATC CGGTTGGAGA AAGGGTTCCA  840
AGTGGTGGTG ACCTTGCGGA GAGAAGATTT TGATGTGGAA GCAGCTGACT CAGCGGGAAA  900
CTGCCTTGAC AGTTTAGTTT TTGTTGCAGG AGATCGGCAA TTTGGTCCTT ACTGTGGTCA  960
```

```
TGGATTCCCT GGGCCTCTAA ATATTGAAAC CAAGAGTAAT GCTCTTGATA TCATCTTCCA 1020
AACTGATCTA ACAGGGCAAA AAAAGGGCTG GAAACTTCGC TATCATGGAG ATCCAATGCC 1080
CTGCCCTAAG GAAGACACTC CCAATTCTGT TTGGGAGCCT GCGAAGGCAA AATATGTCTT 1140
TAGAGATGTG GTGCAGATAA CCTGTCTGGA TGGGTTTGAA GTTGTGGAGG GACGTGTTGG 1200
TGCAACATCT TTCTATTCGA CTTGTCAAAG CAATGGAAAG TGGAGTAATT CCAAACTGAA 1260
ATGTCAACCT GTGGACTGTG GCATTCCTGA ATCCATTGAG AATGGTAAAG TTGAAGACCC 1320
AGAGAGCACT TTGTTTGGTT CTGTCATCCG CTACACTTGT GAGGAGCCAT ATTACTACAT 1380
GGAAAATGGA GGAGGTGGGG AGTATCACTG TGCTGGTAAC GGGAGCTGGG TGAATGAGGT 1440
GCTGGGCCCG GAGCTGCCGA AATGTGTTCC AGTCTGTGGA GTCCCCAGAG AACCCTTTGA 1500
AGAAAAACAG AGGATAATTG GAGGATCCGA TGCAGATATT AAAAACTTCC CCTGGCAAGT 1560
CTTCTTTGAC AACCCATGGG CTGGTGGAGC GCTCATTAAT GAGTACTGGG TGCTGACGGC 1620
TGCTCATGTT GTGGAGGGAA ACAGGGAGCC AACAATGTAT GTTGGGTCCA CCTCAGTGCA 1680
GACCTCACGG CTGGCAAAAT CCAAGATGCT CACTCCTGAG CATGTGTTTA TTCATCCGGG 1740
ATGGAAGCTG CTGGAAGTCC CAGAAGGACG AACCAATTTT GATAATGACA TTGCACTGGT 1800
GCGGCTGAAA GACCCAGTGA AAATGGGACC CACCGTCTCT CCCATCTGCC TACCAGGCAC 1860
CTCTTCCGAC TACAACCTCA TGGATGGGGA CCTGGGACTG ATCTCAGGCT GGGGGCCGAAC 1920
AGAGAAGAGA GATCGTGCTG TTCGCCTCAA GGCGGCCAAGG TTACCTGTAG CTCCTTTAAG 1980
AAAATGCAAA GAAGTGAAAG TGGAGAAACC CACAGCAGAT GCAGAGGCCT ATGTTTTCAC 2040
TCCTAACATG ATCTGTGCTG GAGGAGAGAA GGGCATGGAT AGCTGTAAAG GGGACAGTGG 2100
TGGGGCCTTT GCTGTACAGG ATCCCAATGA CAAGACCAAA TTCTACGCAG CTGGCCTGGT 2160
GTCCTGGGGG CCCCAGTGTG GGACCTATGG GCTCTACACA CGGGTAAAGA ACTATGTTGA 2220
CTGGATAATG AAGACTATGC AGGAAAATAG CACCCCCCGT GAGGACTAAT CCAGATACAT 2280
CCCACCAGCC TCTCCAAGGG TGGTGACCAA TGCATTACCT TCTGTTCCTT ATGATATTCT 2340
CATTATTTCA TCATGACTGA AAGAAGACAC GAGCGAATGA TTTAAATAGA ACTTGATTGT 2400
TGAGACGCCT TGCTAGAGGT AGAGTTTGAT CATAGAATTG TGCTGGTCAT ACATTTGTGG 2460
TCTGACTCCT TGGGGTCCTT TCCCCGGAGT ACCTATTGTA GATAACACTA TGGGTGGGGC 2520
ACTCCTTTCT TGCACTATTC CACAGGGATA CCTTAATTCT TTGTTTCCTC TTTACCTGTT 2580
CAAAATTCCA TTTACTTGAT CATTCTCAGT ATCCACTGTC TATGTACAAT AAAGGATGTT 2640
TATAAGC
```

The first five nucleotides in each exon are underlined to indicate intron–exon boundaries. Exon 2 is only five nucleotides long (203–207) and contains the initiation codon. To avoid confusion, this exon is not indicated with additional underlining, but is indicated by the double-underlined methionine initiation codon (ATG). The termination codon (TAA) and the polyadenylation signal (AATAAA) are indicated.

GENOMIC STRUCTURE

The human *C1S* gene spans 11 kb from the initiation codon to the stop codon and contains 12 exons.[32] The exon/intron structure of the *C1S* gene is very similar to that of the *C1R* gene. The SP domain and the preceding connecting segment are encoded by a single exon (exon 12) (Fig. 11.3). Four ALU elements have been identified in introns 2, 4, 6 and 7.

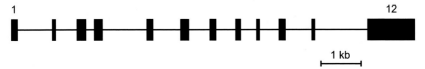

FIGURE 11.3   Diagram indicating the relative position of each exon in the *C1S* gene.

ACCESSION NUMBERS

Human	X06596
	M18767
	J04080
Mouse C1sA	AF459019
Mouse C1sB	AF459020

DEFICIENCY

Eight cases of C1s deficiencies (five families) have been reported. Most C1s-deficient patients had greatly reduced serum levels of C1r, suffered from recurrent pyogenic infections and developed systemic lupus erythematosus or lupus-like disease.[33]

Several deleterious mutations leading to complete C1s deficiency have been identified:

- Y204X (exon 6, nonsense mutation)[34]
- 4-bp deletion (exon 10, frame-shift mutation)+E597X (exon 12, nonsense mutation)[35,36]
- E597X+G630Q (exon 12, nonsense and missense mutations)[37]
- R534X (exon 12, nonsense mutation)[38]

A study has shown that periodontal Ehlers-Danlos syndrome, mainly characterised by early onset periodontitis, easy bruising and pretibial hyperpigmentation, is caused by heterozygous missense or in-frame insertion/deletion mutations in *C1R* (14 variants, 15 families) or *C1S* (2 variants, 2 families).[39]

Pathogenic Human C1s Mutations in Periodontal Ehlers-Danlos Syndrome

C294R (CCP1)	7 patients (1 family)
V301 deletion (CCP1)	9 patients (1 family)

POLYMORPHIC VARIANTS

No polymorphism has been identified at the DNA or protein level. One common and two uncommon variants have been identified by isoelectric focusing of plasma samples.[40]

MUTANT ANIMALS

There are no knockout mice for C1s available.

REFERENCES

1. Garnier G, Circolo A, Xu Y, Volanakis JE. Complement C1r and C1s genes are duplicated in the mouse: differential expression generates alternative isomorphs in the liver and in the male reproductive system. *Biochem J* 2003;**371**(2):631–40.

2. Busby TF, Ingham KC. Domain structure, stability, and interactions of human complement C1s-: characterization of a derivative lacking most of the B chain. *Biochemistry* 1988;**27**(16):6127–35.

3. Medved LV, Busby TF, Ingham KC. Calorimetric investigation of the domain structure of human complement Cl-s: reversible unfolding of the short consensus repeat units. *Biochemistry* 1989;**28**(13):5408–14.

4. Thielens NM, Aude CA, Lacroix MB, Gagnon J, Arlaud GJ. Ca^{2+} binding properties and Ca2(+)-dependent interactions of the isolated NH2-terminal alpha fragments of human complement proteases C1-r and C1-s. *J Biol Chem* 1990;**265**(24):14469–75.

5. Villiers CL, Arlaud GJ, Colomb MG. Domain structure and associated functions of subcomponents C1r and C1s of the first component of human complement. *Proc Natl Acad Sci USA* 1985;**82**(13):4477–81.

6. Bally I, Rossi V, Lunardi T, Thielens NM, Gaboriaud C, Arlaud GJ. Identification of the C1q-binding Sites of Human C1r and C1s: a refined three-dimensional model of the C1 complex of complement. *J Biol Chem* 2009;**284**(29):19340–8.

7. Gregory LA, Thielens NM, Arlaud GJ, Fontecilla-Camps JC, Gaboriaud C. X-ray structure of the Ca2+-binding interaction domain of C1s. Insights into the assembly of the C1 complex of complement. *J Biol Chem* 2003;**278**(34):32157–64.

8. Venkatraman Girija U, Gingras AR, Marshall JE, Panchal R, Sheikh MA, Gal P, et al. Structural basis of the C1q/C1s interaction and its central role in assembly of the C1 complex of complement activation. *Proc Natl Acad Sci USA* 2013;**110**(34):13916–20.

9. Perry AJ, Wijeyewickrema LC, Wilmann PG, Gunzburg MJ, D'Andrea L, Irving JA, et al. A molecular switch governs the interaction between the human complement protease C1s and its substrate, complement C4. *J Biol Chem* 2013;**288**(22):15821–9.

10. Gaboriaud C, Rossi V, Bally I, Arlaud GJ, Fontecilla-Camps JC. Crystal structure of the catalytic domain of human complement c1s: a serine protease with a handle. *EMBO J* 2000;**19**(8):1755–65.

11. Arlaud GJ, Volanakis JE, Thielens NM, Narayana SV, Rossi V, Xu Y. The atypical serine proteases of the complement system. *Adv Immunol* 1998;**69**:249–307.

12. Cooper NR. The classical complement pathway: activation and regulation of the first complement component. *Adv Immunol* 1985;**37**:151–216.

13. Nissen MH, Roepstorff P, Thim L, Dunbar B, Fothergill JE. Limited proteolysis of beta 2-microglobulin at Lys-58 by complement component C1s. *Eur J Biochem* 1990;**189**(2):423–9.

14. Eriksson H, Nissen MH. Proteolysis of the heavy chain of major histocompatibility complex class I antigens by complement component C1s. *Biochim Biophys Acta* 1990;**1037**(2):209–15.

15. Busby Jr WH, Nam TJ, Moralez A, Smith C, Jennings M, Clemmons DR. The complement component C1s is the protease that accounts for cleavage of insulin-like growth factor-binding protein-5 in fibroblast medium. *J Biol Chem* 2000;**275**(48):37638–44.

16. Naito AT, Sumida T, Nomura S, Liu ML, Higo T, Nakagawa A, et al. Complement C1q activates canonical Wnt signaling and promotes aging-related phenotypes. *Cell* 2012;**149**(6):1298–313.

17. Yeo JG, Leong J, Arkachaisri T, Cai Y, Teo BH, Tan JH, et al. Proteolytic inactivation of nuclear alarmin high-mobility group box 1 by complement protease C1s during apoptosis. *Cell Death Discov* 2016;**2**:16069.

18. Bradley K, North J, Saunders D, Schwaeble W, Jeziorska M, Woolley DE, et al. Synthesis of classical pathway complement components by chondrocytes. *Immunology* 1996;**88**(4):648–56.

19. Drouet C, Reboul A. Biosynthesis of C1r and C1s subcomponents. *Behring Inst Mitt* 1989;**84**:80–8.

20. Gulati P, Lemercier C, Guc D, Lappin D, Whaley K. Regulation of the synthesis of C1 subcomponents and C1-inhibitor. *Behring Inst Mitt* 1993;**93**:196–203.

21. Laufer J, Oren R, Farzam N, Goldberg I, Passwell J. Differential cytokine regulation of complement proteins in human glomerular epithelial cells. *Nephron* 1997;**76**(3):276–83.

22. Legoedec J, Gasque P, Jeanne JF, Scotte M, Fontaine M. Complement classical pathway expression by human skeletal myoblasts in vitro. *Mol Immunol* 1997;**34**(10):735–41.

23. Walker DG, Dalsing-Hernandez JE, Lue LF. Human postmortem brain-derived cerebrovascular smooth muscle cells express all genes of the classical complement pathway: a potential mechanism for vascular damage in cerebral amyloid angiopathy and Alzheimer's disease. *Microvasc Res* 2008;**75**(3):411–9.

24. Luo C, Chen M, Madden A, Xu H. Expression of complement components and regulators by different subtypes of bone marrow-derived macrophages. *Inflammation* 2012;**35**(4):1448–61.

25. Bensa JC, Reboul A, Colomb MG. Biosynthesis in vitro of complement subcomponents C1q, C1s and C1 inhibitor by resting and stimulated human monocytes. *Biochem J* 1983;**216**(2):385–92.

26. Kusumoto H, Hirosawa S, Salier JP, Hagen FS, Kurachi K. Human genes for complement components C1r and C1s in a close tail-to-tail arrangement. *Proc Natl Acad Sci USA* 1988;**85**(19):7307–11.

27. Mackinnon CM, Carter PE, Smyth SJ, Dunbar B, Fothergill JE. Molecular cloning of cDNA for human complement component C1s. The complete amino acid sequence. *Eur J Biochem* 1987;**169**(3):547–53.

28. Tosi M, Duponchel C, Meo T, Julier C. Complete cDNA sequence of human complement C1s and close physical linkage of the homologous genes C1s and C1r. *Biochemistry* 1987;**26**(26):8516–24.

29. Petillot Y, Thibault P, Thielens NM, Rossi V, Lacroix M, Coddeville B, et al. Analysis of the N-linked oligosaccharides of human C1s using electrospray ionisation mass spectrometry. *FEBS Lett* 1995;**358**(3):323–8.

30. Thielens NM, Van Dorsselaer A, Gagnon J, Arlaud GJ. Chemical and functional characterization of a fragment of C1-s containing the epidermal growth factor homology region. *Biochemistry* 1990;**29**(14):3570–8.

31. Nguyen VC, Tosi M, Gross MS, Cohen-Haguenauer O, Jegou-Foubert C, de Tand MF, et al. Assignment of the complement serine protease genes C1r and C1s to chromosome 12 region 12p13. *Hum Genet* 1988;**78**(4):363–8.

32. Nakagawa M, Yuasa I, Irizawa Y, Umetsu K. The human complement component C1R gene: the exon-intron structure and the molecular basis of allelic diversity. *Ann Hum Genet* 2003;**67**(3):207–15.

33. Lintner KE, Wu YL, Yang Y, Spencer CH, Hauptmann G, Hebert LA, et al. Early components of the complement classical activation pathway in human systemic autoimmune diseases. *Front Immunol* 2016;**7**:36.

34. Amano MT, Ferriani VP, Florido MP, Reis ES, Delcolli MI, Azzolini AE, et al. Genetic analysis of complement C1s deficiency associated with systemic lupus erythematosus highlights alternative splicing of normal C1s gene. *Mol Immunol* 2008;**45**(6):1693–702.

35. Endo Y, Kanno K, Takahashi M, Yamaguchi K, Kohno Y, Fujita T. Molecular basis of human complement C1s deficiency. *J Immunol* 1999;**162**(4):2180–3.

36. Inoue N, Saito T, Masuda R, Suzuki Y, Ohtomi M, Sakiyama H. Selective complement C1s deficiency caused by homozygous four-base deletion in the C1s gene. *Hum Genet* 1998;**103**(4):415–8.
37. Abe K, Endo Y, Nakazawa N, Kanno K, Okubo M, Hoshino T, et al. Unique phenotypes of C1s deficiency and abnormality caused by two compound heterozygosities in a Japanese family. *J Immunol* 2009;**182**(3):1681–8.
38. Dragon-Durey MA, Quartier P, Fremeaux-Bacchi V, Blouin J, de Barace C, Prieur AM, et al. Molecular basis of a selective C1s deficiency associated with early onset multiple autoimmune diseases. *J Immunol* 2001;**166**(12):7612–6.
39. Kapferer-Seebacher I, Pepin M, Werner R, Aitman T, Nordgren A, Stoiber H, et al. Periodontal Ehlers-Danlos syndrome is caused by mutations in *C1R* and *C1S*, which encode subcomponents C1r and C1s of complement. *Am J Hum Genet* 2016;**99**. http://dx.doi.org/10.1016/j.ajhg.2016.08.019). [in press].
40. Kamboh MI, Ferrell RE. Periodontal Ehlers-Danlos syndrome is caused by mutations in *C1R* and *C1S*, which encode subcomponents C1r and C1s of complement. *J Immunogenet* 1987;**14**(4–5):231–8.

Chapter 12

Factor D

Steven D. Podos, Atul Agarwal, Mingjun Huang
Achillion Pharmaceuticals, Inc., New Haven, CT, United States

OTHER NAMES

Complement factor D, CFD, D component of complement, DF, properdin factor D, PFD, C3 proactivator convertase, C3PAse, adipsin, ADN.[1–3]

PHYSICOCHEMICAL PROPERTIES

Human factor D is produced as a single polypeptide of 253 amino acids (variant 1; 260 amino acids for variant 2) including leader peptide and possibly activation peptide. Sequence numbers in this chapter refer to variant 1 unless indicated otherwise. Factor D circulates primarily as a mature 228 amino acid 24.4 kDa monomer produced by cleavage after amino acid R25; the mature enzyme is identical whether formed from variant 1 or 2. It contains four internal disulphide bonds linking cysteines 51 and 67, 148 and 214, 179 and 195, and 204 and 229. Human factor D is not glycosylated although factor D from another species can be heavily glycosylated.[4,5]

STRUCTURE

Factor D is a single polypeptide chain enzyme that shares a high degree of residue identity with serine proteases (SPs) including bovine pancreatic trypsin and chymotrypsin (Fig. 12.1).[6] Factor D has been shown in crystal structures in multiple ways (Fig. 12.2). A 2 Å resolution X-ray structure (PDB code: 1DSU, Ref. 7) shows an ellipsoidal shape for both factor D molecules. This structure is dominated by antiparallel β-strands that form two shallow β-barrels. The structure contains three very short α-helices. The catalytic triad (His41, Asp89, Ser183) is located in the crevice between the two β-barrels. While the orientation of Ser183 is similar in the two monomers of the 1DSU X-ray structure, Asp89 is oriented away from His41 in monomer A and His41 is oriented away from Ser183 in monomer B. These atypical catalytic triad side-chain orientations disrupt the hydrogen bonding network among the three amino acid side chains and prohibit expression of

The Complement FactsBook. http://dx.doi.org/10.1016/B978-0-12-810420-0.00012-2

FIGURE 12.1   Diagram of factor D protein domain.

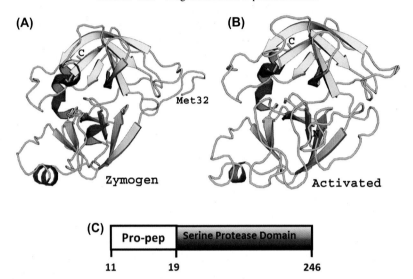

FIGURE 12.2   Crystal structure for complement activating serine protease, factor D. (A) Ribbon diagram of complement factor D crystal zymogen (profactor D) crystal structure (PDB ID: 1FDP). (B) Ribbon diagram of the activated human complement factor D crystal structure (PDB ID: 1DSU). (C) Schematic representation of the single chain factor D.

the catalytic activity of factor D. His41 adopts the 'trans' conformation in monomer B which is unusual for a SP. Four disulphide bridges in factor D contribute to the observed tertiary structure. The factor D molecule does not bind metal ions.

Unlike most proteases in the chymotrypsin family, human factor D circulates in the plasma in a mature 'resting-state' with low proteolytic activity. An extended surface-charge interaction with its physiological substrate C3bB likely releases factor D's seven amino acid self-inhibitory loop (196–202) and promotes its catalytically active conformation, as observed in a cocrystal structure with C3bB (PDB code: 2XWB, 3.5 Å resolution[8]). In the absence of C3bB, Arg202 of the self-inhibitory loop forms a salt-bridge interaction with Asp177, locking up the S1 binding pocket and rendering it inaccessible to the P1 residue of the substrate and destabilising the catalytic triad into a nonproductive arrangement. The 2 Å X-ray structure of a factor D triple mutant (Ser81Tyr, Thr198Ser, and Ser199Trp; PDB code: 1DST), designed to more closely resemble trypsin and demonstrated to have increased catalytic activity, also showed an altered conformation of the self-inhibitory loop, lending further support to this mechanism of conformational activation.[9]

FUNCTION

The factor D SP cleaves its unique substrate, factor B, in Mg^{++}-dependent complex with $C3(H_2O)$ or C3b, to generate the alternative pathway C3 convertases $C3(H_2O)Bb$ and C3bBb.[10,11] Factor B cleavage is an obligatory and rate-limiting step in complement alternative pathway activation. Factor D participates in the amplification loop which contributes significantly to responses elicited by all three complement pathways.[12,13]

Factor D circulates predominantly as a mature enzyme, with no requirement for regulated proteolytic activation and no associated inhibitor subunit. Profactor D retaining a five to seven amino acid activation peptide has been recovered from plasma or from particular recombinant expression systems.[3,14] Interestingly, studies with mice deficient in MASP-1/3 have suggested that these complement lectin pathway enzymes are required for profactor D activation, further connecting these two complement pathways.[15–17] Subsequent studies however in mice and in two human patients have challenged this conclusion, and the physiological significance of profactor D and the enzymes responsible for its activation remain uncertain.[18–20]

Factor D activity is tightly restricted by its high substrate specificity and a dual level of conformational control in which the factor B substrate adopts a susceptible conformation when in complex with its activated C3 cofactor and factor D undergoes induced-fit activation.[8,11] The resulting selectivity is important for preventing unregulated factor D activity and is reflected in its poor activity against synthetic substrates.[9]

In addition to its functions in innate immunity, factor D, also known as adipsin, has long been connected to adipose tissue biology and metabolic disorders.[5,21,22] Evidence has primarily been indirect, resting on its expression in adipose tissue and the metabolic regulation thereof. A report suggested a mechanistic link wherein factor D-dependent generation of complement C3a promotes pancreatic β-cell function and insulin production.[23]

DEGRADATION PATHWAY

Factor D is not processed proteolytically during normal complement activation. Plasma factor D levels are determined largely by the balance between rapid synthesis and a high rate of renal clearance via glomerular filtration and subsequent catabolism or excretion.[24,25] Renal dysfunction can cause significant elevations of circulating factor D levels, as can systemic delivery of an anti-factor D monoclonal antibody that hinders renal elimination.[24,26,27]

TISSUE DISTRIBUTION

Human factor D circulates in blood at approximately $2\,\mu g/mL$, lowest among the core complement proteins.[28] This concentration is not affected by complement activation. Factor D has been detected in cerebrospinal fluid and tears.[29,30]

Unusually among complement proteins, human factor D mRNA is expressed primarily not in liver, but in adipose tissue, from where it is secreted into the blood.[22] Expression has also been detected in the sciatic nerve in mice and in human cells of the monocyte/macrophage lineage.[5,22] Ocular factor D mRNA has been estimated as approximately 40-fold lower than in adipose tissue, and factor D protein has been localised within human donor eyes.[27,31]

REGULATION OF EXPRESSION

Factor D is not an acute-phase protein.[28] Its expression is not regulated by complement activation.[28] Factor D mRNA expression is moderately elevated in fasted mice and substantially decreased in several murine obesity models, under regulation by factors including insulin and neuroendocrine factors such as high glucocorticoid levels.[21,32] Elevated factor D levels have been reported in patients with age-related macular degeneration.[31]

HUMAN PROTEIN SEQUENCE

Variant 1

```
MHSWERLAVL  VLLGAAACAA  PPRGRILGGR  EAEAHARPYM  ASVQLNGAHL   50

CGGVLVAEQW  VLSAAHCLED  AADGKVQVLL  GAHSLSQPEP  SKRLYDVLRA  100

VPHPDSQPDT  IDHDLLLLQL  SEKATLGPAV  RPLPWQRVDR  DVAPGTLCDV  150

AGWGIVNHAG  RRPDSLQHVL  LPVLDRATCN  RRTHHDGAIT  ERLMCAESNR  200

RDSCKGDSGG  PLVCGGVLEG  VVTSGSRVCG  NRKKPGIYTR  VASYAAWIDS  250

VLA
```

Variant 2

```
MHSWERLAVL  VLLGAAACGE  EAWAWAAPPR  GRILGGREAE  AHARPYMASV   50

QLNGAHLCGG  VLVAEQWVLS  AAHCLEDAAD  GKVQVLLGAH  SLSQPEPSKR  100

LYDVLRAVPH  PDSQPDTIDH  DLLLLQLSEK  ATLGPAVRPL  PWQRVDRDVA  150

PGTLCDVAGW  GIVNHAGRRP  DSLQHVLLPV  LDRATCNRRT  HHDGAITERL  200

MCAESNRRDS  CKGDSGGPLV  CGGVLEGVVT  SGSRVCGNRK  KPGIYTRVAS  250

YAAWIDSVLA
```

Leader sequences are underlined. Factor D is found predominantly as mature enzyme, although zymogen forms have been observed with activation peptides of five to seven amino acids, double underlined. Human factor D has no sites of glycosylation or other posttranslational modification.

PROTEIN MODULES

Variant 1

1–25	Leader/activation peptide	exons 1–2
26–248	Trypsin-like serine protease	exons 2–5

Variant 2

1–32	Leader/activation peptide	exons 1–2
33–260	I domain	exons 3–5

CHROMOSOMAL LOCATION

Human	19p13.3
Mouse	10qC1
Rat	7q11

cDNA SEQUENCES

Transcript variant 1

```
GGGTCAGTGT CTCAGCCACA GCGGCTTCAC CATGCACAGC TGGGAGCGCC TGGCAGTTCT    60

GGTCCTCCTA GGAGCGGCCG CCTGCGCGGC GCCGCCCCGT GGTCGGATCC TGGGCGGCAG   120

AGAGGCCGAG GCGCACGCGC GGCCCTACAT GGCGTCGGTG CAGCTGAACG GCGCGCACCT   180

GTGCGGCGGC GTCCTGGTGG CGGAGCAGTG GGTGCTGAGC GCGGCGCACT GCCTGGAGGA   240

CGCCGGCCGAC GGGAAGGTGC AGGTTCTCCT GGGCGCGCAC TCCCTGTCGC AGCCGGAGCC   300

CTCCAAGCGC CTGTACGACG TGCTCCGCGC AGTGCCCCAC CCGGACAGCC AGCCCGACAC   360

CATCGACCAC GACCTCCTGC TGCTACAGCT GTCGGAGAAG GCCACACTGG GCCCTGCTGT   420

GCGCCCCCTG CCCTGGCAGC GCGTGGACCG CGACGTGGCA CCGGGAACTC TCTGCGACGT   480

GGCCGGCTGG GGCATAGTCA ACCACGCGGG CCGCCGCCCG GACAGCCTGC AGCACGTGCT   540

CTTGCCAGTG CTGGACCGCG CCACCTGCAA CCGGCGCACG CACCACGACG GCGCCATCAC   600

CGAGCGCTTG ATGTGCGCGG AGAGCAATCG CCGGGACAGC TGCAAGGGTG ACTCCGGGGG   660

CCCGCTGGTG TGCGGGGGCG TGCTCGAGGG CGTGGTCACC TCGGGCTCGC GCGTTTGCGG   720

CAACCGCAAG AAGCCCGGGA TCTACACCCG CGTGGCGAGC TATGCGGCCT GGATCGACAG   780

CGTCCTGGCC TAGGGTGCCG GGGCCTGAAG GTCAGGGTCA CCCAAGCAAC AAAGTCCCGA   840

GCAATGAAGT CATCCACTCC TGCATCTGGT TGGTCTTTAT TGAGCACCTA CTATATGCAG   900

AAGGGGAGGC CGAGGTGGGA GGATCATTGG ATCTCAGGAG TTCGAGATCA GCATGGGCCA   960

CGTAGCGCGA CTCCATCTCT ACAAATAAAT AAAAAATTAG CTGGGCAATT GGCGGGCATG  1020

GAGGTGGGTG CTTGTAGTTC CAGCTACTCA GGAGGCTGAG GTGGGAGGAT GACTTGAACG  1080

CAGGAGGCTG AGGCTGCAGT GAGTTGTGAT TGCACCACTG CCCT                    1124
```

Transcript variant 2

```
GGGTCAGTGT CTCAGCCACA GCGGCTTCAC CATGCACAGC TGGGAGCGCC TGGCAGTTCT    60

GGTCCTCCTA GGAGCGGCCG CCTGCGGTGA GGAGGCCTGG GCCTGGGCGG CGCCGCCCCG   120

TGGTCGGATC CTGGGCGGCA GAGAGGCCGA GGCGCACGCG CGGCCCTACA TGGCGTCGGT   180

GCAGCTGAAC GGCGCGCACC TGTGCGGCGG CGTCCTGGTG GCGGAGCAGT GGGTGCTGAG   240

CGCGGCGCAC TGCCTGGAGG ACGCGGCCGA CGGGAAGGTG CAGGTTCTCC TGGGCGCGCA   300

CTCCCTGTCG CAGCCGGAGC CCTCCAAGCG CCTGTACGAC GTGCTCCGCG CAGTGCCCCA   360

CCCGGACAGC CAGCCCGACA CCATCGACCA CGACCTCCTG CTGCTACAGC TGTCGGAGAA   420

GGCCACACTG GGCCCTGCTG TGCGCCCCCT GCCCTGGCAG CGCGTGGACC GCGACGTGGC   480

ACCGGGAACT CTCTGCGACG TGGCCGGCTG GGGCATAGTC AACCACGCGG CCGCCGCCC   540

GGACAGCCTG CAGCACGTGC TCTTGCCAGT GCTGGACCGC GCCACCTGCA ACCGGCGCAC   600

GCACCACGAC GGCGCCATCA CCGAGCGCTT GATGTGCGCG GAGAGCAATC GCCGGGACAG   660

CTGCAAGGGT GACTCCGGGG GCCCGCTGGT GTGCGGGGGC GTGCTCGAGG GCGTGGTCAC   720

CTCGGGCTCG CGCGTTTGCG GCAACCGCAA GAAGCCCGGG ATCTACACCC GCGTGGCGAG   780

CTATGCGGCC TGGATCGACA GCGTCCTGGC CTAGGGTGCC GGGGCCTGAA GGTCAGGGTC   840

ACCCAAGCAA CAAAGTCCCG AGCAATGAAG TCATCCACTC CTGCATCTGG TTGGTCTTTA   900

TTGAGCACCT ACTATATGCA GAAGGGGAGG CCGAGGTGGG AGGATCATTG GATCTCAGGA   960

GTTCGAGATC AGCATGGGCC ACGTAGCGCG ACTCCATCTC TACAAATAAA TAAAAAATTA  1020

GCTGGGCAAT TGGCGGGCAT GGAGGTGGGT GCTTGTAGTT CCAGCTACTC AGGAGGCTGA  1080

GGTGGGAGGA TGACTTGAAC GCAGGAGGCT GAGGCTGCAG TGAGTTGTGA TTGCACCACT  1140

GCCCT                                                              1145
```

The first five nucleotides of each exon are underlined to highlight intron/exon boundaries. Initiation and termination codons are double underlined. The two overlapping polyadenylation sites in each variant are highlighted by bold underlines (single site, AATAAA; overlapping sites, AATAAATAAA).

GENOMIC STRUCTURE

The 3.9 kb human factor D gene contains five exons. Exon 1 contains the two indicated alternative 3′ splice sites generating the transcript variants 1 and 2 which differ in leader/activation peptide. The 1.8 kb murine and 3.6 kb cynomolgus monkey genes also comprise five exons of similar arrangements, and alternative splicing generating two differing leader peptides has also been reported in mice (Fig. 12.3).[32]

0.5 kb

FIGURE 12.3   Diagram of genomic structure for factor D.

ACCESSION NUMBERS

Human	Gene	NC_000019.10 (859665..863624)
	mRNA (variants 1, 2)	NM_001928.3, NM_001317335.1
	Protein (variants 1, 2)	NP_001919.2, NP_001304264.1
Mouse	Gene	NC_000076.6 (79890853..79892660)
	mRNA (variants 1, 2, 3)	NM_013459.4, NM_001291915.2, NM_001329541.1
	Protein (variants 1, 2, 3)	NP_038487.1, NP_001278844.1, NP_001316470.1

DEFICIENCY

There are five case reports describing individuals and/or families with factor D deficiency.[33–37] In total, 11 individuals were considered as complete or near-complete factor D-deficient based on their factor D levels being less than 10% of normal ($n=9$) or based on the existence of genetic mutations including a Cys214 to Arg mutation that destroys an internal disulphide bond between two conserved cysteines ($n=2$). Of these 11 individuals, 8 (72.3%) suffered from severe and/or recurrent infections with encapsulated organisms such as *Neisseria meningitidis*. In contrast, family members with partial factor D deficiency (levels >10% of normal) appeared to be unaffected, with no reported instances of severe or recurrent infections. The age at which infections occurred varied from the neonatal period to early adulthood. In addition, no other risks were identified based on review of these case reports including obesity and/or lipodystrophy.

POLYMORPHIC VARIANTS

Rare variants have been identified in recent years as a consequence of the sequencing of whole genomes or exomes of over half a million humans (http://exac.broadinstitute.org/): 4 loss-of-function variants which include nonsense, splice acceptor and splice donor variants caused by single nucleotide changes and 98 missense variants have been identified. Among the 98 missense variants, 76 display a frequency of <1/10,000 including 58 single observations. Only 2 show the maximum frequency of 1/100 to 3/100, and only 4 are present as homozygotes. In addition, all 4 loss-of-function variants exist at a frequency of <1/10,000 and none are present as homozygotes.

MUTANT ANIMALS

No natural occurring factor D-deficient animals have been reported. However, factor D knockout mice have been generated in multiple laboratories to investigate the associated phenotypes and the role of alternative pathway in mouse disease models.[38] In a study investigating the role of factor D in obesity, no differences in body mass and serum lipid profiles were found between factor D-deficient mice and their sufficient littermates on a regular diet.[39] However, these mice exhibited worsened glucose homeostasis when under the metabolic stress of diet-induced obesity, despite showing a mild attenuation in weight gain compared to wild-type mice of the same strain.[23]

REFERENCES

1. HUGO Gene Nomenclature Committee. Available from: http://www.genenames.org/.
2. Müller-Eberhard HJ, Götze. C3 proactivator convertase and its mode of action. *J Exp Med* 1972;**135**:1003–8.
3. Fearon DT, Austen KF, Ruddy S. Properdin factor D: characterization of its active site and isolation of the precursor form. *J Exp Med* 1974;**139**:355–66.
4. Volanakis JE, Barnum SR, Kilpatrick JM. Purification and properties of human factor D. *Methods Enzymol* 1993;**223**:82–97.
5. Cook KS, Min HY, Johnson D, Chaplinsky RJ, Flier JS, Hunt CR, et al. Adipsin: a circulating serine protease homolog secreted by adipose tissue and sciatic nerve. *Science* 1987;**237**:402–5.
6. Greer J. Comparative modeling methods: application to the family of the mammalian serine proteases. *Proteins* 1990;**7**:317–34.
7. Narayana SV, Carson M, el-Kabbani O, Kilpatrick JM, Moore D, Chen X, et al. Structure of human factor D. a complement system protein at 2.0 Å resolution. *J Mol Biol* 1994;**235**:695–708.
8. Forneris F, Ricklin D, Wu J, Tzekou A, Wallace RS, Lambris JD, et al. Structures of C3b in complex with factors B and D give insight into complement convertase formation. *Science* 2010;**330**:1816–20.
9. Kim S, Narayana SV, Volanakis JE. Crystal structure of a complement factor D mutant expressing enhanced catalytic activity. *J Biol Chem* 1995;**270**:24399–405.
10. Lesavre PH, Müller-Eberhard HJ. Mechanism of action of factor D of the alternative complement pathway. *J Exp Med* 1978;**148**:1498–509.
11. Volanakis VE, Narayana SVL. Complement factor D, a novel serine protease. *Protein Sci* 1996;**5**:553–64.
12. Harboe M, Ulvund G, Vien L, Fung M, Mollnes TE. The quantitative role of alternative pathway amplification in classical pathway induced terminal complement activation. *Clin Exp Immunol* 2004;**138**:439–46.
13. Harboe M, Garred P, Karlstrøm E, Lindstad JK, Stahl GL, Mollnes TE. The down-stream effects of mannan-induced lectin complement pathway activation depend quantitatively on alternative pathway amplification. *Mol Immunol* 2009;**47**:373–80.
14. Yamauchi Y, Stevens JW, Macon KJ, Volanakis JE. Recombinant and native zymogen forms of human complement factor D. *J Immunol* 1994;**152**:3645–53.
15. Takahashi M, Ishida Y, Iwaki D, Kanno K, Suzuki T, Endo Y, et al. Essential role of mannose-binding lectin-associated serine protease-1 in activation of the complement factor D. *J Exp Med* 2010;**207**:29–37.

16. Banda NK, Takahashi M, Levitt B, Glogowska M, Nicholas J, Takahashi K, et al. Essential role of complement mannose-binding lectin-associated serine proteases-1/3 in the murine collagen antibody-induced model of inflammatory arthritis. *J Immunol* 2010;**185**:5598–606.

17. Iwaki D, Kanno K, Takahashi M, Endo Y, Matsushita M, Fujita T. The role of mannose-binding lectin-associated serine protease-3 in activation of the alternative complement pathway. *J Immunol* 2011;**187**:3751–8.

18. Degn SE, Jensen L, Hansen AG, Duman D, Tekin M, Jensenius JC, et al. Mannan-binding lectin-associated serine protease (MASP)-1 is crucial for lectin pathway activation in human serum, whereas neither MASP-1 nor MASP-3 is required for alternative pathway function. *J Immunol* 2012;**189**:3957–69.

19. Atik T, Koparir A, Bademci G, Foster II J, Altunoglu U, Mutlu GY, Bowdin S, et al. Novel MASP1 mutations are associated with an expanded phenotype in 3MC1 syndrome. *Orphanet J Rare Dis* 2015;**10**:128.

20. Ruseva MM, Takahashi M, Fujita T, Pickering MC. C3 dysregulation due to factor H deficiency is mannan-binding lectin-associated serine proteases (MASP)-1 and MASP-3 independent *in vivo*. *Clin Exp Immunol* 2014;**176**:84–92.

21. Rosen BS, Cook KS, Yaglom J, Groves DL, Volanakis JE, Damm D, et al. Adipsin and complement factor D activity: an immune-related defect in obesity. *Science* 1989;**244**:1483–7.

22. White RT, Damm D, Hancock N, Rosen BS, Lowell BB, Usher P, et al. Human adipsin is identical to complement factor D and is expressed at high levels in adipose tissue. *J Biol Chem* 1992;**267**:9210–3.

23. Lo JC, Ljubicic S, Leibiger B, Kern M, Leibiger IB, Moede T, et al. Adipsin is an adipokine that improves β cell function in diabetes. *Cell* 2014;**158**:41–53.

24. Volanakis JE, Barnum SR, Giddens M, Galla JH. Renal filtration and catabolism of complement protein D. *N Engl J Med* 1985;**312**:395–9.

25. Sanders PW, Volanakis JE, Rostand SG, Galla JH. Human complement protein D catabolism by the rat kidney. *J Clin Invest* 1986;**77**:1299–304.

26. Pascual M, Steiger G, Estreicher J, Macon K, Volanakis JE, Schifferli JA. Metabolism of complement factor D in renal failure. *Kidney Int* 1988;**34**:529–36.

27. Loyet KM, Good J, Davancaze T, Sturgeon L, Want X, Yang J, et al. Complement inhibition in cynomolgus monkeys by anti–factor D antigen-binding fragment for the treatment of an advanced form of dry age-related macular degeneration. *J Pharmacol Exp Ther* 2014;**351**:527–37.

28. Barnum SR, Niemann MA, Kearney JF, Volanakis JE. Quantitation of complement factor D in human serum by a solid-phase radioimmunoassay. *J Immunol Methods* 1984;**67**:303–9.

29. Hietaharju A, Kuusisto H, Nieminen R, Vuolteenaho K, Elovaara I, Moilanen E. Elevated cerebrospinal fluid adiponectin and adipsin levels in patients with multiple sclerosis: a Finnish co-twin study. *Eur J Neurol* 2010;**17**:332–4.

30. Rentka A, Harsfalvi J, Szucs G, Szekanecz Z, Szodoray P, Koroskenyi K, et al. Membrane array and multiplex bead analysis of tear cytokines in systemic sclerosis. *Immunol Res* 2016;**64**:619–26.

31. Loyet KM, Deforge LE, Katschke Jr KJ, Diehl L, Graham RR, Pao L, et al. Activation of the alternative complement pathway in vitreous is controlled by genetics in age-related macular degeneration. *Invest Ophthalmol Vis Sci* 2012;**53**:6628–37.

32. Lowell BB, Flier JS. Differentiation dependent biphasic regulation of adipsin gene expression by insulin and insulin-like growth factor-1 in 3T3-F442A adipocytes. *Endocrinology* 1990;**127**:2898–906.

33. Min HY, Spiegelman BM. Adipsin, the adipocyte serine protease: gene structure and control of expression by tumor necrosis factor. *Nucleic Acids Res* 1986;**14**:8879–92.

34. Kluin-Nelemans HC, van Velzen-Blad H, van Helden HP, Daha MR. Functional deficiency of complement factor D in a monozygous twin. *Clin Exp Immunol* 1984;**58**:724–30.

35. Hiemstra PS, Langeler E, Compier B, Keepers Y, Leijh PCJ, van den Barselaar MT, et al. Complete and partial deficiencies of complement factor D in a Dutch family. *J Clin Invest* 1989;**84**:1957–61.

36. Weiss SJ, Ahmed AE, Bonagura VR. Complement factor D deficiency in an infant first seen with pneumococcal neonatal sepsis. *J Allergy Clin Immunol* 1998;**102**:1043–4.

37. Biesma DH, Hannema AJ, van Velzen-Blad H, Mulder L, van Zwieten R, Kluijt I, et al. A family with complement factor D deficiency. *J Clin Invest* 2001;**108**:233–40.

38. Sprong T, Roos D, Weemaes C, Neeleman C, Geesing CLM, Mollnes TE, et al. Deficient alternative complement pathway activation due to factor D deficiency by 2 novel mutations in the complement factor D gene in a family with meningococcal infections. *Blood* 2006;**107**:4865–70.

39. Holers VM. The spectrum of complement alternative pathway-mediated diseases. *Immunol Rev* 2008;**223**:300–16.

FURTHER READING

1. Xu Y, Ma M, Ippolito GC, Schroeder Jr HW, Carroll MC, Volanakis JE. Complement activation in factor D-deficient mice. *Proc Natl Acad Sci USA* 2001;**98**:14577–82.

Chapter 13

C2

Kartik Manne, Sthanam V.L. Narayana

University of Alabama at Birmingham, Birmingham, AL, United States

PHYSIOCHEMICAL PROPERTIES

The second component of the complement C2 is a serine protease (SP) (EC 3.4.21.43), synthesised as single chain polypeptide of 752 residues along with a 20 residue leader peptide. It is also a glycoprotein having ~16% carbohydrate.[1]

Mature protein		
pI[2]		~6.0–6.3 (depending on isoform)
M_r (K)	Calculated	82.5
	Observed on SDS-PAGE	90–100
N-linked glycosylation sites	Predicted	9, 92, 270, 313, 447, 451, 601, 631
	Observed[3,4]	270, 313, 447, 451, 601, 631

STRUCTURE

With a modular structure of five domains (Fig. 13.1). C2 is a SP exhibiting 39% sequence identity with factor B. Crystal structures of its N-terminal C2b and C-terminal C2a fragments are available at high resolution (Fig. 13.2). C2b is made of three complement control protein (CCP) domains; two C-terminal CCPs arranged side by side with a large interface, exhibit minimal interaction with the N-terminal CCP domain that extends out.[4] C2a is comprised of a C-terminal SP domain, and an N-terminal von Willebrand type-A (vWA) domain.[3] C2b is the propeptide of C2, and it is clipped by C1s only when C2 is bound to C4b using metal-mediated interactions. In addition to detailing the essential metal binding site in vWA domain, the crystal structure of C2a revealed the distorted 'oxyanion hole' present in the SP domain, which explains the inactivity of C4b-bound C2a and the need for induced conformational changes for the function of C3 convertase C4bC2a.

The Complement FactsBook. http://dx.doi.org/10.1016/B978-0-12-810420-0.00013-4
127

FIGURE 13.1   Diagram of protein domains for C2.

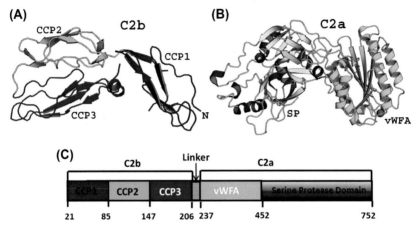

FIGURE 13.2   Crystal Structure of complement serine protease C2: (A) Ribbon diagrams of the N-terminal C2b fragment crystal structure (PDB ID: 3ERB). (B) Ribbon representation of the C-terminal C2a fragment of C2 crystal structure (PDB ID: 2ODP and 2I6S). Bound Mg^{2+} ion is represented as a *orange sphere*. (C) Schematic representation of the single chain complement component C2 and its domains.

FUNCTION

C2 provides the catalytic subunit C2a to the classical/lectin pathway C3- and C5-convertases C4bC2a and C4bC2aC3a, respectively. C2 and its catalytic fragments C2a are poorly reactive but exhibit high specificity. This low activity of C2a is compatible with the mechanism of action where proteolytic activity against its natural substrates C3 and C5 is expressed only in the context of the C4bC2a or C4bC2aC3b complex, respectively. The assembly of the C3 convertase C4bC2a of the classical pathway proceeds in two steps. First C1s cleaves C2 attached to C4b that is deposited on an activating surface. Next, the N-terminal smaller fragment C2b of C2 is released into the fluid phase, while the larger fragment, C2a, remains complexed with C4b to form the C3 convertase, C4bC2a.

TISSUE DISTRIBUTION

Serum protein: 11–35 µg/mL in serum[5]
Primary site of synthesis: liver HepG2 cells[6]
Secondary sites: monocytes/macrophages,[7] fibroblasts,[8] type II alveolar epithelial cells[9] and astroglioma cells[10]

EXPRESSION AND REGULATION

Expression of C2 is restricted to hepatocytes, monocytes and fibroblasts and is stimulated by lipopolysaccharide (LPS) and cytokines (IL-1, IL-4). Interferon IFN-y predominantly enhances the transcription of C2[8,10,11] suggesting C2 expression is upregulated during the acute-phase response. However, the serum levels of C2 remain same during acute phase suggesting decreased stability of the C2 mRNA.[11]

PROTEIN SEQUENCE[12,13]

```
MGPLMVLFCL   LFLYPGLADS   APSCPQNVNI   SGGTFTLSHG   WAPGSLLTYS    50
CPQGLYPSPA   SRLCKSSGQW   QTPGATRSLS   KAVCKPVRCP   APVSFENGIY   100
TPRLGSYPVG   GNVSFECEDG   FILRGSPVRQ   CRPNGMWDGE   TAVCDNGAGH   150
CPNPGISLGA   VRTGFRFGHG   DKVRYRCSSN   LVLTGSSERE   CQGNGVWSGT   200
EPICRQPYSY   DFPEDVAPAL   GTSFSHMLGA   TNPTQKTKES   LGRKIQIQRS   250
GHLNLYLLLD   CSQSVSENDF   LIFKESASLM   VDRIFSFEIN   VSVAIITFAS   300
EPKVLMSVLN   DNSRDMTEVI   SSLENANYKD   HENGTGTNTY   AALNSVYLMM   350
NNQMRLLGME   TMAWQEIRHA   IILLTDGKSN   MGGSPKTAVD   HIREILNINQ   400
KRNDYLDIYA   IGVGKLDVDW   RELNELGSKK   DGERHAFILQ   DTKALHQVFE   450
HMLDVSKLTD   TICGVGNMSA   NASDQERTPW   HVTIKPKSQE   TCRGALISDQ   500
WVLTAAHCFR   DGNDHSLWRV   NVGDPKSQWG   KEFLIEKAVI   SPGFDVFAKK   550
NQGILEFYGD   DIALLKLAQK   VKMSTHARPI   CLPCTMEANL   ALRRPQGSTC   600
RDHENELLNK   QSVPAHFVAL   NGSKLNINLK   MGVEWTSCAE   VVSQEKTMFP   650
NLTDVREVVT   DQFLCSGTQE   DESPCKGESG   GAVFLERRFR   FFQVGLVSWG   700
LYNPCLGSAD   KNSRKRAPRS   KVPPPRDFHI   NLFRMQPWLR   QHLGDVLNFL   750
PL
```

PROTEIN MODULES

1–20	Leader peptide	exon 1
21–85	CCP1	exon 2
86–147	CCP2	exon 3
148–206	CCP3	exon 4
207–236	Linker	exon 5
237–452	von Willebrand type-A domain	exons 6–10
453–466	Linker	exon 11
467–752	Serine protease domain	exons 11–18

Catalytic triad: H507, D561, S679

CHROMOSOMAL LOCATION

Human[14,15]	6p21.3 (localises within the class III region of the MHC)
Mouse[16]	Chromosome 17
Rat	20p12

cDNA SEQUENCE

```
GGCTCTCTAC CTCTCGCCGC CCCTAGGGAG GACACCATGG GCCCACTGAT GGTTCTTTTT    60

TGCCTGCTGT TCCTGTACCC AGGTCTGGCA GACTCGGCTC CCTCCTGCCC TCAGAACGTG   120

AATATCTCGG GTGGCACCTT CACCCTCAGC CATGGCTGGG CTCCTGGGAG CCTTCTCACC   180

TACTCCTGCC CCCAGGGCCT GTACCCATCC CCAGCATCAC GGCTGTGCAA GAGCAGCGGA   240

CAGTGGCAGA CCCCAGGAGC CACCCGGTCT CTGTCTAAGG CGGTCTGCAA ACCTGTGCGC   300

TGTCCAGCCC CTGTCTCCTT TGAGAATGGC ATTTATACCC CACGGCTGGG GTCCTATCCC   360

GTGGGTGGCA ATGTGAGCTT CGAGTGTGAG GATGGCTTCA TATTGCGGGG CTCGCCTGTG   420

CGTCAGTGTC GCCCCAACGG CATGTGGGAT GGAGAAACAG CTGTGTGTGA TAATGGGGCT   480

GGCCACTGCC CCAACCCAGG CATTTCACTG GGCGCAGTGC GGACAGGCTT CCGCTTTGGT   540

CATGGGGACA AGGTCCGCTA TCGCTGCTCC TCGAATCTTG TGCTCACGGG GTCTTCGGAG   600

CGGGAGTGCC AGGGCAACGG GGTCTGGAGT GGAACGGAGC CCATCTGCCG CCAACCCTAC   660

TCTTATGACT TCCCTGAGGA CGTGGCCCCT GCCCTGGGCA CTTCCTTCTC CCACATGCTT   720

GGGGCCACCA ATCCCACCCA GAAGACAAAG GAAAGCCTGG GCCGTAAAAT CCAAATCCAG   780

CGCTCTGGTC ATCTGAACCT CTACCTGCTC CTGGACTGTT CGCAGAGTGT GTCGGAAAAT   840

GACTTTCTCA TCTTCAAGGA GAGCGCCTCC CTCATGGTGG ACAGGATCTT CAGCTTTGAG   900

ATCAATGTGA GCGTTGCCAT TATCACCTTT GCCTCAGAGC CCAAAGTCCT CATGTCTGTC   960

CTGAACGACA ACTCCCGGGA TATGACTGAG GTGATCAGCA GCCTGGAAAA TGCCAACTAT  1020

AAAGATCATG AAAATGGAAC TGGGACTAAC ACCTATGCGG CCTTAAACAG TGTCTATCTC  1080

ATGATGAACA ACCAAATGCG ACTCCTCGGC ATGGAAACGA TGGCCTGGCA GGAAATCCGA  1140

CATGCCATCA TCCTTCTGAC AGATGGAAAG TCCAATATGG GTGGCTCTCC CAAGACAGCT  1200

GTTGACCATA TCAGAGAGAT CCTGAACATC AACCAGAAGA GGAATGACTA TCTGGACATC  1260

TATGCCATCG GGGTGGGCAA GCTGGATGTG GACTGGAGAG AACTGAATGA GCTAGGGTCC  1320

AAGAAGGATG GTGAGAGGCA TGCCTTCATT CTGCAGGACA CAAAGGCTCT GCACCAGGTC  1380

TTTGAACATA TGCTGGATGT CTCCAAGCTC ACAGACACCA TCTGCGGGGT GGGGAACATG  1440

TCAGCAAACG CCTCTGACCA GGAGAGGACA CCCTGGCATG TCACTATTAA GCCCAAGAGC  1500

CAAGAGACCT GCCGGGGGGC CCTCATCTCC GACCAATGGG TCCTGACAGC AGCTCATTGC  1560

TTCCGCGATG GCAACGACCA CTCCCTGTGG AGGGTCAATG TGGGAGACCC CAAATCCCAG  1620

TGGGGCAAAG AATTGCTTAT TGAGAAGGCG GTGATCTCCC CAGGGTTTGA TGTCTTTGCC  1680

AAAAAGAACC AGGGAATCCT GGAGTTCTAT GGTGATGACA TAGCTCTGCT GAAGCTGGCC  1740

CAGAAAGTAA AGATGTCCAC CCATGCCAGG CCCATCTGCC TTCCCTGCAC GATGGAGGCC  1800
```

```
AATCTGGCTC TGCGGAGACC TCAAGGCAGC ACCTGTAGGG ACCATGAGAA TGAACTGCTG  1860

AACAAACAGA GTGTTCCTGC TCATTTTGTC GCCTTGAATG GGAGCAAACT GAACATTAAC  1920

CTTAAGATGG GAGTGGAGTG GACAAGCTGT GCCGAGGTTG TCTCCCAAGA AAAAACCATG  1980

TTCCCCAACT TGACAGATGT CAGGGAGGTG GTGACAGACC AGTTCCTATG CAGTGGGACC  2040

CAGGAGGATG AGAGTCCCTG CAAGGGAGAA TCTGGGGGAG CAGTTTTCCT TGAGCGGAGA  2100

TTCAGGTTTT TTCAGGTGGG TCTGGTGAGC TGGGGTCTTT ACAACCCCTG CCTTGGCTCT  2160

GCTGACAAAA ACTCCCGCAA AAGGGCCCCT CGTAGCAAGG TCCCGCCGCC ACGAGACTTT  2220

CACATCAATC TCTTCCGCAT GCAGCCCTGG CTGAGGCAGC ACCTGGGGGA TGTCCTGAAT  2280

TTTTTACCCC TCTAGCCATG GCCACTGAGC CCTCTGCTGC CCTGCCAGAA TCTGCCGCCC  2340

CTCCATCTTC TACCTCTGAA TGGCCACCCT TAGACCCTGT GATCCATCCT CTCTCCTAGC  2400

TGAGTAAATC CGGGTCTCTA GGATGCCAGA GGCAGCGCAC ACAAGCTGGG AAATCCTCAG  2460

GGCTCCTACC AGCAGGACTG CCTCGCTGCC CCACCTCCCG CTCCTTGGCC TGTCCCCAGA  2520

TTCCTTCCCT GGTTGACTTG ACTCATGCTT GTTTCACTTT CACATGGAAT TTCCCAGTTA  2580

TGAAATTAAT AAAAATCAAT GGTTTCCAC                                    2609
```

GENOMIC STRUCTURE

The *C2* gene has 18 kb of DNA and contains 18 exons.[17] It is made up of some large introns, the largest of which is located upstream of exon 3 and contains a human-specific SINE-type retroposon, called SINE-R.C2.[18] This retroposon, derived from the human endogenous retrovirus HERV-K10, is associated with variable number of tandem repeats (VNTR)[19] locus, which is responsible for the multiallelic restriction fragment length polymorphism (RFLP) of the C2 gene (Fig. 13.3).[20]

ACCESSION NUMBERS

	cDNA	Genomic
Human[12,13,17]	X04481	L09706
	K1236	L09707
	M26301	L09708
Mouse[21]	M57891	M60563–M60579
	J05661	J05661

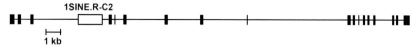

FIGURE 13.3   Diagram of genomic structure for C2.

DEFICIENCY

In addition to recurrent bacterial infections like pneumonia, meningitis and sepsis, C2 deficiency is associated with increased risk of systemic lupus erythematosus (SLE), an autoimmune disorder.[22,23] Less than 1% of the European descent population exhibit C2 deficiency, out of which 60% are asymptotic while close to 30% are at higher risk of SLE.[24–26] Females with C2 deficiency are more prone to SLE than males.[27,28] The human *C2* gene is located on chromosome 6p21.3, and two types of C2 deficiency with mutations are known. The type I is caused by a 28-bp deletion in the *C2* gene, which results in the deletion of exon 6 and no translation of the C2 protein. Hence no protein is detected either in blood or intracellularly. Type II C2 deficiency is caused by a point mutation (Ser^{189}Phe and Gly^{144}Arg), which leads to no secretion of C2 [29] as it is retained in cells.

POLYMORPHIC VARIANTS

Using isoelectric focusing and immunoblotting, nine structural variants of C2 have been identified at the protein level. The variants include the common (C2C), relatively less common (C2B) found in Caucasians, four rare acidic (C2A) types and three rare basic types.[30] With the help of unique restriction enzymes and nine C2 gene haplotypes,[20] several dimorphic RFLPs have been characterised.[31]

MUTANT ANIMALS

C2-deficient mice have not been reported. Factor B-deficient mice, generated by replacing exons 1 to 18 of the gene with a promoterless *lacZ* reporter gene and the neomycin resistance gene, are fully viable when kept under specific pathogen-free conditions, produced no detectable factor B mRNA or protein and had no detectable factor B enzymatic activity or alternative pathway function.[32] However, same mice exhibited two adjacent genes, complement component C2 [33] and *D17H6S45* previously known as *Rd*,[34] downregulated as a result of the above gene disruption. The downregulation of *C2* gene expression was sufficient to cause a complete loss of classical pathway function.

REFERENCES

1. Tomana M, Niemann M, Garner C, Volanakis JE. Carbohydrate composition of the second, third and fifth components and factors B and D of human complement. *Mol Immunol* 1985;**22**:107–11.
2. Alper CA. Inherited structural polymorphism in human C2: evidence for genetic linkage between C2 and Bf. *J Exp Med* 1976;**144**:1111–5.
3. Krishnan V, Xu Y, Macon K, Volanakis JE, Narayana SV. The crystal structure of C2a, the catalytic fragment of classical pathway C3 and C5 convertase of human complement,. *J Mol Biol* 2007;**367**:224–33.

4. Krishnan V, Xu Y, Macon K, Volanakis JE, Narayana SV. The structure of C2b, a fragment of complement component C2 produced during C3 convertase formation. *Acta Crystallogr D Biol Crystallogr* 2009;**65**:266–74.
5. Oglesby TJ, Ueda A, Volanakis JE. Radioassays for quantitation of intact complement proteins C2 and B in human serum. *J Immunol methods* 1988;**110**:55–62.
6. Perlmutter DH, Cole FS, Goldberger G, Colten HR. Distinct primary translation products from human liver mRNA give rise to secreted and cell-associated forms of complement protein C2. *J Biol Chem* 1984;**259**:10380–5.
7. Whaley K. Biosynthesis of the complement components and the regulatory proteins of the alternative complement pathway by human peripheral blood monocytes. *J Exp Med* 1980;**151**:501–16.
8. Katz Y, Cole FS, Strunk RC. Synergism between gamma interferon and lipopolysaccharide for synthesis of factor B, but not C2, in human fibroblasts. *J Exp Med* 1988;**167**:1–14.
9. Strunk RC, Cole FS, Perlmutter DH, Colten HR. γ-Interferon increases expression of class III complement genes C2 and factor B in human monocytes and in murine fibroblasts transfected with human C2 and factor B genes. *J Biol Chem* 1985;**260**:15280–5.
10. Barnum SR, Ishii Y, Agrawal A, Volanakis JE. Production and interferon-gamma-mediated regulation of complement component C2 and factors B and D by the astroglioma cell line U105-MG. *Biochem J* 1992;**287**(Pt 2):595–601.
11. Lappin DF, Guc D, Hill A, McShane T, Whaley K. Effect of interferon-gamma on complement gene expression in different cell types. *Biochem J* 1992;**281**(Pt 2):437–42.
12. Bentley DR. Primary structure of human complement component C2. Homology to two unrelated protein families. *Biochem J* 1986;**239**:339–45.
13. Horiuchi T, Macon KJ, Kidd VJ, Volanakis JE. cDNA cloning and expression of human complement component C2. *J Immunol* 1989;**142**:2105–11.
14. Carroll MC, Campbell RD, Bentley DR, Porter RR. A molecular map of the human major histocompatibility complex class III region linking complement genes C4, C2 and factor B,. *Nature* 1984;**307**:237–41.
15. Carroll MC, Katzman P, Alicot EM, et al. Linkage map of the human major histocompatibility complex including the tumor necrosis factor genes. *Proc Natl Acad Sci USA* 1987;**84**:8535–9.
16. Steinmetz M, Stephan D, Fischer Lindahl K. Gene organization and recombinational hotspots in the murine major histocompatibility complex. *Cell* 1986;**44**:895–904.
17. Ishii Y, Zhu ZB, Macon KJ, Volanakis JE. Structure of the human C2 gene. *J Immunol* 1993;**151**:170–4.
18. Zhu ZB, Jian B, Volanakis JE. Ancestry of SINE-R.C2 a human-specific retroposon. *Hum Genet* 1994;**93**:545–51.
19. Zhu ZB, Hsieh SL, Bentley DR, Campbell RD, Volanakis JE. A variable number of tandem repeats locus within the human complement C2 gene is associated with a retroposon derived from a human endogenous retrovirus. *J Exp Med* 1992;**175**:1783–7.
20. Zhu ZB, Volanakis JE. Allelic associations of multiple RFLPs of the gene encoding complement protein C2. *Am J Hum Genet* 1990;**46**:956–62.
21. Ishikawa N, Nonaka M, Wetsel RA, Colten HR. Murine complement C2 and factor B genomic and cDNA cloning reveals different mechanisms for multiple transcripts of C2 and B. *J Biol Chem* 1990;**265**:19040–6.
22. Pickering MC, Walport MJ. Links between complement abnormalities and systemic lupus erythematosus. *Rheumatology (Oxford)* 2000;**39**:133–41.
23. Pickering MC, Botto M, Taylor PR, Lachmann PJ, Walport MJ. Systemic lupus erythematosus, complement deficiency, and apoptosis. *Adv Immunol* 2000;**76**:227–324.

24. Laich A, Sim RB. Complement C4bC2 complex formation: an investigation by surface plasmon resonance. *Biochim Biophys Acta* 2001;**1544**:96–112.

25. Jonsson G, Sjoholm AG, Truedsson L, Bengtsson AA, Braconier JH, Sturfelt G. Rheumatological manifestations, organ damage and autoimmunity in hereditary C2 deficiency. *Rheumatology (Oxford)* 2007;**46**:1133–9.

26. Macedo AC, Isaac L. Systemic lupus erythematosus and deficiencies of early components of the complement classical pathway. *Front Immunol* 2016;**7**:55.

27. Pons-Estel GJ, Alarcon GS, Scofield L, Reinlib L, Cooper GS. Understanding the epidemiology and progression of systemic lupus erythematosus. *Semin Arthritis Rheum* 2010;**39**:257–68.

28. McCarty DJ, Manzi S, Medsger Jr TA, Ramsey-Goldman R, LaPorte RE, Kwoh CK. Incidence of systemic lupus erythematosus. Race and gender differences. *Arthritis Rheum* 1995;**38**:1260–70.

29. Wetsel RA, Kulics J, Lokki ML, et al. Type II human complement C2 deficiency. Allele-specific amino acid substitutions (Ser189--> Phe; Gly444--> Arg) cause impaired C2 secretion. *J Biol Chem* 1996;**271**:5824–31.

30. Jahn I, Uring-Lambert B, Arnold D, Clemenceau S, Hauptmann G. C2 reference typing report. *Complement Inflamm* 1990;**7**:175–82.

31. Cross SJ, Edwards JH, Bentley DR, Campbell RD. DNA polymorphism of the C2 and factor B genes. Detection of a restriction fragment length polymorphism which subdivides haplotypes carrying the C2C and factor B F alleles. *Immunogenetics* 1985;**21**:39–48.

32. Taylor PR, Nash JT, Theodoridis E, Bygrave AE, Walport MJ, Botto M. A targeted disruption of the murine complement factor B gene resulting in loss of expression of three genes in close proximity, factor B, C2, and D17H6S45,. *J Biol Chem* 1998;**273**:1699–704.

33. Wu LC, Morley BJ, Campbell RD. Cell-specific expression of the human complement protein factor B gene: evidence for the role of two distinct 5'-flanking elements. *Cell* 1987;**48**:331–42.

34. Levi-Strauss M, Carroll MC, Steinmetz M, Meo T. A previously undetected MHC gene with an unusual periodic structure,. *Science* 1988;**240**:201–4.

Chapter 14

Factor B

Jennifer Laskowski, Joshua M. Thurman

University of Colorado School of Medicine, Aurora, CO, United States

OTHER NAMES

CFB, Cfb, C3/C5 convertase, glycine-rich beta glycoprotein (GBG), PBF2, pro-perdin factor B, Al195813, Al255840, B, Bf, C2, Fb, H2-Bf, AHUS4, ARMD14, BFD, CFAB, CFBD, FBI12, PBF2, complement factor B.

PHYSICOCHEMICAL PROPERTIES

Factor B is a single-chain polypeptide of approximately 93 kD.

M_r (K)	Predicted	83
	Observed	90
pI		~7.0668
N-linked glycosylation sites		4; Asn122, Asn142, Asn285, Asn378
Glycation site		Lys291

Note: any references to specific amino acid residues in this chapter are based on the 764-residue protein, which includes the 25-residue leader sequence. (Some sources do not include this region in the protein notation.)

STRUCTURE

Factor B is a single-chain protein that contains five distinct domains (Fig. 14.1). Upon activation of the alternative complement pathway, factor B is cleaved by factor D into a Ba fragment (~33 kD) and a Bb fragment (~60 kD). The N-terminus of intact factor B contains the Ba region (residues 1–259, noncatalytic). The Ba region of the protein encompasses three complement control protein (CCP) domains, which are important for the initial binding of factor B to C3b.[1] CCP modules are also found in several other proteins of the complement cascade, and they contain four conserved cysteine residues that form two internal disulphide bonds.

The Ba portion of the molecule is linked to the Bb region by a 45-residue linker, and the scissile bond that is cleaved by factor D (Arg259-Lys260) is

The Complement FactsBook. http://dx.doi.org/10.1016/B978-0-12-810420-0.00014-6

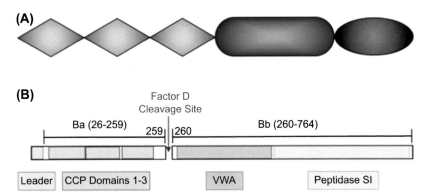

FIGURE 14.1 Structure of intact factor B. (A) Stylised protein domains for factor B. Factor B is generated as a single-chain polypeptide. (B) Cleavage of factor B by factor D generates two fragments, termed Ba and Bb. The Ba fragment contains three complement control protein (CCP) domains. The Bb fragment contains a von Willebrand factor, type A domain (VWA) and a peptidase S1 (serine protease) domain.

contained within the linker region. The scissile bond of unbound factor B is protected from nonspecific cleavage and kept in a 'locked' conformation by two salt bridges, which bind Arg259 to Glu471 and Glu232.[2] Binding of factor B to C3b disrupts the Arg259-Glu471 salt bridge. This induces a conformational change that leads to activation of an Mg^{2+}-dependent metal ion-dependent adhesion site (MIDAS) and stabilisation of the C3bB-Mg^{2+} complex.[2] The conformational change also exposes the scissile bond and enables cleavage of the proteins by factor D.

The Bb portion of the molecule (residues 260–764; catalytic) contains a von Willebrand factor, type A domain (VWA) and a peptidase S1 (PA clan) domain. VWA domains are comprised of ~200 residues that form an α/β open sheet. The MIDAS site in the VWA domain of factor B is also important for binding of the protein to C3b.[3] The peptidase S1 domain lies at the C-terminus of the factor B protein and contains the serine protease (SP) trypsin domain active site. Catalytic activity for SPs belonging to the trypsin family is accomplished through the use of a charge relay system, most often involving a histidine residue hydrogen-bonded to serine and aspartic acid residues. This charge relay system and the amino acid residues that surround it are highly conserved within the trypsin protease family. The crystal structure of factor B is shown below (Fig. 14.2).[4]

FUNCTION

The SP domain of the Bb fragment is the catalytically active component of the alternative pathway C3 and C5 convertases. Once factor B has bound either $C3(H_2O)$ or C3b, it undergoes a conformational shift, and the scissile bond becomes accessible for cleavage by factor D.[5] The Ba fragment is soluble, but the Bb fragment remains attached to C3b. The cleavage of factor B

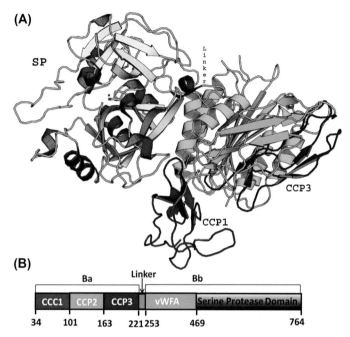

FIGURE 14.2   Crystal structure of factor B. (A) Ribbon diagrams of the complement serine protease factor B crystal structure (PDB ID: 2OK5): The N-terminal CCP1, CCP2 and CCP3 form the Ba fragment and the remaining VWA and serine protease (SP) domains constitute Bb fragment. (B) Schematic representation of the six domain SP factor B, each domain is marked and coloured accordingly and the 'linker' joining Ba and Bb fragments which is a substrate for C3b-bound factor D is marked.

allows rotation of the SP domain, rendering it catalytically active. The C3 convertases cleave C3 at an Arginine–Serine bond in the complement C3 α-chain (residues 726–727), yielding C3a and C3b. The C5 convertase cleaves C5 at an Arginine–Leucine bond in the C5 α-chain (residues 74–75), yielding C5a and C5b. Inactivation of the C3 and C5 convertases is achieved by the dissociation of Bb from C3b.

The half-life of $C3(H_2O)Bb$ and C3bBb are approximately 77 and 90 s, respectively,[6] but decay accelerating factor (DAF or CD55) and factor H accelerate the dissociation of these molecules. Once Bb and C3b have dissociated they cannot rebind.[7] Thus, the activity of the alternative pathway convertases is regulated in several ways. Factor D can only activate factor B when it is bound to $C3(H_2O)$ or C3b. Bb is only catalytically active when in complex with $C3(H_2O)$ or C3b. These convertase complexes are intrinsically unstable, and DAF and factor H make the half-lives of these complexes even shorter.

Direct biologic effects of Ba and Bb have been reported, although the physiologic importance of these effects is not clear. Ba may have chemotactic activity for neutrophils and macrophages, and may limit B cell proliferation.[8] Bb has been reported to affect macrophages and monocyte function, and may promote the growth of B cells.[9]

TISSUE DISTRIBUTION

Factor B is primarily produced in the liver, but studies have also shown that it is produced in monocytes, macrophages, fibroblasts, dendritic cells, intestine and kidneys.[10–13] In the kidneys it is primarily produced in the tubules, and expression is not seen in the glomeruli.[11] Its concentration in plasma is approximately 200 μg/mL.

REGULATION OF EXPRESSION

Factor B is an acute-phase reactant. IL-1β, IL-6, TNFα, IFN-γ and endotoxin increase its expression.[14–18] Conversely, platelet-derived growth factor and epidermal growth factor can decrease the expression of factor B.[19]

PROTEIN SEQUENCES

```
  1   MGSNLSPQLC LMPFILGLLS GGVTTTPWSL ARPQGSCSLE GVEIKGGSFR

 51   LLQEGQALEY VCPSGFYPYP VQTRTCRSTG SWSTLKTQDQ KTVRKAECRA

101   IHCPRPHDFE NGEYWPRSPY YNVSDEISFH CYDGYTLRGS ANRTCQVNGR

151   WSGQTAICDN GAGYCSNPGI PIGTRKVGSQ YRLEDSVTYH CSRGLTLRGS

201   QRRTCQEGGS WSGTEPSCQD SFMYDTPQEV AEAFLSSLTE TIEGVDAEDG

251   HGPGEQQKRK IVLDPSGSMN IYLVLDGSDS IGASNFTGAK KCLVNLIEKV

301   ASYGVKPRYG LVTYATYPKI WVKVSEADSS NADWVTKQLN EINYEDHKLK

351   SGTNTKKALQ AVYSMMSWPD DVPPEGWNRT RHVIILMTDG LHNMGGDPIT

401   VIDEIRDLLY IGKDRKNPRE DYLDVYVFGV GPLVNQVNIN ALASKKDNEQ

451   HVFKVKDMEN LEDVFYQMID ESQSLSLCGM VWEHRKGTDY HKQPWQAKIS

501   VIRPSKGHES CMGAVVSEYF VLTAAHCFTV DDKEHSIKVS VGGEKRDLEI

551   EVVLFHPNYN INGKKEAGIP EFYDYDVALI KLKNKLKYGQ TIRPICLPCT

601   EGTTRALRLP PTTTCQQQKE ELLPAQDIKA LFVSEEEKKL TRKEVYIKNG

651   DKKGSCERDA QYAPGYDKVK DISEVVTPRF LCTGGVSPYA DPNTCRGDSG

701   GPLIVHKRSR FIQVGVISWG VVDVCKNQKR QKQVPAHARD FHINLFQVLP

751   WLKEKLQDED LGFL
```

The leader sequence has been <u>underlined</u>. The four *N*-linked glycosylation sites and one glycation site are shown in **bold** text. The three charge relay system sites (526, 576 and 699) located in the serine protease domain are indicated by **<u>bold and underlined</u>** text. The factor D cleavage site located between residues 259 and 260 is in **bold** and has been shaded.

PROTEIN MODULES

1–25	Leader sequence	exon 1/2
35–100	CCP domain#1	exon 2/3
101–160	CCP domain#2	exon 3/4
163–220	CCP domain#3	exon 4/5
221–269	Linker region	exon 5/6
270–469	VWA domain	exons 6–10
470–476	Connecting segment	exon 11
477–757	Serine protease domain	exons 11–18

CHROMOSOMAL LOCATION

The gene encoding complement factor B resides within the major histocompatibility complex class III region gene cluster on chromosome 6 (6p21.33),[20] just 500 bp away from its classical pathway homologue, complement component 2 (C2).[21] The gene location is 6:31,945,709–31,952,084 (6375 bp). The *CFB* gene is found on chromosome 17 in the mouse (*Mus musculus*) and on chromosome 20 (20p12) in the rat (*Rattus norvegicus*).

CDNA SEQUENCE

The cDNA sequence shown below corresponds to the BF*S allele.

```
   1   ggccttgggg gagggggagg ccagaatgac tccaagagct acaggaaggc aggtcagaga
  61   ccccactgga caaacagtgg ctggactctg caccataaca cacaatcaac aggggagtga
 121   gctggatcct tatttctggt ccctaagtgg gtggtttggg cttactgggg aggagctaag
 181   gccgagagg aggtactgaa ggggagagtc ctggaccttt ggcagcaaag ggtgggactt
 241   ctgcagtttc tgtttccttg actggcagct cagcggggcc ctcccgcttg gatgttccgg
 301   gaaagtgatg tgggtaggac aggcggggcg agccgcaggt gccagaacac agattgtata
 361   aaaggctggg ggctggtggg gagcagggga agggaatgtg accaggtcta ggtctggagt
 421   ttcagcttgg acactgagcc aagcagacaa gcaaagcaag ccaggacaca ccatcctgcc
 481   ccaggcccag cttctctcct gccttccaac gccatgggga gcaatctcag cccccaactc
 541   tgcctgatgc cctttatctt gggcctcttg tctggaggtg tgaccaccac tccatggtct
 601   ttggcccggc cccagggatc ctgctctctg gagggggtag agatcaaagg cggctccttc
 661   cgacttctcc aagagggcca ggcactggag tacgtgtgtc cttctggctt ctacccgtac
 721   cctgtgcaga cacgtacctg cagatctacg gggtcctgga gcaccctgaa gactcaagac
 781   caaaagactg tcaggaaggc agagtgcaga gcaatccact gtccaagacc acacgacttc
 841   gagaacgggg aatactggcc ccggtctccc tactacaatg tgagtgatga gatctctttc
 901   cactgctatg acggttacac tctccggggc tctgccaatc gcacctgcca agtgaatggc
 961   cgatggagtg ggcagacagc gatctgtgac aacggagcgg ggtactgctc caacccgggc
1021   atccccattg gcacaaggaa ggtgggcagc cagtaccgcc ttgaagacag cgtcacctac
1081   cactgcagcc gggggcttac cctgcgtggc tcccagcggc gaacgtgtca ggaaggtggc
1141   tcttggagcg ggacggagcc ttcctgccaa gactccttca tgtacgacac ccctcaagag
1201   gtggccgaag cttttcctgtc ttccctgaca gagaccatag aaggagtcga tgctgaggat
1261   gggcacggcc cagggggaaca acagaagcgg aagatcgtcc tggacccttc aggctccatg
1321   aacatctacc tggtgctaga tggatcagac agcattgggg ccagcaactt cacaggagcc
1381   aaaaagtgtc tagtcaactt aattgagaag gtggcaagtt atggtgtgaa gccaagatat
1441   ggtctagtga catatgccac ataccccaaa atttgggtca aagtgtctga agcagacagc
1501   agtaatgcag actgggtcac gaagcagctc aatgaaatca attatgaaga ccacaagttg
1561   aagtcaggga ctaacaccaa gaaggccctc caggcagtgt acagcatgat gagctggcca
1621   gatgacgtcc ctcctgaagg ctggaaccgc acccgccatg tcatcatcct catgactgat
1681   ggattgcaca acatgggcgg ggacccaatt actgtcattg atgagatccg ggacttgcta
```

```
1741  tacattggca aggatcgcaa aaacccaagg gaggattatc tggatgtcta tgtgtttggg

1801  gtcgggcctt tggtgaacca agtgaacatc aatgctttgg cttccaagaa agacaatgag

1861  caacatgtgt tcaaagtcaa ggatatggaa aacctggaag atgttttcta ccaaatgatc

1921  gatgaaagcc agtctctgag tctctgtggc atggtttggg aacacaggaa gggtaccgat

1981  taccacaagc aaccatggca ggccaagatc tcagtcattc gcccttcaaa gggacacgag

2041  agctgtatgg gggctgtggt gtctgagtac tttgtgctga cagcagcaca ttgtttcact

2101  gtggatgaca aggaacactc aatcaaggtc agcgtaggag gggagaagcg ggacctggag

2161  atagaagtag tcctatttca ccccaactac aacattaatg ggaaaaaaga agcaggaatt

2221  cctgaatttt atgactatga cgttgccctg atcaagctca agaataagct gaaatatggc

2281  cagactatca ggcccatttg tctcccctgc accgagggaa caactcgagc tttgaggctt

2341  cctccaacta ccacttgcca gcaacaaaag gaagagctgc tccctgcaca ggatatcaaa

2401  gctctgtttg tgtctgagga ggagaaaaag ctgactcgga aggaggtcta catcaagaat

2461  ggggataaga aaggcagctg tgagagagat gctcaatatg ccccaggcta tgacaaagtc

2521  aaggacatct cagaggtggt cacccctcgg ttcctttgta ctggaggagt gagtccctat

2581  gctgacccca atacttgcag aggtgattct ggcggcccct tgatagttca caagagaagt

2641  cgtttcattc aagttggtgt aatcagctgg ggagtagtgg atgtctgcaa aaaccagaag

2701  cggcaaaagc aggtacctgc tcacgcccga gactttcaca tcaacctctt tcaagtgctg

2761  ccctggctga aggagaaact ccaagatgag gatttgggtt ttctataagg ggtttcctgc

2821  tggacagggg cgtgggattg aattaaaaca gctgcgacaa ca
```

The methionine initiation codon (atg) is shaded and the termination codon (tag) is shown in **bold** font. The first five nucleotides of each exon are underlined, and the probable polyadenylation site (attaaa) is shown in **bold and underlined** text.

GENOMIC STRUCTURE

The complement factor B gene spans 6.375 kb of genomic DNA (transcript is 2.86 kb) and contains 18 exons, all of which code for protein (Fig. 14.3).

ACCESSION NUMBERS

Species	UniProt	Genomic
Human	P00751	AF019413
Mouse	P04186	M60629–M60646
Rat	Q6MG74	BX883045

FIGURE 14.3 **Genomic structure of the gene for factor B.** The gene spans 6.375 kb of genomic DNA and contains 18 exons (*grey boxes*).

DEFICIENCY

Deficiency of factor B has been associated with an increased risk of infection. Meningococcal and pneumococcal infections have been reported in affected patients.[22,23] Mice with targeted deletion of the gene for factor B have been developed.[24] Exons 3 through 7 of the factor B gene were replaced with a targeting vector, and the mice do not have detectable alternative pathway activity. The mice develop normally, but they are more susceptible to bacterial infections than wild-type mice.

POLYMORPHIC VARIANTS

There are three major allelic variants in the *CFB* gene, collectively referred to as the factor B fast-slow polymorphism, and include variants BF*S ('slow'), BF*FA ('fast' A) and BF*FB ('fast' B). Combined, these three variants occur at a gene frequency greater than 95%, with BF*S being the most common allele.[25] The difference between the three allelic variants occurs at nucleotides 94 and 95 in codon 8 and corresponds to the amino acid at position 32 (with leader sequence) or position 8 (without leader sequence). At this position, the BF*S allele codes for Arg (ccg), BF*FA for Gln (cag) and BF*FB for Trp (tgg).[26]

Disease-associated polymorphic variants of factor B are listed in Table 14.1, and the location of those variants located within significant domains of the factor B protein are shown in Fig. 14.4.

MUTANT ANIMALS

Mice with targeted deletion of the gene for factor B have been developed.[24] Exons 3 through 7 of the factor B gene were replaced with a targeting vector. D3 (129S2/SvPas background) and J1 (129S4/SvJae background) embryonic stem cells were injected into C57BL/6 blastocysts, resulting in chimeric mice. The mice have subsequently been backcrossed onto C57BL/6 and Balb/c backgrounds by investigators. Homozygous deficient mice do not have detectable alternative pathway activity. The mice develop normally, but they are more susceptible to bacterial infections than wild-type mice.[27,28] However, they develop milder renal diseases in a model of lupus[29] and in nephrotoxic serum nephritis.[30] These mice are not yet commercially available, although a similar strain is being developed by the Knockout Mouse Phenotyping Program at The Jackson Laboratory and may become available in the future.

TABLE 14.1 Disease-Associated Polymorphic Variants

Polymorphism	Location/ Molecular Consequence	dbSNP (NCBI)	Disease Association/Major Polymorphic Variant
L9H	T539A, Missense	rs4151667	Reduced risk of age-related macular degeneration (ARMD14)[a]
W28Q	Missense	-	Missense found in FA allele only[b]
W28R	T595C, Missense	-	Missense found in S allele only[b]
R32W	C607T, Missense	rs12614	Predisposition to chronic Hepatitis B[c]
(R8W)		rs12614	fB fast-slow polymorphism | Bf*fB/S
R32Q	G608A, Missense	rs641153	Reduced risk of ARMD14[c]
(R8Q)		rs641153	fB fast-slow polymorphism | Bf *FA/S[b,d]
S166P	C1O10T, Missense	-	aHUS4[e]
R203Q	G1121A, Missense	rs745794224	aHUS4[e]
I242L	A1237C, Missense	rs144812066	aHUS4[e]
G252S	G1267A, Missense	rs4151651	-
Q256Ter	C1279T, Stop gained	rs398123065	Complement factor B deficiency-1 family
F286L	C1371G, Missense	rs117905900	aHUS4; gain of function-mutation leading to increased C3b8b formation[f]
K323Q	A1480C, Missense	-	aHUS4[e]

Continued

TABLE 14.1 Disease-Associated Polymorphic Variants—cont'd

Polymorphism	Location/ Molecular Consequence	dbSNP (NCBI)	Disease Association/Major Polymorphic Variant
K323E	A1480G, Missense	rs121909748	aHUS4; gain of function-mutation leading to increased C3b8b formation[f]
M458I	G1887A, Missense	rs200837114	aHUS4'
K533R	A2111G, Missense	rs149101394	aHUS4*
K565E	A2206G, Missense	rs4151659	[g]
F632Cytst	2407–2410, frame-shift delTTTG	rs398124644	Complement factor B deficiency–1 family
D651E	T2466G, Missense	rs4151660	-
A7365	G2719T, Missense	-	Missense found in FA allele only[b]

aHUS4, atypical hemolytic uremic syndrome, subtype 4; *ARMD14*; age-related macular degeneration, subtype 14.
[a]*Gold et al.*[21]
[b]*Davrinche et al.*[26]
[c]*Jiang et al.*[31]
[d]*Mejia et al.*[25]
[e]*Maga Nishimura et al.*[32]
[f]*Goicoechea de Jorge et al.*[33]
[g]*Mungall et al.*[34]

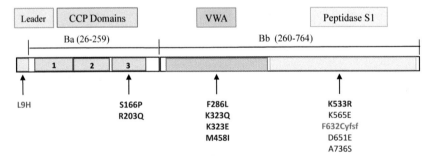

FIGURE 14.4 Location of polymorphic variants in the factor B protein. The locations of the disease-associated factor B variants within the protein are illustrated above. Variants associated with atypical haemolytic uraemic syndrome are shown in red. Variants associated with deficiency of factor B are shown in green. Variants associated with a reduced risk of developing age-related macular degeneration are shown in blue.

REFERENCES

1. Pryzdial EL, Isenman DE. Alternative complement pathway activation fragment Ba binds to C3b. Evidence that formation of the factor B-C3b complex involves two discrete points of contact. *J Biol Chem* 1987;**262**(4):1519–25.
2. Hourcade DE, Mitchell LM. Access to the complement factor B scissile bond is facilitated by association of factor B with C3b protein. *J Biol Chem* 2011;**286**(41):35725–32.
3. Milder FJ, Gomes L, Schouten A, Janssen BJ, Huizinga EG, Romijn RA, et al. Factor B structure provides insights into activation of the central protease of the complement system. *Nat Struct Mol Biol* 2007;**14**(3):224–8.
4. Brenner S. The molecular evolution of genes and proteins: a tale of two serines. *Nature* 1988;**334**(6182):528–30.
5. Forneris F, Ricklin D, Wu J, Tzekou A, Wallace RS, Lambris JD, et al. Structures of C3b in complex with factors B and D give insight into complement convertase formation. *Science* 2010;**330**(6012):1816–20.
6. Fishelson Z, Pangburn MK, Muller-Eberhard HJ. Characterization of the initial C3 convertase of the alternative pathway of human complement. *J Immunol* 1984;**132**(3):1430–4.
7. Pangburn MK, Muller-Eberhard HJ. The C3 convertase of the alternative pathway of human complement. Enzymic properties of the bimolecular proteinase. *Biochem J* 1986;**235**(3):723–30.
8. Ambrus Jr JL, Peters MG, Fauci AS, Brown EJ. The Ba fragment of complement factor B inhibits human B lymphocyte proliferation. *J Immunol* 1990;**144**(5):1549–53.
9. Peters MG, Ambrus Jr JL, Fauci AS, Brown EJ. The Bb fragment of complement factor B acts as a B cell growth factor. *J Exp Med* 1988;**168**(4):1225–35.
10. Passwell J, Schreiner GF, Nonaka M, Beuscher HU, Colten HR. Local extrahepatic expression of complement genes C3, factor B, C2, and C4 is increased in murine lupus nephritis. *J Clin Invest* 1988;**82**(5):1676–84.
11. Welch TR, Beischel LS, Frenzke M, Witte D. Regulated expression of complement factor B in the human kidney. *Kidney Int* 1996;**50**(2):521–5.
12. Strunk RC, Cole FS, Perlmutter DH, Colten HR. gamma-Interferon increases expression of class III complement genes C2 and factor B in human monocytes and in murine fibroblasts transfected with human C2 and factor B genes. *J Biol Chem* 1985;**260**(28):15280–5.
13. Li K, Fazekasova H, Wang N, Sagoo P, Peng Q, Khamri W, et al. Expression of complement components, receptors and regulators by human dendritic cells. *Mol Immunol* 2011;**48**(9–10):1121–7.
14. Colten HR, Dowton SB. Regulation of complement gene expression. *Biochem Soc Symp* 1986;**51**:37–46.
15. Falus A, Beuscher HU, Auerbach HS, Colten HR. Constitutive and IL 1-regulated murine complement gene expression is strain and tissue specific. *J Immunol* 1987;**138**(3):856–60.
16. Perlmutter DH, Dinarello CA, Punsal PI, Colten HR. Cachectin/tumor necrosis factor regulates hepatic acute-phase gene expression. *J Clin Invest* 1986;**78**(5):1349–54.
17. Perlmutter DH, Goldberger G, Dinarello CA, Mizel SB, Colten HR. Regulation of class III major histocompatibility complex gene products by interleukin-1. *Science* 1986;**232**(4752):850–2.
18. Ripoche J, Mitchell JA, Erdei A, Madin C, Moffatt B, Mokoena T, et al. Interferon gamma induces synthesis of complement alternative pathway proteins by human endothelial cells in culture. *J Exp Med* 1988;**168**(5):1917–22.
19. Circolo A, Pierce GF, Katz Y, Strunk RC. Antiinflammatory effects of polypeptide growth factors. Platelet-derived growth factor, epidermal growth factor, and fibroblast growth factor inhibit the cytokine-induced expression of the alternative complement pathway activator factor B in human fibroblasts. *J Biol Chem* 1990;**265**(9):5066–71.

20. Carroll MC, Campbell RD, Bentley DR, Porter RR. A molecular map of the human major histocompatibility complex class III region linking complement genes C4, C2 and factor B. *Nature* 1984;**307**(5948):237–41.

21. Gold B, Merriam JE, Zernant J, Hancox LS, Taiber AJ, Gehrs K, et al. Variation in factor B (BF) and complement component 2 (C2) genes is associated with age-related macular degeneration. *Nat Genet* 2006;**38**(4):458–62.

22. Slade C, Bosco J, Unglik G, Bleasel K, Nagel M, Winship I. Deficiency in complement factor B. *N Engl J Med* 2013;**369**(17):1667–9.

23. Pettigrew HD, Teuber SS, Gershwin ME. Clinical significance of complement deficiencies. *Ann N Y Acad Sci* 2009;**1173**:108–23.

24. Matsumoto M, Fukuda W, Circolo A, Goellner J, Strauss-Schoenberger J, Wang X, et al. Abrogation of the alternative complement pathway by targeted deletion of murine factor B. *Proc Natl Acad Sci USA* 1997;**94**(16):8720–5.

25. Mejia JE, Jahn I, de la Salle H, Hauptmann G. Human factor B. Complete cDNA sequence of the BF*S allele. *Hum Immunol* 1994;**39**(1):49–53.

26. Davrinche C, Abbal M, Clerc A. Molecular characterization of human complement factor B subtypes. *Immunogenetics* 1990;**32**(5):309–12.

27. Mueller-Ortiz SL, Drouin SM, Wetsel RA. The alternative activation pathway and complement component C3 are critical for a protective immune response against *Pseudomonas aeruginosa* in a murine model of pneumonia. *Infect Immun* 2004;**72**(5):2899–906.

28. Li Q, Li YX, Douthitt K, Stahl GL, Thurman JM, Tong HH. Role of the alternative and classical complement activation pathway in complement mediated killing against *Streptococcus pneumoniae* colony opacity variants during acute pneumococcal otitis media in mice. *Microbes Infect/Inst Pasteur* 2012;**14**(14):1308–18.

29. Watanabe H, Garnier G, Circolo A, Wetsel RA, Ruiz P, Holers VM, et al. Modulation of renal disease in MRL/lpr mice genetically deficient in the alternative complement pathway factor B. *J Immunol* 2000;**164**(2):786–94.

30. Thurman JM, Tchepeleva SN, Haas M, Panzer S, Boackle SA, Glogowska MJ, et al. Complement alternative pathway activation in the autologous phase of nephrotoxic serum nephritis. *Am J Physiol Ren Physiol* 2012;**302**(12):F1529–36.

31. Jiang DK, Ma XP, Yu M, Cao G, Ding DL, Chen H, et al. Hepatology 2015;**62**(1):118–28.

32. Maga Nishimura CJ, Weaver AE, Frees KL, Smith RJ. Hum Mutat 2010;**31**(6):E1445 60.

33. Goicoechea de Jorge E, Harris CL, Esparia-Gordillo J, Carreras L, Arranz EA, Garrido CA, et al. Proc Natl Acad Sci USA 2007;**104**(1):240–5.

34. Mungall AJ, Palmer SA, Sims SK, Edward CA, Ashurst JL, Wilming L, et al. Nature 2003;**425**(6960):805–11.

Chapter 15

Factor I

Marcin Okrój[1], Anna M. Blom[2]

[1]*Medical University of Gdańsk, Gdańsk, Poland;* [2]*Lund University, Malmö, Sweden*

OTHER NAMES

Complement factor I, FI, C3b inactivator, C3B/C4B inactivator, C3b-INA, ARMD13, AHUS3, konglutinogen activating factor, EC 3.4.21.45.

PHYSICOCHEMICAL PROPERTIES

Factor I (FI) is the product of a single gene, synthesised as pre-proprotein with 18 aa signal sequence and a 4 aa linker (RRKR), which is excised upon cleavage.

pI	6.69 (Nonglycosylated)		
	Heavy chain	6.83	
	Light chain	6.47	
Amino acids	Signal sequence	18	
	Mature protein	565	
	Heavy chain	317	19–335
	Linker 4		336–339
	Light chain	224	340–583
Molecular mass	Predicted	65, 75 kDa	
	Observed	88 kDa	
Potential N-linked glycosylation sites 6		70, 103, 177, 464, 494, 536	
Interchain disulphide bonds		Cys 327, 453	

STRUCTURE

FI is synthesised as a single-chain precursor, which is posttranslationally processed into the heavy (50 kDa) and the light (38 kDa) chain.[1] Mature protein is a heterodimer of heavy and light chain linked together by a disulphide bond. It is constructed from five domains, four of these (factor I membrane attack complex/FIMAC, CD5 domain, two low-density lipoprotein receptor type A/ LDLRA domains) located in heavy chain and a serine protease (SP) domain,

The Complement FactsBook. http://dx.doi.org/10.1016/B978-0-12-810420-0.00015-8

which contains enzymatic activity and comprises the light chain. According to electron microscopy, FI has a bilobal structure of overall average length of 13.4 ± 1.4 nm and maximal diameters of bulbous parts of 5.8 ± 0.8 and 4.9 ± 0.9 nm.[2]

Analyses of X-ray and neutron scattering data led to conclusion that FIMAC, CD5 domain and one of LDL-receptor class A (LDLRA) domains form one lobe, and the large SP domain forms the other lobe.[3] SP domain is the largest one and contains the catalytic triad: H380, D429 and S525. Single molecule of FI binds two Ca^{2+} ions with binding sites buried within each CD5 domain (Fig. 15.1). Crystal structure of FI is available (pdb: 2XRC) and confirms the bilobal structure.[4] There are 40 cysteine residues, of which 36 are engaged in intradomain disulphide bonds. C33 and C255 bridge FIMAC and LDLRA₁ domains, as well as C327 and C453 bridge heavy chain and SP domains[5] (Fig 15.2).

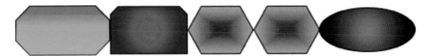

FIGURE 15.1 Scheme presents domain organisation of factor I.

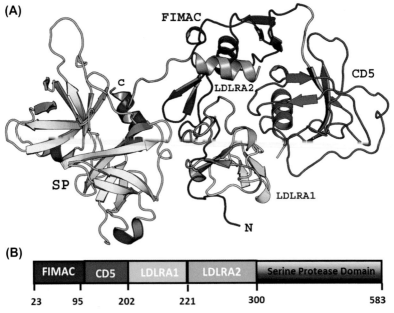

FIGURE 15.2 Crystal structure for complement serine protease (SP), factor I (FI). (A) Ribbon diagram of the human complement SP FI crystal structure (PDB ID: 2XRC). (B) Schematic representation of the single-chain FI domains, coloured similar to the ribbon diagram.

FUNCTION

FI inhibits all pathways of the complement system by cleaving the α′-chains of activated early components C4b and C3b. Interestingly, FI circulates in proteolytically inactive state and achieving enzymatic capacity depends on the presence of a cofactor, which may be factor H (only for C3b), CD46 (MCP), CD35 (CR1) or C4BP.[4,6] FI is an enzyme, which like thrombin, cleaves peptide bonds after arginyl residues but with V_{max} 60,000 to 60 times lower (depending on substrate) when compared to thrombin.[6] Physiological consequences of FI activity are: protection from misguided complement activation and C3 convertase formation on self-surfaces, limiting C3 consumption and formation of C3b breakdown products such as iC3b, which in turn boosts phagocytosis and adaptive immunity upon binding to their specific receptors on immune cells.[7,8] No physiological inhibitor of FI is known, and its zymogen-like form is probably maintained by allosteric modulation exerted by noncatalytic heavy chain.[4] Some synthetic inhibitors such as leupeptin, Pefabloc-Xa, Pefabloc-SC and antipain decrease the activity of FI.[9] Synthetic, low molecular weight substrates for FI have been described, but their cleavage does not need cofactor,[9] and therefore may not reflect physiological conditions when the transient complex of factor I – substrate – cofactor is formed.

TISSUE DISTRIBUTION

Serum/plasma: 35 μg/mL, production supported by hepatocytes in liver.[10]

Secondary sites of synthesis: fibroblasts,[11] keratinocytes,[12] monocytes, myoblasts,[13] endothelial cells,[14] some tumour cells.[15,16]

REGULATION OF EXPRESSION

FI is an acute-phase protein, and its concentration increases during inflammation. Production in hepatocytes is upregulated by IL6.[10,17] In addition, upregulation by IFN-γ was observed in keratinocytes[12] and endothelial cells.[14]

PROTEIN SEQUENCES

Factor I pre-proprotein (NCBI Reference Sequence: NP_000195.2, protein accession: P05156)

Signal sequence is underlined, and **N-linked glycosylation sites** are indicated by bold font. Linker sequence between heavy and light chain is indicated by ~~strikethrough~~.

```
MKLLHVFLLF LCFHLRFCKV TYTSQEDLVE KKCLAKKYTH LSCDKVFCQP WQRCIEGTCV   60

CKLPYQCPKN GTAVCATNRR SFPTYCQQKS LECLHPGTKF LNNGTCTAEG KFSVSLKHGN  120

TDSEGIVEVK LVDQDKTMFI CKSSWSMREA NVACLDLGFQ QGADTQRRFK LSDLSINSTE  180

CLHVHCRGLE TSLAECTFTK RRTMGYQDFA DVVCYTQKAD SPMDDFFQCV NGKYISQMKA  240

CDGINDCGDQ SDELCCKACQ GKGFHCKSGV CIPSQYQCNG EVDCITGEDE VGCAGFASVA  300

QEETEILTAD MDAERRRIKS LLPKLSCGVK NRMHIRRKRI VGGKRAQLGD LPWQVAIKDA  360

SGITCGGIYI GGCWILTAAH CLRASKTHRY QIWTTVVDWI HPDLKRIVIE YVDRIIFHEN  420

YNAGTYQNDI ALIEMKKDGN KKDCELPRSI PACVPWSPYL FQPNDTCIVS GWGREKDNER  480

VFSLQWGEVK LISNCSKFYG NRFYEKEMEC AGTYDGSIDA CKGDSGGPLV CMDANNVTYV  540

WGVVSWGENC GKPEFPGVYT KVANYFDWIS YHVGRPFISQ YNV                    583
```

PROTEIN MODULES

1–18	Signal sequence	exon 1
43–108	FIMAC (Kazal-like domain, Kazal 2 domain)	exon 2
114–215	CD5 domain/scavenger receptor cysteine-rich domain	exon 3/4
224–256	LDLRA domain	exon 5
259–293	LDLRA domain	exon 6
339–569	Trypsin-like SP domain	exons 9–13

CHROMOSOMAL LOCATION

DNA coding for FI is localised on chromosome 4, band 1q25.

Length	61,485 bases[a]/61335[b]
Orientation	Minus strand
Start	109,740,694 bp from pter
End	109,802,179 bp from pter
Exons	13 (13 coding exons)[a]/14 (14)[a]

[a]sequence variant NM_000204.4.
[b]sequence variant NM_001318057.1.

cDNA SEQUENCES AND GENOMIC STRUCTURE

The first five nucleotides in each exon are underlined. The initiation and termination codons (ATG and TAA, respectively) and putative polyadenylation signals AATAAA are double underlined.

Factor I pre-proprotein (NCBI Reference Sequence: NM_000204.4)

```
AGTCTCTTCT GGTGGAAACA AGTTCCTATT GGTCACCTCC CTCAGCTCTT TAATGGGAAG    60

GCCATTTTTT GGTTTCAGTT ACGGTAAATC TCTCTGGATT TCAGCCAAAT TCTTTCAGAG   120

TTCAAAAGTA CAAAGCTCTT TAGGAGGTTT GTTCTTTTGA AACATTTTTT CAGAGGTTAA   180

AAATTAAATT TCAAAGAAT ACCTGGAGTG GAAAAGAGTT CTCAGCAGAG ACAAAGACCC    240

CGAACACCTC CAACATGAAG CTTCTTCATG TTTTCCTGTT ATTTCTGTGC TTCCACTTAA   300

GGTTTTGCAA GGTCACTTAT ACATCTCAAG AGGATCTGGT GGAGAAAAG TGCTTAGCAA    360

AAAAATATAC TCACCTCTCC TGCGATAAAG TCTTCTGCCA GCCATGGCAG AGATGCATTG   420

AGGGCACCTG TGTTTGTAAA CTACCGTATC AGTGCCCAAA GAATGGCACT GCAGTGTGTG   480

CAACTAACAG GAGAAGCTTC CCAACATACT GTCAACAAAA GAGTTTGGAA TGTCTTCATC   540

CAGGGACAAA GTTTTTAAAT AACGGAACAT GCACAGCCGA AGGAAAGTTT AGTGTTTCCT   600

TGAAGCATGG AAATACAGAT TCAGAGGGAA TAGTTGAAGT AAAACTTGTG GACCAAGATA   660

AGACAATGTT CATATGCAAA AGCAGCTGGA GCATGAGGGA AGCCAACGTG GCCTGCCTTG   720

ACCTTGGGTT TCAACAAGGT GCTGATACTC AAAGAAGGTT TAAGTTGTCT GATCTCTCTA   780

TAAATTCCAC TGATGTCTA CATGTGCATT GCCGAGGATT AGAGACCAGT TTGGCTGAAT    840

GTACTTTTAC TAAGAGAAGA ACTATGGGTT ACCAGGATTT CGCTGATGTG GTTTGTTATA   900

CACAGAAAGC AGATTCTCCA ATGGATGACT TCTTTCAGTG TGTGAATGGG AAATACATTT   960

CTCAGATGAA AGCCTGTGAT GGTATCAATG ATTGTGGAGA CCAAAGTGAT GAACTGTGTT  1020

GTAAAGCATG CCAAGGCAAA GGCTTCCATT GCAAATCGGG TGTTTGCATT CCAAGCCAGT  1080

ATCAATGCAA TGGTGAGGTG GACTGCATTA CAGGGGAAGA TGAAGTTGGC TGTGCAGGCT  1140

TTGCATCTGT GGCTCAAGAA GAAACAGAAA TTTTGACTGC TGACATGGAT GCAGAAAGAA  1200

GACGGATAAA ATCATTATTA CCTAAACTAT CTTGTGGAGT TAAAAACAGA ATGCACATTC  1260

GAAGGAAACG AATTGTGGGA GGAAAGCGAG CACAACTGGG AGACCTCCCA TGGCAGGTGG  1320

CAATTAAGGA TGCCAGTGGA ATCACCTGTG GGGGAATTTA TATTGGTGGC TGTTGGATTC  1380

TGACTGCTGC ACATTGTCTC AGAGCCAGTA AAAACTCATCG TTACCAAATA TGGACAACAG  1440

TAGTAGACTG GATACACCCC GACCTTAAAC GTATAGTAAT TGAATACGTG GATAGAATTA  1500

TTTTCCATGA AAACTACAAT GCAGGCACTT ACCAAAATGA CATCGCTTTG ATTGAAATGA  1560

AAAAAGACGG AAACAAAAAA GATTGTGAGC TGCCTCGTTC CATCCCTGCC TGTGTCCCCT  1620

GGTCTCCTTA CCTATTCCAA CCTAATGATA CATGCATCGT TTCTGGCTGG GGACGAGAAA  1680

AAGATAACGA AAGAGTCTTT TCACTTCAGT GGGGTGAAGT TAAACTAATA AGCAACTGCT  1740

CTAAGTTTTA CGGAAATCGT TTCTATGAAA AAGAAATGGA ATGTGCAGGT ACATATGATG  1800
```

```
GTTCCATCGA TGCCTGTAAA GGGGACTCTG GAGGCCCCTT AGTCTGTATG GATGCCAACA 1860

ATGTGACTTA TGTCTGGGGT GTTGTGAGTT GGGGGGAAAA CTGTGGAAAA CCAGAGTTCC 1920

CAGGTGTTTA CACCAAAGTG GCCAATTATT TTGACTGGAT TAGCTACCAT GTAGGAAGGC 1980

CTTTTATTTC TCAGTACAAT GTATAAAATT GTGATCTCTC TCTTCATTCT ATTCTTTTTC 2040

TCTCAAGAGT TCCATTTAAT GGAAATAAAA CGGTATAATT AATAATTCTC TAGGGGGGAA 2100

AAATGAAGCA AATCTCACTG GATATTTTTA AAGGTCTCCA CAGAGTTTAT GCCATATTGG 2160

AATTTTGTTG TATAATTCTC AAATAAAATAT TTTGGTGAAG CATACAAAAA AAAAAAA    2217
```

Sequence NM_001318057.1 encodes 2241 bp transcript, which includes an alternative, in-frame exon in central part. This transcript results in 591 aa pre-proprotein with identical N and C termini, but 7 additional amino acids (Fig. 15.3).

POLYMORPHIC VARIANTS, DEFICIENCY AND DISEASE-RELATED SINGLE NUCLEOTIDE POLYMORPHISMS

Historically, two common polymorphic variants were first identified with iso-electric focusing of neuraminidase-treated EDTA plasma samples followed by western blotting: FI*A and FI*B.[18] Frequency of these alleles depends on population and may vary from 0.893 FI*A and 0.106 FI*B to 0.153 FI*A and 0.847 FI*B.[18,19] Since then many polymorphic variants (such as G119R) were shown to increase the risk of development of age-related macular degeneration[20] or atypical haemolytic uraemic syndrome (aHUS; such as R317W).[21] Over 30 families with complete deficiency of FI have been reported.[6,22] These individuals present with increased rate of infections with Gram-positive and Gram-negative bacteria, autoimmune disorders and glomerulonephritis.[22–25] The identified mutations result either in decreased secretion or loss of function of FI. About 60 point mutations have been reported so far in heterozygous state leading to partial deficiency. These may impair mRNA splicing, folding of protein/secretion, or result in truncated protein (reviewed in Refs. 6,26). Missense mutations related to aHUS (35) are not grouped within a single FI domain, but disseminated through all domains.[6]

FI KNOCKOUT MICE

FI knockout mice have been developed and presented with uncontrolled alternative pathway activation as evidenced by reduced C3, factor B and factor H

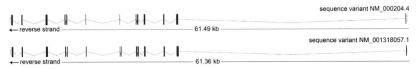

FIGURE 15.3 Genomic organisation of factor I with exon/intron organisation.

levels. Renal histology revealed higher incidence of mesangial expansion and hyalinosis; however, knockout mice did not develop spontaneous glomerulonephritis, in contrast to factor H-deficient mice.[27]

REFERENCES

1. Goldberger G, Arnaout MA, Aden D, Kay R, Rits M, Colten HR. Biosynthesis and postsynthetic processing of human C3b/C4b inactivator (factor I) in three hepatoma cell lines. *J Biol Chem* 1984;**259**:6492–7.

2. DiScipio RG. Ultrastructures and interactions of complement factors H and I. *J Immunol* 1992;**149**:2592–9.

3. Chamberlain D, Ullman CG, Perkins SJ. Possible arrangement of the five domains in human complement factor I as determined by a combination of X-ray and neutron scattering and homology modeling. *Biochemistry* 1998;**37**:13918–29.

4. Roversi P, Johnson S, Caesar JJ, McLean F, Leath KJ, Tsiftsoglou SA, et al. Structural basis for complement factor I control and its disease-associated sequence polymorphisms. *PNAS* 2011;**108**:12839–44.

5. Tsiftsoglou SA, Willis AC, Li P, Chen X, Mitchell DA, Rao Z, et al. The catalytically active serine protease domain of human complement factor I. *Biochemistry* 2005;**44**:6239–49.

6. Nilsson SC, Sim RB, Lea SM, Fremeaux-Bacchi V, Blom AM. Complement factor I in health and disease. *Mol Immunol* 2011;**48**:1611–20.

7. van Lookeren Campagne M, Wiesmann C, Brown EJ. Macrophage complement receptors and pathogen clearance. *Cell Microbiol* 2007;**9**:2095–102.

8. Carroll MC, Isenman DE. Regulation of humoral immunity by complement. *Immunity* 2012;**37**:199–207.

9. Tsiftsoglou SA, Sim RB. Human complement factor I does not require cofactors for cleavage of synthetic substrates. *J Immunol* 2004;**173**:367–75.

10. Morris KM, Aden DP, Knowles BB, Colten HR. Complement biosynthesis by the human hepatoma-derived cell line HepG2. *J Clin Invest* 1982;**70**:906–13.

11. Vyse TJ, Morley BJ, Bartok I, Theodoridis EL, Davies KA, Webster AD, et al. The molecular basis of hereditary complement factor I deficiency. *J Clin Invest* 1996;**97**:925–33.

12. Timar KK, Junnikkala S, Dallos A, Jarva H, Bhuiyan ZA, Meri S, et al. Human keratinocytes produce the complement inhibitor factor I: synthesis is regulated by interferon-gamma. *Mol Immunol* 2007;**44**:2943–9.

13. Schlaf G, Demberg T, Beisel N, Schieferdecker HL, Gotze O. Expression and regulation of complement factors H and I in rat and human cells: some critical notes. *Mol Immunol* 2001;**38**:231–9.

14. Julen N, Dauchel H, Lemercier C, Sim RB, Fontaine M, Ripoche J. In vitro biosynthesis of complement factor I by human endothelial cells. *Eur J Immunol* 1992;**22**:213–7.

15. Okroj M, Holmquist E, Nilsson E, Anagnostaki L, Jirstrom K, Blom AM. Local expression of complement factor I in breast cancer cells correlates with poor survival and recurrence. *Cancer Immunol Immunother* 2015;**64**:468–78.

16. Okroj M, Hsu YF, Ajona D, Pio R, Blom AM. Non-small cell lung cancer cells produce a functional set of complement factor I and its soluble cofactors. *Mol Immunol* 2008;**45**:169–79.

17. Minta JO, Fung M, Paramaswara B. Transcriptional and post-transcriptional regulation of complement factor I (CFI) gene expression in Hep G2 cells by interleukin-6. *Biochim Biophys Acta* 1998;**1442**:286–95.

18. Nakamura S, Abe K. Genetic polymorphism of human factor I (C3b inactivator). *Hum Genet* 1985;**71**:45–8.
19. Zhang L, Stradmann-Bellinghausen B, Rittner C, Schneider PM. Genetic polymorphism of human complement factor I (C3b inactivator) in the Chinese Han population. *Exp Clin Immunogenet* 1999;**16**:30–2.
20. van de Ven JP, Nilsson SC, Tan PL, Buitendijk GH, Ristau T, Mohlin FC, et al. A functional variant in the CFI gene confers a high risk of age-related macular degeneration. *Nat Genet* 2013;**45**:813–7.
21. Kavanagh D, Richards A, Noris M, Hauhart R, Liszewski MK, Karpman D, et al. Characterization of mutations in complement factor I (CFI) associated with hemolytic uremic syndrome. *Mol Immunol* 2008;**45**:95–105.
22. Nilsson SC, Trouw LA, Renault N, Miteva MA, Genel F, Zelazko M, et al. Genetic, molecular and functional analyses of complement factor I deficiency. *Eur J Immunol* 2009;**39**:310–23.
23. Amadei N, Baracho GV, Nudelman V, Bastos W, Florido MP, Isaac L. Inherited complete factor I deficiency associated with systemic lupus erythematosus, higher susceptibility to infection and low levels of factor H. *Scand J Immunol* 2001;**53**:615–21.
24. Sadallah S, Gudat F, Laissue JA, Spath PJ, Schifferli JA. Glomerulonephritis in a patient with complement factor I deficiency. *Am J Kidney Dis* 1999;**33**:1153–7.
25. Bienaime F, Dragon-Durey MA, Regnier CH, Nilsson SC, Kwan WH, Blouin J, et al. Mutations in components of complement influence the outcome of Factor I-associated atypical hemolytic uremic syndrome. *Kidney Int* 2010;**77**:339–49.
26. Rodriguez E, Rallapalli PM, Osborne AJ, Perkins SJ. New functional and structural insights from updated mutational databases for complement factor H, Factor I, membrane cofactor protein and C3. *Biosci Rep* 2014:34.
27. Rose KL, Paixao-Cavalcante D, Fish J, Manderson AP, Malik TH, Bygrave AE, et al. Factor I is required for the development of membranoproliferative glomerulonephritis in factor H-deficient mice. *J Clin Invest* 2008;**118**:608–18.

Part III

C3 Family

Chapter 16

C3

Scott R. Barnum

University of Alabama at Birmingham, Birmingham, AL, United States

PHYSICOCHEMICAL PROPERTIES

C3 is synthesised as a prepro single-chain molecule of 1663 amino acids including a 22-amino acid leader peptide and 4-amino acid cleavage sites.[1]

Mature Protein[2]		
pI	~5.9	
M_r (K)	~190	
	β-chain	α-chain
Amino acids	23–667	672–1663
M_r (K) predicted	~75 (645 aa)	~115 (992 aa)
N-linked glycosylation sites	1 (85)	2 (939, 1617, predicted)
Interchain disulphide bonds	559	816

STRUCTURE

C3 is composed of two polypeptide chains (α- and β-chains) that are disulphide linked with an additional intrachain disulphide linkage in the α-chain. The α-chain has a highly reactive thioester bond formed from neighbouring cysteine and glutamine residues, similar to that found in C4.[3] C3 has a two-domain shape as assessed by scattering solution analysis that consists of a flat ellipsoid ~18 nm long, 2 nm thick and 8–10 nm wide domain, with a smaller flat domain of $2 \times 4 \times 9$ nm (Fig. 16.1).[4] The crystal structure of C3 reveals that it has 13 major domains.[5,6] Eight of the domains have a fibronectin–type-3-core fold and are termed macroglobulin domains (MG). The β-chain is composed of six MG domains in tandem: MG1 through MG6β followed by a linker region (LNK). The α-chain starts with the anaphylatoxin domain (ANA), followed by a linker loop between the ANA domain and the C-terminal half of the MG6 termed MG6α. The α-chain continues with MG7 domain, followed by a CUB domain, then the thioester domain (TED). The TED is a feature common to members of the α2-macroglobulin family of proteins and is found in structurally related proteins in many organisms including insects and nematodes.[7,8] The TED region is followed by another CUB domain, MG8, a short

The Complement FactsBook. http://dx.doi.org/10.1016/B978-0-12-810420-0.00016-X

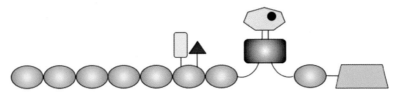

FIGURE 16.1   The protein domain motifs for C3. See Table 1.1 for domain definitions.

anchor region and, finally the C-terminal C345C domain. The C345C domain has a netrin fold and is part of a large family of proteins referred to as NTR (netrin-like module) (Fig. 16.2).[9]

FUNCTION

C3 is the focus of most of the activation and regulatory activities of the complement system. It is cleaved into more biologically active fragments than any other complement protein with the possible exception of C4. Fragments of C3 interact with more proteins in the complement than any other including: C3aR, CR1, CR2, CR3, CR4, factor I, factor H, CHFRs, MCP, DAF, CSMD1, C2 and factor B. As one of the oldest proteins in the complement system, C3 has integrated itself into biology that extends well beyond that of host defence. An exhaustive description of C3 functions is beyond the scope of this text; instead, a brief overview for each of the main fragments is provided. The reader is referred to reviews that highlight C3 functions in greater detail.[10–18]

C3a: Classically viewed as a proinflammatory mediator that contributes to conditions that encompass infection, trauma, ischaemia, autoimmunity and others.[14] C3a along with C5a binds to its receptor on myeloid cells contributing to cellular activation, chemoattraction and degranulation. C3a is also an antimicrobial molecule bactericidal for Gram-positive and Gram-negative organisms.[19] Studies have demonstrated that C3a provides homeostatic signals to T cells and modulates metabolism (reviewed in Ref. 18). Studies have also shown that C3a has antiinflammatory properties including suppression of inflammatory cytokine production and lipopolysaccharide (LPS)-induced shock.[20–23]

C3b/iC3b: Classically viewed as phagocytic opsonin, for clearance of complement-opsonised pathogens by neutrophils, macrophages and dendritic cells, C3b/iC3b may also play a role in synaptic pruning.[11] C3b, through interaction with CD46, also modulates metabolism (reviewed in Ref. 18). Studies have also shown that C3b can target intracellular pathogens to the proteasome for degradation and activation of cytosol-activated mitochondrial antiviral signalling leading to the production of proinflammatory cytokines.[24]

C3f/C3c: No known functions.

C3d/C3dg: Well-established roles in humoral immunity serving to cross-link cell surface antibody and the CD21/35 signalling complex on B cells,[10] although there is evidence that C3d can augment antibody responses in the absence of CD21/35.[25]

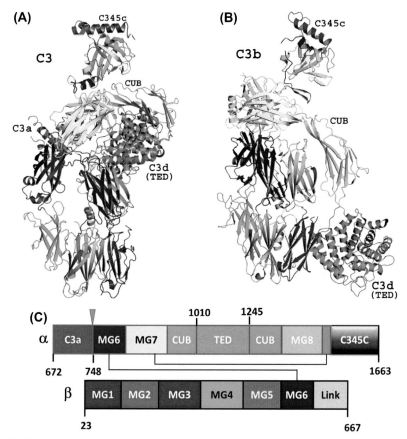

FIGURE 16.2 The crystal structure of C3. (A) Ribbon diagram of the C3 crystal structure (PDB ID: 2A73). The N-terminal C3a and thioester bond hosting C3d domains are marked. (B) Ribbon representation of C3b fragment of C3 (PDB ID: 5FO7) after the loss of C3a. (C) Schematic representation of α- and β-chains of the complement component C3. Domains are coloured similar to the C3 and C3b ribbon representations. The cleavage site between C3a and C3b fragments of C3 is represented by an *inverted triangle* coloured blue.

DEGRADATION PATHWAY

C3 undergoes a series of well-characterised proteolytic cleavages once complement has been activated through any of the pathways (shown below). C3 is also cleaved by several coagulation enzymes including factors IXa, Xa, XIa, plasmin and thrombin through the extrinsic protease pathway (Fig. 16.3).[26,27]

TISSUE DISTRIBUTION

Serum protein: 1–1.5 mg/mL[28]

Primary site of synthesis: Liver[29]

Secondary sites: C3 is produced by nearly all cell types, either at a low constitutive level or when induced by a variety of stimuli.[30] Cells that produce C3 include

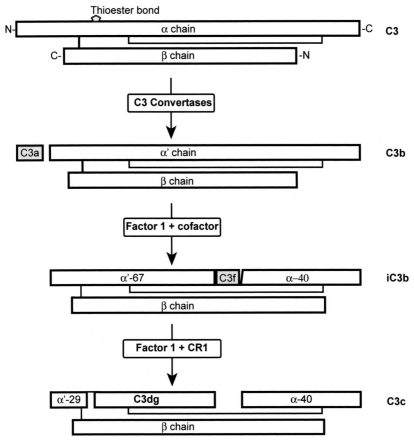

FIGURE 16.3 The degradation pathway for C3 via activation of the complement pathways.

monocytes/macrophages, neutrophils, platelets, fibroblasts, endothelial cells, myoblasts, synovium, T and B cells, astrocytes, microglial cells, neurons, glomerular mesangial cells proximal tubular and glomerular epithelium, pulmonary alveolar type II epithelium, adipocytes, osteoblasts and intestinal epithelial cells.

REGULATION OF EXPRESSION

C3 is an acute-phase protein whose serum levels rise about twofold during infection.[31] The change in C3 concentration in serum during the acute-phase response does not mirror the dramatic increases in C3 synthesis in cells in a tissue-specific fashion, which is mediated by proinflammatory cytokines such as IL-l, IL-6, TNF-α and IFN-γ.[32] Transforming growth factor-β (TGF-β), traditionally considered an antiinflammatory cytokine, upregulates C3 production in monocytes and retinal pigment epithelial cells via PKCδ and ERK1/2, SMAD3-mediated mechanisms, respectively.[33,34] LPS and lipid A also induce the production of C3.[35–37] Hormones such as testosterone, estradiol, progesterone, chorionic gonadotropin and cortico-steroids also regulate C3 production.[38–41] C3 expression is regulated at the level of transcription and translation.[42–45] The C3 promoter has numerous regulatory elements including TATA, CAAT and basic leucine zipper 1 domain among others.[46–49]

HUMAN PROTEIN SEQUENCE

```
MGPTSGPSLL LLLLTHLPLA LGSPMYSIIT PNILRLESEE TMVLEAHDAQ    50

GDVPVTVTVH DFPGKKLVLS SEKTVLTPAT NHMGNVTFTI PANREFKSEK   100

GRNKFVTVQA TFGTQWEKV  VLVSLQSGYL FIQTDKTIYT PGSTVLYRIF   150

TVNHKLLPVG RTVMVNIENP EGIPVKQDSL SSQNQLGVLP LSWDIPELVN   200

MGQWKIRAYY ENSPQQVFST EFEVKEYVLP SFEVIVEPTE KFYYIYNEKG   250

LEVTITARFL YGKKVEGTAF VIFGIQDGEQ RISLPESLKR IPIEDGSGEV   300

VLSRKVLLDG VQNLRAEDLV GKSLYVSATV ILHSGSDMVQ AERSGIPIVT   350

SPYQIHFTKT PKYFKPGMPF DLMVFVTNPD GSPAYRVPVA VQGEDTVQSL   400

TQGDGVAKLS INTHPSQKPL SITVRTKKQE LSEAEQATRT MQALPYSTVG   450

NSNNYLHLSV LRTELRPGET LNVNFLLRMD RAHEAKIRYY TYLIMNKGRL   500

LKAGRQVREP GQDLWLPLS  ITTDFIPSFR LVAYYTLIGA SGQREWADS    550

VWVDVKDSCV GSLWKSGQS  EDRQPVPGQQ MTLKIEGDHG ARWLVAVDK    600

GVFVLNKKNK LTQSKIWDW  EKADIGCTPG SGKDYAGVFS DAGLTFTSSS   650

GQQTAQRAEL QCPQPAARRR RSVQLTEKRM DKVGKYPKEL RKCCEDGMRE   700

NPMRFSCQRR TRFISLGEAC KKVFLDCCNY ITELRRQHAR ASHLGLARSN   750

LDEDIIAEEN IVSRSEFPES WLWNVEDLKE PPKNGISTKL MNIFLKDSIT   800

TWEILAVSMS DKKGICVADP FEVTVMQDFF IDLRLPYSW  RNEQVEIRAV   850

LYNYRQNQEL KVRVELLHNP AFCSLATTKR RHQQTVTIPP KSSLSVPYVI   900

VPLKTGLQEV EVKAAVYHHF ISDGVRKSLK WPEGIRMNK  TVAVRTLDPE   950

RLGREGVQKE DIPPADLSDQ VPDTESETRI LLQGTPVAQM TEDAVDAERL  1000

KHLIVTPSC  GEQNMIGMTP TVIAVHYLDE TEQWEKFGLE KRQGALELIK  1050

KGYTQQLAFR QPSSAFAAFV KRAPSTWLTA YWKVFSLAV  NLIAIDSQVL  1100

CGAVKWLILE KQKPDGVFQE DAPVIHQEMI GGLRNNNEKD MALTAFVLIS  1150

LQEAKDICEE QVNSLPGSIT KAGDFLEANY MNLQRSYTVA IAGYALAQMG  1200

RLKGPLLNKF LTTAKDKNRW EDPGKQLYNV EATSYALLAL LQLKDFDFVP  1250

PWRWLNEQR  YYGGGYGSTQ ATFMVFQALA QYQKDAPDHQ ELNLDVSLQL  1300

PSRSSKITHR IHWESASLLR SEETKENEGF TVTAEGKGQG TLSWTMYHA   1350

KAKDQLTCNK FDLKVTIKPA PETEKRPQDA KNTMILEICT RYRGDQDATM  1400

SILDISMMTG FAPDTDDLKQ LANGVDRYIS KYELDKAFSD RNTLIIYLDK  1450

VSHSEDDCLA FKVHQYFNVE LIQPGAVKVY AYYNLEESCT RFYHPEKEDG  1500

KLNKLCRDEL CRCAEENCFI QKSDDKVTLE ERLDKACEPG VDYVYKTRLV  1550

KVQLSNDFDE YIMAIEQTIK SGSDEVQVGQ QRTFISPIKC REALYKTRLV  1600

HYLMWGLSSD FWGEKPNLSY IIGKDTWVEH WPEEDECQDE ENQKQCQDLG  1650

AFTESMWFG CPN
```

The leader sequence and the cleavage site RRRR between the α- and β-chains are underlined. C1010 and Q1013, whose side chains form the thioester bond are double-underlined and *N*-linked glycosylation sites are indicated as **N**.

PROTEIN MODULES

Amino acid sequences for each relevant section of the protein from leader to the C-terminal end of the protein. List of associated exons for each module:

Sequence	Domain	Exon
1–22	Leader peptide	exon 1
23–126	MG1	exon 2/3
127–231	MG2	exon 3–6
232–350	MG3	exon 7–10
351–451	MG4	exon 10–12
452–556	MG5	exon 12–13
557–599	MG6β	exon 13/14
600–667	LNK	exon 14–16
675–742	ANA	exon 16/17
743–756	α′NT	exon 18
757–828	MG6α	exon 18/19
829–934	MG7	exon 20/21
935–970	CUB1	exon 22–24
980–1323	TED	exon 24–30
1324–1345	CUB2	exon 31
1356–1496	MG8	exon 32–36
1486–1517	Anchor	exon 37
1518–1661	C345 C	exon 38–41

CHROMOSOMAL LOCATION

Human: 19p13.3-p13.2[50]
   Mouse: chromosome 17, 34.3 cM[51]
   Rat: chromosome 9[52]

HUMAN cDNA SEQUENCE

```
ATAAAAAGCC AGCTCCAGCA GGCGCTGCTC ACTCCTCCCC ATCCTCTCCC TCTGTCCCTC    60

TGTCCCTCTG ACCCTGCACT GTCCCAGCAC CATGGGACCC ACCTCAGGTC CCAGCCTGCT   120

GCTCCTGCTA CTAACCCACC TCCCCCTGGC TCTGGGGAGT CCCATGTACT CTATCATCAC   180

CCCCAACATC TTGCGGCTGG AGAGCGAGGA GACCATGGTG CTGGAGGCCC ACGACGCGCA   240

AGGGGATGTT CCAGTCACTG TTACTGTCCA CGACTTCCCA GGCAAAAAAC TAGTGCTGTC   300

CAGTGAGAAG ACTGTGCTGA CCCCTGCCAC CAACCACATG GGCAACGTCA CCTTCACGAT   360

CCCAGCCAAC AGGGAGTTCA AGTCAGAAAA GGGGCGCAAC AAGTTCGTGA CCGTGCAGGC   420

CACCTTCGGG ACCCAAGTGG TGGAGAAGGT GGTGCTGGTC AGCCTGCAGA GCGGGTACCT   480

CTTCATCCAG ACAGACAAGA CCATCTACAC CCCTGGCTCC ACAGTTCTCT ATCGGATCTT   540
```

```
CACCGTCAAC CACAAGCTGC TACCCGTGGG CCGGACGGTC ATGGTCAACA TTGAGAACCC    600

GGAAGGCATC CCGGTCAAGC AGGACTCCTT GTCTTCTCAG AACCAGCTTG GCGTCTTGCC    660

CTTGTCTTGG GACATTCCGG AACTCGTCAA CATGGGCCAG TGGAAGATCC GAGCCTACTA    720

TGAAAACTCA CCACAGCAGG TCTTCTCCAC TGAGTTTGAG GTGAAGGAGT ACGTGCTGCC    780

CAGTTTCGAG GTCATAGTGG AGCCTACAGA GAAATTCTAC TACATCTATA ACGAGAAGGG    840

CCTGGAGGTC ACCATCACCG CCAGGTTCCT CTACGGGAAG AAAGTGGAGG GAACTGCCTT    900

TGTCATCTTC GGGATCCAGG ATGGCGAACA GAGGATTTCC CTGCCTGAAT CCCTCAAGCG    960

CATTCCGATT GAGGATGGCT CGGGGGAGGT TGTGCTGAGC CGGAAGGTAC TGCTGGACGG   1020

GGTGCAGAAC CTCCGAGCAG AAGACCTGGT GGGGAAGTCT TTGTACGTGT CTGCCACCGT   1080

CATCTTGCAC TCAGGCAGTG ACATGGTGCA GGCAGAGCGC AGCGGGATCC CCATCGTGAC   1140

CTCTCCCTAC CAGATCCACT TCACCAAGAC ACCCAAGTAC TTCAAACCAG GAATGCCCTT   1200

TGACCTCATG GTGTTCGTGA CGAACCCTGA TGGCTCTCCA GCCTACCGAG TCCCCGTGGC   1260

AGTCCAGGGC GAGGACACTG TGCAGTCTCT AACCCAGGGA GATGGCGTGG CCAAACTCAG   1320

CATCAACACA CACCCCAGCC AGAAGCCCTT GAGCATCACG GTGCGCACGA AGAAGCAGGA   1380

GCTCTCGGAG GCAGAGCAGG CTACCAGGAC CATGCAGGCT CTGCCCTACA GCACCGTGGG   1440

CAACTCCAAC AATTACCTGC ATCTCTCAGT GCTACGTACA GAGCTCAGAC CCGGGGAGAC   1500

CCTCAACGTC AACTTCCTCC TGCGAATGGA CCGCGCCCAC GAGGCCAAGA TCCGCTACTA   1560

CACCTACCTG ATCATGAACA AGGGCAGGCT GTTGAAGGCG GGACGCCAGG TGCGAGAGCC   1620

CGGCCAGGAC CTGGTGGTGC TGCCCCTGTC CATCACCACC GACTTCATCC CTTCCTTCCG   1680

CCTGGTGGCG TACTACACGC TGATCGGTGC CAGCGGCCAG AGGGAGGTGG TGGCCGACTC   1740

CGTGTGGGTG GACGTCAAGG ACTCCTGCGT GGGCTCGCTG GTGGTAAAAA GCGGCCAGTC   1800

AGAAGACCGG CAGCCTGTAC CTGGGCAGCA GATGACCCTG AAGATAGAGG GTGACCACGG   1860

GGCCCGGGTG GTACTGGTGG CCGTGGACAA GGGCGTGTTC GTGCTGAATA AGAAGAACAA   1920

ACTGACGCAG AGTAAGATCT GGGACGTGGT GGAGAAGGCA GACATCGGCT GCACCCCGGG   1980

CAGTGGGAAG GATTACGCCG GTGTCTTCTC CGACGCAGGG CTGACCTTCA CGAGCAGCAG   2040

TGGCCAGCAG ACCGCCCAGA GGGCAGAACT TCAGTGCCCG CAGCCAGCCG CCCGCCGACG   2100

CCGTTCCGTG CAGCTCACGG AGAAGCGAAT GGACAAAGTC GGCAAGTACC CCAAGGAGCT   2160

GCGCAAGTGC TGCGAGGACG GCATGCGGGA GAACCCCATG AGGTTCTCGT GCCAGCGCCG   2220

GACCCGTTTC ATCTCCCTGG GCGAGGCGTG CAAGAAGGTC TTCCTGGACT GCTGCAACTA   2280

CATCACAGAG CTGCGGCGGC AGCACGCGCG GGCCAGCCAC CTGGGCCTGG CCAGGAGTAA   2340

CCTGGATGAG GACATCATTG CAGAAGAGAA CATCGTTTCC CGAAGTGAGT TCCCAGAGAG   2400

TGGCTGTGG  AACGTTGAGG ACTTGAAAGA GCCACCGAAA AATGGAATCT CTACGAAGCT   2460

CATGAATATA TTTTTGAAAG ACTCCATCAC CACGTGGGAG ATTCTGGCTG TCAGCATGTC   2520
```

```
GGACAAGAAA GGGATCTGTG TGGCAGACCC CTTCGAGGTC ACAGTAATGC AGGACTTCTT 2580

CATCGACCTG CGGCTACCCT ACTCTGTTGT TCGAAACGAG CAGGTGGAAA TCCGAGCCGT 2640

TCTCTACAAT TACCGGCAGA ACCAAGAGCT CAAGGTGAGG GTGGAACTAC TCCACAATCC 2700

AGCCTTCTGC AGCCTGGCCA CCACCAAGAG GCGTCACCAG CAGACCGTAA CCATCCCCCC 2760

CAAGTCCTCG TTGTCCGTTC CATATGTCAT CGTGCCGCTA AAGACCGGCC TGCAGGAAGT 2820

GGAAGTCAAG GCTGCCGTCT ACCATCATTT CATCAGTGAC GGTGTCAGGA AGTCCCTGAA 2880

GGTCGTGCCG GAAGGAATCA GAATGAACAA AACTGTGGCT GTTCGCACCC TGGATCCAGA 2940

ACGCCTGGGC CGTGAAGGAG TGCAGAAAGA GGACATCCCA CCTGCAGACC TCAGTGACCA 3000

AGTCCCGGAC ACCGAGTCTG AGACCAGAAT TCTCCTGCAA GGGACCCCAG TGGCCCAGAT 3060

GACAGAGGAT GCCGTCGACG CGGAACGGCT GAAGCACCTC ATTGTGACCC CCTCGGGCTG 3120

CGGGGAACAG AACATGATCG GCATGACGCC CACGGTCATC GCTGTGCATT ACCTGGATGA 3180

AACGGAGCAG TGGGAGAAGT TCGGCCTAGA GAAGCGGCAG GGGGCCTTGG AGCTCATCAA 3240

GAAGGGGTAC ACCCAGCAGC TGGCCTTCAG ACAACCCAGC TCTGCCTTTG CGGCCTTCGT 3300

GAAACGGGCA CCCAGCACCT GGCTGACCGC CTACGTGGTC AAGGTCTTCT CTCTGGCTGT 3360

CAACCTCATC GCCATCGACT CCCAAGTCCT CTGCGGGGCT GTTAAATGGC TGATCCTGGA 3420

GAAGCAGAAG CCCGACGGGG TCTTCCAGGA GGATGCGCCC GTGATACACC AAGAAATGAT 3480

TGGTGGATTA CGGAACAACA ACGAGAAAGA CATGGCCCTC ACGGCCTTTG TTCTCATCTC 3540

GCTGCAGGAG GCTAAAGATA TTTGCGAGGA GCAGGTCAAC AGCCTGCCAG GCAGCATCAC 3600

TAAAGCAGGA GACTTCCTTG AAGCCAACTA CATGAACCTA CAGAGATCCT ACACTGTGGC 3660

CATTGCTGGC TATGCTCTGG CCCAGATGGG CAGGCTGAAG GGGCCTCTTC TTAACAAATT 3720

TCTGACCACA GCCAAAGATA AGAACCGCTG GGAGGACCCT GGTAAGCAGC TCTACAACGT 3780

GGAGGCCACA TCCTATGCCC TCTTGGCCCT ACTGCAGCTA AAAGACTTTG ACTTTGTGCC 3840

TCCCGTCGTG CGTTGGCTCA ATGAACAGAG ATACTACGGT GGTGGCTATG GCTCTACCCA 3900

GGCCACCTTC ATGGTGTTCC AAGCCTTGGC TCAATACCAA AAGGACGCCC CTGACCACCA 3960

GGAACTGAAC CTTGATGTGT CCCTCCAACT GCCCAGCCGC AGCTCCAAGA TCACCCACCG 4020

TATCCACTGG GAATCTGCCA GCCTCCTGCG ATCAGAAGAG ACCAAGGAAA ATGAGGGTTT 4080

CACAGTCACA GCTGAAGGAA AAGGCCAAGG CACCTTGTCG GTGGTGACAA TGTACCATGC 4140

TAAGGCCAAA GATCAACTCA CCTGTAATAA ATTCGACCTC AAGGTCACCA TAAAACCAGC 4200

ACCGGAAACA GAAAAGAGGC CTCAGGATGC CAAGAACACT ATGATCCTTG AGATCTGTAC 4260

CAGGTACCGG GGAGACCAGG ATGCCACTAT GTCTATATTG GACATATCCA TGATGACTGG 4320

CTTTGCTCCA GACACAGATG ACCTGAAGCA GCTGGCCAAT GGTGTTGACA GATACATCTC 4380

CAAGTATGAG CTGGACAAAG CCTTCTCCGA TAGGAACACC CTCATCATCT ACCTGGACAA 4440

GGTCTCACAC TCTGAGGATG ACTGTCTAGC TTTCAAAGTT CACCAATACT TTAATGTAGA 4500
```

```
GCTTATCCAG CCTGGAGCAG TCAAGGTCTA CGCCTATTAC AACCTGGAGG AAAGCTGTAC  4560

CCGGTTCTAC CATCCGGAAA AGGAGGATGG AAAGCTGAAC AAGCTCTGCC GTGATGAACT  4620

GTGCCGCTGT GCTGAGGAGA ATTGCTTCAT ACAAAAGTCG GATGACAAGG TCACCCTGGA  4680

AGAACGGCTG GACAAGGCCT GTGAGCCAGG AGTGGACTAT GTGTACAAGA CCCGACTGGT  4740

CAAGGTTCAG CTGTCCAATG ACTTTGACGA GTACATCATG GCCATTGAGC AGACCATCAA  4800

GTCAGGCTCG GATGAGGTGC AGGTTGGACA GCAGCGCACG TTCATCAGCC CCATCAAGTG  4860

CAGAGAAGCC CTGAAGCTGG AGGAGAAGAA ACACTACCTC ATGTGGGGTC TCTCCTCCGA  4920

TTTCTGGGGA GAGAAGCCCA ACCTCAGCTA CATCATCGGG AAGGACACTT GGGTGGAGCA  4980

CTGGCCTGAG GAGGACGAAT GCCAAGACGA AGAGAACCAG AAACAATGCC AGGACCTCGG  5040

CGCCTTCACC GAGAGCATGG TTGTCTTTGG GTGCCCCAAC TGACCACACC CCCATTCCAT  5100

GAACCTACAG AGATCCTACA CTGTGGCCAT TGCTGGCTAT GCTCTGGCCC AGATGGGCAG
```

The first five nucleotides in each exon are underlined to indicate the intron–exon boundaries. The methionine initiation codon (ATG) and the termination codon (TGA) are indicated by double-underlining.

GENOMIC STRUCTURE

The gene spans ~43 kb and is encoded by 41 exons (Fig. 16.4).

ACCESSION NUMBERS

Human	K02765
Mouse	K02782

DEFICIENCY

Complete C3 deficiency is rare with only a small number of documented cases to date.[53–55] This deficiency is inherited in an autosomal codominant manner and affected individuals have C3 levels less than 1% of normal.[56] The low level of C3 in these individuals is due, in part, to production by monocytes, indicating differential regulation of C3 production between hepatocytes and monocytes.[57] Affected individuals present with recurrent, life-threatening infections including pneumonia, otitis, meningitis, osteomyelitis, peritonitis and sepsis.[53,54] In addition to infection, some individuals present with immune complex disease, vasculitis, glomerulonephritis or systemic lupus erythematosus.[55,58] Initial studies claimed a sexual disequilibrium, with females more likely to be C3

FIGURE 16.4   The genomic structure of the human C3 gene.

deficient, but this may be due to sample bias. Individuals with heterozygous C3 deficiency have ~50% normal serum levels of C3 and are phenotypically normal.[59] Deficiency is due to either partial gene deletion or single nucleotide polymorphisms.[55,58,60]

POLYMORPHIC VARIANTS

There are two common C3 variants based on electrophoretic mobility (C3S and C3F).[61] Since the original description of these two polymorphisms, over 1600 polymorphic nucleotide variants that represent missense, synonymous, coding sequence, inframe, frame-shift, splice region and stop variants have now been identified for *C3* (Ensembl - Transcript: C3-001 ENST00000245907.10). The most clinically relevant variants are those involved in haemolytic uraemic syndrome,[55,62,63] age-related macular degeneration[64,65] and glomerulopathy.[66]

MUTANT ANIMALS

At least two different C3-deficient mice lines have been developed. The first line was generated using a neomycin-containing cassette that deleted 120 amino acids spanning the junction between the α- and β-chains of *C3*.[67] Constructs were injected into J-1 embryonic stem cells and selected clones were injected into C57BL/6 blastocysts. Male chimeric mice were bred to C57BL/6 mice. A second line of C3-deficient mice was generated by targeting a neomycin-containing construct that deleted the 5' end of the C3 gene down to the fourth intron, thus removing the first three exons of the C3 gene.[68] The construct was injected into RW-4 and D3 embryonic stem cells from 129 SVJ mice. Homologous recombinants were injected into C57BL/6 blastocysts and chimeric males breed to C57BL/6 mice. Both lines of mice grow normally under pathogen-free conditions and display no overt phenotype. C3-deficient mice are highly susceptible to bacterial challenge. Since the generation of these mice, they have been used in infectious and autoimmune disease model systems too numerous to discuss here. These and variations on the original mice are available at The Jackson Laboratory (https://www.jax.org).

REFERENCES

1. de Bruijn MH, Fey GH. Human complement component C3: cDNA coding sequence and derived primary structure. *Proc Natl Acad Sci USA* 1985;**82**:708–12.
2. Tack BD, Prahl JW. Third component of human complement: purification from plasma and physicochemical characterization. *Biochemistry* 1976;**15**:4513–21.
3. Tack BF. The beta-Cys-gamma-Glu thiolester bond in human C3, C4, and alpha 2-macroglobulin. *Springer Semin Immunopathol* 1983;**6**:259–82.
4. Perkins SJ, Sim RB. Molecular modelling of human complement component C3 and its fragments by solution scattering. *Eur J Biochem* 1986;**157**:155–68.
5. Janssen BJ, Huizinga EG, Raaijmakers HC, et al. Structures of complement component C3 provide insights into the function and evolution of immunity. *Nature* 2005;**437**:505–11.

6. Janssen BJ, Christodoulidou A, McCarthy A, Lambris JD, Gros P. Structure of C3b reveals conformational changes that underlie complement activity. *Nature* 2006;**444**:213–6.

7. Blandin S, Levashina EA. Thioester-containing proteins and insect immunity. *Mol Immunol* 2004;**40**:903–8.

8. Budd A, Blandin S, Levashina EA, Gibson TJ. Bacterial alpha2-macroglobulins: colonization factors acquired by horizontal gene transfer from the metazoan genome? *Genome Biol* 2004;**5**:R38.

9. Banyai L, Patthy L. The NTR module: domains of netrins, secreted frizzled related proteins, and type I procollagen C-proteinase enhancer protein are homologous with tissue inhibitors of metalloproteases. *Protein Sci* 1999;**8**:1636–42.

10. Carroll MC, Isenman DE. Regulation of humoral immunity by complement. *Immunity* 2012;**37**:199–207.

11. Stephan AH, Barres BA, Stevens B. The complement system: an unexpected role in synaptic pruning during development and disease. *Annu Rev Neurosci* 2012;**35**:369–89.

12. Mastellos DC, Deangelis RA, Lambris JD. Complement-triggered pathways orchestrate regenerative responses throughout phylogenesis. *Semin Immunol* 2013;**25**:29–38.

13. Phieler J, Garcia-Martin R, Lambris JD, Chavakis T. The role of the complement system in metabolic organs and metabolic diseases. *Semin Immunol* 2013;**25**:47–53.

14. Klos A, Wende E, Wareham KJ, Monk PN. International union of basic and clinical pharmacology. [corrected]. LXXXVII. Complement peptide C5a, C4a, and C3a receptors. *Pharmacol Rev* 2013;**65**:500–43.

15. Zimmer J, Hobkirk J, Mohamed F, Browning MJ, Stover CM. On the functional overlap between complement and anti-microbial peptides. *Front Immunol* 2014;**5**:689.

16. Kolev M, Le Friec G, Kemper C. Complement–tapping into new sites and effector systems. *Nat Rev Immunol* 2014;**14**:811–20.

17. Barbu A, Hamad OA, Lind L, Ekdahl KN, Nilsson B. The role of complement factor C3 in lipid metabolism. *Mol Immunol* 2015;**67**:101–7.

18. Hess C, Kemper C. Complement-mediated regulation of metabolism and basic cellular process. *Immunity* 2016;**45**:240–54.

19. Pasupuleti M, Walse B, Nordahl EA, Morgelin M, Malmsten M, Schmidtchen A. Preservation of antimicrobial properties of complement peptide C3a, from invertebrates to humans. *J Biol Chem* 2007;**282**:2520–8.

20. Takabayashi T, Vannier E, Burke JF, Tompkins RG, Gelfand JA, Clark BD. Both C3a and C3a(desArg) regulate interleukin-6 synthesis in human peripheral blood mononuclear cells. *J Infect Dis* 1998;**177**:1622–8.

21. Takabayashi T, Vannier E, Clark BD, et al. A new biologic role for C3a and C3a desArg: regulation of TNF-alpha and IL-1 beta synthesis. *J Immunol* 1996;**156**:3455–60.

22. Kildsgaard J, Hollmann TJ, Matthews KW, Bian K, Murad F, Wetsel RA. Cutting edge: targeted disruption of the C3a receptor gene demonstrates a novel protective anti-inflammatory role for C3a in endotoxin-shock. *J Immunol* 2000;**165**:5406–9.

23. Boos L, Szalai AJ, Barnum SR. C3a expressed in the central nervous system protects against LPS-induced shock. *Neurosci Lett* 2005;**387**:68–71.

24. Tam JC, Bidgood SR, McEwan WA, James LC. Intracellular sensing of complement C3 activates cell autonomous immunity. *Science* 2014;**345**:1256070.

25. Haas KM, Toapanta FR, Oliver JA, et al. Cutting edge: C3d functions as a molecular adjuvant in the absence of CD21/35 expression. *J Immunol* 2004;**172**:5833–7.

26. Markiewski MM, Nilsson B, Ekdahl KN, Mollnes TE, Lambris JD. Complement and coagulation: strangers or partners in crime? *Trends Immunol* 2007;**28**:184–92.

27. Amara U, Flierl MA, Rittirsch D, et al. Molecular intercommunication between the complement and coagulation systems. *J Immunol* 2010;**185**:5628–36.

28. Kohler PF, Muller-Eberhard HJ. Immunochemical quantitation of the third, fourth and fifth components of human complement: concentrations in the serum of healthy adults. *J Immunol* 1967;**99**:1211–6.

29. Alper CA, Johnson AM, Birtch AG, Moore FD. Human C'3: evidence for the liver as the primary site of synthesis. *Science* 1969;**163**:286–8.

30. Morgan BP, Gasque P. Extrahepatic complement biosynthesis: where, when and why? *Clin Exp Immunol* 1997;**107**:1–7.

31. Gabay C, Kushner I. Acute-phase proteins and other systemic responses to inflammation. *N Engl J Med* 1999;**340**:448–54.

32. Barnum SR, Fey G, Tack BF. Biosynthesis and genetics of C3. *Curr Top Microbiol Immunol* 1990;**153**:23–43.

33. Drouin SM, Kiley SC, Carlino JA, Barnum SR. Transforming growth factor-beta2 regulates C3 secretion in monocytes through a protein kinase C-dependent pathway. *Mol Immunol* 1998;**35**:1–11.

34. Li Y, Song D, Song Y, et al. Iron-induced local complement component 3 (C3) up-regulation via non-canonical transforming growth factor (TGF)-beta signaling in the retinal pigment epithelium. *J Biol Chem* 2015;**290**:11918–34.

35. Nichols WK. LPS stimulation of complement (C3) synthesis by a human monocyte cell line. *Complement* 1984;**1**:108–15.

36. Strunk RC, Whitehead AS, Cole FS. Pretranslational regulation of the synthesis of the third component of complement in human mononuclear phagocytes by the lipid A portion of lipopolysaccharide. *J Clin Invest* 1985;**76**:985–90.

37. Sutton MB, Strunk RC, Cole FS. Regulation of the synthesis of the third component of complement and factor B in cord blood monocytes by lipopolysaccharide. *J Immunol* 1986;**136**:1366–72.

38. Churchill Jr WH, Weintraub RM, Borsos T, Rapp HJ. Mouse complement: the effect of sex hormones and castration on two of the late-acting components. *J Exp Med* 1967;**125**:657–72.

39. Strunk RC, Tashjian Jr AH, Colten HR. Complement biosynthesis in vitro by rat hepatoma cell strains. *J Immunol* 1975;**114**:331–5.

40. Rezvani R, Gupta A, Smith J, et al. Cross-sectional associations of acylation stimulating protein (ASP) and adipose tissue gene expression with estradiol and progesterone in pre- and postmenopausal women. *Clin Endocrinol* 2014;**81**:736–45.

41. Palomino WA, Argandona F, Azua R, Kohen P, Devoto L. Complement C3 and decay-accelerating factor expression levels are modulated by human chorionic gonadotropin in endometrial compartments during the implantation window. *Reprod Sci* 2013;**20**:1103–10.

42. Wilson DR, Juan TS, Wilde MD, Fey GH, Darlington GJ. A 58-base-pair region of the human C3 gene confers synergistic inducibility by interleukin-1 and interleukin-6. *Mol Cell Biol* 1990;**10**:6181–91.

43. Barnum SR, Jones JL. Differential regulation of C3 gene expression in human astroglioma cells by interferon-gamma and interleukin-1 beta. *Neurosci Lett* 1995;**197**:121–4.

44. Drouin SM, Carlino JA, Barnum SR. Transforming growth factor-beta2-mediated regulation of C3 gene expression in monocytes. *Mol Immunol* 1996;**33**:1025–34.

45. Mitchell TJ, Naughton M, Norsworthy P, Davies KA, Walport MJ, Morley BJ. IFN-gamma up-regulates expression of the complement components C3 and C4 by stabilization of mRNA. *J Immunol* 1996;**156**:4429–34.

46. Vik DP, Amiguet P, Moffat GJ, et al. Structural features of the human C3 gene: intron/exon organization, transcriptional start site, and promoter region sequence. *Biochemistry* 1991;**30**:1080–5.

47. Juan TS, Wilson DR, Wilde MD, Darlington GJ. Participation of the transcription factor C/EBP delta in the acute-phase regulation of the human gene for complement component C3. *Proc Natl Acad Sci USA* 1993;**90**:2584–8.

48. Maranto J, Rappaport J, Datta PK. Regulation of complement component C3 in astrocytes by IL-1beta and morphine. *J Neuroimmune Pharmacol* 2008;**3**:43–51.

49. Maranto J, Rappaport J, Datta PK. Role of C/EBP-beta, p38 MAPK, and MKK6 in IL-1beta-mediated C3 gene regulation in astrocytes. *J Cell Biochem* 2011;**112**:1168–75.

50. Whitehead AS, Solomon E, Chambers S, Bodmer WF, Povey S, Fey G. Assignment of the structural gene for the third component of human complement to chromosome 19. *Proc Natl Acad Sci USA* 1982;**79**:5021–5.

51. Natsuume-Sakai S, Hayakawa JI, Takahashi M. Genetic polymorphism of murine C3 controlled by a single co-dominant locus on chromosome 17. *J Immunol* 1978;**121**:491–8.

52. Cox DW, Francke U. Direct assignment of orosomucoid to human chromosome 9 and alpha 2HS-glycoprotein to chromosome 3 using human fetal liver x rat hepatoma hybrids. *Hum Genet* 1985;**70**:109–15.

53. Ross SC, Densen P. Complement deficiency states and infection: epidemiology, pathogenesis and consequences of neisserial and other infections in an immune deficiency. *Medicine* 1984;**63**:243–73.

54. Bitter-Suermann D, Burger R. C3 deficiencies. *Curr Top Microbiol Immunol* 1989;**153**:223–33.

55. Grumach AS, Kirschfink M. Are complement deficiencies really rare? Overview on prevalence, clinical importance and modern diagnostic approach. *Mol Immunol* 2014;**61**:110–7.

56. Alper CA, Colten HR, Rosen FS, Rabson AR, Macnab GM, Gear JS. Homozygous deficiency of C3 in a patient with repeated infections. *Lancet* 1972;**2**:1179–81.

57. Einstein LP, Hansen PJ, Ballow M, et al. Biosynthesis of the third component of complement (C3) in vitro by monocytes from both normal and homozygous C3-deficient humans. *J Clin Invest* 1977;**60**:963–9.

58. Mayilyan KR. Complement genetics, deficiencies, and disease associations. *Protein & Cell* 2012;**3**:487–96.

59. Alper CA, Propp RP, Klemperer MR, Rosen FS. Inherited deficiency of the third component of human complement (C'3). *J Clin Invest* 1969;**48**:553–7.

60. Reis ES, Barbuto JA, Isaac L. Human monocyte-derived dendritic cells are a source of several complement proteins. *Inflamm Res* 2006;**55**:179–84.

61. Alper CA, Propp RP. Genetic polymorphism of the third component of human complement (C'3). *J Clinical Invest* 1968;**47**:2181–91.

62. Vieira-Martins P, El Sissy C, Bordereau P, Gruber A, Rosain J, Fremeaux-Bacchi V. Defining the genetics of thrombotic microangiopathies. *Transfus Apher Sci* 2016;**54**:212–9.

63. Kavanagh D, Richards A, Fremeaux-Bacchi V, et al. Screening for complement system abnormalities in patients with atypical hemolytic uremic syndrome. *Clin J Am Soc Nephrol* 2007;**2**:591–6.

64. Zipfel PF, Lauer N, Skerka C. The role of complement in AMD. *Adv Exp Med Biol* 2010;**703**:9–24.

65. Anderson DH, Radeke MJ, Gallo NB, et al. The pivotal role of the complement system in aging and age-related macular degeneration: hypothesis re-visited. *Prog Retin Eye Res* 2010;**29**:95–112.

66. Prohaszka Z, Nilsson B, Frazer-Abel A, Kirschfink M. Complement analysis 2016: clinical indications, laboratory diagnostics and quality control. *Immunobiology* 2016;**221**:1247–58.

67. Wessels MR, Butko P, Ma M, Warren HB, Lage AL, Carroll MC. Studies of group B streptococcal infection in mice deficient in complement component C3 or C4 demonstrate an essential role for complement in both innate and acquired immunity. *Proc Natl Acad Sci USA* 1995;**92**:11490–4.

68. Circolo A, Garnier G, Fukuda W, et al. Genetic disruption of the murine complement C3 promoter region generates deficient mice with extrahepatic expression of C3 mRNA. *Immunopharmacology* 1999;**42**:135–49.

Chapter 17

C4

David E. Isenman
University of Toronto, Toronto, ON, Canada

OTHER NAMES

None in human, in mouse formerly also known as Ss protein.

PHYSICOCHEMICAL PROPERTIES

Human C4 is synthesised as a prepro single chain of 1744 amino acids including a 19 amino acid leader sequence and two tetrabasic cleavage sites for a furin-like enzyme. The three chains of mature C4 are linked by disulphide bonds, and these chains have the order β-α−γ in the precursor molecule. In primates, sheep and cattle there are two isotypes of C4, C4A and C4B.[1] The A/B designations reflect their more *A*cidic or *B*asic electrophoretic mobilities at alkaline pH.

Mature Protein:
pI ~7.9, varies with allotype.
$E^{1\%}_{280\ nm}$ extinction coefficient 8.2.[2]

	β-Chain	α-Chain	γ-Chain
Amino acids (prepro numbering)	20–675	680–1449 secreted 680–1427 plasma	1454–1744
M_r (K) predicted	71.7	84.6 secreted 82.1 plasma	33.1
Observed	75	93	33
N-linked glycosylation sites (all occupied)[3]	1 (226)	3 (862,1328,1391)	
O-linked glycosylation site		1 (T1244)	
Interchain disulphide bonds: 1 β-α, 2 α-γ	β567	α820, α876, α1394	γ1590, γ1566
Intrachain disulphide bonds: 2 β-chain, 3 α-chain, 5 γ-chain (all disulphide pairings from PDB file 4FXK)[8]	β68–β97 β635–β669	α702–α728 α703–α735	γ1471–γ1535 γ1583–γ1588 γ1595–γ1673 γ1618–γ1742 γ1718–γ1727

The Complement FactsBook. http://dx.doi.org/10.1016/B978-0-12-810420-0.00017-1

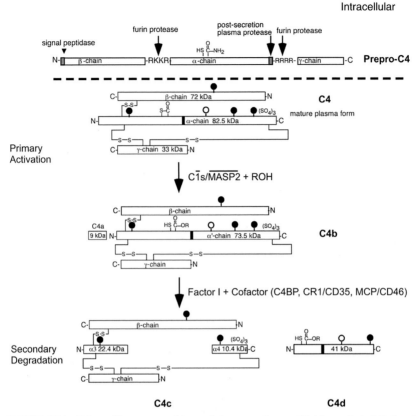

FIGURE 17.1 Polypeptide structure of intracellular prepro human C4 (above *dashed line*) and of mature plasma C4. For the latter, the polypeptide chain composition of its major activation and secondary degradation products are shown. Indicated are the N- and C-termini of chains, the sites of *N*-linked glycosylation (*closed circle*), the site of *O*-glycosylation (*open circle*), the thioester bond formed by the side chains of residues C1010 and Q1013, before and after transacylation to a hydroxyl nucleophile (ROH), the location of the isotypic cluster of residues and the sites of tyrosine sulphation. Only interchain disulphides are shown. The molecular masses indicated are for the polypeptide segments only; approximately 2 kDa should be added for each carbohydrate chain that is present.

Following secretion into the plasma, an elastase-like metalloprotease removes a 22 residue peptide from the C-terminus of α-chain.[3] Other posttranslational modifications include formation of the intramolecular thioester bond,[4] glycosylation[5] and tyrosine sulphation.[6] The chain structures of the biosynthetic intracellular C4 precursor, its mature plasma form and its subsequent degradation products are depicted in Fig. 17.1.

STRUCTURE

C3, C4 and C5 have ~30% sequence identity (mid-40% sequence similarity) and grossly similar overall architectures. The human C4 molecule, which has approximate dimensions of $156 \times 94 \times 64$ Å, consists of 13 major domains and 2 minor

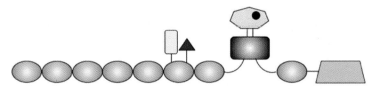

FIGURE 17.2   Diagram of the protein domain composition of C4.

motifs.[8] Eight of the major domains have been dubbed macroglobulin (MG) domains and have an anti-parallel 3-strand/4-strand β-sheet arrangement characteristic of fibronectin type III domains. The additional domains are the thioester-containing TED domain, a CUB domain (complement C1r/C1s, urchin egf, bone morphogenic protein), the ANA domain (corresponding to the C4a fragment), an α-helical connecting segment called Link and a C-terminal domain dubbed C345C that has a netrin-type fold (Fig. 17.2). Separating the Link and the ANA domains is the propeptide segment that is cleaved during biosynthesis, and thus the two halves of MG6 actually reside within two separate chains in mature C4. A similar situation occurs with a propeptide segment within a loop of MG8 and results in MG8 being contributed to by segments within both the α- and γ-chains.

There are currently four X-ray crystal structure entries in the PDB database for human C4. The entries are for C4D (1HZF),[7] native C4 (4FXK) (Fig. 17.3),[8] C4b (4XAM)[9] and a complex between native C4 and fragment of MASP2 consisting of complement control proteins, CCP1 and CCP2, and the serine protease (SP) domains (4FXG).[8]

Domains MG1 through MG6 form a 1.5-turn superhelical entity known as the β-ring, which forms a stable 'base' of the molecule. The TED, CUB and MG8 domains cluster to form a superdomain in native C4, which resides atop the β-chain ring and is 'wedged' in place by the ANA domain. MG7 and C345C, at the top of the native molecule, are peripheral to the TED-CUB-MG8 superdomain. The intact thioester in native C4 is buried at the interface between TED and MG8, and is protected from hydrolysis by the solvent.

Cleavage of C4 by activated C1s liberates the C4a/ANA fragment and results in a dramatic repositioning of several of the domains and sequence motifs in the resulting C4b molecule, while at the same time leaving the arrangement of the domains of the β-ring largely unaltered.[9] Removal of the ANA 'wedge' results in a significant rotation and translation of the MG8 domain. This disrupts the interfaces of MG8 with TED and CUB, allowing them to rotate and move down the side of the β-ring to new positions alongside MG1 and MG2, respectively. Furthermore, the positions of MG7 and MG8 become flipped in C4b, relative to what they were in native C4. These are large conformational changes, with TED moving approximately 60 Å and resulting in the thioester residues become fully exposed and accessible for transacylation onto target nucleophiles. Another major movement is by the newly released α′NT segment, which moves from a sequestered position near MG6 in native C4, to lie fully exposed across the top of the MG7 domain of C4B. This is a highly acidic segment and has been implicated through site-directed mutagenesis as being a contact site for C2.[10] Although there

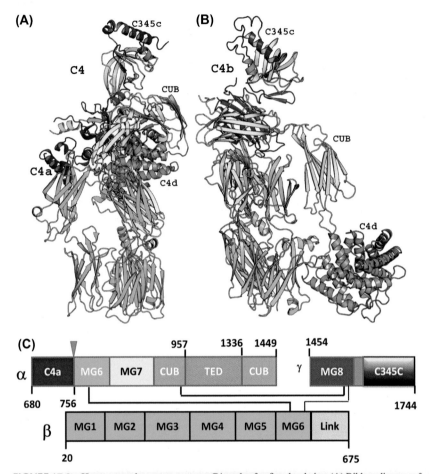

FIGURE 17.3 Human complement component C4 made of α, β and γ chains. (A) Ribbon diagram of the C4 crystal structure (PDB ID: 4FXK). The N-terminal C4a domain and thioester bond hosting C4d are marked. (B) Crystal structure of C4b fragment of C4 (PDB ID: 4XAM), after the loss of C4a, represented as a ribbon diagram. (C) Schematic representation of the complement component C4 domains α, β and γ chains. Domains are coloured similar to the C4 and C4b ribbon representations. The cleavage site between C4a and C4b fragments of C4 is represented by an *inverted triangle*, coloured blue.

are modest changes in the angle at which the C345C domain sits at the top of the molecule, there is evidence that its position is somewhat variable and may even be influenced by crystal packing effects.[9] Finally, as was also the case for their native forms, there is remarkable similarity in the structures of C3b and C4b.[9,11]

FUNCTION

C4 is proteolytically activated through both the classical pathway (by activated C1s) and the lectin pathway (by activated MASP2). The resulting C4b fragment is a modulatory subunit of the classical pathway C3 convertase (C4b2a) and the classical pathway C5 convertase (C4bC2aC3b). C4b has stable binding sites for C2, C5 and nascently activated C3b.[12,13] It also has binding sites for the CCP module containing complement regulatory molecules C4b binding protein (C4BP), complement receptor 1 (CR1/CD35), membrane cofactor protein (MCP/CD46), decay accelerator factor (DAF/CD55) and the regulatory SP factor I. The thioester formed between the side chains of Cys 1010 and Gln 1013 (prepro numbering used throughout) within the C4d region of the α-chain (Fig. 17.1) mediates covalent attachment to the target surface bearing activated forms of C1s and MASP2. The C4a peptide has very weak anaphylatoxin activity (~100–1000 less potent than C3a).[14] This observation combined with the absence of an identified receptor for C4a suggests that C4a should be removed from the list of complement anaphylatoxins.[15,16]

The human C4A and C4B isotypes are defined by a short sequence motif located within the C4d fragment, ~110 amino acids C-terminal to the thioester-forming residues: [1120]**PCPVLD**[1125] for C4A and [1120]**LSPVIH**[1125] for C4B.[17] The C4 isotypes differ not only in their preference of acceptor nucleophiles, amino groups for C4A and hydroxyl groups for C4B, but also in the mechanism through which transacylation is accomplished.[4] In nascently activated C4 of the A isotype, amino group nucleophiles are believed to directly attack the thioester carbonyl. However, due to the presence in C4B of a histidine at residue 1125 in place of the non-nucleophilic residue aspartic acid in C4A, activation first leads to the formation of an intramolecular acyl-imidazole intermediate involving the transacylation of the thioester carbonyl to the ring nitrogen of His 1125. The released thiolate anion can then act as a general base to catalyse the attack of a hydroxyl bearing nucleophile on the reactive acyl-imidazole intermediate. Species that have only one functional isotype of C4 (e.g., mouse) have the C4B isotype equivalent. C4 in the mouse has an isotypic region sequence of [1116]**PCPVIH**[1121], which is a hybrid between the isotypic sequence motifs of human C4A and C4B. However, it has the human C4B-like histidine at the C-terminal position. Thus, it shows preferential transacylation onto hydroxyl groups.[18] Regardless of the isotype, a relatively minor portion (≤10%) of nascently activated C4b fragment transacylates onto the activating surface, the remainder becoming thioester-hydrolysed and is essentially a waste product.[4]

Target-bound C4b is a ligand for CR1, also known as the immune adherence receptor. It has been suggested that C4b of the A isotype is a better ligand for this receptor than is C4b of the B isotype.[19,20] However, this remains controversial. All previously discussed functional interaction sites of C4b are lost when a C4b–cofactor complex (i.e., C4b-C4BP, C4b-MCP or C4b-CR1) is cleaved by factor I C-terminal to residues 956 and 1336 to yield fragments C4c and C4d (Fig. 17.1). A study has shown that fluid-phase C4d, alone or as part of C4, is a ligand for cells (e.g., monocytes) expressing immunoglobulin-like transcript 4 (ILT4). Moreover, binding of C4d is followed by endocytosis, suggesting that ILT4 may act as a scavenger receptor for 'waste' complement split products.[21]

TISSUE DISTRIBUTION

Serum protein: ~0.6 mg/mL[22]
Cerebrospinal fluid: ~3.3 μg/mL[23]
Tears: ~5.6 μg/mL[24]

Primary site of synthesis: liver[25]
Secondary sites: monocytes, macrophages, mammary gland, lung, spleen, kidney, brain, testis, skin fibroblasts and intestinal epithelial cells.[25–28]

REGULATION OF EXPRESSION

C4 is an acute-phase protein that is upregulated by IFN-γ in hepatocytes, monocytes and intestinal epithelial cells.[26,29] IFN-γ increases the stability of the C4 mRNA.[30] Upregulation of C4 synthesis in intestinal epithelial cells is also mediated by IL-6.[26] No other cytokines have been reported to affect C4 expression on their own, although the combination of TNF-α with IFN-γ has been reported to act synergistically in upregulating C4 gene expression.[27] In some, but not all tissues, lipopolysaccharide downregulates the synthesis of C4 and counteracts the upregulation of synthesis produced by IFN-γ.[27,29]

PROTEIN SEQUENCE[31]

```
MRLLWGLIWA SSFFTLSLQK PRLLLFSPSV VHLGVPLSVG VQLQDVPRGQ   50
VVKGSVFLRN PSRNNVPCSP KVDFTLSSER DFALLSLQVP LKDAKSCGLH  100
QLLRGPEVQL VAHSPWLKDS LSRTTNIQGI NLLFSSRRGH LFLQTDQPIY  150
NPGQRVRYRV FALDQKMRPS TDTITVMVEN SHGLRVRKKE VYMPSSIFQD  200
DFVIPDISEP GTWKISARFS DGLESNSSTQ FEVKKYVLPN FEVKITPGKP  250
YILTVPGHLD EMQLDIQARY IYGKPVQGVA YVRFGLLDED GKKTFFRGLE  300
SQTKLVNGQS HISLSKAEFQ DALEKLNMGI TDLQGLRLYV AAAIIEYPGG  350
EMEEAELTSW YFVSSPFSLD LSKTKRHLVP GAPFLLQALV REMSGSPASG  400
IPVKVSATVS SPGSVPEVQD IQQNTDGSGQ VSIPIIIPQT ISELQLSVSA  450
GSPHPAIARL TVAAPPSGGP GFLSIERPDS RPPRVGDTLN LNLRAVGSGA  500
TFSHYYYMIL SRGQIVFMNR EPKRTLTSVS VFVDHHLAPS FYFVAFYYHG  550
DHPVANSLRV DVQAGACEGK LELSVDGAKQ YRNGESVKLH LETDSLALVA  600
LGALDTALYA AGSKSHKPLN MGKVFEAMNS YDLGCGPGGG DSALQVFQAA  650
GLAFSDGDQW TLSRKRLSCP KEKTTRKKRN VNFQKAINEK LGQYASPTAK  700
RCCQDGVTRL PMMRSCEQRA ARVQQLDCRE PFLSCCQFAE SLRKKSRDKG  750
QAGLQRALEI LQEEDLIDED DIPVRSFFPE NWLWRVETVD RFQILTLWLP  800
DSLTTWEIHG LSLSKTKGLC VATPVQLRVF REFHLHLRLP MSVRRFEQLE  850
LRPVLYNYLD KNLTVSVHVS PVEGLCLAGG GGLAQQVLVP AGSARPVAFS  900
VVPTAAAAVS LKVVARGSFE FPVGDAVSKV LQIEKEGAIH REELVYELNP  950
LDHRGRTLEI PGNSDPNMIP DGDFNSYVRV TASDPLDTLG SEGALSPGGV 1000
ASLLRLPRGC GEQTMIYLAP TLAASRYLDK TEQWSTLPPE TKDHAVDLIQ 1050
KGYMRIQQFR KADGSYAAWL SRDSSTWLTA FVLKVLSLAQ EQVGGSPEKL 1100
QETSNWLLSQ QQADGSFQDP CPVLDRSMQG GLVGNDETVA LTAFVTIALH 1150
HGLAVFQDEG AEPLKQRVEA SISKANSFLG EKASAGLLGA HAAAITAYAL 1200
TLTKAPVDLL GVAHNNLMAM AQETGDNLYW GSVTGSQSNA VSPTPAPRNP 1250
SDPMPQAPAL WIETTAYALL HLLLHEGKAE MADQAAAWLT RQGSFQGGFR 1300
STQDTVIALD ALSAYWIASH TTEERGLNVT LSSTGRNGFK SHALQLNNRQ 1350
IRGLEEELQF SLGSKINVKV GGNSKGTLKV LRTYNVLDMK NTTCQDLQIE 1400
VTVKGHVEYT MEANEDYEDY EYDELPAKDD PDAPLQPVTP LQLFEGRRNR 1450
RRREAPKVVE EQESRVHYTV CIWRNGKVGL SGMAIADVTL LSGFHALRAD 1500
LEKLTSLSDR YVSHFETEGP HVLLYFDSVP TSRECVGFEA VQEVPVGLVQ 1550
PASATLYDYY NPERRCSVFY GAPSKSRLLA TLCSAEVCQC AEGKCPRQRR 1600
ALERGLQDED GYRMKFACYY PRVEYGFQVK VLREDSRAAF RLFETKITQV 1650
LHFTKDVKAA ANQMRNFLVR ASCRLRLEPG KEYLIMGLDG ATYDLEGHPQ 1700
YLLDSNSWIE EMPSERLCRS TRQRAACAQL NDFLQEYGTQ GCQV
```

The leader sequence is underlined, as are the cleavage sites (<u>RKKR</u> and <u>RRRR</u>) between the β-α and α-γ chains, respectively. *N*-linked glycosylation sites (all utilised) are indicated by **N** and residues C1010 and Q1013, whose side chains form the thioester bond, are double underlined. The sulphated tyrosines are indicated by **Y**, the site of *O*-glycosylation by **T** and the site of cysteinylation in the C4A isotype by **C**. The site of cleavage by activated C1s C-terminal to R756 is denoted by **R**, and the sites of cleavage by factor I C-terminal to R956 and R1336 are each denoted by *R*.

PROTEIN MODULES

1. Residue boundaries of chains and physiologic fragments.

1–19	Leader peptide	exon 1
20–675	β-chain	exons 1–16
680–1449	α-chain (secreted form)	exons 16–33
1454–1744	γ-chain	exons 33–41
680–756	C4a anaphylatoxin	exons 16–17
757–956	α3-fragment of C4c	exons 18–23
957–1336	C4d fragment	exons 23–30
1337–1427	α4-fragment of C4c	exons 30–33

2. Residue boundaries of structurally defined protein modules.

20–137	MG1	exons 1–3
138–240	MG2	exons 3–7
241–364	MG3	exons 7–10
365–465	MG4	exons 10–12
466–563	MG5	exons 12–13
564–605	MG6β	exons 13–14
606–675	Link	exons 14–16
680–756	ANA	exons 16–17
757–778	α'NT	exon 18
779–832	MG6α	exons 18–20
833–935	MG7	exons 20–22
936–983	CUB$_g$	exons 22–24
984–1323	TED	exons 24–30
1324–1388	CUB$_f$	exons 30–32
1389–1427	MG8α	exons 32–33 (plasma form of α-chain)
1454–1573	MG8γ	exons 33–37
1574–1594	Anchor	exons 37–38
1595–1744	C345C	exons 38–41

The isotype-specific residues, in addition to four other residues that generally, although not absolutely, segregate as a set with either C4A or C4B produce the serological antigenic determinants of human C4 known as Chido (Ch1,2,3,4,5,6) and Rodgers (Rg1,2). Shown in Fig. 17.4 are the residues that have been deduced to form the various Ch and Rg antigenic determinants.[32]

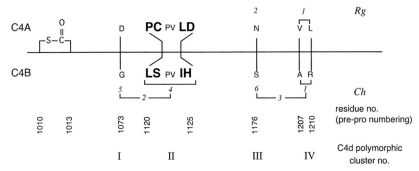

FIGURE 17.4 Isotypic residues (*boldface*) and Ch/Rg antigenic determinants of human C4A and C4B within the C4d fragment.

Rg1, Rg2, Ch1, Ch4, Ch5 and Ch6 are linear sequence determinants, whereas Ch2 and Ch3 are conformational determinants.

CHROMOSOMAL LOCATION[33]

Human: locus 6p21.3 within the class III region of the major histocompatibility complex (MHC) on chromosome 6.
    Telomere...HLA-B..........C2..Bf...**C4A..**CYP21A..**C4B**..CYP21B..........HLA-DR.
Mouse: chromosome 17, locus position 18.29 cM within the class III region of the MHC.
    Telomere...H2-D............C2..Bf...**Slp..**CYP21A..**C4**..CYP21B............H2-Ea.
    Slp is 94% identical to mouse C4. Its function in complement is questionable, because it is not readily cleaved by activated C1s.
Rat: locus 20p12 within the class III region of the MHC on chromosome 20.
    Telomere...RT1-C/E.........C2..Bf...**C4**..CYP21....................RT1-B/D.......RT1-A.

cDNA SEQUENCE[31,34]

```
AGAAGGTAGCAGACAGACAGACGGATCTAACCTCTCTTGGATCCTCCAGCCATGAGGCTG     60
CTCTGGGCATCCAGCTTCTTCACCTTATCTCTGCAGAAGCCCAGGTTGCTCTTGTTCTCT    120
CCTTCTGTGGTTCATCTGGGGGTCCCCCTATCGGTGGGGGTGCAGCTCCAGGATGTGCCC    180
CGAGGACAGGTAGTGAAAGGATCAGTGTTCCTGAGAAACCCATCTCGTAATAATGTCCCC    240
TGCTCCCCAAAGGTGGACTTCACCCTTAGCTCAGAAAGAGACTTCGCACTCCTCAGTCTC    300
CAGGTGCCCTTGAAAGATGCGAAGAGCTGTGTGGCCTCCATCAACTCCTCAGAGGCCCTGAG    360
GTCCAGCTGGTGGCCCATTCGCCATGGCTAAAGGACTCTCTGTCCAGAACGACAAACATC    420
CAGGGTATCAACCTGCTCTTCTCCTCTCGCCGGGGGCACCTCTTTTTGCAGACGGACCAG    480
CCCATTTACAACCCTGGCCAGCGGGTTCGGTACCGGGTCTTTGCTCTGGATCAGAAGATG    540
CGCCCGAGCACTGACACCATCACAGTCATGGTGGAGAACTCTCACGGCCTCCGCGTGCGG    600
AAGAAGGAGGTGTACATGCCCTCGTCCATCTTCCAGGATGACTTTGTGATCCCAGACATC    660
TCAGAGCCAGGGACCTGGAAGATCTCAGCCCGATTCTCAGATGGCCTGGAATCCAACAGC    720
AGCACCCAGTTTGAGGTGAAGAAATATGTCCTTCCCAACTTTGAGGTGAAGATCACCCCT    780
GGAAAGCCCTACATCCTGACGGTGCCAGGCCATCTTGATGAAATGCAGTTAGACATCCAG    840
```

```
GCCAGGTACATCTATGGGAAGCCAGTGCAGGGGGTGGCATATGTGCGCTTTGGGCTCCTA    900
GATGAGGATGGTAAGAAGACTTTCTTTCGGGGGCTGGAGAGTCAGACCAAGCTGGTGAAT    960
GGACAGAGCCACATTTCCCTCTCAAAGGCAGAGTTCCAGGACGCCCTGGAGAAGCTGAAT   1020
ATGGGCATTACTGACCTCCAGGGGCTGCGCCTCTACGTTGCTGCAGCCATCATTGAGTCT   1080
CCAGGTGGGGAGATGGAGGAGGCAGAGCTCACATCCTGGTATTTTGTGTCATCTCCCTTC   1140
TCCTTGGATCTTAGCAAGACCAAGCGACACCTTGTGCCTGGGGCCCCCTTCCTGCTGCAG   1200
GCCTTGGTCCGTGAGATGTCAGGCTCCCCAGCTTCTGGCATTCCTGTCAAAGTTTCTGCC   1260
ACGGTGTCTTCTCCTGGGTCTGTTCCTGAAGCCCAGGACATTCAGCAAAACACAGACGGG   1320
AGCGGCCAAGTCAGCATTCAATAATTATCCCTCAGACCATCTCAGAGCTGCAGCTCTCA    1380
GTATCTGCAGGCTCCCCACATCCAGCGATAGCCAGGCTCACTGTGGCAGCCCCACCTTCA   1440
GGAGGCCCCGGGTTTCTGTCTATTGAGCGGCCGGATTCTCGACCTCCTCGTGTTGGGGAC   1500
ACTCTGAACCTGAACTTGCGAGCCGTGGGCAGTGGGGCCACCTTTTCTCATTACTACTAC   1560
ATGATCCTATCCCGAGGGCAGATCGTGTTCATGAATCGAGAGCCCAAGAGGACCCTGACC   1620
TCGGTCTCGGTGTTTGTGGACCATCACCTGGCACCCTCCTTCTACTTTGTGGCCTTCTAC   1680
TACCATGGAGACCACCCAGTGGCCAACTCCCTGCGAGTGGATGTCCAGGCTGGGGCCTGC   1740
GAGGGCAAGCTGGAGCTCAGCGTGGACGGTGCCAAGCAGTACCGGAACGGGGAGTCCGTG   1800
AAGCTCCACTTAGAAACCGACTCCCTAGCCCTGGTGGCGCTGGGGAGCCTTGGACACAGCT   1860
CTGTATGCTGCAGGCAGCAAGTCCCACAAGCCCCTCAACATGGGCAAGGTCTTTGAAGCT   1920
ATGAACAGCTATGACCTCGGCTGTGGTCCTGGGGGTGGGGACAGTGCCCTTCAGGTGTTC   1980
CAGGCAGCGGGCCTGGCCTTTTCTGATGGAGACCAGTGGACCTTATCCAGAAAGAGACTA   2040
AGCTGTCCCAAGGAGAAGACAACCCGGAAAAAGAGAAACGTGAACTTCCAAAAGGCGATT   2100
AATGAGAAATTGGGTCAGTATGCTTCCCCGACAGCCAAGCGCTGCTGCCAGGATGGGGTG   2160
ACACGTCTGCCCATGATGCGTTCCTGCGAGCAGCGGGCAGCCCGCGTGCAGCAGCCGGAC   2220
TGCCGGGAGCCCTTCCTGTCCTGCTGCCAATTTGCTGAGAGTCTGCGCAAGAAGAGCAGG   2280
GACAAGGGCCAGGCGGGCCTCCAACGAGCCCTGGAGATCCTGCAGGAGGAGGACCTGATT   2340
GATGAGGATGACATTCCCGTGCGCAGCTTCTTCCCAGAGAACTGGCTCTGAGAGTGGAA   2400
ACAGTGGACCGCTTTCAAATATTGACACTGTGGCTCCCCGACTCTCTGACCACGTGGGAG   2460
ATCCATGGCCTGAGCCTGTCCAAAACCAAAGGCCTATGTGTGGCCACCCCAGTCCAGCTC   2520
CGGGTGTTCCGCGAGTTCCACCTGCACCTCCGCCTGCCCATGTCTGTCCGCCGCTTTGAG   2580
CAGCTGGAGCTGCGGGCCTGTCCTCTATAACTACCTGGATAAAAACCTGACTGTGAGCGTC   2640
CACGTGTCCCCAGTGGAGGGGCTGTGCCTGGCTGGGGGCGGAGGGCTGGCCCAGCAGGTG   2700
CTGGTGCCTGCGGGCTCTGCCCGGCCTGTTGCCTTCTCTGTGGTGCCCACGGCAGCCGCC   2760
GCTGTGTCTCTGAAGGTGGTGGCTCGAGGGTCCTTCGAATTCCCTGTGGGAGATGCGGTG   2820
TCCAAGGTTCTGCAGATTGAGAAGGAAGGGCCATCCATAGAGGAGAGCTGGTCTATGAA   2880
CTCAACCCCTTGGACCACCGAGGCCGGACCTTGGAAATACCTGGCAACTCTGATCCCAAT   2940
ATGATCCCTGATGGGGACTTTAACAGCTACGTCAGGGTTACAGCCTCAGATCCATTGGAC   3000
ACTTTAGGCTCTGAGGGGGCCTTGTCACCAGGAGGCGTGGCCTCCCTCTTGAGGCTTCCT   3060
CGAGGCTGTGGGGAGCAAACCATGATCTACTTGGCTCCGACACTGGCTGCTTCCCGCTAC   3120
CTGGACAAGACAGAGCAGTGGAGCACACTGCCTCCCGAGACCAAGGACCACGCCGTGGAT   3180
CTGATCCAGAAAGGCTACATGCGGATCCAGCAGTTTCGGAAGGCGGATGGTTCCTATGCG   3240
GCTTGGTTGTCACGGGACAGCAGCACCTGGCTCACAGCCTTTGTGTTGAAGGTCCTGAGT   3300
TTGGCCCAGGAGCAGGTAGGAGGCTCGCCTGAGAAACTGCAGGAGACATCTAACTGGCTT   3360
CTGTCCCAGCAGCAGGCTGACGGCTCGTTCCAGGACCCCTGTCCAGTGTTAGACAGGAGC   3420
ATGCAGGGGGGGTTTGGTGGGCAATGATGAGACTGTGGCACTCACAGCCTTTGTGACCATC   3480
GCCCTTCATCATGGGCTGGCCGTCTTCCAGGATGAGGGTGCAGAGCCATTGAAGCAGAGA   3540
GTGGAAGCCTCCATCTCAAAGGCAAACTCATTTTTGGGGGAGAAAGCAAGTGCTGGGCTC   3600
CTGGGTGCCCACGCAGCTGCCATCACGGCCTATGCCCTGTCACTGACCAAGGCGCCTGTG   3660
GACCTGCTCGGTGTTGCCCACAACAACCTCATGGCAATGGCCCAGGAGACTGGAGATAAC   3720
CTGTACTGGGGCTCAGTCACTGGTTCTCAGAGCAATGCCGTGTCGCCCACCCCGGCTCCT   3780
CGCAACCCATCCGACCCCATGCCCCAGGCCCCAGCCCTGTGGATTGAAACCACAGCCTAC   3840
GCCCTGCTGCACCTCCTGCTTCACGAGGGCAAAGCAGAGATGGCAGACCAGGCTTCGGCC   3900
TGGCTCACCCGTCAGGGCAGCTTCCAAGGGGGATTCCGCAGTACCCAAGACACGGTGATT   3960
GCCCTGGATGCCCTGTCTGCCTACTGGATTGCCTCCCACACCACTGAGGAGAGGGGTCTC   4020
AATGTGACTCTCAGCTCCACAGGCCGGAATGGGTTCAAGTCCCACGCGCTGCAGCTGAAC   4080
AACCGCCAGATTCGCGGCCTGGAGGAGGAGCTGCAGTTTTTCTTGGGCAGCAAGATCAAT   4140
GTGAAGGTGGGAGGAAACAGCAAAGGAACCCTGAAGGTCCTTCGTACCTACAATGTCCTG   4200
```

cDNA SEQUENCE—*Continued*

```
GACATGAAGAACACGACCTGCCAGGACCTACAGATAGAAGTGACAGTCAAAGGCCACGTC    4260
GAGTACACGATGGAAGCAAACGAGGACTATGAGGACTATGAGTACGATGAGCTTCCAGCC    4320
AAGGATGACCCAGATGCCCCTCTGCAGCCCGTGACACCCCTGCAGCTGTTTGAGGGTCGG    4380
AGGAACCGCCGCAGGAGGGAGGCGCCCAAGGTGGTGGAGGAGCAGGAGTCCAGGGTGCAC    4440
TACACCGTGTGCATCTGGCCGGAACGGCAAGGTGGGGCTGTCTGGCATGGCCATCGCGGAC    4500
GTCACCCTCCTGAGTGGATTCCACGCCCTGCGTGCTGACCTGGAGAAGCTGACCTCCCTC    4560
TCTGACCGTTACGTGAGTCACTTTGAGACCGAGGGGCCCCACGTCCTGCTGTATTTTGAC    4620
TCGGTCCCCACCTCCCGGGAGTGCGTGGGCTTTGAGGCTGTGCAGGAAGTGCCGGTGGGG    4680
CTGGTGCAGCCGGCCAGCGCAACCCTGTACGACTACTACAACCCCGAGCGCAGATGTTCT    4740
GTGTTTTACGGGGCACCAAGTAAGAGCAGACAGACTCTTGGCCACCTTGTGTTCTGCTGAAGTC    4800
TGCCAGTGTGCTGAGGGGAAGTGCCCTCGCCAGCGTCGCGCCCTGGAGCGGGGTCTGCAG    4860
GACGAGGATGGCTACAGGATGAAGTTTGCCTGCTACTACCCCCGTGTGGAGTACGGCTTC    4920
CAGGTTAAGGTTCTCCGAGAAGACAGCAGAGCTGCTTTCCGCCTCTTTGAGACCAAGATC    4980
ACCCAAGTCCTGCACTTCACCAAGGATGTCAAGGCCGCTGCTAATCAGATGCGCAACTTC    5040
CTGGTTCGAGCCTCCTGCCGCCTTCGCTTGGAACCTGGGAAAGAATATTTGATCATGGGT    5100
CTGGATGGGGCCACCTATGACCTCGAGGGACACCCCCAGTACCTGCTGGACTCGAATAGC    5160
TGGATCGAGGAGATGCCCTCTGAACGCCTGTGCCGGAGCACCCGCCAGCGGGCAGCCTGT    5220
GCCCAGCTCAACGACTTCCTCCAGGAGTATGGCACTCAGGGGTGCCAGGTGTGAGGGCTG    5280
CCCTCCCACCTCCGCTGGGAGGAACCTGAACCTGGGAACCATGAAGCTGGAAGCACTGCT    5340
GTGTCCGCTTTCATGAACACAGCCTGGGACCAGGGCATATTAAAGGCTTTTGGCAGCAAA    5400
GTGTCAGTGTTGGC
```

The first five nucleotides in each exon are underlined to indicate the intron/exon boundaries. The methionine initiation codon (<u>ATG</u>), the termination codon (<u>TGA</u>) and the polyadenylation signal (<u>ATTAAA</u>) are double-underlined.

GENOMIC STRUCTURE

Human *C4A* and *C4B* are most commonly tandem loci that are separated by ~10 kb. They are each located within a four gene cluster denoted as RCCX, where R represents the ser/thr nuclear protein kinase RP, the first C represents C4, the second C represents CYP21 steroid hydroxylase and X represents the extracellular matrix protein tenascin-X (TNX). Duplications are of the entire RCCX module, and whereas the duplicated *C4* genes are usually functional, the duplicated *CYP21* genes are often pseudogenes. The *C4A* and *C4B* loci can be either 14.6 or 21 kb in length, with a propensity for the longer gene at *C4A* loci. The size difference is solely due to a larger intron 9 in the long *C4* gene, which in turn is due to the insertion of an ~6.4 kb retroposon element within this intron.[35] In addition to length polymorphism, there is also variability in gene copy number with the two locus models of one *C4A* and one *C4B* gene on each chromosome accounting for only ~55% of the Caucasian population.[36] The number of *C4* genes on each chromosome generally varies between 1 and 3 (up to 4 copies on rare occasions), and in many permutations of *C4A* and *C4B* genes, including the complete absence of either *C4A* or *C4B*. There is also variability in the number of long genes and short genes at *C4* loci. Gene dosage in the diploid genome therefore generally varies between 2 and 6 with 52% of the Caucasian population having a gene dosage of 4. The *C4* gene dosage frequencies for 2, 3, 5 and 6 in the same population are 2%, 25%, 17% and 3%,

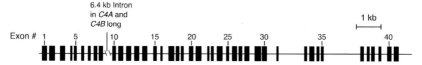

FIGURE 17.5   Exon/intron organisation of the human C4 gene.

respectively.[36] C4 isotypic levels in plasma are, for the most part, determined by gene dosage levels: although there is also some indication that short genes may transcribe faster than long genes.[36] Both human genes, as well as the mouse C4 gene, consist of 41 exons. The mouse gene does not contain the retroposon element and therefore has a size (~15.8 kb) more closely corresponding to the short human gene. Fig. 17.5 depicts the exon/intron organisation within the human C4 gene.[37,38]

ACCESSION NUMBERS

Species	cDNA	Gene
Human C4A	NM_007293.2	NG_011638.1, gene ID 720
Human C4B	NM_001002029.3	NG_011639.1, gene ID 721
Mouse C4 (C57BL/6)	NM_009780.2	NC_000083.6, gene ID 12268
Mouse C4 (B10.WR)	M11729.1	M17440.1
Rat (Norway)	NM_031504.3	NC_005119.4, gene ID 24233

DEFICIENCY

Complete C4 deficiency in humans is extremely rare (28 cases reported[39]), but almost always results in the individual developing systemic lupus erythematosus (SLE) or SLE-like conditions. Increased susceptibility to infections and diseases of the kidney, such as glomerulonephritis, are also commonly seen.[39,40] Partial or complete deficiency of either C4A or C4B occur much more commonly and have a combined frequency of >30% in the normal Caucasian population.[36] Null alleles for *C4A* or *C4B* have three origins: (1) the presence of a monomodular RCCX haplotype with the expression of a single *C4A* or *C4B* gene; (2) the presence of a bimodular RCCX locus with homoexpression of identical C4 isotypes (indicative of a gene conversion process); and (3) the presence of a pseudogene caused by a point mutation. Frequently occurring at a *C4A* locus, the most common origin for a *C4* pseudogene is a two base pair insertion (TC) at codon 1232 of exon 29, thus leading to a frame-shift mutation within exon 30 and the non-expression of the C4 protein.[41,42]

Genetic backgrounds influence the SLE risk associated with partial C4 deficiency states, and partial deficiency states of C4 are one risk factor contributing to the complex etiology of SLE with about half of SLE patients showing no defects in either C4A or C4B protein levels.[39,40,43] Complete C4B deficiency appears also to be associated with infections by encapsulated bacteria in an ethnic-dependent fashion.[44]

Interestingly, whereas high expression levels of C4A appear to be protective with respect to SLE,[43] a report suggests that high expression of C4A in the brain is a significant risk factor for schizophrenia.[45]

POLYMORPHIC VARIANTS

In addition to the non-expressed alleles at both *C4* loci in humans, there are at least 13 allelic polymorphisms of C4A and 16 of C4B.[46] Within an isotype, allotypes are given numbers in increasing order of their anodal migration at alkaline pH. The two most common allotypes in all ethnic groups are C4A3 and C4B1.[46] The isotype-specific polymorphic residues have been identified within the C4d region (see Fig. 17.5) as have some allotype-specific residues in all three chains.[38] In most cases, allotypic amino acid interchanges are without obvious functional consequence, but an exception occurs in the quite rare allotype C4A6. The β-chain residue Arg 477 is replaced by Trp with the result that the molecule loses the ability to bind C5.[47,48] Another rare polymorphism occurs at the neighbouring residue, P478L, and this substitution also results in haemolytically inactive C4B1 allotype.[49] Given the location of the residue substitution, this result is consistent with the allotype having a defect in C5 binding.

MUTANT ANIMALS

Naturally Occurring C4 Deficiency

Guinea pigs that were naturally deficient in C4 were first characterised in the early 1970s and were instrumental in making the case for the existence of the alternative pathway, a controversial topic at that time.[50,51] The diminished antibody response in the C4-deficient guinea pig studies could be rescued not only by the coadministration of a small amount of normal guinea pig serum at the time of primary immunisation, but also by coadministration of purified human C4A and not C4B.[52] Serum from C4-deficient guinea pigs is available from several commercial sources and is used to measure C4 haemolytic activity.[53]

C4 KNOCKOUT MOUSE PHENOTYPES

C4 knockout mice were generated by targeted replacement via homologous recombination of exons 23–29 with a PKG-Neo cassette in 129S4/SvJae embryonic stem (ES) cells.[54] Heterozygous *C4* null ES cells were microinjected into C57BL/6 embryos. Male chimeric mice were first bred with C57BL/6 females, and then brother-sister mating of heterozygous nulls yielded the homozygous *C4*−/− founder mice. These mice are healthy and fertile in routine animal husbandry, but show greater susceptibility to infection by group B *Streptococci*,[55] a type of bacteria that do not directly activate the alternative pathway. The mice were subsequently extensively backcrossed onto the C57BL/6J background and are available from Jackson Laboratories (C4B[tm1Crr]). On certain genetic backgrounds, homozygous C4 knockout mice are predisposed to developing a lupus-like syndrome. These mice spontaneously developed anti-dsDNA antibodies and variable levels of renal histopathology, hallmarks of human SLE, whereas the equivalent C4-sufficient strains of mice did not.[56,57] Studies with C4[−/−] mice also demonstrate that C4 is required to clear self-antigens and that prolonged stimulation by such self-antigens may result in the breaking of B cell tolerance to the self-antigens.[58–60] C4 knockout mice have also been used to confirm the role of the classical complement pathway in the antibody response to T-dependent antigens[54] with a phenotype of low antibody titer, failure to class switch, as well as reduced size and number of germinal centres in the lymph nodes of the immunised mice.

REFERENCES

1. Dodds AW, Law SK. The complement component C4 of mammals. *Biochem J* 1990;**265**:495–502.
2. Isenman DE, Kells DI. Conformational and functional changes in the fourth component of human complement produced by nucleophilic modification and by proteolysis with C1s. *Biochemistry* 1982;**21**:1109–17.
3. Hortin G, Chan AC, Fok KF, Strauss AW, Atkinson JP. Sequence analysis of the COOH terminus of the alpha-chain of the fourth component of human complement. Identification of the site of its extracellular cleavage. *J Biol Chem* 1986;**261**:9065–9.
4. Law SK, Dodds AW. The internal thioester and the covalent binding properties of the complement proteins C3 and C4. *Protein Sci* 1997;**6**:263–74.
5. Chan AC, Atkinson JP. Oligosaccharide structure of human C4. *J Immunol* 1985;**134**:1790–8.
6. Hortin G, Sims H, Strauss AW. Identification of the site of sulfation of the fourth component of human complement. *J Biol Chem* 1986;**261**:1786–93.
7. van den Elsen JM, Martin A, Wong V, Clemenza L, Rose DR, Isenman DE. X-ray crystal structure of the C4d fragment of human complement component C4. *J Mol Biol* 2002;**322**:1103–15.
8. Kidmose RT, Laursen NS, Dobo J, et al. Structural basis for activation of the complement system by component C4 cleavage. *Proc Natl Acad Sci USA* 2012;**109**:15425–30.
9. Mortensen S, Kidmose RT, Petersen SV, Szilagyi A, Prohaszka Z, Andersen GR. Structural basis for the function of complement component C4 within the classical and lectin pathways of complement. *J Immunol* 2015;**194**:5488–96.
10. Pan Q, Ebanks RO, Isenman DE. Two clusters of acidic amino acids near the NH$_2$ terminus of complement component C4 alpha'-chain are important for C2 binding. *J Immunol* 2000;**165**:2518–27.
11. Janssen BJ, Christodoulidou A, McCarthy A, Lambris JD, Gros P. Structure of C3b reveals conformational changes that underlie complement activity. *Nature* 2006;**444**:213–6.
12. Kim YU, Carroll MC, Isenman DE, et al. Covalent binding of C3b to C4b within the classical complement pathway C5 convertase. Determination of amino acid residues involved in ester linkage formation. *J Biol Chem* 1992;**267**:4171–6.
13. Takata Y, Kinoshita T, Kozono H, et al. Covalent association of C3b with C4b within C5 convertase of the classical complement pathway. *J Exp Med* 1987;**165**:1494–507.

14. Hugli TE, Kawahara MS, Unson CG, Molinar-Rode R, Erickson BW. The active site of human C4a anaphylatoxin. *Mol Immunol* 1983;**20**:637–45.

15. Barnum SR. C4a: an anaphylatoxin in name only. *J Innate Immun* 2015;**7**:333–9.

16. Klos A, Wende E, Wareham KJ, Monk PN. International union of basic and Clinical Pharmacology. [corrected]. Lxxxvii. Complement peptide C5a, C4a, and C3a receptors. *Pharmacol Rev* 2013;**65**:500–43.

17. Yu CY, Belt KT, Giles CM, Campbell RD, Porter RR. Structural basis of the polymorphism of human complement components C4A and C4B: gene size, reactivity and antigenicity. *EMBO J* 1986;**5**:2873–81.

18. Dodds AW, Law SK. Structural basis of the binding specificity of the thioester-containing proteins, C4, C3 and alpha-2-macroglobulin. *Complement* 1988;**5**:89–97.

19. Clemenza L, Isenman DE. The C4A and C4B isotypic forms of human complement fragment C4b have the same intrinsic affinity for complement receptor 1 (CR1/CD35). *J Immunol* 2004;**172**:1670–80.

20. Reilly BD, Mold C. Quantitative analysis of C4Ab and C4Bb binding to the C3b/C4b receptor (CR1, CD35). *Clin Exp Immunol* 1997;**110**:310–6.

21. Hofer J, Forster F, Isenman DE, et al. Ig-like transcript 4 as a cellular receptor for soluble complement fragment C4d. *FASEB J* 2016;**30**:1492–503.

22. Holers VM. Complement and its receptors: new insights into human disease. *Annu Rev Immunol* 2014;**32**:433–59.

23. Cova JL, Propp RP, Barron KD. Quantitative relationships of the fourth complement component in human cerebrospinal fluid. *J Lab Clin Med* 1977;**89**:615–21.

24. Willcox MD, Morris CA, Thakur A, Sack RA, Wickson J, Boey W. Complement and complement regulatory proteins in human tears. *Invest Ophthalmol Vis Sci* 1997;**38**:1–8.

25. Cox BJ, Robins DM. Tissue-specific variation in C4 and Slp gene regulation. *Nucleic Acids Res* 1988;**16**:6857–70.

26. Andoh A, Fujiyama Y, Bamba T, Hosoda S. Differential cytokine regulation of complement C3, C4, and factor B synthesis in human intestinal epithelial cell line, Caco-2. *J Immunol* 1993;**151**:4239–47.

27. Kulics J, Circolo A, Strunk RC, Colten HR. Regulation of synthesis of complement protein C4 in human fibroblasts: cell- and gene-specific effects of cytokines and lipopolysaccharide. *Immunology* 1994;**82**:509–15.

28. Whaley K. Biosynthesis of the complement components and the regulatory proteins of the alternative complement pathway by human peripheral blood monocytes. *J Exp Med* 1980;**151**:501–16.

29. Kulics J, Colten HR, Perlmutter DH. Counterregulatory effects of interferon-gamma and endotoxin on expression of the human C4 genes. *J Clin Invest* 1990;**85**:943–9.

30. Mitchell TJ, Naughton M, Norsworthy P, Davies KA, Walport MJ, Morley BJ. IFN-gamma up-regulates expression of the complement components C3 and C4 by stabilization of mRNA. *J Immunol* 1996;**156**:4429–34.

31. Belt KT, Carroll MC, Porter RR. The structural basis of the multiple forms of human complement component C4. *Cell* 1984;**36**:907–14.

32. Yu CY, Campbell RD, Porter RR. A structural model for the location of the Rodgers and the Chido antigenic determinants and their correlation with the human complement component C4A/C4B isotypes. *Immunogenetics* 1988;**27**:399–405.

33. Campbell RD, Law SK, Reid KB, Sim RB. Structure, organization, and regulation of the complement genes. *Annu Rev Immunol* 1988;**6**:161–95.

34. Belt KT, Yu CY, Carroll MC, Porter RR. Polymorphism of human complement component C4. *Immunogenetics* 1985;**21**:173–80.

35. Dangel AW, Mendoza AR, Baker BJ, et al. The dichotomous size variation of human complement C4 genes is mediated by a novel family of endogenous retroviruses, which also establishes species-specific genomic patterns among Old World primates. *Immunogenetics* 1994;**40**:425–36.

36. Blanchong CA, Chung EK, Rupert KL, et al. Genetic, structural and functional diversities of human complement components C4A and C4B and their mouse homologues, Slp and C4. *Int Immunopharmacol* 2001;**1**:365–92.

37. Ulgiati D, Townend DC, Christiansen FT, Dawkins RL, Abraham LJ. Complete sequence of the complement C4 gene from the HLA-A1, B8, C4AQ0, C4B1, DR3 haplotype. *Immunogenetics* 1996;**43**:250–2.

38. Yu CY. The complete exon-intron structure of a human complement component C4A gene. DNA sequences, polymorphism, and linkage to the 21-hydroxylase gene. *J Immunol* 1991;**146**:1057–66.

39. Lintner KE, Wu YL, Yang Y, et al. Early components of the complement classical activation pathway in human systemic autoimmune diseases. *Front Immunol* 2016;**7**:36.

40. Yang Y, Chung EK, Zhou B, et al. The intricate role of complement component C4 in human systemic lupus erythematosus. *Curr Dir Autoimmun* 2004;**7**:98–132.

41. Barba G, Rittner C, Schneider PM. Genetic basis of human complement C4A deficiency. Detection of a point mutation leading to nonexpression. *J Clin Invest* 1993;**91**:1681–6.

42. Boteva L, IMAGEN, Wu YL, et al. Determination of the loss of function complement C4 exon 29 CT insertion using a novel paralog-specific assay in healthy UK and Spanish populations. *PLoS One* 2011;**6**:e22128.

43. Yang Y, Chung EK, Wu YL, et al. Gene copy-number variation and associated polymorphisms of complement component C4 in human systemic lupus erythematosus (SLE): low copy number is a risk factor for and high copy number is a protective factor against SLE susceptibility in European Americans. *Am J Hum Genet* 2007;**80**:1037–54.

44. Bishof NA, Welch TR, Beischel LS. C4B deficiency: a risk factor for bacteremia with encapsulated organisms. *J Infect Dis* 1990;**162**:248–50.

45. Sekar A, Bialas AR, de Rivera H, et al. Schizophrenia risk from complex variation of complement component 4. *Nature* 2016;**530**:177–83.

46. Mauff G, Alper CA, Dawkins R, et al. C4 nomenclature statement (1990). *Complement Inflamm* 1990;**7**:261–8.

47. Anderson MJ, Milner CM, Cotton RG, Campbell RD. The coding sequence of the hemolytically inactive C4A6 allotype of human complement component C4 reveals that a single arginine to tryptophan substitution at beta-chain residue 458 is the likely cause of the defect. *J Immunol* 1992;**148**:2795–802.

48. Ebanks RO, Jaikaran AS, Carroll MC, Anderson MJ, Campbell RD, Isenman DE. A single arginine to tryptophan interchange at beta-chain residue 458 of human complement component C4 accounts for the defect in classical pathway C5 convertase activity of allotype C4A6. Implications for the location of a C5 binding site in C4. *J Immunol* 1992;**148**:2803–11.

49. McLean RH, Niblack G, Julian B, et al. Hemolytically inactive C4B complement allotype caused by a proline to leucine mutation in the C5-binding site. *J Biol Chem* 1994;**269**:27727–31.

50. Ellman L, Green I, Judge F, Frank MM. In vivo studies in C4-deficient Guinea pigs. *J Exp Med* 1971;**134**:162–75.

51. Root RK, Ellman L, Frank MM. Bactericidal and opsonic properties of C4-deficient Guinea pig serum. *J Immunol* 1972;**109**:477–86.
52. Finco O, Li S, Cuccia M, Rosen FS, Carroll MC. Structural differences between the two human complement C4 isotypes affect the humoral immune response. *J Exp Med* 1992;**175**:537–43.
53. Gaither TA, Alling DW, Frank MM. A new one-step method for the functional assay of the fourth component (C4) of human and Guinea pig complement. *J Immunol* 1974;**113**:574–83.
54. Fischer MB, Ma M, Goerg S, et al. Regulation of the B cell response to T-dependent antigens by classical pathway complement. *J Immunol* 1996;**157**:549–56.
55. Wessels MR, Butko P, Ma M, Warren HB, Lage AL, Carroll MC. Studies of group B streptococcal infection in mice deficient in complement component C3 or C4 demonstrate an essential role for complement in both innate and acquired immunity. *Proc Natl Acad Sci USA* 1995;**92**:11490–4.
56. Paul E, Pozdnyakova OO, Mitchell E, Carroll MC. Anti-DNA autoreactivity in C4-deficient mice. *Eur J Immunol* 2002;**32**:2672–9.
57. Chen Z, Koralov SB, Kelsoe G. Complement C4 inhibits systemic autoimmunity through a mechanism independent of complement receptors CR1 and CR2. *J Exp Med* 2000;**192**:1339–52.
58. Walport MJ. Complement. Second of two parts. *N Engl J Med* 2001;**344**:1140–4.
59. Leadbetter EA, Rifkin IR, Hohlbaum AM, Beaudette BC, Shlomchik MJ, Marshak-Rothstein A. Chromatin-IgG complexes activate B cells by dual engagement of IgM and Toll-like receptors. *Nature* 2002;**416**:603–7.
60. Chatterjee P, Agyemang AF, Alimzhanov MB, et al. Complement C4 maintains peripheral B-cell tolerance in a myeloid cell dependent manner. *Eur J Immunol* 2013;**43**:2441–50.

Chapter 18

C5

Rick A. Wetsel

The University of Texas Health Science Center at Houston, Houston, TX, United States

OTHER NAMES

Hc or MuB1 sometimes used to refer to mouse C5.

PHYSICOCHEMICAL PROPERTIES

C5 is synthesised as an intracellular single-chain precursor, pro-C5,[1] of 1676 amino acids including an 18-amino acid leader peptide and an arginine-rich linker sequence (RPRR) located between the N-terminal β-chain and the C-terminal α-chain. The signal and linker peptides are processed from the pro-molecule during secretion, yielding the mature, native two-chain structure that is held together by a single disulphide bond and noncovalent forces.

Mature Protein:		
pI[2]	4.7–5.5	
	6.0 (calculated)	
	β-chain	α-chain
Amino acids[3]	19–673	678–1676
M_r(K)		
Predicted	73.3	112.5
Observed[4]	75.0	115.0
N-linked glycosylation sites[5]		741, 911
Interchain disulphide bond		
1 β–α[6]	567	810
Intrachain disulphide bonds		
1 β-chain		634–669
12 α-chain		698–724
		699–731
		711–732
		856–883
		866–1527
		1101–1159
		1375–1505
		1405–1474
		1520–1525
		1532–1606
		1553–1676
		1654–1657

The Complement FactsBook. http://dx.doi.org/10.1016/B978-0-12-810420-0.00018-3

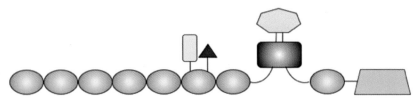

FIGURE 18.1 Protein domains for C5. For domain descriptions see Chapter 1, Table 1.1.

STRUCTURE

The C5 crystal structure has been determined at a resolution of 3.1 Å.[7,8] Similar to its homologous family member C3, C5 contains eight alpha2-macroglobulin-like domains (MG1–MG8), an anaphylatoxin domain (C5a), a C5d domain (structurally homologous to the thioester-containing domain in C3), a CUB domain (between MG8 and C5d) and a C345C C-terminal α-chain domain that is positioned at the top of the C5 molecule (Fig. 18.1).

C5b: On cleavage of C5 to its C5a and C5b fragments, the C5a peptide is released while the C5b remaining molecule undergoes conformational changes resulting in extension of the CUB domain and repositioning of the C5d domain so that the C6 binding site contained in C5d is exposed.[9]

C5a: NMR modelling of the C5a peptide suggested a drumstick or dagger shape core with four helical domains stabilised by three disulphide bonds. The C-terminal segment (residues 70–74) containing the activation domain was seen as a dynamic random coil.[10,11] In contrast, the X-ray crystal structure of C5a has indicated a three-helix core instead of the four-helix bundle indicated in the NMR structure.[12,13] In addition, the N-terminal helices occupy different positions in the NMR and crystal structures. Interestingly, the X-ray structure of C5 shows C5a as having four helical domains. The physiological significance of these structurally different C5a molecules (three- versus four-helical domains) is not known (Fig. 18.2).[14]

FUNCTION

C5 is cleaved during complement activation by Bb or C2b of the alternative or classical/lectin C5 convertase complexes, respectively. Cleavage occurs at a specific site within the C5 α-chain R751-L752 producing the C5a and C5b fragments. The larger C5b fragment via a nascent exposed binding site for C6 initiates the formation of the C5b-9 complex, which causes the cytolysis and destruction of certain bacterial and viral pathogenic organisms, as well as lysis of homologous cells in certain pathological conditions such as paroxysmal nocturnal hemoglobinuria (PNH).[15] The smaller C5a peptide of 74 amino acids is derived from the N-terminus of the α-chain (residues 678–751) and is released from the parent molecule. C5a is often referred to as an anaphylatoxin because of its associated functions in causing smooth muscle contraction, increased vascular permeability and basophil and mast cell degranulation. C5a is also a well-known potent biological chemoattractant that mediates the directed migration of

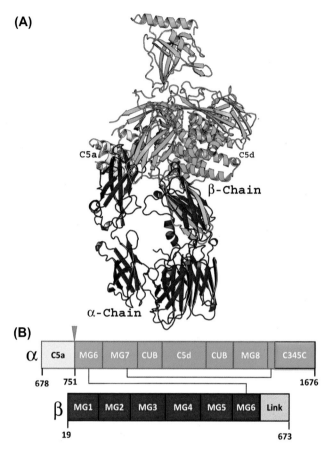

FIGURE 18.2   Crystal structure of human complement component C5. (A) Ribbon diagram of the C5 crystal structure (PDB ID: 3CU7). The N-terminal C5a domain (yellow) and thioester bond hosting C5d (green) are marked. (B) Schematic representation of the complement component C5 made of α- and β-chains and domains are coloured similar to the C5 ribbon representations. The cleavage site between C5a and C5b fragments of C5 is represented by an *inverted triangle*, coloured blue.

neutrophils, eosinophils, basophils and monocytes. It is also important in optimal phagocytosis of pathogens by phagocytic cells via upregulation of Fc receptors on macrophages and neutrophils. C5a has also been shown to have numerous other biological activities in both innate and adaptive immunity, including polarisation of T cell responses, modulation of T cell effector functions and regulation of apoptotic cellular pathways. These effects are primarily thought to occur via C5a modulation of several different cytokines and interleukins, including TNF-α, IL-1β, IL-6, IL-8, IL-12 and IFN-β. In addition to its biological effects on bone marrow-derived cells, C5a has been shown to have direct effects on certain tissue parenchymal cells, including lung epithelial cells, hepatoctyes, vascular endothelial cells, neurons and kidney tubular epithelial cells. Direct C5a-mediated effects on tissue parenchymal cells appear to be primarily important in inflammatory pathological conditions during which expression of the receptor for C5a is significantly upregulated. C5a mediates its biological activities on binding its primary specific receptor C5aR1 (CD88).[16,17] C5aR1 is a classical G-protein-coupled, seven-transmembrane receptor, which on C5a binding results in Ga_i activation and subsequent triggering of multiple intracellular signalling pathways (ERK, Akt, MAPK and P13 K) and β-arrestin recruitment for receptor desensitisation.[18] C5a also binds a second receptor, C5aR2 (also called C5L2).[19] C5aR2 is also a seven-transmembrane receptor, but it is not coupled to G-proteins, because of key amino acid changes where G-protein binding occurs. The biological function of C5aR2 remains controversial.[20] It was originally thought to be simply a decoy receptor that served to temper the inflammatory effects of C5a. However, findings have indicated that it may have both inflammatory and antiinflammatory activities and significant roles in disease pathogenesis.

DEGRADATION PATHWAY

C5 can be degraded by numerous proteolytic enzymes. However, to date no biological activity has been attributed to any degradation fragment of C5 or C5b. Unlike C3 in which C3a can be produced by cleavage of C3 with trypsin and other serine proteases, generation of C5a by proteases other than the C5 convertases is difficult.[21] The crystal structure of C5 indicates that the convertase cleavage site is more ordered in C5 compared to C3, which may explain the differences in nonspecific cleavage and activation of these two molecules.[9.]

Two plasma carboxypeptidases CPN and CPR (also called TAFI and CPU, respectively) remove the C-terminal arginine from the C5a peptide forming the C5a desArg derivative.[22] On a molar basis, human C5a desArg expresses only 1% as much anaphylatoxic activity and 1% chemotactic activity as does native C5a. CPN is constitutively expressed by the liver and is not upregulated during acute phase. CPR circulates as a zymogen that is activated primarily by thrombin during coagulation. In contrast to CPN, CPR is upregulated significantly during the acute-phase response. CPM, a membrane bound carboxypeptidase that is highly expressed on lung alveolar type I epithelial cells, may also be a regulator of C5a.

TISSUE DISTRIBUTION

Serum protein: 75 µg/mL in plasma[23]
Primary site of synthesis: Liver (hepatocytes)
Secondary sites: Lung, spleen, foetal intestine, astrocytes, monocytes, macrophages and type II alveolar epithelial cells.

REGULATION OF EXPRESSION

In the liver, C5 expression is not regulated by inflammatory stimuli to any great degree. Cytokines such as IL-6 and IL-1β that regulate the hepatic expression of numerous acute-phase proteins do not affect C5 expression. At secondary cellular sites of expression, there may be some limited degree of cytokine regulation of C5 expression. Whether small increase in C5 expression by secondary sites has any significant impact on local sites of inflammation and immunity during the host response is not known.

HUMAN PROTEIN SEQUENCE[3]

```
MGLLGILCFL IFLGKTWGQE QTYVISAPKI FRVGASENIV IQVYGYTEAF    50

DATISIKSYP DKKFSYSSGH VHLSSENKFQ NSAILTIQPK QLPGGQNPVS   100

YVYLEVVSKH FSKSKRMPIT YDNGFLFIHT DKPVYTPDQS VKVRVYSLND   150

DLKPAKRETV LTFIDPEGSE VDMVEEIDHI GIISFPDFKI PSNPRYGMWT   200

IKAKYKEDFS TTGTAYFEVK EYVLPHFSVS IEPEYNFIGY KNFKNFEITI   250

KARYFYNKVV TEADVYITFG IREDLKDDQK EMMQTAMQNT MLINGIAQVT   300

FDSETAVKEL SYYSLEDLNN KYLYIAVTVI ESTGGFSEEA EIPGIKYVLS   350

PYKLNLVATP LFLKPGIPYP IKVQVKDSLD QLVGGVPVTL NAQTIDVNQE   400

TSDLDPSKSV TRVDDGVASF VLNLPSGVTV LEFNVKTDAP DLPEENQARE   450

GYRAIAYSSL SQSYLYIDWT DNHKALLVGE HLNIIVTPKS PYIDKITHYN   500

YLILSKGKII HFGTREKFSD ASYQSINIPV TQNMVPSSRL LVYYIVTGEQ   550

TAELVSDSVW LNIEEKCGNQ LQVHLSPDAD AYSPGQTVSL NMATGMDSWV   600

ALAAVDSAVY GVQRGAKKPL ERVFQFLEKS DLGCGAGGGL NNANVFHLAG   650

LTFLTNANAD DSQENDEPCK EILRPRRTLQ KKIEEIAAKY KHSVVKKCCY   700

DGACVNNDET CEQRAARISL GPRCIKAFTE CCVVASQLRA N ISHKDMQLG   750

RLHMKTLLPV SKPEIRSYFP ESWLWEVHLV PRRKQLQFAL PDSLTTWEIQ   800

GVGISNTGIC VADTVKAKVF KDVFLEMNIP YSVVRGEQIQ LKGTVYNYRT   850

SGMQFCVKMS AVEGICTSES PVIDHQGTKS SKCVRQKVEG SSSHLVTFTV   900

LPLEIGLHNI N FSLETWFGK EILVKTLRVV PEGVKRESYS GVTLDPRGIY   950

GTISRRKEFP YRIPLDLVPK TEIKRILSVK GLLVGEILSA VLSQEGINIL  1000

THLPKGSAEA ELMSVVPVFY VFHYLETGNH WNIFHSDPLI EKQKLKKKLK  1050

EGMLSIMSYR NADYSYSVWK GGSASTWLTA FALRVLGQVN KYVEQNQNSI  1100

CNSLLWLVEN YQLDNGSFKE NSQYQPIKLQ GTLPVEAREN SLYLTAFTVI  1150

GIRKAFDICP LVKIDTALIK ADNFLLENTL PAQSTFTLAI SAYALSLGDK  1200

THPQFRSIVS ALKREALVKG NPPIYRFWKD NLQHKDSSVP NTGTARMVET  1250

TAYALLTSLN LKDINYVNPV IKWLSEEQRY GGGFYSTQDT INAIEGLTEY  1300

SLLVKQLRLS MDIDVSYKHK GALHNYKMTD KNFLGRPVEV LLNDDLIVST  1350
```

HUMAN PROTEIN SEQUENCE—*Continued*

```
GFGSGLATVH  VTTVVHKTST  SEEVCSFYLK  IDTQDIEASH  YRGYGNSDYK   1400

RIVACASYKP  SREESSSGSS  HAVMDISLPT  GISANEEDLK  ALVEGVDQLF   1450

TDYQIKDGHV  ILQLNSIPSS  DFLCVRFRIF  ELFEVGFLSP  ATFTVYEYHR   1500

PDKQCTMFYS  TSNIKIQKVC  EGAACKCVEA  DCGQMQEELD  LTISAETRKQ   1550

TACKPEIAYA  YKVSITSITV  ENVFVKYKAT  LLDIYKTGEA  VAEKDSEITF   1600

IKKVTCTNAE  LVKGRQYLIM  GKEALQIKYN  FSFRYIYPLD  SLTWIEYWPR   1650

DTTCSSCQAF  LANLDEFAED  IFLNGC                               1676
```

The signal peptide sequence and linker peptide sequence, <u>RPRR</u>, between the β- and α-chains are underlined. The signal peptide and linker peptide are lost in the mature native protein. The known *N*-linked glycosylation sites are bolded (**N**). The cysteine residues that link the β-and α-chains are double underlined (C̲).

PROTEIN MODULES

1–18	Leader peptide
19–673	β-chain
678–1676	α-chain
678–751	C5a peptide
752–1676	C5a'-chain

CHROMOSOMAL LOCATION

Human[24]: 9q33.2

    Telomere … CNTRL (Centriolin) … **CC5** … TRAF1 (Tumour Necrosis Associated Factor 1) … Centromere

    Mouse[25]: chromosome 2.B

    Telomere … Cntrl… **Hc** … Traf1 … Centromere

    Rat: chromosome 3p.11

    Telomere … Cntrl … **C5** … Traf1 … Centromere

HUMAN cDNA SEQUENCE

```
CTACCTCCAA CCATGGGCCT TTTGGGAATA CTTTGTTTTT TAATCTTCCT GGGGAAAACC     60

TGGGGACAGG AGCAAACATA TGTCATTTCA GCACCAAAAA TATTCCGTGT TGGAGCATCT    120

GAAAATATTG TGATTCAAGT TTATGGATAC ACTGAAGCAT TTGATGCAAC AATCTCTATT    180

AAAAGTTATC CTGATAAAAA ATTTAGTTAC TCCTCAGGCC ATGTTCATTT ATCCTCAGAG    240

AATAAATTCC AAAACTCTGC AATCTTAACA ATACAACCAA AACAATTGCC TGGAGGACAA    300

AACCCAGTTT CTTATGTGTA TTTGGAAGTT GTATCAAAGC ATTTTTCAAA ATCAAAAAGA    360

ATGCCAATAA CCTATGACAA TGGATTTCTC TTCATTCATA CAGACAAACC TGTTTATACT    420

CCAGACCAGT CAGTAAAAGT TAGAGTTTAT TCGTTGAATG ACGACTTGAA GCCAGCCAAA    480

AGAGAAACTG TCTTAACCTT CATAGATCCT GAAGGATCAG AAGTTGACAT GGTAGAAGAA    540

ATTGATCATA TTGGAATTAT CTCTTTTCCT GACTTCAAGA TTCCGTCTAA TCCTAGATAT    600

GGTATGTGGA CGATCAAGGC TAAATATAAA GAGGACTTTT CAACAACTGG AACCGCATAT    660

TTTGAAGTTA AAGAATATGT CTTGCCACAT TTTTCTGTCT CAATCGAGCC AGAATATAAT    720

TTCATTGGTT ACAAGAACTT TAAGAATTTT GAAATTACTA TAAAAGCAAG ATATTTTTAT    780

AATAAAGTAG TCACTGAGGC TGACGTTTAT ATCACATTTG GAATAAGAGA AGACTTAAAA    840

GATGATCAAA AAGAAATGAT GCAAACAGCA ATGCAAAACA CAATGTTGAT AAATGGAATT    900

GCTCAAGTCA CATTTGATTC TGAAACAGCA GTCAAAGAAC TGTCATACTA CAGTTTAGAA    960

GATTTAAACA ACAAGTACCT TTATATTGCT GTAACAGTCA TAGAGTCTAC AGGTGGATTT   1020

TCTGAAGAGG CAGAAATACC TGGCATCAAA TATGTCCTCT CTCCCTACAA ACTGAATTTG   1080

GTTGCTACTC CTCTTTTCCT GAAGCCTGGG ATTCCATATC CCATCAAGGT GCAGGTTAAA   1140

GATTCGCTTG ACCAGTTGGT AGGAGGAGTC CCAGTAATAC TGAATGCACA AACAATTGAT   1200

GTAAACCAAG AGACATCTGA CTTGGATCCA AGCAAAAGTG TAACACGTGT TGATGATGGA   1260

GTAGCTTCCT TTGTGCTTAA TCTCCCATCT GGAGTGACGG TGCTGGAGTT TAATGTCAAA   1320

ACTGATGCTC CAGATCTTCC AGAAGAAAAT CAGGCCAGGG AAGGTTACCG AGCAATAGCA   1380

TACTCATCTC TCAGCCAAAG TTACCTTTAT ATTGATTGGA CTGATAACCA TAAGGCTTTG   1440

CTAGTGGGAG AACATCTGAA TATTATTGTT ACCCCCAAAA GCCCATATAT TGACAAAATA   1500

ACTCACTATA ATTACTTGAT TTTATCCAAG GGCAAAATTA TCCATTTTGG CACGAGGGAG   1560

AAATTTTCAG ATGCATCTTA TCAAAGTATA AACATTCCAG TAACACAGAA CATGGTTCCT   1620

TCATCCCGAC TTCTGGTCTA TTATATCGTC ACAGGAGAAC AGACAGCAGA ATTAGTGTCT   1680

GATTCAGTCT GGTTAAATAT TGAAGAAAAA TGTGGCAACC AGCTCCAGGT TCATCTGTCT   1740

CCTGATGCAG ATGCATATTC TCCAGGCCAA ACTGTGTCTC TTAATATGGC AACTGGAATG   1800
```

HUMAN cDNA SEQUENCE—*Continued*

```
GATTCCTGGG TGGCATTAGC AGCAGTGGAC AGTGCTGTGT ATGGAGTCCA AAGAGGAGCC    1860

AAAAAGCCCT TGGAAAGAGT ATTTCAATTC TTAGAGAAGA GTGATCTGGG CTGTGGGGCA    1920

GGTGGTGGCC TCAACAATGC CAATGTGTTC CACCTAGCTG GACTTACCTT CCTCACTAAT    1980

GCAAATGCAG ATGACTCCCA AGAAAATGAT GAACCTTGTA AAGAAATTCT CAGGCCAAGA    2040

AGAACGCTGC AAAAGAAGAT AGAAGAAATA GCTGCTAAAT ATAAACATTC AGTAGTGAAG    2100

AAATGTTGTT ACGATGGAGC CTGCGTTAAT AATGATGAAA CCTGTGAGCA GCGAGCTGCA    2160

CGGATTAGTT TAGGGCCAAG ATGCATCAAA GCTTTCACTG AATGTTGTGT CGTCGCAAGC    2220

CAGCTCCGTG CTAATATCTC TCATAAAGAC ATGCAATTGG GAAGGCTACA CATGAAGACC    2280

CTGTTACCAG TAAGCAAGCC AGAAATTCGG AGTTATTTTC CAGAAAGCTG GTTGTGGGAA    2340

GTTCATCTTG TTCCCAGAAG AAAACAGTTG CAGTTTGCCC TACCTGATTC TCTAACCACC    2400

TGGGAAATTC AAGGCATTGG CATTTCAAAC ACTGGTATAT GTGTTGCTGA TACTGTCAAG    2460

GCAAAGGTGT TCAAAGATGT CTTCCTGGAA ATGAATATAC CATATTCTGT TGTACGAGGA    2520

GAACAGATCC AATTGAAAGG AACTGTTTAC AACTATAGGA CTTCTGGGAT GCAGTTCTGT    2580

GTTAAAATGT CTGCTGTGGA GGGAATCTGC ACTTCGGAAA GCCCAGTCAT TGATCATCAG    2640

GGCACAAAGT CCTCCAAATG TGTGCGCCAG AAAGTAGAGG GCTCCTCCAG TCACTTGGTG    2700

ACATTCACTG TGCTTCCTCT GGAAATTGGC CTTCACAACA TCAATTTTTC ACTGGAGACT    2760

TGGTTTGGAA AAGAAATCTT AGTAAAAACA TTACGAGTGG TGCCAGAAGG TGTCAAAAGG    2820

GAAAGCTATT CTGGTGTTAC TTTGGATCCT AGGGGTATTT ATGGTACCAT TAGCAGACGA    2880

AAGGAGTTCC CATACAGGAT ACCCTTAGAT TTGGTCCCCA AAACAGAAAT CAAAAGGATT    2940

TTGAGTGTAA AAGGACTGCT TGTAGGTGAG ATCTTGTCTG CAGTTCTAAG TCAGGAAGGC    3000

ATCAATATCC TAACCCACCT CCCCAAAGGG AGTGCAGAGG CGGAGCTGAT GAGCGTTGTC    3060

CCAGTATTCT ATGTTTTTCA CTACCTGGAA ACAGGAAATC ATTGGAACAT TTTTCATTCT    3120

GACCCATTAA TTGAAAAGCA GAAACTGAAG AAAAAATTAA AAGAAGGGAT GTTGAGCATT    3180

ATGTCCTACA GAAATGCTGA CTACTCTTAC AGTGTGTGGA AGGGTGGAAG TGCTAGCACT    3240

TGGTTAACAG CTTTTGCTTT AAGAGTACTT GGACAAGTAA ATAAATACGT AGAGCAGAAC    3300

CAAAATTCAA TTTGTAATTC TTTATTGTGG CTAGTTGAGA ATTATCAATT AGATAATGGA    3360

TCTTTCAAGG AAAATTCACA GTATCAACCA ATAAAATTAC AGGGTACCTT GCCTGTTGAA    3420

GCCCGAGAGA ACAGCTTATA TCTTACAGCC TTTACTGTGA TTGGAATTAG AAAGGCTTTC    3480

GATATATGCC CCCTGGTGAA AATCGACACA GCTCTAATTA AAGCTGACAA CTTTCTGCTT    3540

GAAAATACAC TGCCAGCCCA GAGCACCTTT ACATTGGCCA TTTCTGCGTA TGCTCTTTCC    3600
```

HUMAN cDNA SEQUENCE—*Continued*

```
CTGGGAGATA AAACTCACCC ACAGTTTCGT TCAATTGTTT CAGCTTTGAA GAGAGAAGCT   3660

TTGGTTAAAG GTAATCCACC CATTTATCGT TTTTGGAAAG ACAATCTTCA GCATAAAGAC   3720

AGCTCTGTAC CTAACACTGG TACGGCACGT ATGGTAGAAA CAACTGCCTA TGCTTTACTC   3780

ACCAGTCTGA ACTTGAAAGA TATAAATTAT GTTAACCCAG TCATCAAATG GCTATCAGAA   3840

GAGCAGAGGT ATGGAGGTGG CTTTTATTCA ACCCAGGACA CCATCAATGC CATTGAGGGC   3900

CTGACGGAAT ATTCACTCCT GGTTAAACAA CTCCGCTTGA GTATGGACAT CGATGTTTCT   3960

TACAAGCATA AAGGTGCCTT ACATAATTAT AAAATGACAG ACAAGAATTT CCTTGGGAGG   4020

CCAGTAGAGG TGCTTCTCAA TGATGACCTC ATTGTCAGTA CAGGATTTGG CAGTGGCTTG   4080

GCTACAGTAC ATGTAACAAC TGTAGTTCAC AAAACCAGTA CCTCTGAGGA AGTTTGCAGC   4140

TTTTATTTGA AAATCGATAC TCAGGATATT GAAGCATCCC ACTACAGAGG CTACGGAAAC   4200

TCTGATTACA AACGCATAGT AGCATGTGCC AGCTACAAGC CCAGCAGGGA AGAATCATCA   4260

TCTGGATCCT CTCATGCGGT GATGGACATC TCCTTGCCTA CTGGAATCAG TGCAAATGAA   4320

GAAGACTTAA AAGCCCTTGT GGAAGGGGTG GATCAACTAT TCACTGATTA CCAAATCAAA   4380

GATGGACATG TTATTCTGCA ACTGAATTCG ATTCCCTCCA GTGATTTCCT TTGTGTACGA   4440

TTCCGGATAT TTGAACTCTT TGAAGTTGGG TTTCTCAGTC CTGCCACTTT CACAGTTTAC   4500

GAATACCACA GACCAGATAA ACAGTGTACC ATGTTTTATA GCACTTCCAA TATCAAAATT   4560

CAGAAAGTCT GTGAAGGAGC CGCGTGCAAG TGTGTAGAAG CTGATTGTGG GCAAATGCAG   4620

GAAGAATTGG ATCTGACAAT CTCTGCAGAG ACAAGAAAAC AAACAGCATG TAAACCAGAG   4680

ATTGCATATG CTTATAAAGT TAGCATCACA TCCATCACTG TAGAAAATGT TTTTGTCAAG   4740

TACAAGGCAA CCCTTCTGGA TATCTACAAA ACTGGGGAAG CTGTTGCTGA GAAAGACTCT   4800

GAGATTACCT TCATTAAAAA GGTAACCTGT ACTAACGCTG AGCTGGTAAA AGGAAGACAG   4860

TACTTAATTA TGGGTAAAGA AGCCCTCCAG ATAAAATACA ATTTCAGTTT CAGGTACATC   4920

TACCCTTTAG ATTCCTTGAC CTGGATTGAA TACTGGCCTA GAGACACAAC ATGTTCATCG   4980

TGTCAAGCAT TTTTAGCTAA TTTAGATGAA TTTGCCGAAG ATATCTTTTT AAATGGATGC   5040

TAAAATTCCT GAAGTTCAGC TGCATACAGT TTGCACTTAT GGACTCCTGT TGTTGAAGTT   5100

CGTTTTTTTG TTTTCTTCTT TTTTTAAACA TTCATAGCTG GTCTTATTTG TAAAGCTCAC   5160

TTTACTTAGA ATTAGTGGCA CTTGCTTTTA TTAGAGAATG ATTTCAAATG CTGTAACTTT   5220

CTGAAATAAC ATGGCCTTGG AGGGCATGAA GACAGATACT CCTCCAAGGT TATTGGACAC   5280

CGGAAACAAT AAATTGGAAC ACCTCCTCAA ACCTACCACT CAGGAATGTT TGCTGGGGCC   5340

GAAAGAACAG TCCATTGAAA GGGAGTATTA CAAAAACATG GCCTTTGCTT GAAAGAAAAT   5400

ACCAAGGAAC AGGAAACTGA TCATTAAAGC CTGAGTTTGC TTTC                     5444
```

5kb

FIGURE 18.3   Diagram of genomic structure for human C5.

The first five nucleotides in each exon are underlined to indicate the intron–exon boundaries. The methionine initiation codon (<u>ATG</u>), the termination codon (<u>TAA</u>) and the polyadenylation signal (<u>ATTAAA</u>) are indicated.

Genomic Structure[26,27]

The human C5 structural gene is 96 kb in length (determined from NCBI genomic sequence data) and is highly interrupted by 41 intron–exon boundaries. Although C5 is much larger than the C3 and C4 genes, the intron–exon organisation of all three family members is very similar. Two alternatively processed human C5 polyadenylated transcripts have been cloned and sequenced. However, no in vivo function has been attributed to them (Fig. 18.3).

ACCESSION NUMBERS

	cDNA/Genbank	Genomic/NCBI
Human	M57729	727 or NG_007364
Mouse	M35525	15139 or MGI:MGI:96031
Rat	XM_001079130	362119 or RGD: 2237

DEFICIENCY

Autosomal recessive: Several C5-deficient families have been reported from different ethnic backgrounds and from different geographical regions. Sera from homozygous deficient individuals lack bactericidal activity and have a severely impaired ability to induce chemotaxis. Consequently, all completely C5-deficient individuals display a propensity for severe recurrent infections, particularly to *Neisserial* species, including meningitidis and extragenital gonorrhea. In addition, one individual with complete C5 protein deficiency has systemic lupus erythematosus, but it is not clear if this is directly related to her C5 deficiency.

Human mutations identified: Sixteen different genetic mutations have been identified to date.

The first two genetic mutations identified were two nonsense mutations in two African-American families.[28] Subsequently, these two mutations have been found in Western Cape, South African families, and the exon 1 nonsense mutation has also been observed in a family from Saudi Arabia. The most frequently observed mutation to date appears to be a point mutation resulting in an amino acid substitution in exon 7 (G754>A; p.A252T). This mutation was found in approximately 7% of black African meningococcal disease cases in the Western Cape.

African-American Families[28]

Two nonsense mutations
c.C55>T (exon 1); p.Q19X; Stop codon, truncated protein
c.C4426>T(exon 36); p.R1476X; Stop codon, truncated protein

Western Cape, South African Families[29]

Two nonsense mutations
c.C55>T (exon 1); p.Q19X; Stop codon, truncated protein
c.C4426>T (exon 36); p.R1476X; Stop codon, truncated protein
Point mutation-amino acid substitution
c.G754>A (exon 7); p.A252T

Saudi Family[30]

Nonsense mutation
c.C55>T (exon 1); p.Q19X; Stop codon, truncated protein

Spanish Families[31,32]

Missense/frame-shift mutation
c.4871_73CCC>GC (exon 40); p.A1624fsX1645; truncated protein
Exon splicing enhancer mutation-skipping of exon 15
c.1883_84AG>CTCT; p.E628AfsX649

Italian Family[31]

Nonsense mutation
c.C892>T (exon 9); p.Q298X; stop codon, truncated protein

Brazilian Family[33]

Splice site mutation-skipping of exon 30; 51 amino acid deletion
c.G4017>A; p.1289-1339del

Turkish Family[34]

Exon splicing enhancer-skipping of exon 10
c.A1115>G; p.G335AfsX337

Norwegian Family[35]

Donor splice site mutation-skipping of exon 27
g.G70347 to T; c.3486+1G>T; p.1131-1162del; 32 amino acid deletion

Danish Family[35]

Donor splice site mutation-skipping of exon 19
g.A45286 to G; c.2348+1G>A; p.Q785YfsX789, truncated protein
Single nucleotide mutation resulting in amino acid substitution
c.1775T>G (exon 14); p.M592>R

Dutch Family[36]

4 bp deletion
c.1178_81delAAAC (exon 11); p.T394fsX396; frame-shift, truncated protein
Nonsense mutation
c.4972C>T (exon 41); cp.Q1658X; Stop codon, truncated protein

Morocco Families[37]

2 bp deletion
c.2607_2608del (exon 21); p. Ser870ProfsX3; Frame-shift, truncated protein
3 bp deletion
c.960_962del (Exon 9); p.N320del; 1 amino acid deletion

POLYMORPHIC VARIANTS[38]

A subgroup of Japanese patients with PNH had a poor response to the C5 inhibitor eculizumab. This poor response was due to c.2654G>A variant in exon 21 of the C5 gene, resulting in a p.R885>H substitution. This substitution did not affect the functional activity of C5, but no longer bound eculizumab. Another variant was found in an Argentinian patient of Asian ancestry with PNH who had a poor response to eculizumab. This variant was found in the same region, c.2653C>T; p.R885>C.

MUTANT ANIMALS

39% (28/72) of commonly used inbred strains of mice are C5-deficient.[39] Deficiency of C5 is due to a 2 bp gene deletion in a 5'-exon of the C5 structural gene (exon 7) that results in a frame-shift and premature stop codon.[40] All of the six C5-deficient strains examined contained the 2 bp deletion, indicating that these mice likely inherited the mutation from a common ancestor.

C5-deficient strains A/HeJ, AKR/J, DBA/2J, NZB/B1NJ, SWR/J and B10.D2/oSnJ are all available from Jackson Laboratories. The C5-sufficient strain B10.D2/

nSnJ and C5-deficient strain B10.D2/oSnJ are considered congenic and are often used to examine the effects of C5 deficiency in mouse models of human disease.

REFERENCES

1. Ooi YM, Colten HR. Biosynthesis and post-synthetic modification of a precursor (pro-C5) of the fifth component of mouse complement (C5). *J Immunol* 1979;**123**(6):2494–8.
2. DiScipio RG, Smith CA, Muller-Eberhard HJ, Hugli TE. The activation of human complement component C5 by a fluid phase C5 convertase. *J Biol Chem* 1983;**258**(17):10629–36.
3. Haviland DL, Haviland JC, Fleischer DT, Hunt A, Wetsel RA. Complete cDNA sequence of human complement pro-C5. Evidence of truncated transcripts derived from a single copy gene. *J Immunol* 1991;**146**(1):362–8.
4. Tack BF, Morris SC, Prahl JW. Fifth component of human complement: purification from plasma and polypeptide chain structure. *Biochemistry* 1979;**18**(8):1490–7.
5. Fernandez HN, Hugli TE. Primary structural analysis of the polypeptide portion of human C5a anaphylatoxin. Polypeptide sequence determination and assignment of the oligosaccharide attachment site in C5a. *J Biol Chem* 1978;**253**(19):6955–64.
6. Dolmer K, Sottrup-Jensen L. Disulfide bridges in human complement component C3b. *FEBS Lett* 1993;**315**(1):85–90.
7. Fredslund F, Laursen NS, Roversi P, Jenner L, Oliveira CL, Pedersen JS, et al. Structure of and influence of a tick complement inhibitor on human complement component 5. *Nat Immunol* 2008;**9**(7):753–60.
8. Laursen NS, Gordon N, Hermans S, Lorenz N, Jackson N, Wines B, et al. Structural basis for inhibition of complement C5 by the SSL7 protein from *Staphylococcus aureus*. *Proc Natl Acad Sci USA* 2010;**107**(8):3681–6.
9. Aleshin AE, DiScipio RG, Stec B, Liddington RC. Crystal structure of C5b-6 suggests structural basis for priming assembly of the membrane attack complex. *J Biol Chem* 2012;**287**(23):19642–52.
10. Greer J. Model structure for the inflammatory protein C5a. *Science* 1985;**228**(4703):1055–60.
11. Zuiderweg ER, Henkin J, Mollison KW, Carter GW, Greer J. Comparison of model and nuclear magnetic resonance structures for the human inflammatory protein C5a. *Proteins* 1988;**3**(3):139–45.
12. Cook WJ, Galakatos N, Boyar WC, Walter RL, Ealick SE. Structure of human desArg-C5a. Acta crystallographica Section D. *Biol Crystallogr* 2010;**66**(Pt 2):190–7.
13. Bajic G, Yatime L, Klos A, Andersen GR. Human C3a and C3a desArg anaphylatoxins have conserved structures, in contrast to C5a and C5a desArg. *Protein Sci* 2013;**22**(2):204–12.
14. Schatz-Jakobsen JA, Yatime L, Larsen C, Petersen SV, Klos A, Andersen GR. Structural and functional characterization of human and murine C5a anaphylatoxins. Acta crystallographica Section D. *Biol Crystallogr* 2014;**70**(Pt 6):1704–17.
15. Lambris JDSA, Wetsel RA. The Chemistry and Biology of C3, C4, and C5. In: MFaJ V, editor. *Human complement system in health and disease*. New York, New York, USA: Marcel Dekker; 1998. p. 83–118.
16. Wetsel R, Kildsgaard J, Haviland DL. Complement anaphylatoxins (C3a, C4a, C5a) and their receptors (C3aR, C5aR/CD88) as therapeutic targets in inflammation. In: JLaM H, editor. *Therapeutic interventions in the complement system*. Totowa, New Jersey, USA: Human Press, Inc; 2000. p. 113–54.
17. Wetsel RA. Structure, function and cellular expression of complement anaphylatoxin receptors. *Curr Opin Immunol* 1995;**7**(1):48–53.

18. Gerard C, Gerard NP. C5A anaphylatoxin and its seven transmembrane-segment receptor. *Annu Rev Immunol* 1994;**12**:775–808.

19. Okinaga S, Slattery D, Humbles A, Zsengeller Z, Morteau O, Kinrade MB, et al. C5L2, a non-signaling C5A binding protein. *Biochemistry* 2003;**42**(31):9406–15.

20. Li Rui CL, Wu MCL, Taylor SM, Woodruff TM. C5L2: a controversial receptor of complement anapylatoxin, C5a. *FASEB J* 2016;**27**(3):855–64.

21. Wetsel RA, Kolb WP. Expression of C5a-like biological activities by the fifth component of human complement (C5) upon limited digestion with noncomplement enzymes without release of polypeptide fragments. *J Exp Med* 1983;**157**(6):2029–48.

22. Matthews KW, Mueller-Ortiz SL, Wetsel RA. Carboxypeptidase N: a pleiotropic regulator of inflammation. *Mol Immunol* 2004;**40**(11):785–93.

23. Kohler PF, Muller-Eberhard HJ. Immunochemical quantitation of the third, fourth and fifth components of human complement: concentrations in the serum of healthy adults. *J Immunol* 1967;**99**(6):1211–6.

24. Wetsel RA, Lemons RS, Le Beau MM, Barnum SR, Noack D, Tack BF. Molecular analysis of human complement component C5: localization of the structural gene to chromosome 9. *Biochemistry* 1988;**27**(5):1474–82.

25. D'Eustachio P, Kristensen T, Wetsel RA, Riblet R, Taylor BA, Tack BF. Chromosomal location of the genes encoding complement components C5 and factor H in the mouse. *J Immunol* 1986;**137**(12):3990–5.

26. Haviland DL, Haviland JC, Fleischer DT, Wetsel RA. Structure of the murine fifth complement component (C5) gene. A large, highly interrupted gene with a variant donor splice site and organizational homology with the third and fourth complement component genes. *J Biol Chem* 1991;**266**(18):11818–25.

27. Carney DF, Haviland DL, Noack D, Wetsel RA, Vik DP, Tack BF. Structural aspects of the human C5 gene. Intron/exon organization, 5'-flanking region features, and characterization of two truncated cDNA clones. *J Biol Chem* 1991;**266**(28):18786–91.

28. Wang X, Fleischer DT, Whitehead WT, Haviland DL, Rosenfeld SI, Leddy JP, et al. Inherited human complement C5 deficiency. Nonsense mutations in exons 1 (Gln1 to Stop) and 36 (Arg1458 to Stop) and compound heterozygosity in three African-American families. *J Immunol* 1995;**154**(10):5464–71.

29. Owen EP, Wurzner R, Leisegang F, Rizkallah P, Whitelaw A, Simpson J, et al. A complement C5 gene mutation, c.754G>A:p.A252T, is common in the Western Cape, South Africa and found to be homozygous in seven percent of Black African meningococcal disease cases. *Mol Immunol* 2015;**64**(1):170–6.

30. Arnaout R, Al Shorbaghi S, Al Dhekri H, Al-Mousa H, Al Ghonaium A, Al Saud B, et al. C5 complement deficiency in a Saudi family, molecular characterization of mutation and literature review. *J Clin Immunol* 2013;**33**(4):871–5.

31. Lopez-Lera A, Garrido S, de la Cruz RM, Fontan G, Lopez-Trascasa M. Molecular characterization of three new mutations causing C5 deficiency in two non-related families. *Mol Immunol* 2009;**46**(11–12):2340–7.

32. Delgado-Cervino E, Fontan G, Lopez-Trascasa M. C5 complement deficiency in a Spanish family. Molecular characterization of the double mutation responsible for the defect. *Mol Immunol* 2005;**42**(1):105–11.

33. Aguilar-Ramirez P, Reis ES, Florido MP, Barbosa AS, Farah CS, Costa-Carvalho BT, et al. Skipping of exon 30 in C5 gene results in complete human C5 deficiency and demonstrates the importance of C5d and CUB domains for stability. *Mol Immunol* 2009;**46**(10):2116–23.

34. Pfarr N, Prawitt D, Kirschfink M, Schroff C, Knuf M, Habermehl P, et al. Linking C5 deficiency to an exonic splicing enhancer mutation. *J Immunol* 2005;**174**(7):4172–7.

35. Schejbel L, Fadnes D, Permin H, Lappegard KT, Garred P, Mollnes TE. Primary complement C5 deficiencies – molecular characterization and clinical review of two families. *Immunobiology* 2013;**218**(10):1304–10.

36. Cheng SC, Sprong T, Joosten LA, van der Meer JW, Kullberg BJ, Hube B, et al. Complement plays a central role in *Candida albicans*-induced cytokine production by human PBMCs. *Eur J Immunol* 2012;**42**(4):993–1004.

37. Colobran R, Franco-Jarava C, Martin-Nalda A, Baena N, Gabau E, Padilla N, et al. Novel mutations causing C5 deficiency in three North-African families. *J Clin Immunol* 2016;**36**(4):388–96.

38. Nishimura J, Yamamoto M, Hayashi S, Ohyashiki K, Ando K, Brodsky AL, et al. Genetic variants in C5 and poor response to eculizumab. *New Eng J Med* 2014;**370**:632–9.

39. Cinader B, Dubiski S, Wardlaw AC. Distribution, inheritance, and properties of an antigen, MuB1, and its relation to hemolytic complement. *J Exp Med* 1964;**120**:897–924.

40. Wetsel RA, Fleischer DT, Haviland DL. Deficiency of the murine fifth complement component (C5). A 2-base pair gene deletion in a 5'-exon. *J Biol Chem* 1990;**265**(5):2435–40.

Part IV

Terminal Pathway Components

Chapter 19

C6

Richard G. DiScipio[1,2]

[1]*Torrey Pines Institute for Molecular Studies, San Diego, CA, United States;* [2]*Sanford Burnham Prebys Medical Discovery Institute, La Jolla, CA, United States*

PHYSIOCHEMICAL PROPERTIES

Proform
934 amino acids including a leader peptide of 21 amino acids

Mature Protein
pI = 6.11 (Apoprotein)

pI ~ 5.6 (glycoprotein)

M_r = 102,500 (Apoprotein)

M_r ~ 108,000 (glycoprotein)

913 amino acids

N-linked glycosylation sites: 324 and 855 (proform numbering)

Intrachain disulphide bonds: 32

STRUCTURE

C6 is a single-chain glycoprotein consisting of a single domain [membrane attack complex perforin (MACPF)] of about 250 amino acids and nine cysteine-rich modules (34–77 amino acids) related to those found in thrombospondin (TS), the LDL-receptor (LDL), the epidermal growth factor (EGF), complement control proteins (CCP) and factor I modules (FIM) (Figs 19.1 and 19.2).[1,2]

1–21	Leader	exon 1
22–79	TS1	exons 1/2
80–1315	TS2	exons 2/3
136–174	LDL	exons 3/4
268–519	MACPF	exons 6–10
520–553	EGF	exon 10
565–612	TS3	exon 11
642–701	CCP1	exons 12/13
702–763	CCP2	exon 14
766–842	FIM1	exons 15/16
858–934	FIM2	exons 16/17

The Complement FactsBook. http://dx.doi.org/10.1016/B978-0-12-810420-0.00019-5

FIGURE 19.1 Diagram of protein domains for complement C6.

FIGURE 19.2 Outline scheme for complement C6. The mature protein consists of 913 amino acids blocked into three thrombospondin, one LDL, one epidermal growth factor, two complement control proteins, two factor I modules, and one domain – the MACPF represented by a thickened line. The Ω symbol indicates the place of the disulphide bridged omega-loop. Asparaginyl-linked carbohydrate sites are represented by *hexagons* with CHO inscribed. Calcium ion binds to the LDL module.

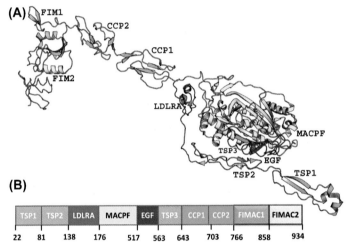

FIGURE 19.3 Human complement component C6. (A) Ribbon diagram of the human complement component C6 crystal structure (PDB ID: 3T5O). (B) Schematic representation of C6 domains, which are coloured similar to the ribbon representation.

Crystal Structure: pdb: 3T5O.[3] Crystal structure of C6 in complex with C5b: pdb: 4E0S,[4] pdb: 4A5W.[5] The overall design of C6 is an elongated molecule with an overall appearance likened to a seahorse. The protein has a maximal length of ~212 Å, and its widest part is ~70 Å (Fig. 19.3).

FUNCTION

C6 binds newly activated C5b and the complex initiates the assembly of the membrane attack complex (MAC) that is a transmembrane pore assembly with a channel size of ~100 Å.[6] The late acting components of complement including C6 can also assemble in soluble complexes referred to as SC5b-9.[7] SC5b-9 complexes are likely to be heterogeneous, since they contain vitronectin and clusterin and are associated with high-density lipoproteins. SC5b-9 may possess multiple functions yet to be discovered, but rapid clearance of the complex is established.[8]

TISSUE DISTRIBUTION

The primary site of synthesis for plasma C6 is the liver hepatocyte.[9] However, other cell types reported to produce C6 in vitro are monocytes,[10] fibroblasts,[11] endothelial cells,[12] neurons and glial cells.[13,14] The normal concentration of C6 in human plasma is 40–60 μg/mL.

REGULATION OF EXPRESSION

C6 is not reported to be a major acute-phase protein.[15]

HUMAN PROTEIN SEQUENCE[1,2]

```
MARRSVLYFI LLNALINKGQ ACFCDHYAWT QWTSCSKTCN SGTQSRHRQI   50
VVDKYYQENF CEQICSKQET RECNWQRCPI NCLLGDFGPW SDCDPCIEKQ  100
SKVRSVLRPS QFGGQPCTAP LVAFQPCIPS KLCKIEEADC KNKFRCDSGR  150
CIARKLECNG ENDCGDNSDE RDCGRTKAVC TRKYNPIPSV QLMGNGFHFL  200
AGEPRGEVLD NSFTGGICKT VKSSRTSNPY RVPANLENVG FEVQTAEDDL  250
KTDFYKDLTS LGHNENQQGS FSSQGGSSFS VPIFYSSKRS ENINHNSAFK  300
QAIQASHKKD SSFIRIHKVM KVLNFTTKAK DLHLSDVFLK ALNHLPLEYN  350
SALYSRIFDD FGTHYFTSGS LGGVYDLLYQ FSSEELKNSG LTEEEAKHCV  400
RIETKKRVLF AKKTKVEHRC TTNKLSEKHE GSFIQGAEKS ISLIRGGRSE  450
YGAALAWEKG SSGLEEKTFS EWLESVKENP AVIDFELAPI VDLVRNIPCA  500
VTKRNNLRKA LQEYAAKFDP CQCAPCPNNG RPTLSGTECL CVCQSGTYGE  550
NCEKQSPDYK SNAVDGQWGC WSSWSTCDAT YKRSRTRECN NPAPQRGGKR  600
CEGEKRQEED CTFSIMENNG QPCINDDEEM KEVDLPEIEA DSGCPQPVPP  650
ENGFIRNEKQ LYLVGEDVEI SCLTGFETVG YQYFRCLPDG TWRQGDVECQ  700
RTECIKPVVQ EVLTITPFQR LYRIGESIEL TCPKGFVVAG PSRYTCQGNS  750
WTPPISNSLT CEKDTLTKLK GHCQLGQKQS GSECICMSPE EDCSHHSEDL  800
CVFDTDSNDY FTSPACKFLA EKCLNNQQLH FLHIGSCQDG RQLEWGLERT  850
RLSSNSTKKE SCGYDTCYDW EKCSASTSKC VCLLPPQCFK GGNQLYCVKM  900
GSSTSEKTLN ICEVGTIRCA NRKMEILHPG KCLA                   934
```

The leader sequence is underlined, and the *N*-linked oligosaccharide sites are shown in bold (**N**), and the mannosyl-tryptophans[16] are presented in italics and underlined (_W_).

CHROMOSOMAL LOCATION

Human: 5p13.1
   Murine: 15A1, 15 1.97 cM
   Rat: 2q16
   The genes encoding C6 and C7 are tightly linked being separated by 160 kbp and are arranged in opposite orientations on human chromosome 5.[17,18]

HUMAN cDNA SEQUENCE[1,2]

```
AATCATTTGC AAAGTTCAAG TTCCAGAGAA CATTTATTTT GACAACCCTC TAGGTGTTGC   60
TAGGCTTCTG GGATATGACA GCATTGCCTT GTGTTAGCTA GCAATAAGAA AAGAAGCTTT  120
GTTTGGATTA ACATATATAC CCTCTTCATT CTGCATACCT ATTTTTTCCC CAATAATTTG  180
CAGCTTAGGT CCGAGGACAC CACAAACTCT GCTTAAAGGG CCTGGAGGCT CTCAAGGCAT  240
GGCCAGACGC TCTGTCTTGT ACTTCATCCT GCTGAATGCT CTGATCAACA AGGGCCAAGC  300
CTGCTTCTCT GATCACTATG CATGGACTCA GTGGACCAGC TGCTCAAAAA CTTGCAATTC  360
TGGAACCCAG AGCAGACACA GACAAATAGT AGTAGATAAG TACTACCAGG AAAACTTTTG  420
TGAACAGATT TGCAGCAAGC AGGAGACTAG AGAATGTAAC TGGCAAAGAT GCCCCATCAA  480
CTGCCTCCTG GGAGATTTTG GACCATGGTC AGACTGTGAC CCTTGTATTG AAAAACAGTC  540
TAAAGTTAGA TCTGTCTTGC GTCCCAGTCA GTTTGGGGGA CAGCCATGCA CTGCGCCTCT  600
GGTAGCCTTT CAACCATGCA TTCCATCTAA GCTCTGCAAA ATTGAAGAGG CTGACTGCAA  660
GAATAAATTT CGCTGTGACA GTGGCCGCTG CATTGCCAGA AAGTTAGAAT GCAATGGAGA  720
AAATGACTGT GGAGACAATT CAGATGAAAG GGACTGTGGG AGGACAAAGG CAGTATGCAC  780
ACGGAAGTAT AATCCCATCC CTAGTGTACA GTTGATGGGC AATGGGTTTC ATTTTCTGGC  840
AGGAGAGCCC AGAGGAGAAG TCCTTGATAA CTCTTTCACT GGAGGAATAT GTAAAACTGT  900
CAAAAGCAGT AGGACAAGTA ATCCATACCG TGTTCCGGCC AATCTGGAAA ATGTCGGCTT  960
TGAGGTACAA ACTGCAGAAG ATGACTTGAA AACAGATTTC TACAAGGATT TAACTTCTCT 1020
TGGACACAAT GAAAATCAAC AAGGCTCATT CTCAAGTCAG GGGGGGAGCT CTTTCAGTGT 1080
ACCAATTTTT TATTCCTCAA GAGAGAAGTGA AAATATCAAC CATAATTCTG CCTTCAAACA 1140
AGCCATTCAA GCCTCTCACA AAAAGGATTC TAGTTTTATT AGGATCCATA AAGTGATGAA 1200
AGTCTTAAAC TTCACAACGA AAGCTAAAGA TCTGCACCTT TCTGATGTCT TTTTGAAAGC 1260
ACTTAACCAT CTGCCTCTAG AATACAACTC TGCTTTGTAC AGCCGAATAT TCGATGACTT 1320
TGGGACTCAT TACTTCACCT CTGGCTCCCT GGGAGGCGTG TATGACCTTC TCTATCAGTT 1380
TAGCAGTGAG GAACTAAAGA ACTCAGGTTT AACCGAGGAA GAAGCCAAAC ACTGTGTCAG 1440
GATTGAAACA AAGAAACGCG TTTTTATTTGC TAAGAAAACA AAAGTGGAAC ATAGGTGCAC 1500
CACCAACAAG CTGTCAGAGA AACATGAAGG TTCATTTATA CAGGGAGCAG AGAAATCCAT 1560
ATCCCTGATT CGAGGTGGAA GGAGTGAATA TGGAGCAGCT TTGGCATGGG AGAAAGGGAG 1620
CTCTGGTCTG GAGGAGAAGA CATTTTCTGA GTGGTTAGAA TCAGTGAAGG AAAATCCTGC 1680
TGTGATTGAC TTTGAGCTTG CCCCCATCGT GGACTTGGTA AGAAACATCC CCTGTGCAGT 1740
GACAAAACGG AACAACCTCA GGAAAGCTTT GCAAGAGTAT GCAGCCAAGT TCGATCCTTG 1800
CCAGTGTGCT CCATGCCCTA ATAATGGCCG ACCCACCCTC TCAGGGACTG AATGTCTGTG 1860
TGTGTGTCAG AGTGGCACCT ATGGTGAGAA CTGTGAGAAA CAGTCTCCAG ATTATAAATC 1920
CAATGCAGTA GACGGACAGT GGGGTTGTTG GTCTTCCTGG AGTACCTGTG ATGCTACTTA 1980
TAAGAGATCG AGAACCCGAG AATGCAATAA TCCTGCCCCC CAACGAGGAG GGAAACGCTG 2040
TGAGGGGGAG AAGCGACAAG AGGAAGACTG CACATTTTCA ATCATGGAAA ACAATGGACA 2100
ACCATGTATC AATGATGACG AAGAAATGAA AGAGGTCGAT CTTCCTGAGA TAGAAGCAGA 2160
TTCCGGGTGT CCTCAGCCAG TTCCTCCAGA AAATGGATTT ATCCGGAATG AAAAGCAACT 2220
ATACTTGGTT GGAGAAGATG TTGAAATTTC ATGCCTTACT GGCTTTGAAA CTGTTGGATA 2280
CCAGTACTTC AGATGCTTAC CAGACGGGAC CTGGAGACAA GGGGATGTGG AATGCCAACG 2340
GACGGAGTGC ATCAAGCCAG TTGTGCAGGA AGTCCTGACA ATTACACCAT TTCAGAGATT 2400
GTATAGAATT GGTGAATCCA TTGAGCTAAC TTGCCCCAAA GGCTTTGTTG TTGCTGGGCC 2460
ATCAAGGTAC ACATGCCAGG GGAATTCCTG GACACCACCC ATTTCAAACT CTCTCACCTG 2520
TGAAAAAGAT ACTCTAACAA AATTAAAAGG CCATTGTCAG CTGGGACAGA AACAATCAGG 2580
ATCTGAATGC ATTTGTATGT CTCCAGAAGA AGACTGTAGC CATCATTCAG AAGATCTCTG 2640
TGTGTTTGAC ACAGACTCCA ACGATTACTT TACTTCACCC GCTTGTAAGT TTTTGGCTGA 2700
GAAATGTTTA AATAATCAGC AACTCCATTT TCTACATATT GGTTCCTGCC AAGCAGGCCG 2760
CCAGTTAGAA TGGGGTCTTG AAAGGACAAG ACTTTCATCC AACAGCACAA AGAAAGAATC 2820
CTGTGGCTAT GACACCTGCT ATGACTGGGA AAAATGTTCA GCCTCCACTT CCAAATGTGT 2880
CTGCCTATTG CCCCCACAGT GCTTCAAGGG TGGAAACCAA CTCTACTGTG TCAAAATGGG 2940
ATCATCAACA AGTGAGAAAA CATTGAACAT CTGTGAAGTG GAACTATAAA GATGTGCAAA 3000
CAGGAAGATG GAAATACTGC ATCCTGGAAA GTGTTTGGCC TAGCACAATT ACTGCTAGGC 3060
CCAGCACAAT GAACAGATTT ACCATCCCGA AGAACCAACT CCTACAAATG AGAATTCTTG 3120
CACAAACAGC AGACTGGCAT GCTCAAAGTT ACTGACAAAA ATTATTTTCT GTTAGTTTGA 3180
GATCATTATT CTCCCCTGAC TCTCCTGTTT GGGCATGTCT TATTCAGTTC CAGCTCATGA 3240
CGCCCTGTAG CATACCCCTA GGTACCAACT TCCACAGCAG TCTCGTAAAT TCTCCTGTTC 3300
ACATTGTACA AAAATAATGT GACTTCTGAG GCCCTTAAGT AGCCTGTGAC ATTAAGCATT 3360
CTCGCAATTA GAAATAAGAA TAAAACCCAT AATTTTCTTC AATGAGTTAA TAAACAGAAA 3420
TCTCCAGAAC CTCTGAAACA CATTCTTGAA GCCCAGCTTT CATATCTTCA TTCAACAAAT 3480
AATTTCTGAG TGTGTATACA GGATGTCAAG TACTGACCAA AGTCCTGAGA ACTCGGCAGA 3540
TAATAAAACA GACAAAGCC TTTGCCTTCA TGAAGCATAC ATTCATTCAG GGGTAGACAC 3600
ACAAAAAATG AAATAAACAG GTAAAATATG TAGCATGTTT GATGGTGATA ATTGTTTTGG 3660
AGAAAATACA ACACGGAAGG GTGAGTCCCA TCCTTAATTA TCATCTTATG CAAGGAAATA 3720
TT
```

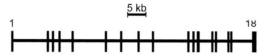

FIGURE 19.4 Gene map for complement C6. The gene for C6 is ~80 kb containing 18 exons (*vertical lines*). The scale bar is 5 kb.

GENOMIC STRUCTURE

The C6 gene is encoded by 18 exons that span ~80 kb.[19] The genes for C6 and C7 are closely linked in opposite orientations (Fig. 19.4).[18]

ACCESSION NUMBERS

Protein
Human: NP_000056   NP_001108603
Murine: NP_057913

MRNA
Human: NM_001115131   NM_000065
Murine: NM_016704

Gene
Human: NG_011582
Murine: Gene ID: 109828

DEFICIENCY

C6 deficiency OMIM: 217050; 612446.

Total C6 deficiencies are known mainly from South Africans and Afro-Americans in which single base pair deletions result in premature stop codons: 879delG in exon 6, 1195delC in exon 7 and 1936delG in exon 12.[20–22] One patient from Japan also had a complete C6 deficiency as a result of single base pair deletion at exon 2: 291,292,293,294delC.[22] Four other deficiencies of C6 from South Africans have been identified: 821delA, 828delG, 1138delC and 1879delG.[23] An individual with a heterozygous alteration in C6 has been reported in which an abnormal splice site occurred at exon 16 along with a missense mutation (C867R) in exon 17 resulting in a truncated molecule that retained some activity.[24] A subtotal C6 deficiency is also known as a consequence of an abnormal 5′-splice donor site at intron 15 that results in a truncated protein lacking the last FIM.[25] Individuals with combined subtotal deficiencies in both C6 and C7 are reported in which the gene for C6 has an abnormal donor splice site at intron 15, and that of C7 has a codon substitution in exon 11 resulting in an R521S.[26] Autosomal recessive traits deficiencies of C6 are associated with recurrent *Neisserial meningitidis* infections.[22,23,27]

POLYMORPHIC VARIANTS

Although isoelectric focusing suggests at least 10 natural variants, the most common C6 variants are C6A (0.65) and C6B (0.35), which result from an E119A substitution. The two forms are not functionally defective.[28,29]

MUTANT ANIMALS

A naturally occurring PVG strain of rats is C6 deficient having 31 bp deletion in exon 10.[30,31] C6-deficient rats exhibit a lower level of pathology in experimental allergic encephalomyelitis and myasthenia gravis relative to wild types.[32,33]

Embryos and sperm from a C6-deficient mouse (C6^{Q0}) (C3.Cg$C6^{Q0}$/ Mmmh) caused by a spontaneous mutation derived from Peru-Coppock strain are available from Mutant Mouse Resource & Research Center (MMRRC) (MGI:88233).

REFERENCES

1. DiScipio RG, Hugli TE. The molecular architecture of human complement component C6. *J Biol Chem* 1989;**264**:16197–206.
2. Haefliger J-A, Tschopp J, Vial N, Jenne DE. Complete primary structure and functional characterization of the sixth component of the human complement system. *J Biol Chem* 1989;**264**:18041–51.
3. Aleshin AE, Schraufstatter IU, Stec B, Bankston LA, Liddington RC, DiScipio RG. Structure of complement C6 suggests a mechanism for initiation and unidirectional, sequential assembly of the membrane attack complex (MAC). *J Biol Chem* 2012;**287**:10210–22.
4. Aleshin AE, DiScipio RG, Stec B, Liddington RC. Crystal structure of C5b-6 suggests structural basis for priming assembly of the membrane attack complex. *J Biol Chem* 2012;**287**:19642–52.
5. Hadders MA, Bubeck D, Roversi P, Hakobyan S, Forneris F, Morgan BP, et al. Assembly and regulation of the membrane attack complex based on structures of C5b6 and sC5b9. *Cell Rep* 2012;**1**:200–7.
6. Sonnen AF, Henneke P. Structural biology of the membrane attack complex. *Subcell Biochem* 2014;**80**:83–116.
7. Kolb WP, Muller-Eberhard HJ. The membrane attack mechanism of complement. Isolation and subunit composition of the C5b-9 complex. *J Exp Med* 1975;**141**:724–35.
8. Hugo F, Berstecher C, Krämer S, Fassbender W, Bhakdi S. In vivo clearance studies of the terminal fluid-phase complement complex in rabbits. *Clin Exp Immunol* 1989;**77**:112–6.
9. Hobart MJ, Lachmann PJ, Calne Y. C6: synthesis by the liver in vivo. *J Exp Med* 1977;**146**:629–30.
10. Hetland G, Johnson E, Falk RJ, Eskeland T. Synthesis of complement components C5, C6, C7, C8 and C9 in vitro by human monocytes and assembly of the terminal complement complex. *Scand J Immunol* 1986;**24**:421–8.
11. Garred P, Hetland G, Mollnes TE, Stoervold G. Synthesis of C3, C5, C6, C7, C8, and C9 by human fibroblasts. *Scand J Immunol* 1990;**32**:555–60.
12. Johnson E, Hetland G. Human umbilical vein endothelial cells synthesize functional C3, C5, C6, C8 and C9 in vitro. *Scand J Immunol* 1991;**33**:667–71.

13. Gasque P, Fontaine M, Morgan BP. Complement expression in human brain. Biosynthesis of terminal pathway components and regulators in human glial cells and cell lines. *J Immunol* 1995;**154**:4726–33.

14. Terai K, Walker DG, McGeer EG, McGeer PL. Neurons express proteins of the classical complement pathway in Alzheimer disease. *Brain Res* 1997;**769**:385–90.

15. Kilicarslan A, Uysal A, Roach EC. Acute phase reactants. *Acta Medica* 2013;**2**:2–7.

16. Hofsteenge J, Blommers M, Hess D, Furmanek A, Miroshnichenko O. The four terminal components of the complement system are C-mannosylated on multiple tryptophan residues. *J Biol Chem* 1999;**274**:32786–94.

17. Hobart MJ, Fernie BA, DiScipio RG, Lachmann PJ. A physical map of the C6 and C7 complement gene region on chromosome 5p13. *Hum Mol Genet* 1993;**2**:1035–6.

18. Hobart MJ, Fernie BA, DiScipio RG. Structure of the human C7 gene and comparison with C6, C8A, C8B, and C9 genes. *J Immunol* 1995;**154**:5188–94.

19. Hobart MJ, Fernie B, DiScipio RG. Structure of the human C6 gene. *Biochemistry* 1993;**32**:6198–205.

20. Hobart MJ, Fernie BA, Fijen KA, Orren A. The molecular basis of C6 deficiency in the Western Cape, South Africa. *Hum Genet* 1998;**103**:506–12.

21. Zhu ZB, Totemchokchyakarn K, Atkinson TP, Volanakis JE. Molecular defects leading to human complement component C6 deficiency in an African-American family. *Clin Exp Immunol* 1998;**111**:91–6.

22. Nishizaka H, Horiuchi T, Zhu Z-B, Fukumori Y, Nagasawa K, Hayashi K, et al. Molecular bases for inherited human complement component C6 deficiency in two unrelated individuals. *J Immunol* 1996;**156**:2309–15.

23. Orren A, Owen EP, Henderson HE, van der Merwe L, Leisegang F, Stassen C, et al. Complete deficiency of the sixth complement component (C6Q0), susceptibility to *Neisseria meningitidis* infections and analysis of the frequencies of C6Q0 gene defects in South Africans. *Clin Exp Immunol* 2012;**167**:459–71.

24. Westra D, Kurvers RA, van den Heuvel LP, Würzner R, Hoppenreijs EP, van der Flier M, et al. Compound heterozygous mutations in the C6 gene of a child with recurrent infections. *Mol Immunol* 2014;**58**:201–5.

25. Wurzner R, Hobart MJ, Fernie BA, Mewar D, Potter PC. Molecular basis of subtotal complement C6 deficiency. A carboxy-terminally truncated but functionally active C6. *J Clin Invest* 1995;**95**:1877–83.

26. Fernie BA, Wurzner R, Orren A, Morgan BP, Potter PC, Platonov AE, et al. Molecular bases of combined subtotal deficiencies of C6 and C7. *J Immunol* 1996;**157**:3648–57.

27. Mayilyan KR. Complement genetics, deficiencies, and disease associations. *Protein Cell* 2012;**3**:487–96.

28. Fernie BA, Delbridge G, Hobart MJ. Correlation of a Glu/Ala substitution at position 98 with the complement C6 A/B phenotypes. *Hum Mol Genet* 1993;**2**:591–2.

29. Würzner R, Witzel-Schlömp K, Tokunaga K, Fernie BA, Hobart MJ, Orren A. Reference typing report for complement components C6, C7 and C9 including mutations leading to deficiencies. *Exp Clin Immunogenet* 1998;**15**:268–85.

30. Leenaerts PL, Stad RK, Hall BM, Van Damme BJ, Vanrenterghem Y, Daha MR. Hereditary C6 deficiency in a strain of PVG/c rats. *Clin Exp Immunol* 1994;**97**:478–82.

31. Bhole D, Stahl GL. Molecular basis for complement component 6 (C6) deficiency in rats and mice. *Immunobiology* 2004;**209**:559–68.

32. Chamberlain-Banoub J, Neal JW, Mizuno M, Harris CL, Morgan BP. Complement membrane attack is required for endplate damage and clinical disease in passive experimental myasthenia gravis in Lewis rats. *Clin Exp Immunol* 2006;**146**:278–86.

33. Tran GT, Hodgkinson SJ, Carter N, Killingsworth M, Spicer ST, Hall BM. Attenuation of experimental allergic encephalomyelitis in complement component 6-deficient rats is associated with reduced complement C9 deposition, P-selectin expression, and cellular infiltrate in spinal chords. *J Immunol* 2002;**168**:4293–300.

Chapter 20

C7

Richard G. DiScipio[1,2]

[1]*Torrey Pines Institute for Molecular Studies, San Diego, CA, United States;* [2]*Sanford Burnham Prebys Medical Discovery Institute, La Jolla, CA, United States*

PHYSIOCHEMICAL PROPERTIES

Proform
843 amino acids including a leader peptide of 22 amino acids

Mature Protein

pI = 6.09	(Apoprotein)	pI ~5.6 (glycoprotein)
M_r = 91,100	(Apoprotein)	M_r ~97,000 (glycoprotein)

821 amino acids
N-linked glycosylation sites: 202 and 754 (proform numbering)
Intrachain disulphide bonds: 28

STRUCTURE

Complement component C7 is a single-chain glycoprotein consisting of a single domain (MACPF) of about 240 amino acids and eight cysteine-rich modules (34–75 amino acids) (Fig. 20.1) homologous to those found in thrombospondin (TS), the LDL-receptor (LDL), the epidermal growth factor (EGF), complement control proteins (CCP) and factor I modules (FIM) (Fig. 20.2).[1]

1–22	Leader	exons 0/1
23–77	TS1	exons 2/3
78–119	LDL	exons 3/4
214–454	MACPF	exons 6/10
454–487	EGF	exon 10
496–549	TS2	exon 11
569–628	CCP1	exons 12/13
629–690	CCP2	exon 14
695–768	FIM1	exons 15/16
769–843	FIM2	exons 16/17

NMR structure of the C7 FIM pair (pdb, 2WCY) indicated a compact structure exhibiting pseudo-twofold rotational symmetry.[2] Electron microscopy has imaged C7 as an elongated molecule with an overall appearance not dissimilar

The Complement FactsBook. http://dx.doi.org/10.1016/B978-0-12-810420-0.00020-1

213

FIGURE 20.1 Diagram of protein domains for C7.

FIGURE 20.2 A summary diagram for the covalent structure of C7. The mature protein is of 821 amino acids and includes two thrombospondin (TS), one LDL, one epithelial growth factor (EDGF), two complement control proteins (CCP), two factor I modules (FIM) along with a single MACPF domain (*thick line*). N-linked carbohydrates are shown by the hexagons labelled as CHO. The disulphide linked Ω-loop is represented.

from a parsnip. The protein has a maximal length of ~150 Å, and at its widest part is ~50 Å (Fig. 20.3).[1]

FUNCTION

C7 binds the C5b6 complex and continues the assembly of the membrane attack complex (MAC), which is a multiprotein transmembrane pore with a mean channel size of ~100 Å.[3] The late acting components of complement including C7 can also assemble into soluble complexes, containing C5 through C9, referred to as sC5b-9.[4] Vitronectin, clusterin and high-density lypoproteins are found in association with sC5b-9 complexes, and thus these assemblies are likely to be heterogeneous suggesting multiple activities; however, the only clearly established function is clearance.[5] C7 has been reported to bind endothelial cells and exert an antiinflammatory effect.[6] C7 binds to plasminogen through the lysine-binding sites of plasminogen potentially targeting plasmin activity to sites of complement deposition.[7]

TISSUE DISTRIBUTION

C7 is not synthesised by hepatocytes but rather by cells of monocyte origin such as Kupffer cells. C7 is normally in the range of 50–90 µg/mL in human plasma.[8] Other cell types reported to produce C7 in vitro include fibroblasts,[9] neurons[10] and glial cells.[11]

REGULATION OF EXPRESSION

C7 is not listed as a major acute-phase protein[12]; however, INF-γ induced C7 expression in astrocyte cell lines.[11]

HUMAN PROTEIN SEQUENCE[1]

```
MKVISLFILV GFIGEFQSFS SASSPVNCQW DFYAPWSECN GCTKTQTRRR   50
SVAVYGQYGG QPCVGNAFET QSCEPTRGCP TEECCCERFR CFGGQCISKS  100
LVCNGDSDCD EDSADEDRCE DSERRPSCDI DKPPPNIELT GNGYNELTGQ  150
FRNRVINTKS FGGQCRKVFS GDGKDFYRLS GNVLSYTFQV KINNDFNYEF  200
YNSTWSYVKH TSTEHTSSSR KRSFFRSSSS SSRSYTSHTN EIHKGKSYQL  250
LVVENTVEVA QFINNNPEFL QLAEPFWKEL SHLPSLYDYS AYRRLIDQYG  300
THYLQSGSLG GEYRVLFYVD SEKLKQNDFN SVEEKKCKSS GWHFVVKFSS  350
HGCKELENAL KAASGTQNNV LRGEPFIRGG GAGFISGLSY LELDNPAGNK  400
RRYSAWAESV TNLPQVIKQK LTPLYELVKE VPCASVKKLY LKWALEEYLD  450
EFDPCHCRPC QNGGLATVEG THCLCHCKPY TFGAACEQGV LVGNQAGGVD  500
GGWSCWSSWS PCVQGKKTRS RECNNPPPSG GGRSCVGETT ESTQCEDEEL  550
EHLRLLEPHC FPLSLVPTEF CPSPPALKDG FVQDEGPMFP VGKNVVYTCN  600
EGYSLIGNPV ARCGEDLRWL VGEMHCQKIA CVLPVLMDGI QSHPQKPFYT  650
VGEKVTVSCS GGMSLEGPSA FLCGSSLKWS PEMKNARCVQ KENPLTQAVP  700
KCQRWEKLQN SRCVCKMPYE CGPSLDVCAQ DERSKRILPL TVCKMHVLHC  750
QGRNYTLTGR DSCTLPASAE KACGACPLWG KCDAESSKCV CREASECEEE  800
GFSICVEVNG KEQTMSECEA GALRCRGQSI SVTSIRPCAA ETQ         843
```

The leader sequence is underlined, and the *N*-linked glycosylation sites are shown in bold (**N**). Sites of C-mannosyl tryptophans[13] are in italics and underlined (W).

CHROMOSOMAL LOCATION

Human: 5p13.1
Murine: 15A1; 15 1.98 cM
Rat: 2q16

The human genes for C6 and C7 are separated by only 160 kbp on chromosome 5 and are positioned in a 'head-to-tail' manner relative to each other (Fig. 20.3).[14,15]

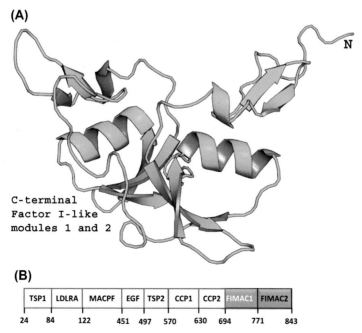

(A)

N

C-terminal
Factor I-like
modules 1 and 2

(B)

TSP1	LDLRA	MACPF	EGF	TSP2	CCP1	CCP2	FIMAC1	FIMAC2

24   84   122   451   497   570   630   694   771   843

FIGURE 20.3 Crystal structure for human complement component C7. (A) Ribbon diagram of the C-terminal factor I-like modules of C7, whose structure is determined with the help of solution NMR (PDB ID: 2WCY). (B) Schematic representation of C7 domains.

HUMAN cDNA SEQUENCE[1]

```
AGGGAGAGGC AGAGAGGCAG GCAGCCTGCT GGGCTCTTCC TGCTGTTGAA AACTTACCCG   60
GCCCTTACAG AGGAAATCTT CCTCCTCTCT TCTGCCCTGA ATGTTTTCCC AAACATGAAG  120
GTGATAAGCT TATTCATTTT GGTGGGATTT ATAGGAGAGT TCCAAAGTTT TTCAAGTGCC  180
TCCTCTCCAG TCAACTGCCA GTGGGACTTC TATGCCCCTT GGTCAGAATG CAATGGCTGT  240
ACCAAGACTC AGACTCGCAG GCGGTCAGTT GCTGTGTATG GGCAGTATGG AGGCCAGCCT  300
TGTGTTGGAA ATGCTTTTGA AACACAGTCC TGTGAACCTA CAAGAGGATG TCCAACAGAG  360
GAGGGATGTG GAGAGCGTTT CAGGTGCTTT TCAGGTCAGT GCATCAGCAA ATCATTGGTT  420
TGCAATGGGG ATTCTGACTG TGATGAAGAC AGTGCTGATG AAGACAGATG TGAGGACTCA  480
GAAAGGAGAC CTTCCTGTGA TATCGATAAA CCTCCTCCTA ACATAGAACT TACTGGAAAT  540
GGTTACAATG AACTCACTGG CCAGTTTAGG AACAGAGTCA TCAATACCAA AAGTTTTGGT  601
GGTCAATGTA GAAAGGTGTT TAGTGGGGAT GGAAAAGATT TCTACAGGCT GAGTGGAAAT  660
GTCCTGTCCT ATACATTCCA GGTGAAAATA AATAATGATT TTAATTATGA ATTTTACAAT  720
AGTACTTGGT CTTATGTAAA ACATACGTCG ACAGAACACA CATCATCTAG TCGGAAGCGC  780
TCCTTTTTTA GATCTTCATC ATCTTCTTCA CGCAGTTATA CTTCACATAC CAATGAAATC  840
CATAAAGGAA AGAGTTACCA ACTGCTGGTT GTTGAGAACA CTGTTGAAGT GGCTCAGTTC  900
ATTAATAACA ATCCAGAATT TTTACAACTT GCTGAGCCAT TCTGGAAGGA GCTTTCCCAC  960
CTCCCCTCTC TGTATGACTA CAGTGCCTAC CGAAGATTAA TCGACCAGTA CGGGACACAT 1020
TATCTGCAAT CTGGGTCGTT AGGAGGAGAA TACAGAGTTC TATTTTATGT GGACTCAGAA 1080
AAATTAAAAC AAAATGATTT TAATTCAGTC GAAGAAAAGA AATGTAAATC CTCAGGTTGG 1140
CATTTTGTCG TTAAATTTTC AAGTCATGGA TGCAAGGAAC TGGAAAACGC TTTAAAAGCT 1200
GCTTCAGGAA CCCAGAACAA TGTATTGCGA GGAGAACCGT TCATCAGAGG GGGAGGTGCA 1260
GGCTTCATAT CTGGCCTTAG TTACCTAGAG CTGGACAATC CTGCTGGAAA CAAAAGGCGA 1320
TATTCTGCCT GGGCAGAATC TGTGACTAAT CTTCCTCAAG TCATAAAACA AAAGCTGACA 1380
CCTTTATATG AGCTGGTAAA GGAAGTACCT TGTGCCTCTG TGAAAAAACT ATACCTGAAA 1440
TGGGCTCTTG AAGAGTATCT GGATGAATTT GACCCCTGTC ATTGCCGGCC TTGTCAAAAT 1500
GGTGGTTTGG CTACTGTTGA GGGGACCCAT TGTCTGTGCC ATTGCAAACC GTACACATTT 1560
GGTGCGGCGT GTGAGCAAGG AGTCCTCGTA GGGAATCAAG CAGGAGGGGT TGATGGAGGT 1620
TGGAGTTGCT GGTCCTCTTG GAGCCCCTGT GTCCAAGGGA AGAAAACAAG AAGCCGTGAA 1680
TGCAATAACC CACCTCCCAG TGGGGGTGGG AGATCCTGCG TTGGAGAAAC GACAGAAAGC 1740
ACACAATGCG AAGATGAGGA GCTGGAGCAC TTGAGGTTGC TTGAACCACA TTGCTTTCCT 1800
TTGTCTTTGG TTCCAACAGA ATTCTGTCCA TCACCTCCTG CCTTGAAAGA TGGGATTTGT 1860
CAAGATGAAG GTACAATGTT TCCTGTGGGG AAAAATGTAG TGTACACTTG CAATGAAGGA 1920
TACTCTCTTA TTGGAAACCC AGTGGCCAGA TGTGGAGAAG ATTTACGGTG GCTTGTTGGG 1980
GAAATGCATT GTCAGAAAAT TGCCTGTGTT CTACCGTGAC TGATGGATGG CATACAGAGT 2040
CACCCCCAAA AACCTTTCTA CACAGTTGGT GAGAAGGTGA CTGTTTCCTG TTCAGGTGAC 2100
ATGTCCTTAG AAGGTCCTTC AGCATTTCTC TGTGGCTCCA GCCTTAAGTG GAGTCCTGAG 2160
ATGAAGAATG CCCGCTGTGT ACAAAAAGAA AATCCGTTAA CACAGGCAGT GCCTAAATGT 2220
CAGCGCTGGG AGAAACTGCA GAATTCAAGA TGTGTTTGTA AAATGCCCTA CGAATGTGGA 2280
CCTTCCTTGG ATGTATGTGC TCAAGATGAG AGAAGCAAAA GGATACTGCC TCTGACAGTT 2340
TGCAAGATGC ATGTTCTCCA CTGTCAGGGT AGAAATTACA CCCTTACTGG TAGGGACAGC 2400
TGTACTCTGC CTGCCTCAGC TGAGAAAGCT TGTGGTGCCT GCCCACTGTG GGGAAAATGT 2460
GATGCTGAGA GCAGCAAATG TGTCTGCCGA GAAGCATCGG AGTGCGAGGA AGAAGGGTTT 2520
AGCATTTGTG TGGAAGTGAA CGGCAAGGAG CAGACGATGT CTGAGTGTGA GGCGGGCGCT 2580
CTGAGATGCA GAGGGCAGAG CATCTCTGTC ACCAGCATAA GGCCTTGTGC TGCGGAAACC 2640
CAGTAGGCTC CTGGAGGCCC TGGTCAGCTT GCTTGGAATC CAGCAGGCAG CTGGGGCTGA 2700
GTGAAAACAT CTGCACAACT GGGCACTGGA CAGCTTTTCC TTCTTCTCCA GTGTCTACCT 2760
TCCTCCTCAA CTCCCAGCCA TCTGTATAAA CACAATCCTT TGTTCTCCCA AATCTGAATC 2820
GAATTACTCT TTTTGCCTCCT TTTTAATGTC AGTAAGGATA TGAGCCTTTG CACAGGCTGG 2880
CTGCGTGTTC TTGAAATAGG TGTTACCTTC TCTGGGCCTT GGTTTTTTAA AATCTGTAAA 2940
ATTAGAGGAT TGCACTAGAG AAACTTGAAT GCTCCATTCA GGCCTATCAT TTTATTAAGT 3000
ATGATTGACA CAGCCCATGG GCCAGAACAC ACTCTACAAA ATGACTAGGA TAACAGAAAG 3060
AACGTGATCT CCTGATTAGA GAGGGTGGTT TTCCTCAATG GAACCAAATA TAAAGAGGAC 3120
TTGAACAAAA ATGACAGATA CAAACTATTT CTATCCTGAG TAGTAATCTC ACACTTCATC 3180
CTATAGAGTC AACCACCACA GATAGGAATT CCTTATTCTT TTTTTAATTT TTTTAAGACA 3240
GAGTCTCACT TTGTTGCCCA GGCTGGAGCG CAGTGGGGTG ATCTCATCTC CCTGCAACCT 3300
CCGCCTCCTG GGTTCAAGCG ATTCTTGTGC CTCAGCTTCC CAAGCAGCTG GGATTACAGG 3360
```

```
TGCCCGCCAC CACGCCCAGC TAATTTTTGC ATTTTTAGTA GAGATGGGGT TTCACCATGT 3420
TGGCCACGCT CGTCTCCAAC TCCTGACCTC AGGTAATCCG CCTGCCTTGG CCTCCCAAAG 3480
TGCTGGGATT ACAGACATGA ACCACCACGC CTGGCTGGAA TACTTACTCT TGTCGGGAGA 3540
TTGAACCACT AAAATGTTAG AGCAGAATTC ATTATGCTGT GGTCACAGGG GTGTCTTGTC 3600
TGAGAACAAA TACAATTCAG TCTTCTCTTT GGGGTTTTAG TATGTGTCAA ACATAGGACT 3660
GGAAGTTTGC CCCTGTTCTT TTTTCTTTTG AAAGAACATC AGTTCATGCC TGAGGCATGA 3720
GTGACTGTGC ATTTGAGAAT AGTTTTCCCT ATTCTGTGGA TACAGTCCCA GAGTTTTCAG 3780
GGAGTACACA GGTAGATTAG TTTGAAGCAT TGACCTTTTA TTTATTCCTT ATTTCTCTTT 3840
CATCAAAACA AAACAGCAGC TGTGGGAGGA GAAATGAGAG GGCTTAAATG AAATTTAAAA 3900
TAAGCTATAT TATACAAATA CTATCTCTGT ATTGTTCTGA CCCTGGTAAA TATATTTCAA 3960
AACTTCAGAT GACAAGGATT AGAACACTCA TTAAAGATGC TATTCTTCAG AAAAAAAAAA 4020
AAAAAAAAAA AAAA
```

The intron–exon boundaries are indicated by underlining the first five nucleotides. The initiation codon is double-underlined as are the termination codon and polyadenylation signal.

GENOMIC STRUCTURE

The gene for C7 spans ~80 kb and includes 18 exons (Fig. 20.4). It is linked closely to the gene for C6, but in opposite orientations.

ACCESSION NUMBERS

Protein
Human NP_000578
Murine: NP_001230766

mRNA
Human: NM_000587
Murine: NM_001243837

Gene
Human: NG_011692
Murine: Gene ID: 109828

DEFICIENCY

C7 deficiency: OMIM: 610102, 217070.

Several mutations are reported in which base pair deletions cause formation of truncated C7 polypeptides. These include a mutation in exon 10 leading to a premature stop codon C464X; an 11 base pair deletion in exon 6; and a 2 base pair deletion in exon 14.[16] Deletions of exons 8 and 9 in the gene for C7 causes

FIGURE 20.4 A scheme for the gene of C7. The 18 exons are shown as *vertical lines*, and the scale is represented by the *bar* (5 kb). The gene is ~80 kb.

a deficiency resulting in an increased susceptibility to neisserial infections.[17] In a Japanese study, a subject was found to have a transversion in exon 16 resulting in a stop codon (C728X), and in another person a 2 base pair deletion also resulted in a truncated protein.[18] A combined subtotal deficiency in both C6 and C7 is reported in which a defect at a 5′ donor splice site in intron 15 of the gene for C6 results in a truncated protein, along with a single base change in C7 in exon 11 (R521S). The latter alteration results in a C7 protein of normal size but of low circulation concentration.[19] Deficiencies in C7 are associated with susceptibility to meningococcal infections.[20,21]

POLYMORPHIC VARIANTS

By isoelectric focusing, Caucasians have one common allele for C7, designated C7*1 (0.99), along with several very rare alleles. Within the Japanese population four alleles have been described, although C7*1 is predominant (0.84) in this group as well.[22,23] Two common isoelectrically similar C7 alleles are C7M (0.8) and C7N (0.2) T565P. C7N has slightly lower specific haemolytic activity than C7M.[24,25]

Other polymorphic variants include T576S,[26] S389T,[27] R220Q, E682Q, R687H[28] and G379R.[29]

MUTANT ANIMALS

C7-deficient mice were generated in C57BL/6J ES cells using a targeted insertion of a gene trap vector in the intronic region between exons 6 and 7 of the *C7* gene. Single insertion of the vector was confirmed by FISH analysis and targeting position confirmed by sequence analysis. Heterozygous chimeric mice were backcrossed to C57BL/6J mice. Successful deletion of C7 was verified by RT-PCR and haemolytic assay. C7-deficient mice when aerosol challenged with *Mycobacterium tuberculosis* had reduced bacterial burden, lower lung occlusion, increased alveolar lymphocytic infiltration and increased levels of lung IFN-γ and TNF-α compared to wild-type mice.[30]

REFERENCES

1. DiScipio RG, Chakravarti DN, Muller-Eberhard HJ, Fey GH. The structure of human complement component C7 and the C5b-7 complex. *J Biol Chem* 1988;**263**:549–60.
2. Phelan MM, Thai CT, Soares DC, Ogata RT, Barlow PN, Bramham J. Solution structure of factor I-like modules from complement C7 reveals a pair of follistatin domains in compact pseudosymmetric arrangement. *J Biol Chem* 2009;**284**:19637–49.
3. Sonnen AF, Henneke P. Structural biology of the membrane attack complex. *Sub-cellular Biochem* 2014;**80**:83–116.
4. Kolb WP, Muller-Eberhard HJ. The membrane attack mechanism of complement. Isolation and subunit composition of the C5b-9 complex. *J Exp Med* 1975;**141**:724–35.
5. Hugo F, Berstecher C, Kramer S, Fassbender W, Bhakdi S. In vivo clearance studies of the terminal fluid-phase complement complex in rabbits. *Clin Exp Immunol* 1989;**77**:112–6.

6. Bossi F, Rizzi L, Bulla R, et al. C7 is expressed on endothelial cells as a trap for the assembling terminal complement complex and may exert anti-inflammatory function. *Blood* 2009;**113**:3640–8.

7. Reinartz J, Hansch GM, Kramer MD. Complement component C7 is a plasminogen-binding protein. *J Immunol* 1995;**154**:844–50.

8. Wurzner R, Joysey VC, Lachmann PJ. Complement component C7. Assessment of in vivo synthesis after liver transplantation reveals that hepatocytes do not synthesize the majority of human C7. *J Immunol* 1994;**152**:4624–9.

9. Garred P, Hetland G, Mollnes TE, Stoervold G. Synthesis of C3, C5, C6, C7, C8, and C9 by human fibroblasts. *Scand J Immunol* 1990;**32**:555–60.

10. Terai K, Walker DG, McGeer EG, McGeer PL. Neurons express proteins of the classical complement pathway in Alzheimer disease. *Brain Res* 1997;**769**:385–90.

11. Gasque P, Fontaine M, Morgan BP. Complement expression in human brain. Biosynthesis of terminal pathway components and regulators in human glial cells and cell lines. *J Immunol* 1995;**154**:4726–33.

12. kilicarslan A, Uysal A, Roach EC. Acute phase reactants. *Acta Medica* 2013;**2**:2–7.

13. Hofsteenge J, Blommers M, Hess D, Furmanek A, Miroshnichenko O. The four terminal components of the complement system are C-mannosylated on multiple tryptophan residues. *J Biol Chem* 1999;**274**:32786–94.

14. Hobart MJ, Fernie BA, DiScipio RG, Lachmann PJ. A physical map of the C6 and C7 complement component gene region on chromosome 5p13. *Hum Mol Genet* 1993;**2**:1035–6.

15. Hobart MJ, Fernie BA, DiScipio RG. Structure of the human C7 gene and comparison with the C6, C8A, C8B, and C9 genes. *J Immunol* 1995;**154**:5188–94.

16. Barroso S, Rieubland C, Jose alvarez A, et al. Molecular defects of the C7 gene in two patients with complement C7 deficiency. *Immunology* 2006;**118**:257–60.

17. Thomas AD, Orren A, Connaughton J, Feighery C, Morgan BP, Roberts AG. Characterization of a large genomic deletion in four Irish families with C7 deficiency. *Mol Immunol* 2012;**50**:57–9.

18. Nishizaka H, Horiuchi T, Zhu ZB, Fukumori Y, Volanakis JE. Genetic bases of human complement C7 deficiency. *J Immunol* 1996;**157**:4239–43.

19. Fernie BA, Wurzner R, Orren A, et al. Molecular bases of combined subtotal deficiencies of C6 and C7: their effects in combination with other C6 and C7 deficiencies. *J Immunol* 1996;**157**:3648–57.

20. Barroso S, Lopez-Trascasa M, Merino D, Alvarez AJ, Nunez-Roldan A, Sanchez B. C7 deficiency and meningococcal infection susceptibility in two Spanish families. *Scand J Immunol* 2010;**72**:38–43.

21. Mayilyan KR. Complement genetics, deficiencies, and disease associations. *Protein & Cell* 2012;**3**:487–96.

22. Hobart MJ, Joysey V, Lachmann PJ. Inherited structural variation and linkage relationships of C7. *J Immunogenet* 1978;**5**:157–63.

23. Nakamura S, Ooue O, Abe K. Genetic polymorphism of the seventh component of complement in a Japanese population. *Hum Genet* 1984;**66**:279–81.

24. Wurzner R, Fernie BA, Jones AM, Lachmann PJ, Hobart MJ. Molecular basis of the complement C7 M/N polymorphism. A neutral amino acid substitution outside the epitope of the allospecific monoclonal antibody WU 4-15. *J Immunol* 1995;**154**:4813–9.

25. Wurzner R, Orren A, Potter P, et al. Functionally active complement proteins C6 and C7 detected in C6- and C7-deficient individuals. *Clin Exp Immunol* 1991;**83**:430–7.

26. Dewald G, Nothen MM, Ruther K. A common Ser/Thr polymorphism in the perforin-homologous region of human complement component C7. *Hum Hered* 1994;**44**:301–4.

27. Fernie BA, Wurzner R, Unsworth DJ, Tuxworth RI, Hobart MJ. DNA polymorphisms of the complement C6 and C7 genes. *Ann Hum Genet* 1995;**59**:163–81.

28. Fernie BA, Hobart MJ. Complement C7 deficiency: seven further molecular defects and their associated marker haplotypes. *Hum Genet* 1998;**103**:513–9.

29. Fernie BA, Orren A, Schlesinger M, et al. DNA haplotypes of the complement C6 and C7 genes associated with deficiencies of the seventh component; and a new DNA polymorphism in C7 exon 13. *Ann Hum Genet* 1997;**61**:287–98.

30. Welsh KJ, Lewis CT, Boyd S, Braun MC, Actor JK. Complement factor C7 contributes to lung immunopathology caused by *Mycobacterium tuberculosis*. *Clin Dev Immunol* 2012;**2012**:429675.

Chapter 21

C8

Richard G. DiScipio[1,2]
[1]*Torrey Pines Institute for Molecular Studies, San Diego, CA, United States;* [2]*Sanford Burnham Prebys Medical Discovery Institute, La Jolla, CA, United States*

PHYSIOCHEMICAL PROPERTIES

	C8α	C8β	C8γ
pI apoprotein	5.74	7.85	8.52
pI glycosylated	~5.4	~7.5	
M_r apoprotein	61,700	61,000	20,400
M_r glycosylated	~64,000	~63,500	
Amino acids (preform)	584	591	202
(Mature)	554	537	182
N-linked glycosylation sites (proform numbering)	43,*437*	101,243,553	

Asparagines designated by italicised residue numbers are probably not glycosylated.

Mature Protein		
M_r	143,100 (apoprotein)	~150,000 (glycoprotein)
pI	~6.9 (apoprotein)	~6.6 (glycoprotein)

1273 amino acids

Intrachain disulphide bond: C8α (Cys194) to C8γ (Cys60) (proform numbering)

STRUCTURE

Complement C8 consists of three polypeptide chains two of which (C8α and C8β) are noncovalently associated, whereas the C8γ is disulphide bridged to C8α. Both C8α and C8β are homologues of C6, C7 and C9 (Fig. 21.1). Linearly, these are arranged starting with a module (54–55 amino acids) that is related to those found in thrombospondin (TS), followed by an LDL-related module (LDL), the MACPF domain of about ~260 amino acids, followed by an epidermal growth factor module (EGF), and ending with another TS module. In contrast C8γ is a single domain polypeptide that is a member of the lipocalin family (Fig. 21.2).

The Complement FactsBook. http://dx.doi.org/10.1016/B978-0-12-810420-0.00021-3

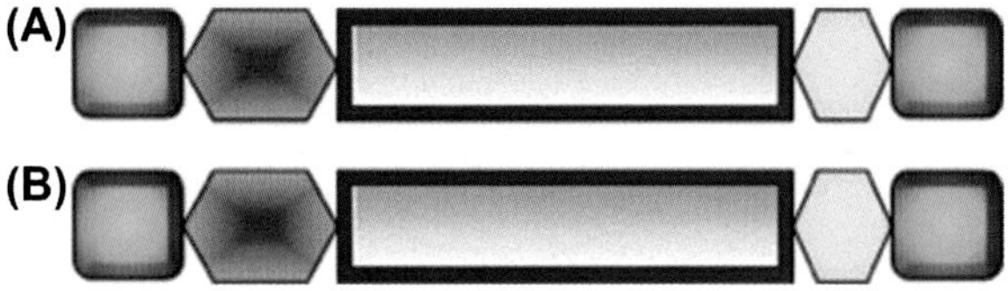

(A)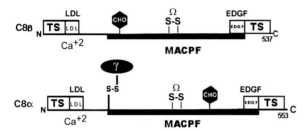

(B)

FIGURE 21.1   Diagram of protein domains. (For domain definitions, see key in front of book.)

FIGURE 21.2   Sketches of the polypeptide chains for complement C8. The polypeptide chains of C8α and C8β are homologous, each containing two thrombospondin (TS), one LDL, one epidermal growth factor (EDGF) modules, along with a single domain, the MACPF, represented by a thick line. Locations of the known N-linked carbohydrate sites are represented by CHO-labelled hexagons. The locations of the disulphide bonded Ω-loops are shown. The γ-chain is disulphide bridged to the α-subunit as displayed.

C8α

1–30	Leader	exons 1/2
34–88	TS1	exons 2/3
89–130	LDL	exons 3/4
233–496	MACPF	exons 6–10
497–528	EGF	exon 10
534–583	TS2	exon 11

C8β

1–54	Leader	exons 1/2
60–114	TS1	exons 2/3
115–155	LDL	exons 3/4
248–500	MACPF	exons 6–10
501–534	EGF	exons 10/11
544–591	TS2	exon 12

C8γ

1–20	Leader	exon 1
21–202	Lipocalin domain	exons 1–5

CRYSTAL STRUCTURES

C8α MACPF domain pdb: 2QQH,[1] pdb: 2RD7 [2]

C8α pdb: 3OJY [3]

C8γ pdb: 2OVE,[4] pdb: 1LF7 [5], pdb: 3OJY,[3] pdb: 2QOS [6]

C8 pdb: 3OJY [3]

Fig. 21.3 shows the crystal structure of C8.

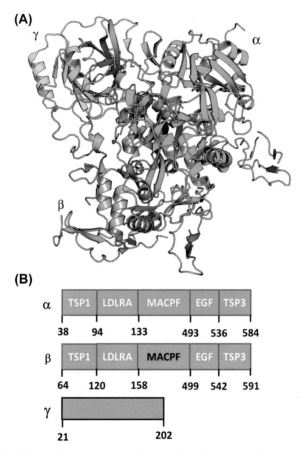

FIGURE 21.3 Diagram of crystal structure for C8. (A) Ribbon diagram of the heterotrimeric α-, β- and γ-chains of C8 crystal structure (PDB ID: 3OJY). (B) Schematic representation of C8 α-, β- and γ- chains, each of their domains coloured according to their ribbon representation.

FUNCTION

Complement component C8 binds to C5b7 and integrates into the growing membrane attack complex (MAC), a ~100 Å transmembrane channel structure that destroys the functional integrity of cells.[7] Through the C8 subunit, a binding site is created for the first C9 molecule, and subsequently C9 molecules accumulate circularly to form the final assembly characterised by a transmembrane pore of about 100 Å. C8 is also a subunit of the soluble terminal complex component sC5b-9 that also contains vitronectin and clusterin.[8] The functions of this macromolecular assembly are not comprehensively established, but may include clearance[9] and wound healing mediated through the vitronectin subunit.[10]

TISSUE DISTRIBUTION

Normal plasma concentration of human C8 is 55–80 μg/mL, and the major cell type for synthesis for plasma C8 is the liver hepatocyte.[11] However, several cell types in vitro (including monocytes,[12] glial cells[13] and neurons,[14] endothelial cells[15] and fibroblasts[16]) are reported to produce this protein.

REGULATION OF EXPRESSION

C8αγ and C8β are synthesised and secreted independently. Although C8 is not reported to be a major acute-phase reactant,[17] HepG2 cells increase expression of C8αγ in response to IL-1β, IL-6 and IFN-γ, whereas only IFN-γ upregulates C8β production.[18]

HUMAN PROTEIN SEQUENCE

C8α[19]

MFAVVFFILS	LMTCQPGVTA	QEKVNQRVRR	AATPAAVTCQ	LS**N**WSEWTDC	50
FPCQDKKYRH	RSLLQPNKFG	GTICSGDIWD	QASCSSSTTC	VRQAQCGQDF	100
QCKETGRCLK	RHLVCNGDQD	CLDGSDEDDC	EDVRAIDEDC	SQYEPIPGSQ	150
KAALGYNILT	QEDAQSVYDA	SYYGGQCETV	YNGEWRELRY	DSTCERLYYG	200
DDEKYFRKPY	NFLKYHFEAL	ADTGISSEFY	DNANDLLSKV	KKDKSDSFGV	250
TIGIGPAGSP	LLVGVGVSHS	QDTSFLNELN	KYNEKKFIFT	RIFTKVQTAH	300
FKMRKDDIML	DEGMLQSLME	LPDQYNYGMY	AKFINDYGTH	YITSGSMGGI	350
YEYILVIDKA	KMESLGITSR	DITTCFGGSL	GIQYEDKINV	GGGLSGDHCK	400
KFGGGKTERA	RKAMAVEDII	SRVRGGSSGW	SGGLAQ**N**RST	ITYRSWGRSL	450
KYNPVVIDFE	MQPIHEVLRH	TSLGPLEAKR	QNLRRALDQY	LMEFNACRCG	500
PCFNNGVPIL	EGTSCRCQCR	LGSLGAACEQ	TQTEGAKADG	S<u>W</u>SC<u>W</u>SS<u>W</u>SV	550
CRAGIQERRR	ECDNPAPQNG	GASCPGRKVQ	TQAC		580

C8β[20, 21]

MKNSRTWAWR	APVELFLLCA	ALGCLSLPGS	RGERPHSFGS	NAVNKSFAKS	50
RQMRSVDVTL	MPIDCELSS<u>W</u>	SS<u>W</u>TTCDPCQ	KKRYRYAYLL	QPSQFHGEPC	100
NFSDKEVEDC	VTNRPCRSQV	RCEGFVCAQT	GRCVNRRLLC	NGDNDCGDQS	150
DEANCRRIYK	KCQHEMDQYW	GIGSLASGIN	LFTNSFEGPV	LDHRYYAGGC	200
SPHYILNTRF	RKPYNVESYT	PQTQGKYEFI	LKEYESYSDF	ER**N**VTEKMAS	250
KSGFSFGFKI	PGIFELGISS	QSDRGKHYIR	RTKRFSHTKS	VFLHARSDLE	300
VAHYKLKPRS	LMLHYEFLQR	VKRLPLEYSY	GEYRDLFRDF	GTHYITEAVL	350
GGIYEYTLVM	NKEAMERGDY	TLNNVHACAK	NDFKIGGAIE	EVYVSLGVSV	400
GKCRGILNEI	KDRNKRDTMV	EDLVVLVRGG	ASEHITTLAY	QELPTADLMQ	450
EWGDAVQYNP	AIIKVKVEPL	YELVTATDFA	YSSTVRQNMK	QALEEFQKEV	500
SSCHCAPCQG	NGVPVLKGSR	CDCICPVGSQ	GLACEVSYRK	NTPIDGKWNC	550
<u>W</u>S**N**<u>W</u>SSCSGR	RKTRQRQCNN	PPPQNGGSPC	SGPASETLDC	S	591

C8γ[22]

MLPPGTATLL	TLLLAAGSLG	QKPQRPRRPA	SPISTIQPKA	NFDAQQFAGT	50
WLLVAVGSAC	RFLQEQGHRA	EATTLHVAPQ	GTAMAVSTFR	KLDGICWQVR	100
QLYGDTGVLG	RFLLQARDAR	GAVHVVVAET	DYQSFAVLYL	ERAGQLSVKL	150
YARSLPVSDS	VLSGFEQRVQ	EAHLTEDQIF	YFPKYGFCEA	ADQFHVLDEV	200
RR					202

Leader sequences are identified by underlining, and potential *N*-linked glycosylation sites are represented in bold (**N**); however, the asparagines (bold/italics) in C8α at position 43 and those in C8β at positions 101 and 553 are probably not glycosylated. C-mannosylated tryptophans[23] are represented in italics and underlined (*W*).

CHROMOSOMAL LOCATION

	C8α	C8β	C8γ
Human	1p32	1p32	9q34.3
Murine	4, 48.69 cM	4, 48.61 cM	2A3, 2 17.31 cM
Rat	5q34	5q34	3p13

HUMAN cDNA SEQUENCE

C8α[19]

```
TGTGTATCTG GGTGAGTTTC CAACATCAGA TAGATCTTAC AGGTCCCAGC CTGTAGACAT   60
CTTTTACTCC AATTTCCTGA ATAGATAGCT TTATTCCTTC AAGGTAATAT AGTGCGGTGG  120
CTTCTGGCTG AGATGTTTGC TGTTGTTTTC TTCATCTTGT CTTTGATGAC TTGTCAGCCT  180
GGGGTAACTG CACAGGAGAA GGTGAACCAG AGAGTAAGAC GGGCAGCTAC ACCCGCAGCA  240
GTTACCTGCC AGCTGAGCAA CTGGTCAGAG TGGACAGATT GCTTTCCGTG CCAGGACAAA  300
AAGTACCGAC ACCGGAGCCT CTTGCAGCCA AACAAGTTTG GGGGAACCAT CTGCAGTGGT  360
GACATCTGGG ATCAAGCCAG CTGCTCCAGT TCTACAACTT GTGTAAGGCA AGCACAGTGT  420
GGACAGGATT TCCAGTGTAA GGAGACAGGT CGCTGCCTGA AACGCCACCT TGTGTGTAAT  480
GGAGACCAGG ACTGCCTTGA TGGCTCTGAT GAGGACGACT GTGAAGATGT CAGGGCCATT  540
GACGAAGACT GCAGCCAGTA TGAACCAATT CCAGGATCAC AGAAGGCAGC CTTGGGGTAC  600
AATATCCTGA CCCAGGAAGA TGCTCAGAGT GTGTACGATG CCAGTTATTA TGGGGGCCAG  660
TGTGAGACGG TATACAATGG GGAATGGAGG GAGCTTCGAT ATGACTCCAC CTGTGAACGT  720
CTCTACTATG GAGATGATGA GAAATACTTT CGGAAACCCT ACAACTTTCT GAAGTACCAC  780
TTTGAAGCCC TGGCAGATAC TGGAATCTCC TCAGAGTTTT ATGATAATGC AAATGACCTT  840
CTTTCCAAAG TTAAAAAAGA CAAGTCTGAC TCATTTGGAG TGACCATCGG CATAGGCCCA  900
GCCGGCAGCC CTTTATTGGT GGGTGTAGGT GTATCCCACT CACAAGACAC TTCATTCTTG  960
AACGAATTAA ACAAGTATAA TGAGAAGAAA TTCATTTTCA CAAGAATCTT CACAAAGGTG 1020
CAGACTGCAC ATTTTAAGAT GAGGAAGGAT GACATTATGC TGGATGAAGG AATGCTGCAG 1080
TCATTAATGG AGCTTCCAGA TCAGTACAAT TATGGCATGT ATGCCAAGTT CATCAATGAC 1140
TATGGCACCC ATTACATCAC ATCTCGGATCC ATGGGTGGCA TTTATGAATA TATCCTGGTG 1200
ATTGACAAAG CAAAAATGGA ATCCCTTGGT ATTACCAGCA GAGATATCAC GACATGTTTT 1260
GGAGGCTCCT TGGGCATTCA ATATGAAGAC AAAATAAATG TTGGTGGAGG TTTATCAGGA 1320
GACCATTGTA AAAAATTGG AGGTGGCAAA ACTGAAAGGG CCAGGAAGGC CATGGCTGTG 1380
GAAGACATTA TTTCTCGGGT GCGAGGTGGC AGTTCTGGCT GGAGCGGTGG CTTGGCACAG 1440
AACAGGAGCA CCATTACATA CCGTTCCTGG GGGAGGTCAT TAAAGTATAA TCCTGTTGTT 1500
ATCGATTTTG AGATGCAGCC TATCCACGAG GTGCTGCGGC ACACAAGCCT GGGGCCTCTG 1560
GAGGCCAAGC GCCAGAACCT GCGCCGCGCC TTGGACCAGT ATCTGATGGA ATTCAATGCC 1620
TGCCGATGTG GGCCTTGCTT CAACAATGGG GTGCCCATCC TCGAGGGCAC CAGCTGCAGG 1680
TGCCAGTGCC GCCTGGGTAG CTTGGGTGCT GCCTGTGAGC AAACACAGAC AGAAGGAGCC 1740
AAAGCAGATG GGAGCTGGAG TTGCTGGAGC TCCTGGTCTG TATGCAGAGC AGGCATCCAG 1800
GAAAGGAGAA GAGAGTGTGA CAATCCAGCA CCTCAGAATG GAGGGCCTC GTGTCCAGGG 1860
CGGAAAGTAC AGACGCAGGC TTGCTGAGGG CCTCTGGACA CAGGCTGGAC CAGATGCTGT 1920
GGATGTCGAC CCCTGCACTG ACTATTGGAT AAAGACTTCT TTCAACTAAG AGAAGATGCA 1980
```

```
AATCAGCACA CTTTTTTCTT TGTTCTGCCA GCTTCCAGGC CTAAGACTAG GTTTTGCTGT 2040
CTACAGCCAA CTATTCTATT AGTTACAAAA CTCAATCATT TTATTCAGCA ACTGGATGTT 2100
GACTGTTAAC TAGAAGCTCT GTCCTACTTA CAGCACTTTG GATCATCAAA AAAATAAAGT 2160
AAAATAGAAA ACTGAGAAAA CTCAATCCAT GACCAGGGAG AACTTACAGG ATGTTAGAGA 2220
CAAAACAAGC AGACACCTGA AACAATCAAC GCCCAATAAA ACAAAGTAGG ATGAAAATTC 2280
TCTTAGTTCT TTGATAACAA TTTGTTCACT CATAGAAACA TTATTAATTG GTAGGGTAAG 2340
CAGACACTCT GAAACAATGA GAAAAATACT AAAAATTGAC TTGAGTTATT TC
```

C8β[20, 21]

```
ATTATTCTGG TTGGTTTCCA GAGTGACAGG TAAGTTTTTG GTCTGTGCAA AGTCTGTTTC   60
CAGTCACTAG TGGCTTTCTG TTTACTTTGC AGAGCTATTT GCTCTTGGGG ACAGAAGCTG  120
ACAGTGGCAC TCACAGCACA GGCTTGTTAT GGGTCTAGCA GCCTCTGTGG CATCTCCTGT  180
CACATTGGGA AAATGAAGAA TTCCAGGACA TGGGCTTGGA GGGCGCCGGT GGAGCTATTT  240
CTTCTCTGTG CTGCCCTGGG CTGTCTCAGT TTGCCTGGCT CCAGAGGTGA AAGGCCACAT  300
TCCTTTGGGT CAAATGCAGT CAACAAGAGC TTTGCTAAGA GCAGACAGAT GCGGAGTGTG  360
GATGTTACCC TGATGCCCAT TGATTGTGAG CTGTCTAGTT GGTCCTCTTG GACCCACATGT 420
GACCCCTGTC AGAAGAAAAG GTACAGGTAT GCCTACTTGC TCCAGCCCTC TCAGTTCCAT  480
GGGGAACCGT GCAACTTCTC TGACAAGGAA GTCGAAGACT GTGTTACCAA CAGACCATGC  540
AGAAGTCAAG TGCGATGTGA AGGCTTTGTG TGTGCACAGA CAGGAAGGTG TGTAAACCGC  600
AGACTTCTTT GCAATGGGGA CAATGACTGT GGAGACCAGT CAGATGAAGC AAACTGTAGA  660
AGGATTTATA AAAAATGTCA GCATGAAATG GACCAATACT GGGGAATTGG CAGTCTGGCC  720
AGTGGGATAA ATTTGTTCAC AAACAGTTTT GAGGGCCCAG TTCTTGATCA CAGGTATTAT  780
GCAGGTGGAT GCTCCCCGCA TTACATCCTG AACACGAGGT TTAGGAAGCC CTACAATGTG  840
GAAAGCTACA CGCCACAGAC CCAAGGCAAA TACGAATTCA TATTAAAAGA GTATGAATCA  900
TACTCAGATT TTGAACGCAA TGTCACAGAG AAAATGGCAA GCAAGTCTGG TTTCAGTTTT  960
GGTTTTAAAA TACCTGGAAT ATTTGAACTT GGCATCAGTA GTCAAAGTGA TCGAGGCAAA 1020
CACTATATTA GGAGAACCAA ACGATTCTCT CATACTAAAA GCGTATTTCT GCATGCACGC 1080
TCTGACCTTG AAGTAGCACA TTACAAGCTG AAACCCAGAA GCCTCATGCT CCATTACGAG 1140
TTCCTTCAGA GAGTTAAGCG GCTGCCCCTG GAGTACAGCT ACGGGGAATA CAGAGATCTC 1200
TTCCGTGATT TTGGGACCCA CTACATCACA GAGGCTGTGC TTGGGGGCAT TTATGAATAC 1260
ACCCTCGTTA TGAACAAAGA GGCCATGGAG AGAGGAGATT ATACTCTTAA CAACGTCCAT 1320
GCCTGTGCCA AAAATGATTT TAAAATTGGT GGTGCCATTG AAGAGGTCTA CGTCAGTCTG 1380
GGTGTGTCTG TAGGCAAATG CAGAGGTATT CTGAATGAAA TAAAAGACAG AAACAAGAGG 1440
GACACCATGG TGGAGGACTT GGTGGTCCTG GTACGAGGAG GGGCAAGTGA GCACATCACC 1500
ACCCTGGCAT ACCAGGAGCT GCCGACGGCG GACCTGATGC AGGAGTGGGG AGACGCTGTG 1560
CAGTACAACC CAGCCATCAT CAAAGTTAAG GTGGAGCCTG TGTATGAACT AGTGACAGCC 1620
ACAGATTTTG CCTATTCCAG CACAGTGAGG CAGAACATGA AGCAGGCACT GGAGGAGTTC 1680
CAGAAGGAAG TTAGTTCCTG CCACTGTGCT CCCTGCCAAG GAAATGGAGT CCCTGTCCTG 1740
AAAGGATCAC GCTGTGACTG CATCTGTCCT GTTGGATCCC AAGGCCTAGC CTGTGAGGTC 1800
TCCTATCGGA AGAATACCCC CATTGATGCG AAGTGGAATT GCTGGTCAAA TTGGTCTTCA 1860
TGCTCTGGAA GACGTAAGAC AAGACAAAGG CAGTGTAACA ATCCACCTCC TCAAAATGGG 1920
GGTAGCCCCT GTTCAGGCCC TGCTTCAGAA ACACTTGACT GCTCCTAGCA GATGATACAG 1980
CAGTGGGCTA CATACAATGA GAGCCCTGAG CCCTCAAGAA CTCATGCCAG CTCAGCCCTA 2040
CACCAGTTTC CACCTGGAGT TCATGCAAGG GCAAAAGGCA GTGCCATGCA AGCTGTTTAA 2100
AATAAAGATG TTACCTTGTA AAATGCAAGT TGATTTAAAT AAATACTGAG TTAAAGGCTT 2160
AAAAAAAAAA AAAAAAAA
```

C8γ[22]

```
AGAGTAGACT CTGTCCTGGG ACTTGGTGGT GCTACCCTTG GCCTCCCACA GTCCTGCCAC   60
CCTGCTGCCG CCACCATGCT GCCCCCTGGG ACTGCGACCC TCTTGACTCT GCTCCTGGCA  120
GCTGGCTCGC TGGGCCAGAA GCCTCAGAGG CCACGCCGGC CCGCATCCCC CATCAGCACC  180
ATCCAGCCCA AGGCCAATTT TGATGCTCAG CAGTTTGCAG GGACCTGGCT CCTTGTGGCT  240
GTGGGCTCCG CTTGCCGTTT CCTGCAGGAG CAGGGCCACC GGGCCGAGGC CACCACACTG  300
CATGTGGCTC CCCAGGGCAC AGCCATGGCT GTCAGTACCT TCCGAAAGCT GGATGGGATC  360
TGCTGGCAGG TGCGCCAGCT CTATGGAGAC ACAGGGGTCC TCGGCCGCTT CCTGCTTCAA  420
GCCCGAGACG CCCGAGGGGC TGTGCACGTG GTTGTCGCTG AGACCGACTA CCAGAGTTTC  480
GCTGTCCTGT ACCTGGAGCG GGCGGGGCAG CTGTCAGTGA AGCTCTACGC CCGCTCGCTC  540
CCTGTGAGCG ACTCGGTCCT GAGTGGGTTT GAGCAGCGGG TCCAGGAGGC CCACCTGACT  600
GAGGACCAGA TCTTCTACTT CCCCAAGTAC GGCTTCTGCG AGGCTGCAGA CCAGTTCCAC  660
GTCCTGACGC AAGTGAGGAG GTGAGGCCGG CACACAGCTC CAGTGCTGAG AAGTCAGTGC  720
CCCGAGAGAC GACCCCACCA GTGGGGTGCC CGCTGCCTGT CCTCCGTGAA ACCAGCCTCA  780
GATCAGGGCC CTGCCACCCA GGGCAGGGGA TCTTCTGCCG GCTGCCCCAG AGGACAGTGG  840
GTGGAGTGGT ACCTACTTAT TAAAGTCTC AGACCCCAAA AAAAAAAA
```

The initiation and termination codons along with polyadenylation signal are double-underlined. The start of exon boundaries is shown by a single underline.

GENOMIC STRUCTURE

C8α[24]

The gene for C8α is ~70 kb consisting of 11 exons (Fig. 21.4).

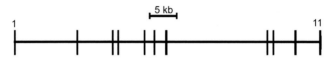

FIGURE 21.4   Schematic of the gene for the α-subunit. The gene is ~70 kb and is made of 11 exons (*vertical bars*). A scale marker (5 kb) is shown.

C8β[25]

The gene for C8β is ~40 kb and consists of 12 exons (Fig. 21.5). The genes for C8α and C8β are closely linked in a head-to-tail orientation on chromosome 1p.

FIGURE 21.5   Schematic of the gene for the β-subunit. This gene is ~40 kb and contains 12 exons (*vertical bars*). The scale bar is 5 kb.

C8γ[26]

The gene for C8γ is 1.8 kb and has 7 exons (Fig. 21.6). C8γ is in a region encoding other genes for members of the lipocalin family including α1-microglobulin and α1-acid glycoprotein.[26]

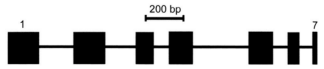

FIGURE 21.6   Schematic of the gene for the γ-chain of C8. This gene is 1.8 kb long and includes 7 exons (*vertical lines*). The scale line is 200 bp.

ACCESSION NUMBERS

	C8α	C8β	C8γ
Protein			
Human	NP_000553	NP_000057	NP_000597
Murine	NP_666260	NP_598643	NP_081338
mRNA			
Human	NM_000562	NM_000606	NM_000066
		NM_001278543	
		NM_001278544	
Murine	AB077298	AB077306	BC019967
			NM_027062.2
Gene			
Human	NG_012049	NG_007285	NG_029580
Murine	ID: 230558	ID: 110382	ID: 69379

DEFICIENCY

C8α:
OMIM: 120950
OMIM: 613790
C8β:
OMIM: 120960
OMIM: 613789

A total deficiency in C8αγ activity was reported to coincide with a near deficiency of functionally active C8β, but the genetic defect was not reported.[27] Two Japanese subjects were found to have C8α deficiencies with differing underlying defects. In one case, a base pair change at the intron–exon 2 boundary was found to be causative. In the other case, both a 5'-splice mutation and a base pair alteration causing a premature stop codon R394X resulted in the deficiency.[28] In another C8β deficiency, a premature stop codon in exon 9 was found to be responsible.[29] C8α and C8β deficiencies are autosomal recessive traits that are associated with an increased susceptibility to recurrent neisserial infections.[29,30]

Polymorphic Variants

C8αγ: Two common variants Q93 K: A (0.59) and B (0.39), along with five rare variants are known.[31,32]

C8β: Two common variants exist: A (0.94) and B (0.05) generating G117R. In addition two rare variants are reported.[31,33]

C8γ: A polymorphism is reported in exon 1 (T207G).[34]

MUTANT ANIMALS

Embryos and sperm of C8β-targeted mutant mice (C57BL/6N-C8btm1a-(EUCOMM)Hmgu/H) are available from MRC Harwell (HAR) and Mutant Mouse Regional Resource Centers (MMRRC) (MGI: 88236).

REFERENCES

1. Hadders MA, Beringer DX, Gros P. Structure of C8alpha-MACPF reveals mechanism of membrane attack in complement immune defense. *Science* 2007;**317**:1552–4.

2. Slade DJ, Lovelace LL, Chruszcz M, Minor W, Lebioda L, Sodetz JM. Crystal structure of the MACPF domain of human complement protein C8 alpha in complex with the C8 gamma subunit. *J Mol Biol* 2008;**379**:331–42.

3. Lovelace LL, Cooper CL, Sodetz JM, Lebioda L. Structure of human C8 protein provides mechanistic insight into membrane pore formation by complement. *J Biol Chem* 2011;**286**:17585–92.

4. Chiswell B, Lovelace LL, Brannen C, Ortlund EA, Lebioda L, Sodetz JM. Structural features of the ligand binding site on human complement protein C8gamma: a member of the lipocalin family. *Biochim Biophys Acta* 2007;**1774**:637–44.

5. Ortlund E, Parker CL, Schreck SF, Ginell S, Minor W, Sodetz JM, et al. Crystal structure of human complement protein C8gamma at 1.2 A resolution reveals a lipocalin fold and a distinct ligand binding site. *Biochemistry* 2002;**41**:7030–7.

6. Lovelace LL, Chiswell B, Slade DJ, Sodetz JM, Lebioda L. Crystal structure of complement protein C8gamma in complex with a peptide containing the C8gamma binding site on C8alpha: implications for C8gamma ligand binding. *Mol Immunol* 2008;**45**:750–6.

7. Sonnen AF, Henneke P. Structural biology of the membrane attack complex. *Subcell Biochem* 2014;**80**:83–116.

8. Kolb WP, Muller-Eberhard HJ. The membrane attack mechanism of complement. Isolation and subunit composition of the C5b-9 complex. *J Exp Med* 1975;**141**:724–35.

9. Hugo F, Berstecher C, Krämer S, Fassbender W, Bhakdi S. In vivo clearance studies of the terminal fluid-phase complement complex in rabbits. *Clin Exp Immunol* 1989;**77**:112–6.

10. Biesecker G. The complement SC5b-9 complex mediates cell adhesion through a vitronectin receptor. *J Immunol* 1980;**145**:209–14.

11. Ng SC, Sodetz JM. Biosynthesis of C8 by hepatocytes. Differential expressionand intracellular association of the α – γ and β subunits. *J Immunol* 1987;**139**:3021–7.

12. Hetland G, Johnson E, Falk RJ, Eskeland T. Synthesis of complement components C5, C6, C7, C8 and C9 in vitro by human monocytes and assembly of the terminal complement complex. *Scand J Immunol* 1986;**24**:421–8.

13. Gasque P, Fontaine M, Morgan BP. Complement expression in human brain. Biosynthesis of terminal pathway components and regulators in human glial cells and cell lines. *J Immunol* 1995;**154**:4726–33.

14. Terai K, Walker DG, McGeer EG, McGeer PL. Neurons express proteins of the classical complement pathway in Alzheimer disease. *Brain Res* 1997;**769**:385–90.

15. Johnson E, Hetland G. Human umbilical vein endothelial cells synthesize functional C3, C5, C6, C8 and C9 in vitro. *Scand J Immunol* 1991;**33**:667–71.

16. Garred P, Hetland G, Mollnes TE, Stoervold G. Synthesis of C3, C5, C6, C7, C8, and C9 by human fibroblasts. *Scand J Immunol* 1990;**32**:555–60.

17. Kilicarslan A, Uysal A. E.C. R. Acute phase reactants. *Acta Medica* 2013;**2**:2–7.

18. Scheurer B, Rittner C, Schneider PM. Expression of the human complement C8 subunits is independently regulated by interleukin 1 beta, interleukin 6, and interferon gamma. *Immunopharmacology* 1997;**38**:167–75.

19. Rao AG, Howard OMZ, Ng S, Whitehead AS, Colten HR, Sodetz JM. Complementary DNA and derived amino acid sequence of the α subunit of human complement protein C8: evidence for the existence of a separate α subunit messenger RNA. *Biochemistry* 1987;**26**:3556–64.

20. Haefliger J-A, Tschopp J, Nardelli D, Wahli W, Kocher H-P, Tosi M, et al. Complementary DNA cloning of complement C8β and its sequence homology to C9. *Biochemistry* 1987;**26**: 3551–6.

21. Howard OMZ, Rao AG, Sodetz JM. Complementary DNA and derived amino acid sequence of the β subunit of human complement protein C8: identification of a close structural and ancestral relationship to the α subunit and C9. *Biochemistry* 1987;**26**:3565–70.

22. Ng SC, Rao AG, Howard OMZ, Sodetz JM. The eighth component of human complement: evidence that it is an oligomeric serum protein assembled from products of three different genes. *Biochemistry* 1987;**26**:5229–33.

23. Hofsteenge J, Blommers M, Hess D, Furmanek A, Miroshnichenko O. The four terminal components of the complement system are C-mannosylated on multiple tryptophan residues. *J Biol Chem* 1999;**274**:32786–94.

24. Michelotti GA, Snider JV, Sodetz JM. Genomic organization of human complement protein C8 alpha and further examination of its linkage to C8 beta. *Hum Genet* 1995;**95**:513–8.

25. Kaufmann T, Rittner C, Schneider PM. The human complement component C8B gene: structure and phylogenetic relationship. *Hum Genet* 1993;**92**:69–75.

26. Kaufmann KM, Sodetz JM. Genomic structure of the human complement protein C8γ: homology to the lipocalin gene family. *Biochemistry* 1994;**33**:5162–6.

27. Tedesco F, Roncelli BH, Agnello V, Sodetz JM. Two distinct abnormalities in patients with C8 alpha-gamma deficiency. Low level of C8 beta chain and the presence of dysfunctional C8 alpha-gamma subunit. *J Clin Invest* 1990;**86**:884–8.

28. Kojima T, Horiuchi T, Nishizaka H, Fukumori Y, Amano T, Nagasawa K, et al. Genetic basis of human complement C8 alpha-gamma deficiency. *J Immunol* 1998;**161**:3762–6.

29. Kaufmann T, Hansch G, Rittner C, Spath P, Tedesco F, Schneider PM. Genetic basis of human complement C8β deficiency. *J Immunol* 1993;**150**:4943–7.

30. Mayilyan KR. Complement genetics, deficiencies, and disease associations. *Protein Cell* 2012;**3**:487–96.

31. Rogde S, Mevag B, Teisberg P, Gedde-Dahl T, Tedesco F, Olaisen B. Genetic polymorphism of complement component C8. *Hum Genet* 1985;**70**:211–6.

32. Rittner C, Stradmann-Bellinghausen B. Human complement C81 (C8 A) polymorphism: detection and segregation of new variants. *Hum Genet* 1993;**94**:413–6.

33. Dewald G, Hemmer S, Nöthen MM. Human complement component C8. Molecular basis of the beta-chain polymorphism. *FEBS Lett* 1994;**340**:211–5.

34. Dewald G, Cichon S, Bryant SP, Hemmer S, Nöthen MM, Spurr NK. The human complement C8G gene, a member of the lipocalin gene family: polymorphisms and mapping to chromosome 9q34.3. *Ann Hum Genet* 1996;**60**:281–91.

Chapter 22

C9

Paul Morgan
Cardiff University, Cardiff, United Kingdom

PHYSICOCHEMICAL PROPERTIES

Immature protein: C9 is synthesised as a single-chain precursor of 559 amino acids, including a leader peptide of 21 amino acids. The mature protein comprises 538 amino acids. For all amino acid numbering, the translation initiator methionine is numbered as +1.

pI	4.7
M_r (kDa) (predicted)	60.7 (nonglycosylated)
M_r (kDa) (observed)	69–73 kDa (glycosylated)
N-Glycosylation sites	N277; N415 (both occupied)
C-Mannosylation sites	W48; W51
Intrachain disulphides[1]	43↔78, 54↔57, 88↔94, 101↔112, 107↔125, 119↔134, 142↔181, 254↔255, 380↔405, 510↔526, 513↔528, 530↔539

STRUCTURE

C9 is a single-chain plasma glycoprotein. It is a member of the membrane attack complex/perforin-like (MACPF)/CDC superfamily that have in common a set of four structural domains, thrombospondin-1 (TSP1), LDL-R-associated (LDLRA) structural domain, MACPF and epidermal growth factor (EGF). C9 is the simplest of the complement proteins in the MACPF/CDC superfamily, comprising one each of these four domains in sequence from the amino-terminus: TSP1, amino acids 42–95; LDLRA, amino acids 99–136; MACPF, amino acids 138–509; EGF, amino acids 510–540 (Fig. 22.1).

   C9 is cleaved by thrombin at a single site, after amino acid 244, to yield two fragments that have been termed C9a and C9b. In addition, trypsin cleaves C9 after amino acid 391; however, the fragments remaining are associated, because they are disulphide-linked.[2] Although helpful in ex-vivo functional studies, these cleavages are not of physiological relevance. C9 is glycosylated (~8% by weight). Two N-linked glycans occupy N277 and

The Complement FactsBook. http://dx.doi.org/10.1016/B978-0-12-810420-0.00022-5

FIGURE 22.1 Schematic of protein domain structure for C9.

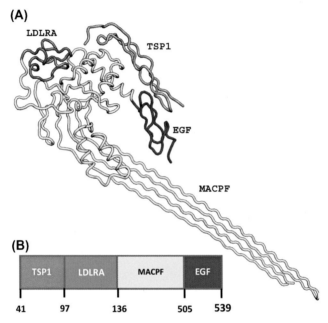

FIGURE 22.2 Crystal structure for human C9. (A) Ribbon representation of single-chain C9 molecule (PDB ID: 5FMW), from the poly-C9 component of the complement membrane attack complex structure determined with the help of electron microscopy. (B) Schematic representation of the C9 domains coloured similar to the ribbon representation.

N415. W48 and/or W51 are *C*-mannosylated.[3] By electron microscopic (EM) imaging, C9 appears as a globular ellipsoid of 7–8 nm diameter.[4] By solution scattering, monomeric C9 in solution has an apparent diameter of 12 nm.[5] Purified C9 has a propensity to autopolymerise, particularly when incubated in the presence of Zn^{2+} ions, undergoing major conformational change to form a ring polymer (poly-C9)[6] High-resolution EM analyses reveal that poly-C9 comprises 22 C9 monomers unfolded, rotated and aligned to form a barrel-like structure with an internal diameter of ~12 nm, external diameter of ~22 nm and height of ~16 nm.[7] The crystal structure of C9: structural details have been extrapolated from crystals of other complement proteins in the MACPF/CDC superfamily, as well as electron microscopy (Fig. 22.2).

FUNCTION

The sole function of C9 is to form the cytolytic membrane attack complex (MAC), an important antibacterial defense. The membrane-bound MAC precursor complex, comprising C5b, C6, C7 and C8, recruits C9 from the fluid phase and facilitates its unfolding and insertion into the membrane, aligned with the C8 α-chain. Unfolding of the first C9 exposes binding sites for the next C9 molecule which binds, unfolds, inserts and aligns alongside the previous molecule. This sequential recruitment then continues until the ring structure of the MAC is complete. High-resolution EM studies have revealed that the complete MAC replicates the 22-fold symmetry of poly-C9, the barrel comprising 18 copies of C9 with one "stave" each from C6, C7, C8β and C8α.[8] The resultant MAC pore has an internal diameter of ~11 nm. Unlike poly-C9, the MAC is not a complete ring but rather resembles a flexible split-washer.

MAC formation is inefficient, and the majority of precursor complexes do not attach to the membrane, but rather remain in the fluid phase. Soluble MAC contains C5b, C6, C7, C8 and one or more C9 molecules together with the complement chaperones S-protein and clusterin (sC5b-9). These soluble complexes are biologically inert, but provide a useful clinical measure of ongoing complement activation.

TISSUE DISTRIBUTION

C9 is present in human plasma at a concentration of 46–66 mg/L (mean ± sd from local control cohort) and in cerebrospinal fluid (0.2 mg/L mean in local control cohort).

Primary site of synthesis: liver (hepatocyte); reported secondary sites of synthesis including monocytes, macrophages, fibroblasts, microglia, astroglia and neurons. The contributions of these secondary sites of synthesis to plasma levels is likely slight, but the local effects may be large.

REGULATION OF EXPRESSION

C9 is a moderate acute-phase reactant, and levels increase 1.5- to 2-fold in response to acute inflammatory triggers. This is regulated by the well-described acute-phase cytokines, including IL-6 and IFN-γ.

HUMAN PROTEIN SEQUENCE

The leader peptide is highlighted. Sites of *C*-mannosylation are in bold. *N*-glycosylation sites are in bold and underlined.

```
  1  MSACRSFAVA ICILEISILT AQYTTSYDPE LTESSGSASH IDCRMSPWSE

 51  WSQCDPCLRQ MFRSRSIEVF GQFNGKRCTD AVGDRRQCVP TEPCEDAEDD

101  CGNDFQCSTG RCIKMRLRCN GDNDCGDFSD EDDCESEPRP PCRDRVVEES

151  ELARTAGYGI NILGMDPLST PFDNEFYNGL CNRDRDGNTL TYYRRPWNVA

201  SLIYETKGEK NFRTEHYEEQ IEAFKSIIQE KTSNFNAAIS LKFTPTETNK

251  AEQCCEETAS SISLHGKGSF RFSYSKNETY QLFLSYSSKK EKMFLHVKGE

301  IHLGRFVMRN RDVVLTTTFV DDIKALPTTY EKGEYFAFLE TYGTHYSSSG

351  SLGGLYELIY VLDKASMKRK GVELKDIKRC LGYHLDVSLA FSEISVGAEF

401  NKDDCVKRGE GRAVNITSEN LIDDVVSLIR GGTRKYAFEL KEKLLRGTVI

451  DVTDFVNWAS SINDAPVLIS QKLSPIYNLV PVKMKNAHLK KQNLERAIED

501  YINEFSVRKC HTCQNGGTVI LMDGKCLCAC PFKFEGIACE ISKQKISEGL

551  PALEFPNEK
```

PROTEIN MODULES

Leader peptide (exon 1): amino acids 1–21.
TSP1 domain (exons 2/3): amino acids 42–95.
LDLRA domain (exons 3/4): amino acids 99–136.
MACPF domain (exons 4–10): amino acids 138–509.
EGF domain (exon 10): amino acids 510–540.

CHROMOSOMAL LOCATION

The gene encoding human C9 is at Chr5p14-p12. The gene spans ~87 kb and contains 11 exons. The C9 gene is linked with the genes for C6 and C7.
The gene encoding mouse C9 is at Chr15 (15 A1) and comprises 11 exons.
The gene encoding rat C9 is at Chr2q16 and comprises 11 exons.

HUMAN cDNA SEQUENCE

```
CAGCATGTCA GCCTGCCGGA GCTTTGCAGT TGCAATCTGC NTTTTAGAAA TAANCATCCT    60

CACAGCACAG TACACGACCA GTTATGACCC AGAGCTAACA GAAAGCAGTG GCTCTGCATC   120

ACACATAGAC CGCAGAATGA GCCCCTGGAG TGAATGGTCA CAATGCGATC CTTGTCTCAG   180

ACAAATGTTT CGTTCAAGAA GCATTGAGGT CTTTGGACAA TTTAATGGGA AAAGATGCAC   240

CGACGCTGTG GGAGACAGAC GACAATGTGT GCCCACAGAG CCCTGTGAGG ATGCTGAGGA   300

TGACTGCGGA AATGACTTTC AATGCAGTAC AGGCAGATGC ATAAAGATGC GACTTCGGTG   360
```

```
TAATGGTGAC AATGACTGCG GAGACTTTTC AGATGAGGAT GATTGTGAAA GTGAGCCCCG   420

TCCCCCCTGC AGAGACAGAG TGGTAGAAGA GTCTGAGCTG GCACGAACAG CAGGCTATGG   480

GATCAACATT TTAGGGATGG ATCCCCTAAG CACACCTTTT GACAATGAGT TCTACAATGG   540

ACTCTGTAAC CGGGATCGGG ATGGAAACAC TCTGACATAC TACCGAAGAC CTTGGAACGT   600

GGCTTCTTTG ATCTATGAAA CCAAAGGCGA GAAAAATTTC AGAACCGAAC ATTACGAAGA   660

ACAAATTGAA GCATTTAAAA GTATCATCCA AGAGAAGACA TCAAATTTTA ATGCAGCTAT   720

ATCTCTAAAA TTTACACCCA CTGAAACAAA TAAAGCTGAA CAATGTTGTG AGGAAACAGC   780

CTCCTCAATT TCTTTACATG GCAAGGGTAG TTTTCGGTTT TCATATTCCA AAAATGAAAC   840

TTACCAACTA TTTTTGTCAT ATTCTTCAAA GAAGGAAAAA ATGTTTCTGC ATGTGAAAGG   900

AGAAATTCAT CTGGGAAGAT TTGTAATGAG AAATCGCGAT GTGCTCACAA CAACTTTTGT   960

GGATGATATA AAAGCTTTGC CAACTACCTA TGAAAAGGGA GAATATTTTG CCTTTTTGGA  1020

AACCTATGGA ACTCACTACA GTAGCTCTGG GTCTCTAGGA GGACTCTATG AACTAATATA  1080

TGTTTTGGAT AAAGCTTCCA TGAAGCGGAA AGGTGTTGAA CTAAAAGACA TAAAGAGATG  1140

CCTTGGGTAT CATCTGGATG TATCTCTGGC TTTCTCTGAA ATCTCTGTTG GAGCTGAATT  1200

TAATAAAGAT GATTGTGTAA AGAGGGGAGA GGGTAGAGCT GTAAACATCC CCAGTGAAAA  1260

CCTCATAGAT GATGTTGTTT CACTCATAAG AGGTGGAACC AGAAAATATG CATTTGAACT  1320

GAAAGAAAAG CTTCTCCGAG GAACCGTGAT TGATGTGACT GACTTTGTCA ACTGGGCCTC  1380

TTCCATAAAT GATGCTCCTG TTCTCATTAG TCAAAAACTG TCTCCTATAT ATAATCTGGT  1440

TCCAGTGAAA ATGAAAAATG CACACCTAAA GAAACAAAAC TTGGAAAGAG CCATTGAAGA  1500

CTATATCAAT GAATTTAGTG TAAGAAAATG CCACACATGC CAAAATGGAG GTACAGTGAT  1560

TCTAATGGAT GGAAAGTGTT TGTGTGCCTG CCCATTCAAA TTTGAGGGAA TTGCCTGTGA  1620

AATCAGTAAA CAAAAAATTT CTGAAGGATT GCCAGCCCTA GAGTTCCCCA ATGAAAAATA  1680

GAGCTGTTGG CTTCTCTGAG CTCCAGTGGA AGAAGAAAAC ACTAGTACCT TCAGATCCTA  1740

CCCCTGAAGA TAATCTTAGC TGCCAAGTAA ATAGCAACAT GCTTCATGAA AATCCTACCA  1800

ACCTCTGAAG TCTCTTCTCT CTTAGGTCTA TAATTTTTTT TTAATTTTTC TTCCTTAAAC  1860

TCCTGTGATG TTTCCATTTT TTGTTCCCTA ATGAGAAGTC AACAGTGAAA TACGCGAGAA  1920

CTGCTTTATC CCACGGAAAA AGCCAATCTC TTCTAAAAAA AAAACAAAAT TAAATTAAAA  1980

ACAGAATGTT GGTTAAAAAA ACTTCAAAGT AATTTTCAAA CGGCTTTGTA TGGTTAACAT  2040

ATTCTGCCAG GTCCATGACC ACACGTCTGT ACCATGCAAT TTAACTCTTA TTTACATTGT  2100

TATGTTTAGT TTGGTTATTT GCTTAGGTGT GCATACATTC ATTCAGCAAA TGCTGAGCAC  2160

CAGCCACGTG CACAGCAGTT GCTTTTACTA GTCTTAGCTC TACGATTTAA ATCCATGTGT  2220

CCAAGGGGGA AAACATATTA TATTTGTAAC CAAAAACTAC TAGTTTACCA GAGGACTGAA  2280

GGGAGATAAA GAGGAGTTGG TTAATGGGTA CAAAAATCCA GTTAGATGAA AGGAATAATA  2340

TAGATAGTGT TCAGTAGCAG AATAGAATGA ACATAAACTA TTAGTTTAAA TTATGTGAAA  2400

TTCCTTCTAT TTGATCATAT TTTACAAGAA AAAACATCAA TTTTATATAG TCCAACTTAA  2460

TACCTAGC
```

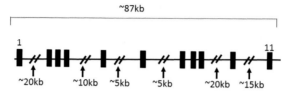

FIGURE 22.3   Diagram of gene structure for C9.

Intron/exon boundaries are shown by underlining the first five nucleotides in each exon; the initiation (ATG) and termination (TAA) sequences are double-underlined.

GENOMIC STRUCTURE

The *C9* gene spans ~87 kb and contains 11 exons.[9] Exon 1 is of indeterminate length (>300bp) and includes numerous putative regulatory elements. Exons 2, 3 and 4 (148bp) are clustered and, together, encode the TSP1 and LDLRA domains. The MACPF domain is encoded across exons 4–10. Exon 10 also encodes the EGF domain (Fig. 22.3).

ACCESSION NUMBERS

Human C9	UniProtKB: P02748
	HGNC: 1358
	Entrez gene: 735
	Ensembl: ENSG00000113600
	OMIM: 120940
Mouse C9	UniProtKB: K3BKN4

DEFICIENCY

C9 deficiency is common in Japan with a homozygous frequency of ~1:1000 (carrier frequency of ~6%). It is the most common inherited immunodeficiency in this population. Deficiency is caused by a nonsense mutation in exon 4, a C > T transition that converts the codon for R95 to a TGA stop (R95*), leading to premature stop and no protein expression.[10] This same mutation is found, albeit less commonly, in Korea and China.[11] In other races C9 deficiency is rare; several isolated cases and affected families have been described, usually with stop mutations, including C54* and R154*.

Deficiency of C9, in common with deficiencies of other MAC components, predisposes people to infection with *Neisseria*, most often meningococcal meningitis or septicaemia.

POLYMORPHIC VARIANTS

A study in a Korean cohort identified 20 genetic variants in the *C9* gene that resolved into seven haplotypes.[12] The P167S polymorphism is associated with increased risk of age-related macular degeneration (AMD) through an unknown mechanism.[13] The R95* mutation associated with C9 deficiency in Japanese, in heterozygosity, is linked with reduced risk for AMD.[14]

MUTANT ANIMALS

No naturally C9-deficient or C9 gene-targeted rodent has been described.

REFERENCES

1. Lengweiler S, Schaller J, Rickli EE. Identification of disulfide bonds in the ninth component (C9) of human complement. *FEBS Lett* 1996;**380**:8–12.
2. Shiver JW, Dankert JR, Donovan JJ, Esser AF. The ninth component of human complement (C9). Functional activity of the b fragment. *J Biol Chem* July 25, 1986;**261**(21):9629–36.
3. Hofsteenge J, Blommers M, Hess D, Furmanek A, Miroshnichenko O. The four terminal components of the complement system are C-mannosylated on multiple tryptophan residues. *J Biol Chem* November 12, 1999;**274**(46):32786–94.
4. DiScipio RG. The size, shape and stability of complement component C9. *Mol Immunol* August 1993;**30**(12):1097–106.
5. Smith KF, Harrison RA, Perkins SJ. Molecular modeling of the domain structure of C9 of human complement by neutron and X-ray solution scattering. *Biochemistry* January 28, 1992;**31**(3):754–64.
6. Podack ER, Tschopp J. Polymerization of the ninth component of complement (C9): formation of poly(C9) with a tubular ultrastructure resembling the membrane attack complex of complement. *Proc Natl Acad Sci U.S.A* 1982;**79**:574–8.
7. Dudkina NV, Spicer BA, Reboul CF, Conroy PJ, Lukoyanova N, Elmlund H, et al. Structure of the poly-C9 component of the complement membrane attack complex. *Nat Commun* 2016;**7**:10588.
8. Serna M, Giles JL, Morgan BP, Bubeck D. Structural basis of complement membrane attack complex formation. *Nat Commun* 2016;**7**:10587.
9. Witzel-Schlömp K, Späth PJ, Hobart MJ, Fernie BA, Rittner C, Kaufmann T, et al. The human complement C9 gene: identification of two mutations causing deficiency and revision of the gene structure. *J Immunol* 1997;**158**:5043–9.
10. Kira R, Ihara K, Takada H, Gondo K, Hara T. Nonsense mutation in exon 4 of human complement C9 gene is the major cause of Japanese complement C9 deficiency. *Hum Genet* 1998;**102**:605–10.
11. Khajoee V, Ihara K, Kira R, Takemoto M, Torisu H, Sakai Y, et al. Founder effect of the C9 R95X mutation in orientals. *Hum Genet* 2003;**112**:244–8.
12. Bae JS, Pasaje CF, Park BL, Cheong HS, Kim JH, Park TJ, et al. Genetic analysis of complement component 9 (C9) polymorphisms with clearance of hepatitis B virus infection. *Dig Dis Sci* 2011;**56**:2735–41.
13. Seddon JM, Yu Y, Miller EC, Reynolds R, Tan PL, Gowrisankar S, et al. Rare variants in CFI, C3 and C9 are associated with high risk of advanced age-related macular degeneration. *Nat Genet* 2013;**45**:1366–70.
14. Nishiguchi KM, Yasuma TR, Tomida D, Nakamura M, Ishikawa K, Kikuchi M, et al. C9-R95X polymorphism in patients with neovascular age-related macular degeneration. *Invest Ophthalmol Vis Sci* 2012;**53**:508–12.

Part V

Regulatory Proteins

Chapter 23

C1 Inhibitor

Christian Drouet, Denise Ponard, Arije Ghannam
Université Grenoble Alpes, Grenoble, France

PHYSICOCHEMICAL PROPERTIES

Immature Protein

C1 Inhibitor (C1Inh) is secreted as a single chain of M_r 100,000.[1] In the cell lysates, two forms are immunoprecipitated, a 100,000-M_r C1Inh, as observed in the culture medium, and an 80,000-M_r species. C1Inh is susceptible to both endoglycosidase F and H, with two products of respective M_r 68,000 and 71,000,[1] in agreement with the known glycosylation pattern. The cell-free translation product of C1Inh mRNA exhibits a 60,000-M_r species.[2] C1Inh is heavily glycosylated with approximately 50% of its total weight contributed by carbohydrates. Biosynthesis of rodent and human C1Inh is similar.

Mature Protein

C1Inh is a single polypeptide chain containing 500 amino acid residues, including a 22-residue signal peptide.

pI	2.7–2.8
Predicted mol. wt.	
Predicted	52,900
	71,100 (glycosylated)
Observed	104,000 (under nonreducing), 85,000–93,000 under reducing conditions
N-glycosylation sites	7 (25, 69, 81, 238, 253, 272 and 352) on the mature protein, with core 1 or possibly core 8 glycans.[3,4] N-glycan heterogeneity is found at Asn[25]: Hex6HexNAc5 (major), Hex5HexNAc4 (minor), dHex1Hex5HexNAc4 (minor) and dHex1Hex6HexNAc5 (minor).
O-glycosylation sites	14 (7 verified by carbohydrate analysis at positions 48, 64, 71, 83, 88, 92, 96.[3,4])
Intrachain disulphide bonds	2; 123–428, 130–205

STRUCTURE

C1Inh is a protease inhibitor belonging to the serpin superfamily. C1Inh has a two-domain structure:

1. The N-terminal domain (residues 23–134; also sometimes referred to as the N-terminal tail), not essential for C1Inh to inhibit proteases. This domain has no similarity to other serpins and carries the majority of glycosylation sites with a 7× repeating Glx-Pro-Thr-Thr/Gln peptide unit (85–88, 89–92, 93–96, 97–100, 101–104, 105–108, 116–119). It is highly divergent in C1Inh proteins for different species (man, mouse, bovine and rabbit).
2. The C-terminal serpin domain (residues 135–500), arranged in a globular serpin structure, which is the part of C1Inh that provides the inhibitory activity (Fig. 23.1).

The crystal structure of the latent serpin domain (119-500; PDB ID 2OAY [5]) features nine α-helices and three β-sheets. A mobile reactive site loop (RCL), exposed for interaction with target serine proteases and typical for serpins is uncleaved and incorporated into sheet A, forming s4A. A 3D model of the native serpin domain of C1Inh (138–500) has been generated from α1-antitrypsin structure (PDB ID 1M6Q [6]). Domains with conserved sequence among serpins can be found at structurally crucial positions, with five regions ascribed for serpin function (Fig. 23.2).

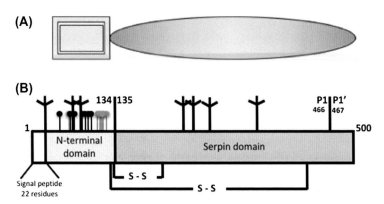

FIGURE 23.1 Two schematic representations of C1 inhibitor protein. (A) Diagram of protein domains for C1 Inhibitor, see key in front of book. (B) Diagram with residue numbering conforms to the mature protein. *Branched symbols*, *N*-glycosylation (n=7); *black circles*, verified *O*-glycosylation sites (n=7); *grey circles*, potential *O*-glycosylation (n=7). S—S, disulphide bonds (n=2).

(A)

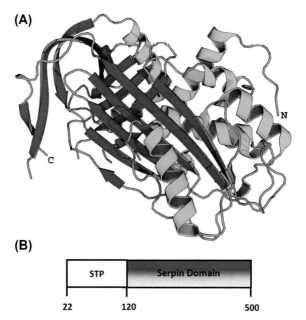

(B)

FIGURE 23.2 Crystal structure for complement C1 Inhibitor. (A) Ribbon diagram of C1 Inhibitor crystal structure (PDB ID: 2OAY), representing the Serpin domain (latent form) and a small portion of the N-terminal domain (residues 97–120). The overall structure resembles those of other serpins, with highly conserved nine α-helices (coloured *cyan*) and three β-sheets (coloured *magenta*). (B) Schematic representation of two-domain C1 Inhibitor.

FUNCTIONS

C1Inh activities are divided into two main categories:

1. Vascular permeability is controlled through inhibition of two proteases of the contact system involved in bradykinin (BK) generation, factor XIIa ($k_{on} = 3.7 \times 10^4 \, M^{-1} \cdot s^{-1}$) and plasma kallikrein ($k_{on} = 1.7 \times 10^4 \, M^{-1} \cdot s^{-1}$).
2. Antiinflammatory function is supported via control of complement (C1r, C1s, with $k_{on} = 43 \times 10^4 \, M^{-1} \cdot s^{-1}$, and MASP2 with $k_{on} = 0.16 \times 10^4 \, M^{-1} \cdot s^{-1}$) and the plasma contact system proteases, in addition to a competition mechanism involving its sialyl Lewis^x and cell selectins, compromising interactions between inflammatory cells and endothelium.[7] This is also consistent with the binding of C1Inh to endothelial cells.

C1Inh also partially controls the intrinsic coagulation protease factor XI and thrombin, and the fibrinolytic proteases plasmin and tissue plasminogen activator, where C1Inh does not appear to play a significant role. Control of C1s is enhanced 30- to 60-fold by heparin, but heparin has no effect on the inhibition of factor XIIa or kallikrein.

DEGRADATION PATHWAY

The target protease recognises the RCL exposed at the surface of C1Inh, which mimics the substrate specificity of the protease and cleaves the Arg^{466}-Thr^{467} P1-P1′ peptide bond. Cleavage triggers a molecular rearrangement, with a fourth strand in β-sheet A, and translocates the covalently linked enzyme from the top to the bottom pole of the serpin.[8] The serpin and the protease become irreversibly distorted with covalent bond formation between the inhibitor and the active site serine of the protease.[9] The protease within the complex, because of the distortion, is more susceptible to proteolytic attack. The failure to trap a noncognate protease as a complex results in cleavage of the RCL, with subsequent C1Inh cleavage as demonstrated by a circulating 95,000-M_r species. Serpin–protease complexes are removed from the circulation primarily via the low-density lipoprotein receptor-related protein. These complexes are removed from the circulation much more rapidly than is the native, active C1Inh.

Carbohydrate, more precisely sialic acid, does appear to play a role in the in vivo stability of C1Inh. Removal of sialic acid enhances the clearance of C1Inh in the rabbit, presumably via binding to the asialoglycoprotein receptor.

TISSUE DISTRIBUTION

Basal serum concentration is 210–345 μg/L.

Primary site of synthesis (cell type): The hepatocyte is the primary source of C1Inh for circulating C1Inh protein.

Secondary sites of synthesis (cell type): Cells of the monocyte/macrophage lineage are the predominant extrahepatic sources of C1Inh. Kupffer cells, peritoneal macrophages, microglial cells, fibroblasts, endothelial cells, placenta and megakaryocytes synthesise and secrete the protein both in vivo and in vitro. The in vivo local production has been demonstrated in brain, spleen, liver, heart, kidney and lung.[10] In the spleen, production of C1Inh was colocalised with that of a specific marker for white pulp follicular dendritic cells. The distribution of human C1Inh-expressing cells was qualitatively indistinguishable from that of its mouse counterpart.

REGULATION OF EXPRESSION

C1Inh is an acute-phase protein. Inflammation and tissue injury positively modulate C1Inh expression at extrahepatic sites, with the serum level rising ~2.5-fold during inflammation. Biosynthesis of C1Inh by macrophages can be stimulated by cytokines, particularly by interferon-γ, IL-6 and TNF-α.[11]

HUMAN PROTEIN SEQUENCE

```
MASRLTLLTL LLLLLAGDRA SSNPNATSSS SQDPESLQDR GEGKVATTVI  50
SKMLFVEPIL EVSSLPTTNS TTNSATKITA NTTDEPTTQP TTEPTTQPTI 100
QPTQPTTQLP TDSPTQPTTG SFCPGPVTLC SDLESHSTEA VLGDALVDFS 150
LKLYHAFSAM KKVETNMAFS PFSIASLLTQ VLLGAGENTK TNLESILSYP 200
KDFTCVHQAL KGFTTKGVTS VSQIFHSPDL AIRDTFVNAS RTLYSSSPRV 250
LSNNSDANLE LINTWVAKNT NNKISRLLDS LPSDTRLVLL NAIYLSAKWK 300
TTFDPKKTRM EPFHFKNSVI KVPMMNSKKY PVAHFIDQTL KAKVGQLQLS 350
HNLSLVILVP QNLKHRLEDM EQALSPSVFK AIMEKLEMSK FQPTLLTLPR 400
IKVTTSQDML SIMEKLEFFD FSYDLNLCGL TEDPDLQVSA MQHQTVLELT 450
ETGVEAAAAS AISVARTLLV FEVQQPFLFV LWDQQHKFPV FMGRVYDPRA 500
```

The leader peptide is underlined; *N*-glycosylation sites are in bold letters and underlined, *O*-glycosylation sites are in bold characters. The Arg[466] P1 residue within the RCL is in bold characters and double-underlined. UniProt ID P05155. MEROPS ID I04.024.

PROTEIN MODULES

1–22	Signal peptide	exon 1
23–134	Mucin-like N-terminal domain	exon 2
135–500	Serpin domain, including RCL (451–471)	exons 3–8

CHROMOSOMAL LOCATION

The human C1Inh gene (*SERPING1*) is located on chromosome 11 (11q12. q13.1).[12]

CDNA SEQUENCE

```
ctgatttaca ggaactcaca ccagcgatca atcttcctta atttgtaact gggcagtgtc  60
ccgggccagc caatagctaa gactgccccc cccgcacccc accctccctg accctggggg 120
actctctact cagtctgcac tggagctgcc tggtgaccag aagtttggag tccgctgacg 180
tcgccgccca gatggcctcc aggctgaccc tgctgaccct cctgctgctg ctgctggctg 240
gggatagagc ctcctcaaat ccaaatgcta ccagctccag ctcccaggat ccagagagtt 300
tgcaagacag aggcgaaggg aaggtcgcaa caacagttat ctccaagatg ctattcgttg 360
aacccatcct ggaggtttcc agcttgccga caaccaactc aacaaccaat tcagccacca 420
aaataacagc taataccact gatgaaccca ccacacaacc caccacagag cccaccaccc 480
aacccaccat ccaacccacc caaccaacta cccagctccc aacagattct cctacccagc 540
```

```
ccactactgg gtccttctgc ccaggacctg ttactctctg ctctgacttg gagagtcatt  600
caacagaggc cgtgttgggg gatgctttgg tagatttctc cctgaagctc taccacgcct  660
tctcagcaat gaagaaggtg gagaccaaca tggccttttc cccattcagc atcgccagcc  720
tccttaccca ggtcctgctc gggggctgggg agaacaccaa aacaaacctg gagagcatcc  780
tctcttaccc caaggacttc acctgtgtcc accaggccct gaagggcttc acgaccaaag  840
gtgtcacctc agtctctcag atcttccaca gcccagacct ggccataagg gacacctttg  900
tgaatgcctc tcggaccctg tacagcagca gccccagagt cctaagcaac aacagtgacg  960
ccaacttgga gctcatcaac acctgggtgg ccaagaacac caacaacaag atcagccggc 1020
tgctagacag tctgccctcc gatacccgcc ttgtcctcct caatgctatc tacctgagtg 1080
ccaagtggaa gacaacattt gatcccaaga aaaccagaat ggaacccttt cacttcaaaa 1140
actcagttat aaaagtgccc atgatgaata gcaagaagta ccctgtggcc catttcattg 1200
accaaacttt gaaagccaag gtggggcagc tgcagctctc ccacaatctg agtttggtga 1260
tcctggtacc ccagaacctg aaacatcgtc ttgaagacat ggaacaggct ctcagccctt 1320
ctgttttcaa ggccatcatg gagaaactgg agatgtccaa gttccagccc actctcctaa 1380
cactaccccg catcaaagtg acgaccagcc aggatatgct ctcaatcatg gagaaattgg 1440
aattcttcga tttttcttat gaccttaacc tgtgtgggct gacagaggac ccagatcttc 1500
aggtttctgc gatgcagcac cagacagtgc tggaactgac agagactggg gtggaggcgg 1560
ctgcagcctc cgccatctct gtggcccgca ccctgctggt ctttgaagtg cagcagccct 1620
tcctcttcgt gctctgggac cagcagcaca agttccctgt cttcatgggg cgagtatatg 1680
accccagggc ctgagacctg caggatcagg ttagggcgag cgctacctct ccagcctcag 1740
ctctcagttg cagccctgct gctgcctgcc tggacttggc ccctgccacc tcctgcctca 1800
ggtgtccgct atccaccaaa agggctccct gagggtctgg gcaagggacc tgcttctatt 1860
agcccttctc catggccctg ccatgctctc caaaccactt tttgcagctt tctctagttc 1920
aagttcacca gactctataa ataaaacctg acagaccatg actttcaaaa aaaaaaaaaa 1980
aaaa
```

The first five nucleotides in each exon are underlined to indicate the intron–exon boundaries. The Met initiation codon (atg), the termination codon (tga) and the probable polyadenylation signal (aataaa) are indicated.

GENOMIC STRUCTURE

SERPING1 gene (serpin peptidase inhibitor, clade G, member 1) extends over a 17,340 bp. Identification No: MIM ID 606860; RefSeq NM_000062.2; ENSEMBL ID ENST00000278407.8.

The gene, comprising eight exons and seven introns, presents an unusual promoter with no TATA sequence, but a terminal deoxynucleotidyl transferase (TdT)-like initiator and a pyrimidine-rich region of potential H-DNA structure centred around the position normally associated with the TATA box and potentially acting as positive transcriptional regulator. *SERPING1* gene gives rise to a transcription product of 1827 bp excluding the poly(A) tail. Three sequence elements potentially involved in determining the transcriptional response to IFN-γ[11] are present at −120, and in the intron 1 (375, 394). Two relaxed CpG

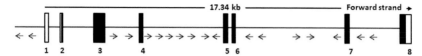

FIGURE 23.3 *SERPING1* gene. C1 Inhibitor encoded gene is organised in eight exons and seven introns. Enhancers are found within 5′ and intron 1. Translation starts at exon 2. Repetitive *AluI* (*arrows*) occupy more than one-third of the total length of the intervening sequences.

islands are present; the proximal one (exon 2) is constitutively unmethylated, while the most distant one (5.7 Kb upstream the transcriptional start site) is fully methylated, in both hereditary angioedema (HAE) patients and controls (Fig. 23.3). Observation that demethylating agents strongly increase transcription suggests an upstream indirect epigenetic regulation of trans-acting factors like transcription factors or microRNAs in the hepatocyte.[13]

DEFICIENCY

Hereditary angioedema (HAE; phenotype MIM No. 106100) is a rare inherited disease, which is clinically characterised by recurrent acute swelling episodes resulting from increased vascular permeability.[14] The prevalence is 1:50,000 to 1:75,000, with essentially heterozygous conditions (only five homozygous individuals). *SERPING1* mutations display a dominant negative effect usually resulting in protein levels far below the expected 50%. Consumption of C1Inh is also observed in rare situations, resulting in acquired angioedema. Excessive BK formation due to pathological activation of the factor XII-driven plasma contact system is a consistent finding in acute episodes of HAE.[15] BK belongs to the kinin family and initiates signalling cascades that increase vascular permeability and induce tissue swelling. Patients have recurrent and unpredictable episodes of localised edema that may affect the skin, gastrointestinal tract or larynx. Also, C4 and C2 cleavage goes uncontrolled, resulting in complement autoactivation.

More than 500 variants have been recorded as associated with HAE, with 43.2% of missense/nonsense variants, 35.6% of small insertions/deletions, 11.5% of large deletions/insertions and 9.7% of splice variants. Pathogenic amino acid substitutions were found distributed over the entire length of the coding sequence, except for the glycosylated amino-terminal extension, whose sequence tolerates extensive variation, as indicated by comparisons across species. Up to 15% of these changes are found in situations without a family history of angioedema and represent de novo mutations. Over 150 mutations of the *SERPING1* gene associated with HAE are described in a public database (URL http://hae.biomembrane.hu/).

POLYMORPHIC VARIANTS

Significant Examples

5′ sequence	rs2244169,[b] rs2511990,[b] rs2509897[b]
Intron 2	rs199473715, rs1005510[b]
Exon 3	rs11229062 (p.L11R), rs11229062 (p.D39E), rs11546660 (p.V56A), rs200534715 (p.T118A), rs281875168 (p.Y154C), rs281875169 (p.S170F), rs281875170 (p.G184R)
Intron 4	rs11229063, rs141593943
Exon 5	rs281875171 (p.L230P), rs281875172 (p.I232K)
Intron 6	rs2511988,[a,b] rs2511989,[b] rs1005510[b]
Exon 6	rs281875173 (p.W299R), rs1803212 (p.T308S)
Exon 8	rs281875174 (p.L430Q), rs281875175 (p.M441T), rs281875176 (p.L447P), rs281875177 (p.V473G), rs4926 (p.V480M), rs281875178 (p.D497D)

[a]When present in C1Inh-deficient families, it was shown to be associated with more severely affected HAE individuals.
[b]Associated with a risk for age-related macular degeneration.[16]

MUTANT ANIMALS

Disruption of the *SERPING1* gene by gene trapping enabled the generation of homozygous deficient mice. Two insertional strategies have been developed. The *SERPING1* gene was targeted either into a splice acceptor site in intron 6, 210 bp upstream of exon 7,[17] or at position +1 in the donor splice site of intron 2.[18] These mice presented with increased vascular permeability compared to wild-type littermates.

REFERENCES

1. Prandini MH, Reboul A, Colomb MG. Biosynthesis of complement C1 inhibitor by Hep G2 cells. Reactivity of different glycosylated forms of the inhibitor with C1s. *Biochem J* 1986;**237**:93–8.
2. Tosi M, Duponchel C, Bourgarel P, Colomb M, Meo T. Molecular cloning of human C1 inhibitor: sequence homologies with alpha 1-antitrypsin and other members of the serpins superfamily. *Gene* 1986;**42**:265–72.
3. Bock SC, Skriver K, Nielsen E, Thogersen HC, Wiman B, Donaldson VH, et al. Human C1 inhibitor: primary structure, cDNA cloning, and chromosomal localization. *Biochemistry* 1986;**25**:4292–301.
4. Perkins SJ, Smith KF, Amatayakul S, Ashford D, Rademacher TW, Dwek RA, et al. Two-domain structure of the native and reactive centre cleaved forms of C1 inhibitor of human complement by neutron scattering. *J Mol Biol* 1990;**214**:751–63.
5. Beinrohr L, Harmat V, Dobó J, Lörincz Z, Gál P, Závodszky P. C1 Inhibitor serpin domain structure reveals the likely mechanism of heparin potentiation and conformational disease. *J Biol Chem* 2007;**282**:21100–9.
6. Bos IG, Hack CE, Abrahams JP. Structural and functional aspects of C1-inhibitor. *Immunobiology* 2002;**205**:518–33.

7. Davis III AE, Lu F, Mejia P. C1 inhibitor, a multi-functional serine protease inhibitor. *Thromb Haemost* 2010;**104**:886–93.

8. Gooptu B, Lomas DA. Conformational pathology of the serpins: themes, variations, and therapeutic strategies. *Annu Rev Biochem* 2009;**78**:147–76.

9. Huntington JA, Read RJ, Carrell RW. Structure of a serpin-protease complex shows inhibition by deformation. *Nature* 2000;**407**:923–6.

10. Vinci G, Lynch NJ, Duponchel C, Lebastard T-M, Milon G, Stover C, et al. In vivo biosynthesis of endogenous and of human C1 Inhibitor in transgenic mice: tissue distribution and colocalization of their expression. *J Immunol* 2002;**169**:5948–54.

11. Zahedi K, Prada AE, Davis III AE. Transcriptional regulation of the C1 inhibitor gene by γ-interferon. *J Biol Chem* 1994;**269**:9669–74.

12. Carter PE, Duponchel C, Tosi M, Fonthergill JE. Complete nucleotide sequence of the gene for human C1 inhibitor with an unusually high density of *Alu* elements. *Eur J Biochem* 1991;**197**:301–8.

13. López-Lera A, Pernia O, López-Trascasa M, Ibanez de Caceres I. Expression of the *SERPING1* gene is not regulated by promoter hypermethylation in peripheral blood mononuclear cells from patients with hereditary angioedema due to C1-inhibitor deficiency. *Orphanet J Rare Dis* 2014;**9**:103e5.

14. Cicardi M, Aberer W, Banerji A, Bas M, Bernstein JA, Bork K, et al. Classification, diagnosis and approach to treatment of angioedema: consensus report from the Hereditary Angioedema International Working Group. *Allergy* 2014;**69**:602–16.

15. Björkqvist J, Sala-Cunill A, Renné T. Hereditary angioedema: a bradykinin-mediated swelling disorder. *Thromb Haemost* 2013;**109**:368–74.

16. Ennis S, Jomary C, Mullins R, Cree A, Chen X, MacLeod A, et al. Association between the *SERPING1* gene and age-related macular degeneration: a two-stage case–control study. *Lancet* 2008;**372**:1828–34.

17. Han ED, MacFarlane RC, Mulligan AN, Scafidi J, Davis 3rd AE. Increased vascular permeability in C1 inhibitor-deficient mice is mediated by the bradykinin type 2 receptor. *J Clin Invest* 2002;**109**:1057–63.

18. Oschatz C, Maas C, Lecher B, Lecher B, Jansen T, Björkqvist J, et al. Mast cells increase vascular permeability by heparin-initiated bradykinin formation in vivo. *Immunity* 2011;**34**:258–68.

Chapter 24

C4b-Binding Protein

Marcin Okrój[1], Anna M. Blom[2]

[1]*Medical University of Gdańsk, Gdańsk, Poland;* [2]*Lund University, Malmö, Sweden*

OTHER NAMES

Proline-rich protein, PRP; complement component 4 binding protein, C4-bp, C4b-bp, C4BP.

PHYSICOCHEMICAL PROPERTIES

Human C4b-binding protein (C4BP) is an oligomeric glycoprotein of molecular mass around 500 kDa, depending on the isoform and glycosylation.[1] Three major isoforms are found in plasma, including (1) the major isoform α7/β1, consisting of seven α-chains (75 kDa each) and one β-chain (45 kDa) (Fig. 24.1), (2) α7/β0 isoform, consisting of seven α-chains and (3) α6/β1 isoform, consisting of six α-chains and one β-chain. The relative proportions of these three isoforms are determined by genetic factors[2] and are regulated by acute-phase cytokines.[3] C4BPα- and C4BPβ-chains (if present) are linked together via their C-terminal parts by disulphide bonds.

	α-chain	β-chain
Calculated pI (nonglycosylated)	6.48	4.91
Amino Acids		
Signal sequence	48	17
Mature protein	549	235
M_r		
Predicted	61.67 kDa	26.42 kDa
Observed	70 kDa	45 kDa
Potential N-linked glycosylation sites	3; 221, 506, 528	5; 64, 71, 98, 117, 154
Interchain disulphide bonds	Cys 546, 558	Cys 202, 216

The Complement FactsBook. http://dx.doi.org/10.1016/B978-0-12-810420-0.00024-9

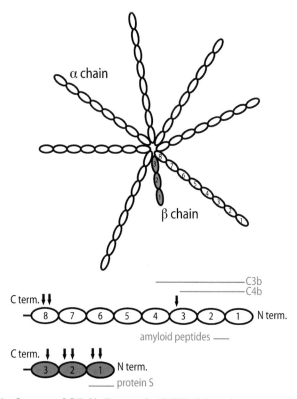

FIGURE 24.1 Structure of C4b-binding protein (C4BP). Schematic representation of the α7β1 isoform of C4BP. Complement control protein domains are labelled with appropriate numbers and potential *N*-glycosylation sites are indicated with *arrows*. Fragments, which confer binding sites for C4b, C3b, amyloid peptides and S protein, are shown with *grey lines*.

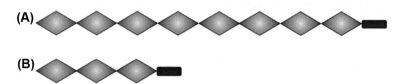

FIGURE 24.2 Diagram of protein domains (for domain definitions, see key in front of the book).

STRUCTURE

C4BP is dominated by complement control protein domains (CCP or short consensus repeats) with eight and three CCP domains in α- and β-chains, respectively (Fig. 24.2). Moreover, C-terminal regions of α- and β-chains contain c.a. 60 aa-long extensions with two cysteine residues and amphipathic regions, which are engaged in intracellular polymerisation of the entire molecule.[4]

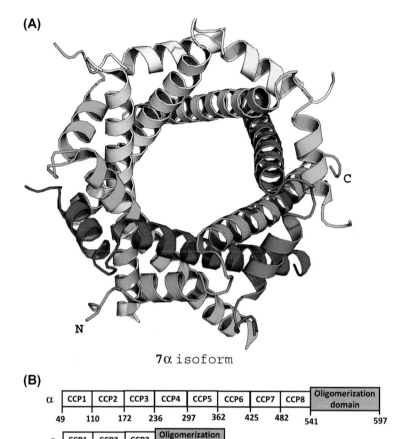

FIGURE 24.3 Crystal structure of C4b-binding protein (C4BP). (A) Ribbon diagram (PDB ID: 4BOF) of 7α isoform heptameric core oligomerisation domain complex of the human C4BP protein. (B) Schematic representation of α- and β-chains of the C4BP protein, made of multiple complement control protein domains.

Electron microscopy reveals an octopus-like shape with tentacles formed by N-terminal regions of α-chains.[5,6] N-terminal CCP1 domain of β-chain contains a high-affinity binding site for anticoagulant protein S.[7,8] Three-dimensional structure of the entire C4BP molecule has not been determined yet, but structures of human CCP1-2 domains of α-chain and central core domain[8] have been obtained (pdb entries 2A55 and 4B0F, Fig. 24.3).

FUNCTION

C4BP is the main soluble inhibitor of the classical and the lectin pathways.[9] Additionally it acts as a soluble inhibitor of the alternative pathway, however to a much lesser extent compared to factor H.[10] In detail, C4BP serves as cofactor in factor I-mediated proteolysis of cell-bound and soluble C4b, as well as soluble C3b.[11,12] This activity may be enhanced by binding of monomeric C-reactive protein.[13] C4BP also binds nascent C4b molecules, thus preventing assembly of the classical C3 convertases[14] and accelerates the decay of the classical C3 convertase.[15] Residues critical for complement inhibition are localised within the N-terminus of the α-chain (CCP1-3). All β-chain-containing C4BP molecules circulate in plasma bound to protein S, thus abolishing its anticoagulant function.[11,15] Conversely, binding of protein S does not affect complement inhibitory function of C4BP.[16] Presence of protein S enables binding of C4BP to phosphatidylserine exposed on the surface of dying cells and thus limits excessive complement activation.[17–19] Another protective role of C4BP stems from binding to islet amyloid polypeptide, which triggers complement activation when amyloid deposits aggregate in pancreatic islets.[20] Binding of C4BP to other amyloid proteins such as Alzheimer's Aβ or prions was also reported.[21,22] In general, C4BP tends to bind self-molecules, which, in some circumstances, are recognised by the activator of the classical pathway C1q. These are amyloid, C-reactive protein and some extracellular matrix proteins.

Many pathogens utilise C4BP to evade immune response. Recruiting of human C4BP onto the surfaces of Gram-positive and Gram-negative bacteria, viruses and fungi has been described,[8] and the in vivo importance of such interaction for group A streptococci has been proven using transgenic mouse expressing human C4BPα-chains.[23]

TISSUE DISTRIBUTION

Serum protein: 150–300 μg/mL
  Primary site of synthesis: Liver
  Secondary sites of synthesis: Pancreas (islets), lungs, some cancer cells

REGULATION OF EXPRESSION

C4BP is an acute-phase protein, and its concentration is elevated upon increase of proinflammatory cytokines. Importantly, excess β-chains may be potentially harmful in circulation since they influence the coagulation cascade by inactivating protein S. Therefore, expression of α- and β-chains is regulated differently, and the α7β0 isoform is primarily upregulated on inflammation.[3,24] HNF1 transcription factor is essential for activity of the *C4BPA* promoter,[25] whereas cooperation between NF-I/CTF and HNF3 transcription factors is necessary for activity of the *C4BPB* promoter. In vitro studies showed that IL-6, IL-1β and IFN-γ increase production of both C4BPα- and C4BPβ-mRNAs, and TNF-α has the opposite effect.[3] However, synergistic action of IFN-γ and TNF-α leads to 10-fold increase of C4BPα-mRNA and only marginal increase of C4BPβ-mRNA. Secretion of C4BPβ depends on coexpression of protein S and the proteins form a complex intracellularly.[26]

PROTEIN SEQUENCES

C4BPα (NCBI Reference Sequence: NP_000706.1, protein accession P04003).

```
MHPPKTPSGA LHRKRKMAAW PFSRLWKVSD PILFQMTLIA ALLPAVLGNC   50
GPPPTLSFAA PMDITLTETR FKTGTTLKYT CLPGYVRSHS TQTLTCNSDG  100
EWVYNTFCIY KRCRHPGELR NGQVEIKTDL SFGSQIEFSC SEGFFLIGST  150
TSRCEVQDRG VGWSHPLPQC EIVKCKPPPD IRNGRHSGEE NFYAYGFSVT  200
YSCDPRFSLL GHASISCTVE NETIGVWRPS PPTCEKITCR KPDVSHGEMV  250
SGFGPIYNYK DTIVFKCQKG FVLRGSSVIH CDADSKWNPS PPACEPNSCI  300
NLPDIPHASW ETYPRPTKED VYVVGTVLRY RCHPGYKPTT DEPTTVICQK  350
NLRWTPYQGC EALCCPEPKL NNGEITQHRK SRPANHCVYF YGDEISFSCH  400
ETSRFSAICQ GDGTWSPRTP SCGDICNFPP KIAHGHYKQS SSYSFFKEEI  450
IYECDKGYIL VGQAKLSCSY SHWSAPAPQC KALCRKPELV NGRLSVDKDQ  500
YVEPENVTIQ CDSGYGVVGP QSITCSGNRT WYPEVPKCEW ETPEGCEQVL  550
TGKRLMQCLP NPEDVKMALE VYKLSLEIEQ LELQRDSARQ STLDKEL     597
```

C4BPβ (NCBI Reference Sequence: NP_000707.1, protein accession P20851).

```
MFFWCACCLM VAWRVSASDA EHCPELPPVD NSIFVAKEVE GQILGTYVCI   50
KGYHLVGKKT LFCNASKEWD NTTTECRLGH CPDPVLVNGE FSSSGPVNVS  100
DKITFMCNDH YILKGSNRSQ CLEDHTWAPP FPICKSRDCD PPGNPVHGYF  150
EGNNFTLGST ISYYCEDRYY LVGVQEQQCV DGEWSSALPV CKLIQEAPKP  200
ECEKALLAFQ ESKNLCEAME NFMQQLKESG MTMEELKYSL ELKKAELKAK  250
LL                                                      252
```

Signal sequences are underlined and *N*-linked glycosylation sites are indicated by bold font.

PROTEIN MODULES

C4BPα

1–48		Signal sequence	exon 2
49–109	(Uniprot)	CCP domain 1	exon 3
111–172	(Uniprot)	CCP domain 2	exon 4/5
173–236	(Uniprot)	CCP domain 3	exon 6
237–296	(Uniprot)	CCP domain 4	exon 7
297–362	(Uniprot)	CCP domain 5	exon 8
363–424	(Uniprot)	CCP domain 6	exon 9
425–482	(Uniprot)	CCP domain 7	exon 10
483–540	(Uniprot)	CCP domain 8	exon 11
541–597		C-terminal oligomerisation domain	exon 12

C4BPβ

1–17		Signal sequence	exon 1
21–78	(Uniprot)	CCP domain 1	exon 1/2
79–136	(Uniprot)	CCP domain 2	exon 3
137–193	(Uniprot)	CCP domain 3	exon 4/5
193–252		C-terminal oligomerisation domain	exon 5/6

CHROMOSOMAL LOCATION AND GENOMIC STRUCTURE

DNA coding for both α- and β-chains is localised on chromosome 1, band 1q32, within the so-called RCA (regulators of complement activation) cluster (Fig. 24.4).

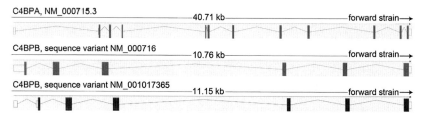

FIGURE 24.4 Genomic organisation of C4b-binding protein (C4BP). Exon/intron organisation of genes coding for C4BPα and C4BPβ is shown. Coding exons are indicated by *filled rectangles* and noncoding exons are shown as *open rectangles*.

	α-chain	β-chain
Length	40,735 bases	10,755* bases/11,152 bases**
Orientation	Plus strand	Plus strand
Start	207,104,238 bp from pter	207,088,842 bp from pter
End	207,144,972 bp from pter	207,099,993 bp from pter
exons	12 (11 coding exons)	6 (6 coding exons)*/7 (6)**

* sequence variant NM_000716.3,** sequence variant NM_001017365

CDNA SEQUENCES

The first five nucleotides in each exon are underlined. The initiation and termination codons (ATG and TAA, respectively) and putative polyadenylation signals (AATAAA) are <u>double-underlined.</u>

C4BPα (NCBI Reference Sequence: NM_000715.3)

```
TGAAATTCAG GGACTCTTTG GTGGAGCAAT TACCAGTCAA CTTCAGGGTA TTATGATAAA    60
CTCTGATCTG GGGAGGAACC AGGACTACAT AGATCAAGGC AGTTTTCTTC TTTGAGAAAC   120
TATCCCAGAT ATCATCATAG AGTCTTCTGC TCTTCCTCAA CTACCAAAGA AAAACATCAG   180
CGAAGCAGCA GGCCATGCAC CCCCCAAAAA CTCCATCTGG GGCTCTTCAT AGAAAAAGGA   240
AAATGGCAGC CTGGCCCTTC TCCAGGCTGT GGAAAGTCTC TGATCCAATT CTCTTCCAAA   300
TGACCTTGAT CGCTGCTCTG TTGCCTGCTG TTCTTGGCAA TTGTGGTCCT CCACCCACTT   360
TATCATTTGC TGCCCCGATG GATATTACGT TGACTGAGAC ACGCTTCAAA ACTGGAACTA   420
CTCTGAAATA CACCTGCCTC CCTGGCTACG TCAGATCCCA TTCAACTCAG ACGCTTACCT   480
GTAATTCTGA TGGCGAATGG GTGTATAACA CCTTCTGTAT CTACAAACGA TGCAGACACC   540
CAGGAGAGTT ACGTAATGGG CAAGTAGAGA TTAAGACAGA TTTATCTTTT GGATCACAAA   600
TAGAATTCAG CTGTTCAGAA GGATTTTTCT TAATTGGCTC AACCACTAGT CGTTGTGAAG   660
TCCAAGATAG AGGAGTTGGC TGGAGTCATC CTCTCCCACA ATGTGAAATT GTCAAGTGTA   720
AGCCTCCTCC AGACATCAGG AATGGAAGGC ACAGCGGTGA AGAAAATTTC TACGCATACG   780
GCTTTTCTGT CACCTACAGC TGTGACCCCG GCTTCTCACT CTTGGGCCAT GCCTCCATTT   840
CTTGCACTGT GGAGAATGAA ACAATAGGTG TTTGGAGACC AAGCCCTCCT ACCTGTGAAA   900
AAATCACCTG TCGCAAGCCA GATGTTTCAC ATGGGGAAAT GGTCTCTGGA TTTGGACCCA   960
TCTATAATTA CAAAGACACT ATTGTGTTTA AGTGCCAAAA AGGTTTTGTT CTCAGAGGCA  1020
GCAGTGTAAT TCATTGTGAT GCTGATAGCA AATGGAATCC TTCTCCTCCT GCTTGTGAGC  1080
CCAATAGTTG TATTAATTTA CCAGACATTC CACATGCTTC CTGGGAAACA TATCCTAGGC  1140
CGACAAAAGA GGATGTGTAT GTTGTTGGGA CTGTGTTAAG GTACCGCTGT CATCCTGGCT  1200
ACAAACCCAC TACAGATGAG CCTACGACTG TGATTTGTCA GAAAAATTTG AGATGGACCC  1260
CATACCAAGG ATGTGAGGCG TTATGTTGCC CTGAACCAAA GCTAAATAAT GGTGAAATCA  1320
CTCAACACAG GAAAAGTCGT CCTGCCAATC ACTGTGTTTA TTTCTATGGA GATGAGATTT  1380
CATTTTCATG TCATGAGACC AGTAGGTTTT CAGCTATATG CCAAGGAGAT GGCACGTGGA  1440
GTCCCCGAAC ACCATCATGT GGAGACATTT GCAATTTTCC TCCTAAAATT GCCCATGGGC  1500
ATTATAAACA ATCTAGTTCA TACAGCTTTT TCAAAGAAGA GATTATATAT GAATGTGATA  1560
AAGGCTACAT TCTGGTCGGA CAGGCGAAAC TCTCCTGCAG TTATTCACAC TGGTCAGCTC  1620
CAGCCCCTCA ATGTAAAGCT CTGTGTCGGA AACCAGAATT AGTGAATGGA AGGTTGTCTG  1680
TGGATAAGGA TCAGTATGTT GAGCCTGAAA ATGTCACCAT CCAATGTGAT TCTGGCTATG  1740
GTGTGGTTGG TCCCCAAAGT ATCACTTGCT CTGGGAACAG AACCTGGTAC CCAGAGGTGC  1800
```

```
CCAAGTGTGA GTGGGAGACC CCCGAAGGCT GTGAACAAGT GCTCACAGGC AAAAGACTCA   1860
TCCACTCTCT CCCAAACCCA GAGGATGTCA AAATGGCCCT GGAGGTATAT AAGCTGTCTC   1920
TGGAAATTGA ACAACTGGAA CTACAGAGAG ACAGCGCAAG ACAATCCACT TTGGATAAAG   1980
AACTATAATT TTTCTCAAAA GAAGGAGGAA AAGGTGTCTT GCTGGCTTGC CTCTTGCAAT   2040
TCAATACAGA TCAGTTTAGC AAATCTACTG TCAATTTGGC AGTGATATTC ATCATAATAA   2100
ATATCTAGAA ATGATAATTT GCTAAAGTTT AGTGCTTTGA GATTGTGAAA TTATTAATCA   2160
TCCTCTGTGT GGCTCATGTT TTTGCTTTTC AACACACAAA GCACAAATTT TTTTTCGATT   2220
AAAAATGTAT GTATAAAAAA CTAAAAAAAA AAAAAA                             2256
```

C4BPβ (NCBI Reference Sequence: NM_000716.3)

This variant, also known as A19, represents the longest transcript and encodes the longer isoform α7/β1. The same isoform is encoded by NM_001017365.2 (previously known as A12) and NM_001017367.1 transcripts, which differ in the 5′ UTR region. Another isoform α7/β0 lacking single aa in signal sequence as a result of alternate, in-frame splicing is encoded by transcripts NM_001017366.2 and NM_001017364.1.

```
ATTCCCAGAA ACCAGGAAAT TTCTGAAACT GAGTTTTAAA CCTGGCAACT CTTTTACCAT     60
CCCTTTGTAT TTTTGCAGTT CCTCCCTATA CCAGCTTCCT TACAGCTGTT CTGATTTCCT    120
GTCGTAAGAT TTTTTCTTTT CATTTTCTCC AGGGAGGGTA AATCAATATA ACCTCTTCTC    180
TAGAGAGGAG GTGGTTAGGT TGGTCTTAAG CAGTGTTAGA AGATCTATTT TTTTTCAAAC    240
CAGGTGTCTG AGCTGGGTGA ATTCCAGCCT GGGGAGAGGA CTTTGATCAC CAGATGTTTT    300
TTTGGTGTGC GTGCTGTCTT ATGGTTGCGT GGCGAGTTTC TGCTTCAGAT GCAGAGCACT    360
GTCCAGAGCT TCCTCCAGTG GACAATAGCA TATTTGTCGC AAAGGAGGTG GAAGGACAGA    420
TTCTGGGGAC TTACGTTTGT ATCAAGGGCT ACCACCTGGT AGGAAGAAG ACCCTTTTTT    480
GCAATGCCTC TAAGGAGTGG GATAACACCA CTACTGAGTG CCGCTTGGGC CACTGTCCTG    540
ATCCTGTGCT GGTGAATGGA GAGTTCAGTT CTTCAGGGCC TGTGAATGTA AGTGACAAAA    600
TCACGTTTAT GTGCAATGAC CACTACATCC TCAAGGGCAG CAATCGGAGC CAGTGTCTAG    660
AGGACCACAC CTGGGCACCT CCCTTTCCCA TCTGCAAAAG TAGGGACTGT GACCCTCCTG    720
GGAATCCAGT TCATGGCTAT TTTGAAGGAA ATAACTTCAC CTTAGGATCC ACCATTAGTT    780
ATTACTGTGA AGACAGGTAC TACTTAGTGG GCGTGCAGGA GCAGCAATGC GTTGATGGGG    840
AGTGGAGCAG TGCACTTCCA GTCTGCAAGT TGATCCAGGA AGCTCCCAAA CCAGATGTG    900
AGAAGGCACT TCTTGCCTTT CAGGAGAGTA AGAACCTCTG CGAAGCCATG GAGAACTTTA    960
TGCAACAATT AAAGGAAAGT GGCATGACAA TGGAGGAGCT AAAATATTCT CTGGAGCTGA   1020
AGAAAGCTGA GTTGAAGGCA AAATTGTTGT AACACTACAG CTGAGCAGAT GTAATAGAAA   1080
TAAACCTATG AATAAATTTT CTTCTTGGTT CTGAAAAAAA AAAAAAAAA A             1131
```

DEFICIENCY

Total deficiency of C4BP has not been reported. Familial low levels of C4BP were reported in one patient, whose symptoms clinically resembled that of Behçet disease.[27]

POLYMORPHIC VARIANTS AND DISEASE-RELATED SINGLE NUCLEOTIDE POLYMORPHISMS

The rare nonsynonymous polymorphism in the α-chain of C4BP (R240H) was weakly associated with atypical haemolytic uraemic syndrome in two independent cohorts,[28] while no significant association was found in the third cohort.[29] Mutations I126T and R120H in α-chain and T232A in β-chain were found in patients with recurrent, spontaneous pregnancy loss, but not in control subjects.[30]

C4BP KNOCKOUT AND TRANSGENIC MICE

Mouse C4BP is composed exclusively of α-chains, which lack CCP5-6 and both cysteines in the C-terminal extensions compared to the human protein. Nonetheless, mouse C4BP forms noncovalent polymers. Mouse *C4BPB* evolved into pseudogene. C4BP knockout mice in which *C4BPA* was inactivated do not present with any clear phenotype in absence of challenge.[31] Transgenic mice expressing human C4BPα-chains are also available.[23]

REFERENCES

1. Sanchez-Corral P, Criado Garcia O, Rodriguez de Cordoba S. Isoforms of human C4b-binding protein. I. Molecular basis for the C4BP isoform pattern and its variations in human plasma. *J Immunol* 1995;**155**:4030–6.
2. Esparza-Gordillo J, Soria JM, Buil A, et al. Genetic determinants of variation in the plasma levels of the C4b-binding protein (C4BP) in Spanish families. *Immunogenetics* 2003;**54**:862–6.
3. Criado Garcia O, Sanchez-Corral P, Rodriguez de Cordoba S. Isoforms of human C4b-binding protein. II. Differential modulation of the C4BPA and C4BPB genes by acute phase cytokines. *J Immunol* 1995;**155**:4037–43.
4. Kask L, Hillarp A, Ramesh B, Dahlback B, Blom AM. Structural requirements for the intracellular subunit polymerization of the complement inhibitor C4b-binding protein. *Biochemistry* 2002;**41**:9349–57.
5. Dahlback B, Smith CA, Muller-Eberhard HJ. Visualization of human C4b-binding protein and its complexes with vitamin K-dependent protein S and complement protein C4b. *PNAS* 1983;**80**:3461–5.
6. Mohlin FC, Blom AM. Purification and functional characterization of C4b-binding protein (C4BP). *Meth Mol Biol* 2014;**1100**:169–76.
7. Blom AM, Covell DG, Wallqvist A, Dahlback B, Villoutreix BO. The C4b-binding protein-protein S interaction is hydrophobic in nature. *Biochim Biophys Acta* 1998;**1388**:181–9.
8. Ermert D, Blom AM. C4b-binding protein: the good, the bad and the deadly. Novel functions of an old friend. *Immunol Lett* 2016;**169**:82–92.
9. Blom AM, Villoutreix BO, Dahlback B. Functions of human complement inhibitor C4b-binding protein in relation to its structure. *Arch Immunol Ther Exp (Warsz)* 2004;**52**:83–95.
10. Seya T, Holers VM, Atkinson JP. Purification and functional analysis of the polymorphic variants of the C3b/C4b receptor (CR1) and comparison with H, C4b-binding protein (C4bp), and decay accelerating factor (DAF). *J Immunol* 1985;**135**:2661–7.
11. Blom AM, Kask L, Dahlback B. CCP1-4 of the C4b-binding protein alpha-chain are required for factor I mediated cleavage of complement factor C3b. *Mol Immunol* 2003;**39**:547–56.
12. Fujita T, Gigli I, Nussenzweig V. Human C4-binding protein. II. Role in proteolysis of C4b by C3b-inactivator. *J Exp Med* 1978;**148**:1044–51.
13. Mihlan M, Blom AM, Kupreishvili K, et al. Monomeric C-reactive protein modulates classic complement activation on necrotic cells. *FASEB J* 2011;**25**:4198–210.
14. Blom AM, Foltyn-Zadura A, Villoutreix BO, Dahlbäck B. Positively charged amino acids at the interface between a-chain CCP1 and CCP2 of C4BP are required for regulation of the classical C3-convertase. *Mol Immunol* 2000;**37**:445–53.
15. Gigli I, Fujita T, Nussenzweig V. Modulation of the classical pathway C3 convertase by plasma proteins C4 binding protein and C3b inactivator. *PNAS* 1979;**76**:6596–600.

16. Dahlbäck B, Hildebrand B. Degradation of human complement component C4b in the presence of the C4b-binding protein-protein S complex. *Biochem J* 1983;**209**:857–63.

17. Webb JH, Blom AM, Dahlbäck B. Vitamin K-dependent protein S localizing complement regulator C4b-binding protein to the surface of apoptotic cells. *J Immunol* 2002;**169**:2580–6.

18. Trouw LA, Nilsson SC, Goncalves I, Landberg G, Blom AM. C4b-binding protein binds to necrotic cells and DNA, limiting DNA release and inhibiting complement activation. *J Exp Med* 2005;**201**:1937–48.

19. Trouw LA, Bengtsson AA, Gelderman KA, Dahlback B, Sturfelt G, Blom AM. C4b-binding protein and factor H compensate for the loss of membrane-bound complement inhibitors to protect apoptotic cells against excessive complement attack. *J Biol Chem* 2007;**282**:28540–8.

20. Sjolander J, Westermark GT, Renstrom E, Blom AM. Islet amyloid polypeptide triggers limited complement activation and binds complement inhibitor C4b-binding protein, which enhances fibril formation. *J Biol Chem* 2012;**287**:10824–33.

21. Trouw LA, Nielsen HM, Minthon L, et al. C4b-binding protein in Alzheimer's disease: binding to Abeta1-42 and to dead cells. *Mol Immunol* 2008;**45**:3649–60.

22. Sjoberg AP, Nystrom S, Hammarstrom P, Blom AM. Native, amyloid fibrils and beta-oligomers of the C-terminal domain of human prion protein display differential activation of complement and bind C1q, factor H and C4b-binding protein directly. *Mol Immunol* 2008;**45**:3213–21.

23. Ermert D, Shaughnessy J, Joeris T, et al. Virulence of group a streptococci is enhanced by human complement inhibitors. *PLoS Pathog* 2015;**11**:e1005043.

24. Garcia de Frutos P, Alim RI, Härdig Y, Zöller B, Dahlbäck B. Differential regulation of alpha and beta chains of C4b-binding protein during acute-phase response resulting in stable plasma levels of free anticoagulant protein S. *Blood* 1994;**84**:815–22.

25. Arenzana N, Rodriguez de Cordoba S, Rey Campos J. Expression of the human gene coding for the alpha-chain of C4b-binding protein, C4BPA, is controlled by an HNF1-dependent hepatic-specific promoter. *Biochem J* 1995;**308**:613–21.

26. Carlsson S, Dahlback B. Dependence on vitamin K-dependent protein S for eukaryotic cell secretion of the beta-chain of C4b-binding protein. *J Biol Chem* 2010;**285**:32038–46.

27. Trapp RG, Fletcher M, Forristal J, West CD. C4 binding protein deficiency in a patient with atypical Behcet's disease. *J Rheumatol* 1987;**14**:135–8.

28. Blom AM, Bergstrom F, Edey M, et al. A novel non-synonymous polymorphism (p.Arg240His) in C4b-binding protein is associated with atypical hemolytic uremic syndrome and leads to impaired alternative pathway cofactor activity. *J Immunol* 2008;**180**:6385–91.

29. Martinez-Barricarte R, Goicoechea de Jorge E, Montes T, Layana AG, Rodriguez de Cordoba S. Lack of association between polymorphisms in C4b-binding protein and atypical haemolytic uraemic syndrome in the Spanish population. *Clin Exp Immunol* 2009;**155**:59–64.

30. Mohlin FC, Mercier E, Fremeaux-Bacchi V, et al. Analysis of genes coding for CD46, CD55, and C4b-binding protein in patients with idiopathic, recurrent, spontaneous pregnancy loss. *Eur J Immunol* 2013;**43**:1617–29.

31. Wenderfer SE, Soimo K, Wetsel RA, Braun MC. Analysis of C4 and the C4 binding protein in the MRL/lpr mouse. *Arthritis Res Ther* 2007;**9**:R114.

Chapter 25

Decay-Accelerating Factor

Joseph M. Christy[1], Christopher B. Toomey[2], David M. Cauvi[2], Kenneth M. Pollard[1]

[1]The Scripps Research Institute, La Jolla, CA, United States; [2]University of California, San Diego, La Jolla, CA, United States

PHYSIOCHEMICAL PROPERTIES

The immature form of DAF is a 381 amino acid single-chain prepropeptide.[1] The immature form includes a leader peptide of 34 amino acids, as well as a 28 amino acid C-terminal signal peptide.[2] DAF is cleaved posttranslationally at the C-terminal signal peptide (located at Ser352) and a glycosylphosphatidylinositol (GPI) anchor is added.[3] The mature glycoprotein has a molecular weight of approximately 70 kDa under non-reduced conditions and 75 kDa under reduced conditions.[4,5] It consists of four complement control protein (CCP) domains of approximately 60 amino acids each and contains an N-linked glycosylation site at amino acid 95 (Fig. 25.1).[1,2,6]

pl   ~6.7–7.8

STRUCTURE

DAF consists of four CCP domains with a single N-linked glycan between CCP 1 and 2.[1,2] DAF contains a heavily O-glycosylated membrane proximal domain rich in serines, threonines and prolines (STP), as well as a membrane-bound GPI anchor.[3,7] The crystal structure is shown in Fig. 25.2.

FUNCTION

DAF intrinsically protects host cells from autologous complement attack by preventing the formation and accelerating the decay of the classical and alternative C3 and C5 convertases, thereby inhibiting the cleavage of C3 and C5.[4,5,9,10] When purified DAF is added to cells, it is able to incorporate into their membranes and display functional activity.[5]

DAF has also been found to regulate T cell tolerance and thus negatively regulate several animal models of autoimmune diseases.[11–15]

FIGURE 25.1   Diagram of protein domains for DAF.[8]

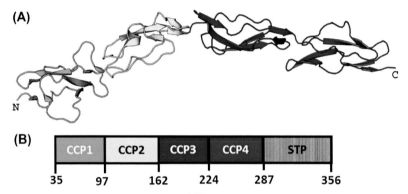

FIGURE 25.2   Crystal structure of DAF.[1–3,7,8] (A) Ribbon diagram of DAF (PDB ID: 1OJY). (B) Schematic representation of DAF CCP domains.

TISSUE DISTRIBUTION

DAF is widely distributed on most cell types. It is present on all cell types that are in close contact with serum complement proteins, including major circulating blood elements and cells that line extravascular compartments.[16,17] Soluble variants of DAF protein are found in plasma, cerebrospinal fluid, saliva, urine, synovial fluid, tears, aqueous humor, and vitreous humor, with concentrations ranging 4–400 ng/mL.[17,18]

REGULATION OF EXPRESSION

In normal individuals, constitutive DAF expression on peripheral blood elements ranges 2,100–85,000 molecules/cell.[19] By contrast, expression on endothelial and epithelial cells is >100,000 molecules/cell.[17] Constitutive DAF expression also varies by tissue type.[20] DAF expression can be induced by phorbol ester, various acute and chronic inflammatory cytokines such as IL-1β, IL-6, TNF-α, TGF-β1 and IFN-γ, tissue-specific factors and prostaglandins.[21–26] Modulation of DAF expression is primarily at the transcriptional level,[24–26] and the promoter has a cAMP response element.[27,28] Expression of the mouse *DAF1* gene is controlled by the Sp1 transcription factor.[29]

PROTEIN SEQUENCE[1,2]

```
MTVARPSVPAALPLLGELPR  LLLLVLLCLP AVWGDCGLPP DVPNAQPALE    50
GRTSFPEDTV ITYKCEESFV KIPGEKDSVI CLKGSQWSDI EEFCNRSCEV   100
PTRLNSASLK QPYITQNYFP VGTVVEYECR PGYRREPSLS PKLTCLQNLK   150
WSTAVEFCKK KSCPNPGEIR NGQIDVPGGI LFGATISFSC NTGYKLFGST   200
SSFCLISGSS VQWSDPLPEC REIYCPAPPQ IDNGIIQGER DHYGYRQSVT   250
YACNKGFTMI GEHSIYCTVN NDEGEWSGPP PECRGKSLTS KVPPTVQKPT   300
TVNVPTTEVS PTSQKTTTKT TTPNAQATRS TPVSRTTKHF HETTPNKGSG   350
TTSGTTRLLS GHTCFTLTGL LGTLVTMGLL T
```

Leader sequence is underlined, *N*-linked glycosylation site is indicated in bold, and the site of cleavage for GPI anchor attachment is double-underlined. Amino acid differences between publications: 80 Ile↔Thr, 85 Ser↔Met.

PROTEIN MODULES

1–34	Leader Sequence[1,2]	exon 1[30]
35–95	CCP	exon 2
97–159	CCP	exon 3
162–221	CCP	exons 4/5
224–284	CCP	exon 6
287–356	STP	exons 7–9

CHROMOSOMAL LOCATION

Human: 1q32, located adjacent to the genes comprising the RCA gene family, which includes MCP, CR1, CR2, factor H and C4bp.[30,31]

Mouse: chromosome 1, co-localised with C4bp gene.[32]

cDNA SEQUENCE[1,2,27]

```
ACTGCAACTC GCTCCGGCCG CTGGGCGTAG CTGCGACTCG GCGGAGTCCC GGCGGCGCGT    60
CCTTGTTCTA ACCCGGCGCG CCATGACCGT CGCGCGGCCG AGCGTGCCCG CGGCGCTGCC   120
CCTCCTCGGG GAGCTGCCCC GGCTGCTGCT GCTGGTGCTG TTGTGCCTGC CGGCCGTGTG   180
GGGTGACTGT GGCCTTCCCC CAGATGTACC TAATGCCCAG CCAGCTTTGG AAGGCCGTAC   240
AAGTTTTCCC GAGGATACTG TAATAACGTA CAAATGTGAA GAAAGCTTTG TGAAAATTCC   300
TGGCGAGAAG GACTCAGTGA TCTGCCTTAA GGGCAGTCAA TGGTCAGATA TTGAAGAGTT   360
CTGCAATCGT AGCTGCGAGG TGCCAACAAG GCTAAATTCT GCATCCCTCA AACAGCCTTA   420
TATCACTCAG AATTATTTTC CAGTCGGTAC TGTTGTGGAA TATGAGTGCC GTCCAGGTTA   480
CAGAAGAGAA CCTTCTCTAT CACCAAAACT AACTTGCCTT CAGAATTTAA AATGGTCCAC   540
AGCAGTCGAA TTTTGTAAAA AGAAATCATG CCCTAATCCG GGAGAAAATAC GAAATGGTCA   600
GATTGATGTA CCAGGTGGCA TATTATTTGG TGCAACCATC TCCTTCTCAT GTAACACAGG   660
GTACAAATTA TTTGGCTCGA CTTCTAGTTT TTGTCTTATT TCAGGCAGCT CTGTCCAGTG   720
GAGTGACCCG TTGCCAGAGT GCAGAGAAAT TTATTGTCCA GCACCACCAC AAATTGACAA   780
```

cDNA SEQUENCE—*Continued*

```
TGGAATAATT CAAGGGGAAC GTGACCATTA TGGATATAGA CAGTCTGTAA CGTATGCATG      840

TAATAAAGGA TTCACCATGA TTGGAGAGCA CTCTATTTAT TGTACTGTGA ATAATGATGA      900

AGGAGAGTGG AGTGGCCCAC CACCTGAATG CAGAGGAAAA TCTCTAACTT CCAAGGTCCC      960

ACCAACAGTT CAGAAACCTA CCACAGTAAA TGTTCCAACT ACAGAAGTCT CACCAACTTC     1020

TCAGAAAACC ACCACAAAAA CCACCACACC AAATGCTCAA GCAACACGGA GTACACCTGT     1080

TTCCAGGACA ACCAAGCATT TTCATGAAAC AACCCCAAAT AAAGGAAGTG GAACCACTTC     1140

AGGTACTACC CGTCTTCTAT CTGGGCACAC GTGTTTCACG TTGACAGGTT TGCTTGGGAC     1200

GCTAGTAACC ATGGGCTTGC TGACTTAGCC AAAGAAGAGT TAAGAAGAAA ATACACACAA     1260

GTATACAGAC TGTTCCTAGT TTCTTAGACT TATCTGCATA TTGGATAAAA TAAATGCAAT     1320

TGTGCTCTTC ATTTAGGATG CTTTCATTGT CTTTAAGATG TGTTAGGAAT GTCAACAGAG     1380

CAAGGAGAAA AAAGGCAGTC CTGGAATCAC ATTCTTAGCA CACCTACACC TCTTGAAAAT     1440

AGAACAACTT GCAGAATTGA GAGTGATTCC TTTCCTAAAA GTGTAAGAAA GCATAGAGAT     1500

TTGTTCGTAT TTAGAATGGG ATCACGAGGA AAAGAGAAGG AAAGTGATTT TTTTCCACAA     1560

GATCTGTAAT GTTATTTCCA CTTATAAAGG AAATAAAAAA TGAAAAACAT TATTTGGATA     1620

TCAAAAGCA ATAAAAACCC AATTCAGTCT CTTCTAAGCA AAATTGCTAA AGAGAGATGA     1680

ACCACATTAT AAAGTAATCT TTGGCTGTAA GGCATTTTCA TCTTTCCTTC GGGTTGGCAA     1740

AATATTTTAA AGGTAAAACA TGCTGGTGAA CCAGGGGTGT TGATGGTGAT AAGGGAGGAA     1800

TATAGAATGA AAGACTGAAT CTTCCTTTGT TGCACAAATA GAGTTTGGAA AAAGCCTGTG     1860

AAAGGTGTCT TCTTTGACTT AATGTCTTTA AAAGTATCCA GAGATACTAC AATATTAACA     1920

TAAGAAAAGA TTATATATTA TTTCTGAATC GAGATGTCCA TAGTCAAATT TGTAAATCTT     1980

ATTCTTTTGT AATATTTATT TATATTTATT TATGACAGTG AACATTCTGA TTTTACATGT     2040

AAAACAAGAA AAGTTGAAGA AGATATGTGA AGAAAAATGT ATTTTTCCTA AATAGAAATA     2100

AATGATCCCA TTTTTTGGT
```

Position 1 is the transcriptional start site. The first five nucleotides in each exon are underlined. Exon 10 encodes an *Alu* type sequence (not in the cDNA sequence above; shown below) which is not found in GPI-anchored DAF mRNA.[1,30] The initiation codon (ATG), termination codon (TAG) and the four polyadenylation signals (AATAAA) are indicated. Nucleotide differences between the published sequences: 321T↔C, 336G↔T, 337T↔G.

EXON 10[1]

```
1164   GTTCTCGTCC TGTCACCCAG GCTGGTATGC GGTGGTGTGA TCGTAGCTCA CTGCAGTCTC
       GAACTCCTGG GTTCAAGCGA TCCTTCCACT TCAGCCTCCC AAGTAGCTGG TACTACAG
```

GENOMIC STRUCTURE

The DAF gene is approximately 40 kb long and contains 11 exons, with exon 10 excluded from GPI-anchored DAF mRNA.[30] Unspliced DAF mRNA containing exon 10 is thought to encode the secreted soluble variant of DAF protein.[1] Exons range 21–956 bp, while introns range 0.5–19.8 kb(Fig. 25.3).[30]

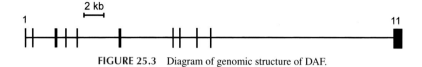

FIGURE 25.3   Diagram of genomic structure of DAF.

ACCESSION NUMBERS (EMBL/GENBANK)

Human	M31516	M64356 (promoter)
	M15799	
	M64653	
	S72858	
Mouse[32,33]	L41365–6	
	D63679	
Orangutan[34]	S67775	
Guinea pig[35]	D49416–D49421	
Rat[36]	AF039583–4	

DEFICIENCY

Deficient expression of GPI-anchored proteins including DAF is associated with paroxysmal nocturnal haemoglobinuria, which involves complement-mediated erythrocyte lysis.[16,37,38] Mutation of the phosphatidylinositol glycan A (*PIG-A*) gene in bone marrow stem cells disrupts GPI anchor synthesis (G1cNAc-PI synthesis), which prevents GPI-anchored protein expression on peripheral blood elements.[39–42] DAF-deficient cells exhibit increased uptake of C3b.[43] By contrast, the Inab phenotype is characterised by a complete absence of DAF expression only and does not significantly affect the sensitivity of cells to complement-mediated lysis.[44–46] Decreased expression of DAF has been notably linked to several diseases involving immune dysregulation, as well as anaemias and proliferative diseases.[47–54] Deficient complement regulatory proteins in systemic lupus erythematosus (SLE) patients have been associated with auto-immune haemolytic anaemia (AHIA) and lymphopenia.[48,49] In particular, SLE patients with AHIA exhibit decreased levels of DAF-expressing peripheral lymphocytes, red blood cells and granulocytes.[49,55]

POLYMORPHIC VARIANTS[56–61]

The Cromer blood group system consists of 12 high-incidence (reported below) and 3 low-incidence antigens residing on DAF, and each phenotype is the result of a single amino acid change in DAF.

G237T; R52L
G237C; R52P
T327G; L82R
C678T; S199L
G761C; A227P

T321A; I80N
C831T; T250M
G801A; R240H
C729T; P216L
T808G; H242Q
G548A; E156K
A822G; Q247R

MUTANT ANIMALS

BCD55[tm1Song]: Parent ES cell line is TL-1m derived from 129S6/SvEvTac, inserted into C57BL/6 background mice. Exons 2 and 3 of the *DAF1* gene encoding CCP1 and CCP2, respectively, were replaced with a neomycin selection cassette.[62] These mice exhibit increased sensitivity to experimental autoimmune myasthenia gravis[63] and are commercially available through the Mutant Mouse Resource & Research Centers.

CD55[tm1Mme]: Parent ES cell line is GK129 derived from 129P2/OlaHsd, inserted into C57BL/6 background mice. Exons 3 and 5 of the *DAF1* gene encoding CCP2 and CCP3, respectively, were removed via a Cre-loxP recombination.[20] These mice exhibit increased sensitivity to experimental autoimmune myasthenia gravis[64] and increased complement-mediated podocyte injury in acute nephrotoxic nephritis.[65] It is not commercially available.

Tg(CD55)[66]: Generated via inserting DAF1 expression cassette into C57BL/6 background mice. *DAF1* Tg mouse DAF1 expression on peripheral blood cells increased approximately 40% compared to wild-type mice. *DAF1* Tg mice exhibit stronger complement regulation through decreased C3b deposition and C5a/C3a levels, as well as protection against experimental autoimmune encephalomyelitis via reduced central nervous system inflammation and demyelination.

REFERENCES

1. Caras IW, Davitz MA, Rhee L, Weddell G, Martin Jr DW, Nussenzweig V. Cloning of decay-accelerating factor suggests novel use of splicing to generate two proteins. *Nature* 1987;**325**(6104):545–9.
2. Medof ME, Lublin DM, Holers VM, Ayers DJ, Getty RR, Leykam JF, et al. Cloning and characterization of cDNAs encoding the complete sequence of decay-accelerating factor of human complement. *Proc Natl Acad Sci USA* 1987;**84**(7):2007–11.
3. Medof ME, Walter EI, Roberts WL, Haas R, Rosenberry TL. Decay accelerating factor of complement is anchored to cells by a C-terminal glycolipid. *Biochemistry* 1986;**25**(22):6740–7.
4. Nicholson-Weller A, Burge J, Fearon DT, Weller PF, Austen KF. Isolation of a human erythrocyte membrane glycoprotein with decay-accelerating activity for C3 convertases of the complement system. *J Immunol* 1982;**129**(1):184–9.

5. Medof ME, Kinoshita T, Nussenzweig V. Inhibition of complement activation on the surface of cells after incorporation of decay-accelerating factor (DAF) into their membranes. *J Exp Med* 1984;**160**(5):1558–78.

6. Lublin DM, Krsek-Staples J, Pangburn MK, Atkinson JP. Biosynthesis and glycosylation of the human complement regulatory protein decay-accelerating factor. *J Immunol* 1986;**137**(5):1629–35.

7. Davitz MA, Low MG, Nussenzweig V. Release of decay-accelerating factor (DAF) from the cell membrane by phosphatidylinositol-specific phospholipase C (PIPLC). Selective modification of a complement regulatory protein. *J Exp Med* 1986;**163**(5):1150–61.

8. Lukacik P, Roversi P, White J, Esser D, Smith GP, Billington J, et al. Complement regulation at the molecular level: the structure of decay-accelerating factor. *Proc Natl Acad Sci USA* 2004;**101**(5):1279–84.

9. Pangburn MK. Differences between the binding sites of the complement regulatory proteins DAF, CR1, and factor H on C3 convertases. *J Immunol* 1986;**136**(6):2216–21.

10. Fujita T, Inoue T, Ogawa K, Iida K, Tamura N. The mechanism of action of decay-accelerating factor (DAF). DAF inhibits the assembly of C3 convertases by dissociating C2a and Bb. *J Exp Med* 1987;**166**(5).1221–8.

11. Liu J, Miwa T, Hilliard B, Chen Y, Lambris JD, Wells AD, et al. The complement inhibitory protein DAF (CD55) suppresses T cell immunity in vivo. *J Exp Med* 2005;**201**(4):567–77.

12. Lin F, Kaminski HJ, Conti-Fine BM, Wang W, Richmonds C, Medof ME. Markedly enhanced susceptibility to experimental autoimmune myasthenia gravis in the absence of decay-accelerating factor protection. J Clin Invest;110(9):1269–74.

13. Miwa T, Maldonado MA, Zhou L, Sun X, Luo HY, Cai D, et al. Deletion of decay-accelerating factor (CD55) exacerbates autoimmune disease development in MRL/lpr mice. *Am J Pathol* 2002;**161**(3):1077–86.

14. Sogabe H, Nangaku M, Ishibashi Y, Wada T, Fujita T, Sun X, et al. Increased susceptibility of decay-accelerating factor deficient mice to anti-glomerular basement membrane glomerulonephritis. *J Immunol* 2001;**167**(5):2791–7.

15. Toomey CB, Cauvi DM, Song W-C, Pollard KM. Decay-accelerating factor 1 (Daf1) deficiency exacerbates xenobiotic-induced autoimmunity. *Immunology* 2010;**131**(1):99–106.

16. Nicholson-Weller A, March JP, Rosen CE, Spicer DB, Austen KF. Surface membrane expression by human blood leukocytes and platelets of decay-accelerating factor, a regulatory protein of the complement system. *Blood* 1985;**65**(5):1237–44.

17. Medof ME, Walter EI, Rutgers JL, Knowles DM, Nussenzweig V. Identification of the complement decay-accelerating factor (DAF) on epithelium and glandular cells and in body fluids. *J Exp Med* 1987;**165**(3):848–64.

18. Lass JH, Walter EI, Burris TE, Grossniklaus HE, Roat MI, Skelnik DL, et al. Expression of two molecular forms of the complement decay-accelerating factor in the eye and lacrimal gland. *Invest Ophthalmol Vis Sci* 1990;**31**(6):1136–48.

19. Kinoshita T, Medof ME, Silber R, Nussenzweig V. Distribution of decay-accelerating factor in the peripheral blood of normal individuals and patients with paroxysmal nocturnal hemoglobinuria. *J Exp Med* 1985;**162**(1):75–92.

20. Lin F, Fukuoka Y, Spicer A, Ohta R, Okada N, Harris CL, et al. Tissue distribution of products of the mouse decay-accelerating factor (DAF) genes. Exploitation of a Daf1 knock-out mouse and site-specific monoclonal antibodies. *Immunology* 2001;**104**(2):215–25.

21. Bryant RW, Granzow CA, Siegel MI, Egan RW, Billah MM. Phorbol esters increase synthesis of decay-accelerating factor, a phosphatidylinositol-anchored surface protein, in human endothelial cells. *J Immunol* 1990;**144**(2):593–8.

22. Spiller OB, Criado-García O, Rodríguez De Córdoba S, Morgan BP. Cytokine-mediated up-regulation of CD55 and CD59 protects human hepatoma cells from complement attack. *Clin Exp Immunol* 2000;**121**(2):234–41.

23. Cocuzzi ET, Bardenstein DS, Stavitsky A, Sundarraj N, Medof ME. Upregulation of DAF (CD55) on orbital fibroblasts by cytokines. Differential effects of TNF-ß and TNF-a. *Curr Eye Res* 2001;**23**(2):86–92.

24. Ahmad SR, Lidington EA, Ohta R, Okada N, Robson MG, Davies KA, et al. Decay-accelerating factor induction by tumour necrosis factor-α, through a phosphatidylinositol-3 kinase and protein kinase C-dependent pathway, protects murine vascular endothelial cells against complement deposition. *Immunology* 2003;**110**(2):258–68.

25. Holla VR, Wang D, Brown JR, Mann JR, Katkuri S, DuBois RN. Prostaglandin E2 regulates the complement inhibitor CD55/decay-accelerating factor in colorectal cancer. *J Biol Chem* 2005;**280**(1):476–83.

26. Andoh A, Kinoshita K, Rosenberg I, Podolsky DK. Intestinal trefoil factor induces decay-accelerating factor expression and enhances the protective activities against complement activation in intestinal epithelial cells. *J Immunol* 2001;**167**(7):3887–93.

27. Ewulonu UK, Ravi L, Medof ME. Characterization of the decay-accelerating factor gene promoter region. *Proc Natl Acad Sci USA* 1991;**88**(11):4675–9.

28. Thomas DJ, Lublin DM. Identification of 5′-flanking regions affecting the expression of the human decay accelerating factor gene and their role in tissue-specific expression. *J Immunol* 1993;**150**(1):151–60.

29. Cauvi DM, Cauvi G, Pollard KM. Constitutive expression of murine decay-accelerating factor 1 is controlled by the transcription factor Sp1. *J Immunol* 2006;**177**(6):3837–47.

30. Post TW, Arce MA, Liszewski MK, Thompson ES, Atkinson JP, Lublin DM. Structure of the gene for human complement protein decay accelerating factor. *J Immunol* 1990;**144**(2):740–4.

31. Lublin DM, Lemons RS, Le Beau MM, Holers VM, Tykocinski ML, Medof ME, et al. The gene encoding decay-accelerating factor (DAF) is located in the complement-regulatory locus on the long arm of chromosome 1. *J Exp Med* 1987;**165**(6):1731–6.

32. Spicer AP, Seldin MF, Gendler SJ. Molecular cloning and chromosomal localization of the mouse decay-accelerating factor genes. Duplicated genes encode glycosylphosphatidylinositol-anchored and transmembrane forms. *J Immunol* 1995;**155**(6):3079–91.

33. Fukuoka Y, Yasui A, Okada N, Okada H. Molecular cloning of murine decay accelerating factor by immunoscreening. *Int Immunol* 1996;**8**(3):379–85.

34. Nickells MW, Alvarez JI, Lublin DM, Atkinson JP. Characterization of DAF-2, a high molecular weight form of decay-accelerating factor (DAF; CD55), as a covalently cross-linked dimer of DAF-1. *J Immunol* 1994;**152**(2):676–85.

35. Nonaka M, Miwa T, Okada N, Nonaka M, Okada H. Multiple isoforms of guinea pig decay-accelerating factor (DAF) generated by alternative splicing. *J Immunol* 1995;**155**(6):3037–48.

36. Hinchliffe SJ, Spiller OB, Rushmere NK, Morgan BP. Molecular cloning and functional characterization of the rat analogue of human decay-accelerating factor (CD55). *J Immunol* 1998;**161**(10):5695–703.

37. Pangburn MK, Schreiber RD, Muller-Eberhard HJ. Deficiency of an erythrocyte membrane protein with complement regulatory activity in paroxysmal nocturnal hemoglobinuria. *Proc Natl Acad Sci USA* 1983;**80**(17):5430–4.

38. Nicholson-Weller A, March JP, Rosenfeld SI, Austen KF. Affected erythrocytes of patients with paroxysmal nocturnal hemoglobinuria are deficient in the complement regulatory protein, decay accelerating factor. *Proc Natl Acad Sci USA* 1983;**80**(16):5066–70.

39. Armstrong C, Schubert J, Ueda E, Knez JJ, Gelperin D, Hirose S, et al. Affected paroxysmal nocturnal hemoglobinuria T lymphocytes harbor a common defect in assembly of N-acetyl-D-glucosamine inositol phospholipid corresponding to that in class A Thy-1- murine lymphoma mutants. *J Biol Chem* 1992;**267**(35):25347–51.

40. Takahashi M, Takeda J, Hirose S, Hyman R, Inoue N, Miyata T, et al. Deficient biosynthesis of N-acetylglucosaminyl-phosphatidylinositol, the first intermediate of glycosyl phosphatidylinositol anchor biosynthesis, in cell lines established from patients with paroxysmal nocturnal hemoglobinuria. *J Exp Med* 1993;**177**(2):517–21.

41. Hidaka M, Nagakura S, Horikawa K, Kawaguchi T, Iwamoto N, Kagimoto T, et al. Impaired glycosylation of glycosylphosphatidylinositol-anchor synthesis in paroxysmal nocturnal hemoglobinuria leukocytes. *Biochem Biophys Res Commun* 1993;**191**(2):571–9.

42. Takeda J, Miyata T, Kawagoe K, Iida Y, Endo Y, Fujita T, et al. Deficiency of the GPI anchor caused by a somatic mutation of the PIG-A gene in paroxysmal nocturnal hemoglobinuria. *Cell* 1993;**73**(4):703–11.

43. Medof ME, Kinoshita T, Silber R, Nussenzweig V. Amelioration of lytic abnormalities of paroxysmal nocturnal hemoglobinuria with decay accelerating factor. *Proc Natl Acad Sci USA* 1985;**82**(9):2980–4.

44. Reid M, Mallinson G, Sim R, Poole J, Pausch V, Merry A, et al. Biochemical studies on red blood cells from a patient with the Inab phenotype (decay-accelerating factor deficiency). *Blood* 1991;**78**(12):3291–7.

45. Telen M, Green A. The Inab phenotype: characterization of the membrane protein and complement regulatory defect. *Blood* 1989;**74**(1):437–41.

46. Daniels GL, Mallinson, Okubo, Hori, Kataoka, et al. Decay-accelerating factor (CD55) deficiency phenotypes in Japanese. *Transfus Med* 1998;**8**(2):141–7.

47. żupańska B, Bogdanik I, Fabijanska-Mitek J, Pyl H. Autoimmune haemolytic anaemia with a paroxysmal nocturnal haemoglobinuria-like defect. *Eur J Haematol* 1999;**62**(5):346–9.

48. Richaud-Patin Y, Pérez-Romano B, Carrillo-Maravilla E, Rodriguez AB, Simon AJ, Cabiedes J, et al. Deficiency of red cell bound CD55 and CD59 in patients with systemic lupus erythematosus. *Immunol Lett* 2003;**88**(2):95–9.

49. García-Valladares I, Atisha-Fregoso Y, Richaud-Patin Y, Jakez-Ocampo J, Soto-Vega E, Elías-López D, et al. Diminished expression of complement regulatory proteins (CD55 and CD59) in lymphocytes from systemic lupus erythematosus patients with lymphopenia. *Lupus* 2006;**15**(9):600–5.

50. Ruiz-Argüelles A, Llorente L. The role of complement regulatory proteins (CD55 and CD59) in the pathogenesis of autoimmune hemocytopenias. *Autoimmun Rev* 2007;**6**(3):155–61.

51. Meletis J, Terpos E, Samarkos M, Meletis C, Apostolidou E, Komninaka V, et al. Detection of CD55- and/or CD59-deficient red cell populations in patients with lymphoproliferative syndromes. *Hematol J* 2001;**2**(1):33–7.

52. Kaiafa G, Papadopoulos A, Ntaios G, Saouli Z, Savopoulos C, Tsesmeli N, et al. Detection of CD55- and CD59-deficient granulocytic populations in patients with myelodysplastic syndrome. *Ann Hematol* 2008;**87**(4):257–62.

53. Meletis J, Terpos E, Samarkos M, Meletis C, Apostolidou E, Komninaka V, et al. Detection of CD55- and/or CD59-deficient red cell populations in patients with plasma cell dyscrasias. *Int J Hematol* 2002;**75**(1):40–4.

54. Venneker GT, Das PK, Naafs B, Tigges AJ, Bos JD, Asghar SS. Morphoea lesions are associated with aberrant expression of membrane cofactor protein and decay accelerating factor in vascular endothelium. *Br J Dermatol* 1994;**131**(2):237–42.

55. Alegretti AP, Mucenic T, Merzoni J, Faulhaber GA, Silla LM, Xavier RM. Expression of CD55 and CD59 on peripheral blood cells from systemic lupus erythematosus (SLE) patients. *Cell Immunol* 2010;**265**(2):127–32.

56. Stafford HA, Tykocinski ML, Lublin DM, Holers VM, Rosse WF, Atkinson JP, et al. Normal polymorphic variations and transcription of the decay accelerating factor gene in paroxysmal nocturnal hemoglobinuria cells. *Proc Natl Acad Sci USA* 1988;**85**(3):880–4.

57. Lublin DM, Thompson ES, Green AM, Levene C, Telen MJ. Dr(a-) polymorphism of decay accelerating factor. Biochemical, functional, and molecular characterization and production of allele-specific transfectants. *J Clin Invest* 1991;**87**(6):1945–52.

58. Telen MJ, Rao N, Udani M, Thompson ES, Kaufman RM, Lublin DM. Molecular mapping of the Cromer blood group Cra and Tca epitopes of decay accelerating factor: toward the use of recombinant antigens in immunohematology. *Blood* 1994;**84**(9):3205–11.

59. Storry JR, Sausais L, Hue-Roye K, Mudiwa F, Ferrer Z, Blajchman MA, et al. GUTI: a new antigen in the Cromer blood group system. *Transfusion* 2003;**43**(3):340–4.

60. Banks J, Poole J, Ahrens N, Seltsam A, Salama A, Hue-Roye K, et al. SERF: a new antigen in the Cromer blood group system. *Transfus Med* 2004;**14**(4):313–8.

61. Hue-Roye K, Lomas-Francis C, Belaygorod L, Lublin DM, Barnes J, Chung A, et al. Three new high-prevalence antigens in the Cromer blood group system. *Transfusion* 2007;**47**(9):1621–9.

62. Sun X, Funk CD, Deng C, Sahu A, Lambris JD, Song WC. Role of decay-accelerating factor in regulating complement activation on the erythrocyte surface as revealed by gene targeting. *Proc Natl Acad Sci USA* 1999;**96**(2):628–33.

63. Morgan BP, Chamberlain-Banoub J, Neal JW, Song W, Mizuno M, Harris CL. The membrane attack pathway of complement drives pathology in passively induced experimental autoimmune myasthenia gravis in mice. *Clin Exp Immunol* 2006;**146**(2):294–302.

64. Lin F, Kaminski HJ, Conti-Fine BM, Wang W, Richmonds C, Medof ME. Markedly enhanced susceptibility to experimental autoimmune myasthenia gravis in the absence of decay-accelerating factor protection. *J Clin Invest* 2002;**110**(9):1269–74.

65. Lin F, Emancipator SN, Salant DJ, Medof ME. Decay-accelerating factor confers protection against complement-mediated podocyte injury in acute nephrotoxic nephritis. *Lab Invest* 2002;**82**(5):563–9.

66. Li Q, Huang D, Nacion K, Bu H, Lin F. Augmenting DAF levels in vivo ameliorates experimental autoimmune encephalomyelitis. *Mol Immunol* 2009;**46**(15):2885–91.

Chapter 26

Membrane Cofactor Protein

M. Kathryn Liszewski, John P. Atkinson
Washington University School of Medicine, St. Louis, MO, United States

OTHER NAMES

MCP, CD46.

PHYSICOCHEMICAL PROPERTIES

Membrane cofactor protein (MCP) is a type 1 transmembrane glycoprotein expressed on most cells as four common isoforms that arise by alternative splicing of a single gene.[1] Each isoform bears a 34-amino acid (aa) signal peptide followed by 328–350 aa (the four mature common isoforms are C1, 328 aa; C2, 335 aa; BC1, 343 aa; BC2 350 aa).[2–4]

pI	3.9–5.8

Higher molecular weight species possess more O-linked sugars, which correlates with a more acidic pI.

M_r (kDa) predicted	~39
Observed	51–58 (C isoforms)
	59–68 (BC isoforms)
N-linked glycosylation sites	3 (83, 114 and 273) all sites are occupied[2,4]

STRUCTURE

The N-terminus of all commonly expressed human isoforms consists of four contiguous complement control protein (CCP) modules (Fig. 26.1). Following this is an alternatively spliced segment enriched in serines, threonines and prolines (STP domain) that is heavily O-glycosylated.[4,5] The gene contains three such STP exons termed A, B and C (reviewed in Ref. 6). The STP segment of two of the regularly expressed isoforms consists of B+C (29 amino acids) and the other two consist of C alone (14 amino acids). STP exon A is uncommon.[2,3,7] Following the STP domain and common to all isoforms is a juxtamembranous sequence of 12 aa of unknown function encoded by a single exon. This

The Complement FactsBook. http://dx.doi.org/10.1016/B978-0-12-810420-0.00026-2
271

FIGURE 26.1   Diagram of protein domains for membrane cofactor protein.

is followed by the hydrophobic transmembrane domain and a charged intracytoplasmic anchor. The carboxyl-terminal tail is also alternatively spliced. Thus, MCP contains one of two nonhomologous cytoplasmic tails of 16 or 23 amino acids, termed CYT-1 or CYT-2.

FUNCTION

MCP serves as a cofactor for the factor-I-mediated cleavage of C3b and C4b that deposits on self-tissue. Regulation occurs intrinsically in that MCP protects the cell on which it is anchored, but not neighbouring cells, and does not engage in either fluid-phase C3b or C3b bound to immune complexes in the fluid phase.[8]

Jagged1, a NOTCH family protein member, was identified as a physiological ligand for MCP.[9,10] MCP tethers Jagged1 on the surface of resting T cells, an interaction that is critical for human Th1 immunity.

MCP has been called a 'pathogen magnet'[11] (reviewed in Refs. 12, 13). Pathogens 'hijack' MCP as a receptor, coopt its complement regulatory activity and utilise its signalling capabilities and/or internalisation mechanisms. Nine human-specific pathogens target MCP including four viruses (measles, adenovirus groups B and D, herpes virus 6) and five bacterial species (*Neisseria gonorrhoeae*, *Neisseria meningitidis*, *Streptococcus pyogenes*, *Escherichia coli* and *Fusobacterium nucleatum*). Additionally, MCP is a receptor for bovine viral diarrhoea virus.

Reproduction. MCP plays a role in reproduction and is expressed as a hypoglycosylated isoform (C isoform) only on the inner acrosomal membrane (IAM) that is exposed following binding of spermatozoa to the zona pellucida (the acrosome reaction).[14] Interestingly, New World monkeys express MCP that lacks CCP1 on most cells, yet the full-length form is retained on spermatozoa. In mice, rats and guinea pigs, MCP is expressed on the IAM of spermatozoa. Mice also express MCP in the retina.[15] A reduction in or lack of MCP on spermatozoa has been associated with human idiopathic male infertility. Furthermore, MCP is a target for antisperm antibodies (reviewed in Ref. 14).

Epithelial cells. On activation, MCP interacts with the cytoskeleton of epithelial cells to induce changes, including promoting autophagy, suggesting a possible role in normal epithelial cell homoeostasis (reviewed in Ref. 13). Upregulation occurs in many epithelial cancers, including those of the ovary, colon, prostate and bladder, as well as in leukaemias.[6,16,17]

CD4+ T cells. Over the past 15 years, it has become increasingly clear that MCP plays important immunomodulatory roles. These studies have been extensively reviewed, especially relative to human T cell biology.[13,18,19] Cross-linking of CD3 and MCP on human CD4+ T cells regulates both the induction of interferon-γ-secreting effector T helper-type 1 (Th1) responses and their subsequent contraction into interleukin 10 (IL-10)-producing regulatory T cells (reviewed in Ref. 13). Costimulation of CD4+ T cells also activates intracellular stores of C3 that can be cleaved in a convertase-independent mechanism (i.e., cathepsin L in CD4+ T cells).[20] The generated C3b can transfer to the cell surface to engage MCP where autocrine stimulation regulates nutrient influx and metabolic reprogramming pathways.[21] As a measles virus (MV) receptor, activation of MCP with MV hemagglutinin on monocytes downregulates IL-12 production, a finding of potential significance for the immune suppressive sequelae of MV infection.[22]

Intracytoplasmic tail signalling. The two alternatively spliced cytoplasmic tails for MCP each bear signalling and trafficking consensus sequences.[5,23] The CYT-1 isoform precursor processing time is about four-fold faster with a t ½ of 10–13 min (in Chinese hamster ovary transfectants), while CYT-2 isoform precursor t ½ is ~35–40 min.[5,23] While CYT-1 has consensus sequences for phosphorylation by both protein kinase C and casein kinase-2, CYT-2 bears motifs for the Src-family kinases, casein kinase-2 and for nuclear targeting.[1]

Interaction with natural or pathogenic ligands induces varied types of signalling and associations with intracellular binding partners in epithelial cells, macrophages and T cells, as well as lymphocytes of MCP transgenic mice, among others (reviewed in Refs. 1, 13, 24). Tails may drive different functions. For example, engagement of CYT-1 isoforms by pathogenic ligands directly induces autophagosome formation.[25] CYT-1 is also critical for adherence of *N. gonorrhoeae* to host epithelial cells.[26] The literature on MCP signalling and T cell biology is particularly substantial and is rapidly expanding.[13,27,28] CYT-driven signalling may serve as a molecular rheostat in activated T cells in which CYT-1 processing is required to turn 'on' T cell activation while CYT-2 processing is needed to turn it off (reviewed in Ref. 27). Further, a model is proposed in which MCP engagement drives metabolic reprogramming during Th1 responses.[21] TCR activation, which induces generation of MCP ligand C3b,[20] increases expression of isoforms bearing CYT-1 that upregulate glucose and amino acid influx into CD4+ T cells. In the Th1 cell contraction phase, MCP isoform expression in CD4+ T cells reverts to a CYT-2 predominant pattern with concomitant downregulation of this system.

TISSUE DISTRIBUTION

Most cell types express the same genetically regulated ratio of each of the four isoforms of MCP. Tissue-specific isoform expression, though, occurs in the kidney and salivary glands (BC2 isoforms predominate), as well as brain (C2),[29] foetal heart[30] and spermatozoa (hypoglycosylated C2 isoforms predominate[31,32]).

REGULATION OF EXPRESSION

Expression levels on peripheral blood mononuclear cells and granulocytes are ~10,000/cell, haematopoietic-derived cell lines 20,000–50,000, and epithelial-derived cell lines from 100,000 to 250,000 (reviewed in Ref. 1). MCP levels are increased in certain haematologic malignancies, on most solid tumour cell lines and following SV40 transformation (reviewed in Refs. 1, 17).

MCP expression is upregulated in glomerular capillary walls and in mesangial regions of diseased kidney tissues and astrocytes following cytomegalovirus infection.[33] MCP is downregulated following ligand binding and antibody cross-linking (reviewed by Ref. 13). In epithelial cells, MCP engagement also induces shedding of the extracellular domain as a result of cleavage by matrix metalloproteinases and the ADAMs (a disintegrin and metalloproteinase) family (reviewed in Ref. 13). In T cells, modulation of MCP expression has also been described, but often in the setting of activated cells and suggesting the importance of feedback loops for integration of autocrine and paracrine signals for such regulation (reviewed in Ref. 13).

Further, surface expression can be reduced via anti-MCP antibodies or following MV infection. In contrast, a decrease in cell surface expression on epithelial cells following interaction with piliated *N. gonorrhoeae* occurs via shedding. Finally, MCP can be released by tumour cells as a soluble form (reviewed in Ref. 1)

HUMAN PROTEIN SEQUENCE

Isoform ABC1 392 aa or ABC2 399 aa.

```
MEPPGRRECP FPSWRFPGLL LAAMVLLLYS FSDACEEPPT FEAMELIGKP   50

KPYYEIGERV DYKCKKGYFY IPPLATHTIC DRNHTWLPVS DDACYRETCP  100

YIRDPLNGQA VPANGTYEFG YQMHFICNEG YYLIGEEILY CELKGSVAIW  150

SGKPPICEKV LCTPPPKIKN GKHTFSEVEV FEYLDAVTYS CDPAPGPDPF  200

SLIGESTIYC GDNSVWSRAA PECKVVKCRF PVVENGKQIS GFGKKFYYKA  250

TVMFECDKGF YLDGSDTIVC DSNSTWDPPV PKCLKVLPPS STKPPALSHS  300

VSTSSTTKSP ASSASGPRPT YKPPVSNYPG YPKPEEGILD SLDVWVIAVI  350

VIAIVVGVAV ICVVPYRYLQ RRKKKGTYLT DETHREVKFT SL

            or CYT-2      KADG GAEYATYQTK STTPAEQRG
```

The leader sequence is underlined, *N*-linked glycosylation sites are bolded (**N**), rare alternatively spliced STP-A is underlined with dashed line, commonly occurring STP segments are underlined, and cytoplasmic tails (also alternatively spliced) at C-terminus are underlined with a wavy line.

PROTEIN MODULES

1–34	Leader peptide	exon 1
35–95	CCP1	exon 2
96–158	CCP2	exons 3/4
159–224	CCP3	exon 5
225–285	CCP4	exon 6
286–329	STP	exons 7–9
	A domain: VLPPSSTKPPALSHS	exon 7
	B domain: VSTSSTTKSPASSAS	exon 8
	C domain: GPRPTYKPPVSNYP	exon 9
330–342	Undefined segment	exon 10
343–366	Transmembrane domain	exons 11/12
367–376	Intracytoplasmic anchor	exon 12
377–392	Cytoplasmic tail one	
	TYLTDETHREVKFTSL	exon 13
377–399	Cytoplasmic tail two	
	KADGGAEYATYQTKSTTPAEQRG	exon 14

CHROMOSOMAL LOCATION

Human: 1q3.2.[4–35]

A nonexpressed *MCP-like* genomic element includes sequences 93% homologous at the nucleotide level with the MCP 5′ terminus (i.e., exons 1–3) and is located within 60 kb of MCP.[6]

Mouse: chromosome 1, closely linked to the complement receptor 2 (*Cr2*) gene.[36,37]

Rat: chromosome 13.[37–39]

HUMAN cDNA SEQUENCE (ABC1)

```
  1 attgttgcgt cccatatctg gacccagaag ggacttccct gctcggctgg ctctcggttt
 61 ctctgctttc ctccggagaa ataacagcgt cttccgcgcc gcgcatggag cctcccggcc
121 gccgcgagtg tccctttcct tcctggcgct ttcctgggtt gcttctggcg gccatggtgt
181 tgctgctgta ctccttctcc gatgcctgtg aggagccacc aacatttgaa gctatggagc
241 tcattggtaa accaaaaccc tactatgaga ttggtgaacg agtagattat aagtgtaaaa
301 aaggatactt ctatatacct cctcttgcca cccatactat ttgtgatcgg aatcatacat
361 ggctacctgt ctcagatgac gcctgttata gagaaacatg tccatatata cgggatcctt
421 taaatggcca agcagtccct gcaaatggga cttacgagtt tggttatcag atgcactta
481 tttgtaatga gggttattac ttaattggtg aagaaattct atattgtgaa cttaaaggat
541 cagtagcaat ttggagcggt aagcccccaa tatgtgaaaa ggttttgtgt acaccacctc
601 caaaaataaa aaatggaaaa cacaccttta gtgaagtaga agtatttgag tatcttgatg
661 cagtaactta tagttgtgat cctgcacctg gaccagatcc attttcactt attggagaga
721 gcacgattta ttgtggtgac aattcagtgt ggagtcgtgc tgctccagag tgtaaagtgg
781 tcaaatgtcg atttccagta gtcgaaaatg gaaaacagat atcaggattt ggaaaaaaat
841 tttactacaa agcaacagtt atgtttgaat gcgataaggg tttttacctc gatggcagcg
901 acacaattgt ctgtgacagt aacagtactt gggatcccc agttccaaag tgtcttaaag
961 tgctgcctcc atctagtaca aaacctccag ctttgagtca ttcagtgtcg acttcttcca
```

HUMAN cDNA SEQUENCE (ABC1)—*Continued*

```
1021 ctacaaaatc tccagcgtcc agtgcctcag gtcctaggcc tacttacaag cctccagtct
1081 caaattatcc aggatatcct aaacctgagg aaggaatact tgacagtttg gatgtttggg
1141 tcattgctgt gattgttatt gccatagttg ttggagttgc agtaatttgt gttgtcccgt
1201 acagatatct tcaaaggagg aagaagaaag gcacatacct aactgatgag acccacagag
1261 aagtaaaatt tacttctctc tgagaaggag agatgagaga aaggtttgct tttatcatta
1321 aaaggaaagc agatggtgga gctgaatatg ccacttacca gactaaatca accactccag
1381 cagagcagag aggctgaata gattccacaa cctggttttgc cagttcatct tttgactcta
1441 ttaaaatctt caatagttgt tattctgtag tttcactctc atgagtgcaa ctgtggctta
1501 gctaatattg caatgtggct tg
```

The first five nucleotides in each exon are underlined to indicate the intron–exon boundaries. The methionine initiation codon (ATG) and the termination codon (TGA) are double underlined.

GENOMIC STRUCTURE

The *MCP* gene spans a minimum of 43 kb and is encoded by 14 exons. There are two sites for alternative splicing: exons 7, 8 and 9 encode the STP domain commonly expressed without exon 7 and commonly including exons 8 and 9 (isoforms B+C) or exon 9 (isoform C) alone; exons 13 and 14 encode the alternatively spliced cytoplasmic tails, CYT-1 and CYT-2. Since exon 13 contains an in-frame stop codon, its expression as CYT-1 converts exon 14 into the 3′ untranslated region of *MCP* (Fig. 26.2).

ACCESSION NUMBERS

The *MCP* gene is conserved in chimpanzee, Rhesus monkey, dog, cow, mouse and rat. At least 73 organisms have orthologues with the human gene for MCP (see: http://www.ncbi.nlm.nih.gov/gene?Db=gene&Cmd=DetailsSearch&Term=4179).

Human		
	MCP-BC2	Y00651
	MCP-BC1	X59405
	MCP-C1	X59406
	MCP-C2	X59407
	MCP-ABC2	X59409
	MCP-ABC1	X59410
Mouse		AB001566

DEFICIENCY/DISEASES

More than 60 disease-associated mutations in *MCP* have been identified. The majority are linked to a rare thrombotic microangiopathic-based disease, atypical

FIGURE 26.2 Diagram of genomic structure of membrane cofactor protein.

haemolytic uraemic syndrome (aHUS), but new putative links to systemic lupus erythematosus, glomerulonephritis and pregnancy-related disorders, among others, have also been identified (reviewed in Refs. 12, 40, 41). Further, activated CD4[+] T cells from MCP-deficient patients fail to mount Th1 and other responses (reviewed in Refs. 13, 28). Most mutations occur in the CCP domains, although others are in the STP, TM or CYT segments. Most are also heterozygous leading to a state of haploinsufficiency, although complete deficiencies have also been described.[42,43]

Other diseases implicated in MCP dysregulation include recurrent infections in MCP-deficient patients, autoimmune conditions such as rheumatoid arthritis or multiple sclerosis and asthma (reviewed in Refs. 13, 27).

POLYMORPHIC VARIANTS

A *Hind*III RFLP correlates with the phenotypic polymorphism of MCP.[44] Although this size polymorphism results from variable splicing of exon 8,[2] the *Hind*III RFLP occurs in an intron between exon 1 (5′ UT/signal peptide) and exon 2 (CCP1). The higher molecular weight species of MCP correlates to the presence of the *Hind*III site, while the lower molecular weight species lack the site. *Pvu*II and *Bgl*II RFLPs also have been described.[44,45]

Regulation of MCP expression may also be, at least in part, due to allelic variation. A variant allele of MCP in the 3′ UT (the CC genotype at the rs7144 locus) accounts for a higher level of expression on lymphocytes and other cells.[46] Specific polymorphisms also appear to influence the strength of MCP signalling, especially as relates to antibody responses to measles vaccination.[46–48]

The *MCP* gene contains a specific single nucleotide polymorphism haplotype block termed the *MCPggaac* haplotype in the *MCP* promoter region that may be associated in vitro with reduced transcriptional activity (Refs. 49–51 and reviewed in Ref. 41). It is linked to the development of aHUS, but only in the setting of a causative variant in another alternative pathway protein.

MUTANT ANIMALS

An MCP transgenic pig was created from a minigene using genomic DNA from the 5′ terminus of a >40 kb gene that was restricted to include the promoter sequence (343 bp upstream of the major transcriptional start site) through intron 3, and spliced in-frame to MCP exons 3–14 encoding isoform BC1.[52] High-level expression closely recapitulated humans except that pig E expressed MCP. This model has been utilised for xenotransplantation of pig kidney to baboon and porcine islets into cynomolgus monkeys.[52,53] The pigs are available from Revivicor Inc. (Blacksburg, Virginia).

A commercially available transgenic mouse model, MYII (JAX Labs), was developed using a 402-kb yeast artificial chromosome (YAC) that included the complete genomic sequence of *MCP*. MYII mimics human expression[54,55] except that, in contrast to humans, but similar to most other primates, the *MCP* is expressed on murine erythrocytes.[55] MYII was utilised in studies for xeno-transplantation[54] for infection by adenovirus 35 (Adv35 vectors) (potentially useful for testing potency and safety of therapeutic recombinant Ad35 vectors)[56] and for measles.[57] Reportedly, the MYII strain had aggressive behaviour and parental litter killing; hCD46Ge became obese as they aged; and the UAB line showed some YAC clone instability.[55]

The hCD46Ge CD46 murine model was utilised to study *N. meningitidis* infection[58] and to study measles by crossing to a strain that inactivated the alpha/beta interferon receptor.[59]

Transgenic mice also have been made bearing specific isoforms, for example, C2,[60] BC1 [61] and BC2.[62]

A mouse model for age-related macular degeneration (AMD) has also been developed by the creation of a *Cd46⁻/⁻* mouse.[63] Wild-type mice express *MCP* only on the IAM of spermatozoa and in the eye (neurosensory retina, retinal pigment epithelium and choroid).[14,15] Upon aging, the *Cd46⁻/⁻* mouse develops changes in the eye consistent with the dry type of AMD. It also develops a more severe pathology following laser-induced retinal injury.[63] Surprisingly, fertility of these mice is intact.

REFERENCES

1. Liszewski MK, Kemper C, Price JD, Atkinson JP. Emerging roles and new functions of CD46. *Springer Semin Immunopathol* 2005;**27**:345–58.
2. Post TW, Liszewski MK, Adams EM, Tedja I, Miller EA, Atkinson JP. Membrane cofactor protein of the complement system: alternative splicing of serine/threonine/proline-rich exons and cytoplasmic tails produces multiple isoforms that correlate with protein phenotype. *J Exp Med* 1991;**174**:93–102.
3. Russell SM, Sparrow RL, McKenzie IFC, Purcell DFJ. Tissue-specific and allelic expression of the complement regulator CD46 is controlled by alternative splicing. *Eur J Immunol* 1992;**22**:1513–8.
4. Lublin DM, Liszewski MK, Post TW, Arce MA, Le Beau MM, Rebentisch MB. Molecular cloning and chromosomal localization of human membrane cofactor protein (MCP). Evidence for inclusion in the multigene family of complement-regulatory proteins. *J Exp Med* 1988;**168**:181–94.
5. Ballard LL, Bora NS, Yu GH, Atkinson JP. Biochemical characterization of membrane cofactor protein of the complement system. *J Immunol* 1988;**141**:3923–9.
6. Liszewski MK, Post TW, Atkinson JP. Membrane cofactor protein (MCP or CD46): newest member of the regulators of complement activation gene cluster. *Annu Rev Immunol* 1991;**9**:431–55.
7. Seya T, Hirano A, Matsumoto M, Nomura M, Ueda S. Human membrane cofactor protein (MCP, CD46): multiple isoforms and functions. *Int J Biochem Cell Biol* 1999;**31**:1255–60.
8. Oglesby TJ, Allen CJ, Liszewski MK, White DJG, Atkinson JP. Membrane cofactor protein (MCP;CD46) protects cells from complement-mediated attack by an intrinsic mechanism. *J Exp Med* 1992;**175**:1547–51.
9. Heeger PS, Kemper C. Novel roles of complement in T effector cell regulation. *Immunobiology* 2012;**217**:216–24.

10. Le Friec G, Sheppard D, Whiteman P, et al. The CD46-Jagged1 interaction is critical for human TH1 immunity. *Nat Immunol* 2012;**13**:1213–21.

11. Cattaneo R. Four viruses, two bacteria, and one receptor: membrane cofactor protein (CD46) as pathogens' magnet. *J Virol* 2004;**78**:4385–8.

12. Liszewski MK, Atkinson JP. Complement regulator CD46: genetic variants and disease associations. *Hum Genomics* 2015;**9**:7.

13. Yamamoto H, Fara AF, Dasgupta P, Kemper C. CD46: the 'multitasker' of complement proteins. *Int J Biochem Cell Biol* 2013;**45**:2808–20.

14. Riley-Vargas RC, Lanzendorf S, Atkinson JP. Targeted and restricted complement activation on acrosome-reacted spermatozoa. *J Clin Invest* 2005;**115**:1241–9.

15. Lyzogubov V, Wu X, Jha P, et al. Complement regulatory protein CD46 protects against choroidal neovascularization in mice. *Am J Pathol* 2014;**184**:2537–48.

16. Hara T, Kojima A, Fukuda H, et al. Levels of complement regulatory proteins, CD35 (CR1), CD46 (MCP) and CD55 (DAF) in human haematological malignancies. *Br J Haematol* 1992;**82**:368–73.

17. Fishelson Z, Donin N, Zell S, Schultz S, Kirschfink M. Obstacles to cancer immunotherapy: expression of membrane complement regulatory proteins (mCRPs) in tumors. *Mol Immunol* 2003;**40**:109–23.

18. Cardone J, Le Friec G, Kemper C. CD46 in innate and adaptive immunity: an update. *Clin Exp Immunol* 2011;**164**:301–11.

19. Kemper C, Atkinson JP. T-cell regulation: with complements from innate immunity. *Nat Rev Immunol* 2007;**7**:9–18.

20. Liszewski MK, Kolev M, Le Friec G, et al. Intracellular complement activation sustains T cell homeostasis and mediates effector differentiation. *Immunity* 2013;**39**:1143–57.

21. Kolev M, Dimeloe S, Le Friec G, et al. Complement regulates nutrient influx and metabolic reprogramming during Th1 cell responses. *Immunity* 2015;**42**:1033–47.

22. Karp CL, Wysocka M, Wahl LM, et al. Mechanism of suppression of cell-mediated immunity by measles virus. [erratum appears in Science 1997 Feb 21;275(5303):1053] *Science* 1996;**273**:228–31.

23. Liszewski MK, Tedja I, Atkinson JP. Membrane cofactor protein (CD46) of complement: processing differences related to alternatively spliced cytoplasmic domains. *J Biol Chem* 1994;**269**:10776–9.

24. Kolev M, Friec GL, Kemper C. Complement – tapping into new sites and effector systems. *Nat Rev Immunol* 2014;**14**:811–20.

25. Joubert PE, Meiffren G, Gregoire IP, et al. Autophagy induction by the pathogen receptor CD46. *Cell Host Microbe* 2009;**6**:354–66.

26. Kallstrom H, Blackmer Gill D, Albiger B, Liszewski MK, Atkinson JP, Jonsson AB. Attachment of *Neisseria gonorrhoeae* to the cellular pilus receptor CD46: identification of domains important for bacterial adherence. *Cell Microbiol* 2001;**3**:133–43.

27. Ni Choileain S, Astier AL. CD46 processing: a means of expression. *Immunobiology* 2012;**217**:169–75.

28. Hess C, Kemper C. Complement-mediated regulation of metabolism and basic cellular processes. *Immunity* 2016;**45**:240–54.

29. Johnstone RW, Russell SM, Loveland BE, McKenzie IF. Polymorphic expression of CD46 protein isoforms due to tissue-specific RNA splicing. *Mol Immunol* 1993;**30**:1231–41.

30. Gorelick A, Oglesby TJ, Rashbaum W, Atkinson JP, Buyon JP. Ontogeny of membrane cofactor protein: phenotypic divergence in the fetal heart. *Lupus* 1995;**4**:293–6.

31. Cervoni F, Oglesby TJ, Adams EM, et al. Identification and characterization of membrane cofactor protein (MCP) on human spermatozoa. *J Immunol* 1992;**148**:1431–7.

32. Riley RC, Kemper C, Leung M, Atkinson JP. Characterization of human membrane cofactor protein (MCP; CD46) on spermatozoa. *Mol Reprod Dev* 2002;**62**:534–46.

33. Spiller OB, Morgan BP, Tufaro F, Devine DV. Altered expression of host-encoded complement regulators on human cytomegalovirus-infected cells. *Eur J Immunol* 1996;**26**:1532–8.

34. Bora NS, Lublin DM, Kumar BV, Hockett RD, Holers VM, Atkinson JP. Structural gene for human membrane cofactor protein (MCP) of complement maps to within 100 kb of the 3' end of the C3b/C4b receptor gene. *J Exp Med* 1989;**169**:597–602.

35. Hourcade D, Garcia AD, Post TW, et al. Analysis of the regulators of complement activation (RCA) gene cluster with yeast artificial chromosomes (YACs). *Genomics* 1992;**12**:289–300.

36. Tsujimura A, Shida K, Kitamura M, et al. Molecular cloning of a murine homologue of membrane cofactor protein (CD46): preferential expression in testicular germ cells. *Biochem J* 1998;**330**(Pt 1):163–8.

37. Miwa T, Nonaka M, Okada N, Wakana S, Shiroishi T, Okada H. Molecular cloning of rat and mouse membrane cofactor protein (MCP, CD46): preferential expression in testis and close linkage between the mouse Mcp and Cr2 genes on distal chromosome 1. *Immunogenetics* 1998;**48**:363–71.

38. Mead R, Hinchliffe SJ, Morgan BP. Molecular cloning, expression and characterization of the rat analogue of human membrane cofactor protein (MCP/CD46). *Immunology* 1999;**98**:137–43.

39. Mizuno M, Harris CL, Morgan BP. Immunohistochemical analysis of membrane complement regulatory proteins in rat testis: unique roles for DAF and MCP in spermatozoal function? *Mol Immunol* 2004;**41**:280.

40. Rodriguez de Cordoba S, Hidalgo MS, Pinto S, Tortajada A. Genetics of atypical hemolytic uremic syndrome (aHUS). *Semin Thromb Hemost* 2014;**40**:422–30.

41. Kavanagh D, Goodship TH, Richards A. Atypical hemolytic uremic syndrome. *Semin Nephrol* 2013;**33**:508–30.

42. Fremeaux-Bacchi V, Moulton EA, Kavanagh D, et al. Genetic and functional analyses of membrane cofactor protein (CD46) mutations in atypical hemolytic uremic syndrome. *J Am Soc Nephrol* 2006;**17**:2017–25.

43. Couzi L, Contin-Bordes C, Marliot F, et al. Inherited deficiency of membrane cofactor protein expression and varying manifestations of recurrent atypical hemolytic uremic syndrome in a sibling pair. *Am J Kidney Dis* 2008;**52**:e5–9.

44. Bora NS, Post TW, Atkinson JP. Membrane cofactor protein of the complement system: a Hind III restriction fragment length polymorphism that correlates with the expression polymorphism. *J Immunol* 1991;**146**:2821–5.

45. Wilton AN, Johnstone RW, McKenzie IF, Purcell DF. Strong associations between RFLP and protein polymorphisms for CD46. *Immunogenetics* 1992;**36**:79–85.

46. Clifford HD, Hayden CM, Khoo SK, Zhang G, Le Souef PN, Richmond P. CD46 measles virus receptor polymorphisms influence receptor protein expression and primary measles vaccine responses in naive Australian children. *Clin Vaccine Immunol* 2012;**19**:704–10.

47. Ovsyannikova IG, Haralambieva IH, Vierkant RA, O'Byrne MM, Jacobson RM, Poland GA. The association of CD46, SLAM and CD209 cellular receptor gene SNPs with variations in measles vaccine-induced immune responses: a replication study and examination of novel polymorphisms. *Hum Hered* 2011;**72**:206–23.

48. Dhiman N, Poland GA, Cunningham JM, et al. Variations in measles vaccine-specific humoral immunity by polymorphisms in SLAM and CD46 measles virus receptors. *J Allergy Clin Immunol* 2007;**120**:666–72.

49. Fremeaux-Bacchi V, Fakhouri F, Garnier A, et al. Genetics and outcome of atypical hemolytic uremic syndrome: a nationwide French series comparing children and adults. *Clini Journal Am Soc Nephrol* 2013;**8**:554–62.

50. Esparza-Gordillo J, Goicoechea de Jorge E, Buil A, et al. Predisposition to atypical hemolytic uremic syndrome involves the concurrence of different susceptibility alleles in the regulators of complement activation gene cluster in 1q32. CORRIGENDUM *Hum Mol Genet* 2005;**14**:1107.

51. Fremeaux-Bacchi V, Kemp EJ, Goodship JA, et al. The development of atypical haemolytic-uraemic syndrome is influenced by susceptibility factors in factor H and membrane cofactor protein: evidence from two independent cohorts. *J Med Genet* 2005;**42**:852–6.

52. Loveland BE, Milland J, Kyriakou P, et al. Characterization of a CD46 transgenic pig and protection of transgenic kidneys against hyperacute rejection in non-immunosuppressed baboons. *Xenotransplantation* 2004;**11**:171–83.

53. van der Windt DJ, Bottino R, Casu A, et al. Long-term controlled normoglycemia in diabetic non-human primates after transplantation with hCD46 transgenic porcine islets. *Am J Transpl* 2009;**9**:2716–26.

54. Yannoutsos N, Ijzermans JN, Harkes C, et al. A membrane cofactor protein transgenic mouse model for the study of discordant xenograft rejection. *Genes Cell* 1996;**1**:409–19.

55. Kemper C, Leung M, Stephensen CB, et al. Membrane cofactor protein (MCP; CD46) expression in transgenic mice. *Clin Exp Immunol* 2001;**124**:180–9.

56. Verhaagh S, de Jong E, Goudsmit J, et al. Human CD46-transgenic mice in studies involving replication-incompetent adenoviral type 35 vectors. *J Gen Virol* 2006;**87**:255–65.

57. Oldstone MB, Lewicki H, Thomas D, et al. Measles virus infection in a transgenic model: virus-induced immunosuppression and central nervous system disease. *Cell* 1999;**98**:629–40.

58. Johansson L, Rytkonen A, Wan H, et al. Human-like immune responses in CD46 transgenic mice. *J Immunol* 2005;**175**:433–40.

59. Mrkic B, Pavlovic J, Rulicke T, et al. Measles virus spread and pathogenesis in genetically modified mice. *J Virol* 1998;**72**:7420–7.

60. Horvat B, Rivailler P, Varior-Krishnan G, Cardoso A, Gerlier D, Rabourdin-Combe C. Transgenic mice expressing human measles virus (MV) receptor CD46 provide cells exhibiting different permissivities to MV infections. *J Virol* 1996;**70**:6673–81.

61. Rall GF, Manchester M, Daniels LR, Callahan EM, Belman AR, Oldstone MB. A transgenic mouse model for measles virus infection of the brain. *Proc Natl Acad Sci USA* 1997;**94**:4659–63.

62. Miyagawa S, Mikata S, Tanaka H, et al. The regulation of membrane cofactor protein (CD46) expression by the 3′ untranslated region in transgenic mice. *Biochem Biophys Res Commun* 1997;**233**:829–33.

63. Lyzogubov VV, Bora PS, Wu X, et al. The complement regulatory protein (CD46) deficient mouse – a tractable model of dry type age-related macular degeneration. *Am J Pathol* 2016;**186**:2088–104.

Chapter 27

Properdin

Viviana P. Ferreira
University of Toledo, Toledo, OH, United States

PHYSIOCHEMICAL PROPERTIES

Properdin is synthesised as a single-chain molecule of 469 amino acids, which includes a 27-amino acid leader sequence[1] (mature protein is 442 amino acids). Each properdin monomer is composed of six complete thrombospondin type 1 repeat (TSR) domains labelled TSR1–6[2,3] and a truncated N-terminal domain (TSR0) containing key conserved residues.[3]

pI[4]		>9.5
		8.32 (theoretical)
M_r (K)	Predicted[2]	53.3 (glycosylated)
		48.5 (unglycosylated)
	Observed	50–55
Glycosylation sites[5,6]		
(C-Mannosylated)		14 tryptophans (83, 86, 139, 142, 145, 196, 199, 260, 263, 321, 324, 382, 385, 388)
(O-Fucosylated)		4 (92, 151, 208, 272)
(N-Glycosylated)		1 (428; occupied)

STRUCTURE

Under physiological conditions, properdin forms mainly cyclic dimers (P_2), trimers (P_3) and tetramers (P_4) in a 26:54:20 (P_2:P_3:P_4) ratio, via head-to-tail associations of its monomers.[7,8] Higher order oligomers are found during prolonged storage and freeze/thaw cycles of pure unfractionated properdin[8,9] and have the ability to bind nonspecifically to surfaces[10] (likely due to the high positively charged nature of properdin that increases nonspecific ionic interactions), but have not been proven to be physiological.

Each monomer of properdin is 26 nm (length) × 2.5 nm (diameter).[7,11] Each TSR domain is ~4 nm × 1.7 nm × 1.7 nm.[11] Atomic structures for properdin have not been resolved, although thrombospondin TSR domains have served as models.[12,13] Electron microscopy[14] and X-ray scattering and modelling data[3] for

The Complement FactsBook. http://dx.doi.org/10.1016/B978-0-12-810420-0.00027-4

properdin have proposed distinct models for properdin structure at the vertices where the monomers interact, whereby TSR0-1 and 5–6 participate in forming the vertices of properdin oligomers,[3,14] and potentially TSR2 and TSR4 (Fig. 27.1). [14]

FUNCTION

Properdin is a positive regulator of complement activity by binding to C3b and factor B or Bb[14–16] and stabilising the alternative pathway C3 and C5 convertases. The half-life of alternative pathway convertases is ~90 s,[17] thus properdin function, by increasing convertase stability 5 to10-fold,[4] is essential for effective complement amplification. Properdin binds to membrane-bound C3b with higher affinity than to its fluid-phase counterparts.[15] In addition, within the past decade, properdin was shown to act as an initiator of alternative pathway activity (reviewed in Refs. 18, 19), as originally proposed by Pillemer in 1954, by acting as a pattern-recognition molecule and selectively binding to certain surfaces, recruiting C3b or C3(H$_2$O) and factor B, generating new functional convertases (Fig. 27.2). As indicated in later sections, properdin is essential for protection against gonococcal infections (by facilitating complement-dependent killing of the pathogen) and for mediating complement-dependent inflammation and disease pathogenesis in various models.

Properdin oligomers can be detected when unfractionated properdin preparations, which contain P$_2$, P$_3$, P$_4$ and nonphysiological aggregated P$_n$, are separated by size exclusion or ion exchange chromatography.[10] The nonphysiological

FIGURE 27.1   Diagram of protein domain structure for properdin.

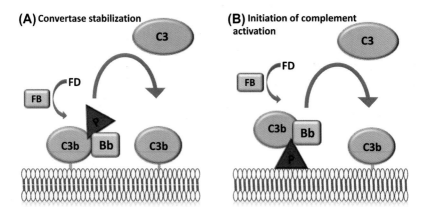

FIGURE 27.2   Properdin as a convertase stabiliser (A) and as a complement initiator (B).

P_n are referred to as 'aggregated' or 'activated' properdin, given their ability to activate and consume complement in solution.[8,9] Thus, studies using purified properdin to understand its biological roles need to eliminate the P_n fraction. Unfractionated/pure properdin from plasma and each isolated oligomer are visualised as ~53 kDa monomers in SDS-PAGE under both reduced and non-reduced conditions. Properdin oligomers/aggregates are not observed in SDS-PAGE due to the noncovalent nature of the interactions. Properdin complexed with other molecules such as $C3(H_2O)$, C3 and immunoglobulins have been observed migrating at higher molecular weights in SDS-PAGE and Immuno Western blot.[20,21]

Physiological properdin binds to sulphatide [Gal(3-SO_4)βl-lCer], while dextran sulphate (500,000 mw) and fucoidan inhibit this interaction.[22] The binding of purified physiological properdin to certain surfaces can be inhibited in the presence of serum.[23,24] Whether properdin produced locally by activated pro-inflammatory cells[25] (leading to higher local concentration) can bypass potential regulatory mechanisms in serum remains undetermined.

Studies using mutant forms of properdin lacking single TSR domains[6] revealed that (1) lack of domains 4 (partially) or 5 or 6 (completely) abrogates binding to C3b and sulphatides; (2) TSR3 deletion does not affect binding to C3b or sulphatides nor oligomer formation; (3) TSR4 and 5 deletion mutants do not form trimers and tetramers, but retain the ability to dimerise; and (4) deletion of TSR6 inhibits oligomerisation.

TISSUE DISTRIBUTION[26]

Serum protein concentration: 4–25 µg/mL in plasma.

Primary sites of synthesis: Unlike most complement proteins, properdin is produced primarily by leucocytes rather than hepatocytes. Leucocyte subsets that synthesise properdin constitutively include human T cells, monocytes, dendritic cells and mast cells. Mature neutrophils contain large stores of properdin in secondary granules, which are rapidly released upon neutrophil stimulation.

Secondary sites of synthesis: adipocytes and stimulated endothelial cells. Various cell lines also express properdin including H-9, HuT78, Jurkat, T-ALL, HL-60, U-937, MonoMac6 and 3T3-L1 adipocytes.

REGULATION OF EXPRESSION

Properdin release from granulocyte granules is upregulated by various agonists including C5a, fMLP, PMA, TNF-α and IL-8.[25,27] Properdin synthesis can be stimulated in certain monocytic cell lines (HL-60, U937, Mono Mac6) by PMA, bacterial LPS, IL-1β, IFN-γ and TNF-α (reviewed in Ref. 26), and by laminar shear stress on endothelial cells.[28] Properdin is also released constitutively from other cells (see previous section). Specific signalling mechanisms regulating expression remain to be determined.

HUMAN PROTEIN SEQUENCE[29]

<u>**MITEGAQAPR LLLPPLLLLL TLPATG**</u>S DPV LCFTQYEESS GKCKGLLGGG   50

VSVEDCCLNT AFAYQKRSGG LCQPCRSPRW SLWSTWAPCS VTCSEGSQLR 100

YRRCVGWNGQ CSGKVAPGTL EWQLQACEDQ QCCPEMGGWS GWGPWEPCSV 150

TCSKGTRTRR RACNHPAPKC GGHCPGQAQE SEACDTQQVC PTHGAWATWG 200

PWTPCSASCH GGPHEPKETR SRKCSAPEPS QKPPGKPCPG LAYEQRRCTG 250

LPPCPVAGGW GPWGPVSPCP VTCGLGQTME QRTCNHPVPQ HGGPFCAGDA 300

TRTHICNTAV PCPVDGEWDS WGEWSPCIRR NMKSISCQEI PGQQSRGRTC 350

RGRKFDGHRC AGQQQDIRHC YSIQHCPLKG SWSEWSTWGL CMPPCGPNPT 400

RARQRLCTPL LPKYPPTVSM VEGQGEK**N**VT FWGRPLPRCE ELQGQKLVVE 450

EKRPCLHVPA CKDPEEEEL

The leader sequence is underlined and bolded, the change of exons is in larger letters, the single *N*-linked glycosylation site is bolded (N), and the 14 C-mannosylated tryptophans are underlined.

PROTEIN MODULES[3,6,11,30]

Protein modules were obtained by running the amino acid sequence on the ScanProsite website (http://prosite.expasy.org/scanprosite/).[30]

	AA sequence	exon
	None	exon 1
Signal peptide 1–27	27	exon 2
TSP type-1 0	28–76	exon 3
TSP type-1 1	77–134	exon 4
TSP type-1 2	136–191	exon 5
TSP type-1 3	193–255	exon 6
TSP type-1 4	257–313	exon 7
TSP type-1 5	315–377	exon 8
TSP type-1 6	379–462	exons 9/10

CHROMOSOMAL LOCATION

The human properdin gene is located at the short arm of the X chromosome (Xp11.3-p11.23[31]) or more specifically Xp11.4 based on the National Center for Biotechnology Information (NCBI).[32]

The mouse properdin gene is located in chromosome X, 16.44 cM, cytoband A3.[33]

HUMAN cDNA SEQUENCE[34]

GAGCACCGCA CACTCACTTC ACCCTGGTTC AACACCCCCA CGAGGTTGAC CCCGTCATTA 60

TGTTAGAGAT TATGCATTTT CCACATAGGG AAACTGAGGC TCAGGGGTGT TAAGTGACTC 120

ACCCAAGGTC ACACGGCTAG GAAGTTGCTG CACGCTCCTA TGCTCCATTT CCTCTGGGAG 180

CCTATCAACC CAGATAAAGC GGGACCTCCT CTCTGGTAGA GGTGCAGGGG GCAGTACTCA 240

AC**ATG**ATCAC AGAGGGAGCG CAGGCCCCTC GATTGTTGCT GCCGCCGCTG CTCCTGCTGC 300

TCACCCTGCC AGCCACAGGC TCAGACCCCG TGCTCTGCTT CACCCAGTAT GAAGAATCCT 360

CCGGCAAGTG CAAGGGCCTC CTGGGGGGTG GTGTCAGCGT GGAAGACTGC TGTCTCAACA 420

CTGCCTTTGC CTACCAGAAA CGTAGTGGTG GGCTCTGTCA GCCTTGCAGG TCCCCACGAT 480

GGTCCCTGTG GTCCACATGG GCCCCCTGTT CGGTGACGTG CTCTGAGGGC TCCCAGCTGC 540

GGTACCGGCG CTGTGTGGGC TGGAATGGGC AGTGCTCTGG AAAGGTGGCA CCTGGGACCC 600

TGGAGTGGCA GCTCCAGGCC TGTGAGGACC AGCAGTGCTG TCCTGAGATG GGCGGCTGGT 660

CTGGCTGGGG GCCCTGGGAG CCTTGCTCTG TCACCTGCTC CAAAGGGACC CGGACCCGCA 720

GGCGAGCCTG TAATCACCCT GCTCCCAAGT GTGGGGGCCA CTGCCCAGGA CAGGCACAGG 780

AATCAGAGGC CTGTGACACC CAGCAGGTCT GCCCCACACA CGGGGCCTGG GCCACCTGGG 840

GCCCCTGGAC CCCCTGCTCA GCCTCCTGCC ACGGTGGACC CCACGAACCT AAGGAGACAC 900

GAAGCCGCAA GTGTTCTGCA CCTGAGCCCT CCCAGAAACC TCCTGGGAAG CCCTGCCCGG 960

GGCTAGCCTA CGAGCAGCGG AGGTGCACCG GCCTGCCACC CTGCCCAGTG GCTGGGGGCT 1020

GGGGGCCTTG GGGCCCTGTG AGCCCCTGCC CTGTGACCTG TGGCCTGGGC CAGACCATGG 1080

AACAACGGAC GTGCAATCAC CCTGTGCCCC AGCATGGGGG CCCCTTCTGT GCTGGCGATG 1140

CCACCCGGAC CCACATCTGC AACACAGCTG TGCCCTGCCC TGTGGATGGG GAGTGGGACT 1200

CGTGGGGGGA GTGGAGCCCC TGTATCCGAC GGAACATGAA GTCCATCAGC TGTCAAGAAA 1260

TCCCGGGCCA GCAGTCACGC GGGAGGACCT GCAGGGGCCG CAAGTTTGAC GGACATCGAT 1320

GTGCCGGGCA ACAGCAGGAT ATCCGGCACT GCTACAGCAT CCAGCACTGC CCCTTGAAAG 1380

GATCATGGTC AGAGTGGAGT ACCTGGGGGC TGTGCATGCC CCCCTGTGGA CCTAATCCTA 1440

CCCGTGCCCG CCAGCGCCTC TGCACACCCT TGCTCCCCAA GTACCCGCCC ACCGTTTCCA 1500

TGGTCGAAGG TCAGGGCGAG AAGAACGTGA CCTTCTGGGG GAGACCGCTG CCACGGTGTG 1560

AGGAGCTACA AGGGCAGAAG CTGGTGGTGG AGGAGAAACG ACCATGTCTA CACGTGCCTG 1620

CTTGCAAAGA CCCTGAGGAA GAGGAACTCT AACACTTCTC TCCTCCACTC TGAGCCCCCT 1680

GACCTTCCAA ACCTCAATAA ACTAGCCTCT TCGAAAAAAA AAAAAAAAAA AAAAAAAAAA 1740

AAAAAAAAAA AAAAAAAAAA AAAAAAAAAA AAAAAAAAAA AAAAA

The first five nucleotides in each exon are underlined to indicate the intron–exon boundaries. First methionine ATG is in bold. Stop codon is indicated <u>TAA</u>. The polyadenylation signal (AATAAA) is also indicated. The sequence above is from accession number NM_002621.2 (The NM_002621 sequence was derived from AL009172, X57748 [1,35,36] and BC015756.). There are two transcript variants that encode for the same protein.[37]

GENOMIC STRUCTURE

The human properdin gene is composed of 10 exons spanning approximately 6 kb[36,38] (Fig. 27.3). The mouse gene is composed of nine exons.[39]

ACCESSION NUMBERS

Human[1,29,35,36]	NM_002621 (sequence derived from AL009172, X57748 and BC015756)
	M83652
	S49355
	DB146591
	AK122955
	AK310695
	KU178249
	KU178250
Mouse[39–41]	NM_008823 (sequence derived from BY347370, AK159291 and AL671853)
	NP_032849
	BC138306
	AK150212
	AK159291
	X12905
Guinea pig[42]	S81116
Rat[43]	NM_001106757
Zebrafish[44]	NM_001160126

DEFICIENCY[45]

Properdin deficiency is X-linked, recessive and leads to defective alternative pathway function. Over 25 families have properdin deficiency and most patients are male.[46] Properdin is critical for facilitating complement-dependent killing

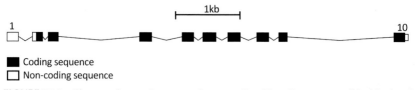

■ Coding sequence
☐ Non-coding sequence

FIGURE 27.3 Diagram of genomic structure for properdin with coding sequence (*black box*) and non-coding sequence (*white box*).

of gonococcal strains,[47] and patients with properdin deficiency are highly susceptible (~3000-fold higher than normal) to fulminant meningococcal infections. Deficiency has also been associated with recurrent otitis media and pneumonia infections.[48] Type I (complete deficiency) is defined by a complete lack of circulating properdin.[46] Mutations identified include R79 to stop,[49] R161 to stop, S206 to stop, W388 to stop,[50] G298V, W321G,[49] W321S, among others. Type II (partial deficiency) is characterised by very low levels of circulating properdin.[46] Mutations identified include R100W,[51] Q343R,[52] among others. Properdin recovered from these affected individuals have oligomerisation defects and primarily consist of dimers.[52] Type III (dysfunctional protein) is characterised by severe functional impairment despite normal plasma levels of properdin. Mutations identified include Y414D substitution in TSR6, which abrogates the ability of properdin to bind C3b and regulate the alternative pathway.[38] The numbering for the mutants listed includes signal peptide.

POLYMORPHIC VARIANTS[45,53]

Multiple properdin polymorphic variants have been identified and many are not yet classified in their functional consequences and thus the potential for pathogenic (mutant) phenotypes exists. Some examples of properdin polymorphic variants include rs8177068, rs8177076, rs8177077, rs909523, rs7060246, rs61737993, rs1048118. Polymorphism rs090523 has a suggestive, but not definitive, association with placental malaria. No associations with age-related macular degeneration exist.

MUTANT ANIMALS

Three properdin mutant mice have been generated. Two are complete properdin knockouts[54,55] and one is properdin deficient in the myeloid lineage.[56] One of the properdin-deficient strains was generated by conventional gene targeting.[54] A targeting construct contained the 5' promoter and exon for TSR1 of the properdin gene and the 3' part of the gene (exons for TSR5 and 6) flanking a positive neomycin selection gene (neo^r). Homologous recombination resulted in replacement of exons 3–7 (coding for TSR2–4) by neo^r of the targeting vector in embryonic stem cells derived from male 129/Ola mice (E14.1a). Embryonic stem cell clones containing the disrupted allele of the targeted gene were microinjected into C57BL/6 and the resulting chimeras were intercrossed resulting in hemizygous males that were backcrossed onto the C57BL/6 background.

Another properdin null ($Cfp^{-/-}$)[55] mouse was generated, unintentionally, using Cre-Lox recombination technology containing neo^r and diphtheria toxin genes as positive and negative selection markers, respectively. Detection of the 5'lox properdin gene failed due to neo^R insertion into the small intron (201 bp) between properdin exons 5 and 6, disrupting the properdin gene. Properdin-deficient ($properdin^{-/-}$) mice developed, grew and reproduced normally.[55]

Properdin deficiency in knockout mice leads to (1) significant protection of mice from arthritis, asthma, abdominal aortic aneurism development, zymosan-induced nonseptic shock and early atherosclerotic lesions in male mice fed a low-fat diet (but this protective function was eliminated upon feeding mice a high-fat diet); (2) increased severity of polymicrobial septic peritonitis in vivo, colitis in an IL-10$^{-/-}$ model of inflammatory bowel disease and infectious colitis utilising *Citrobacter rodentium*; and (3) worsened C3 glomerulopathy when factor H function is either partially or completely impaired (reviewed in Ref. 18).

Gene targeting with the Cre/lox system was used to generate myeloid-specific properdin conditional knockout mice[56] by flanking exons 3–5 of the mouse properdin gene with two loxP sites, creating mice (*Cfp$^{fl/fl}$*) that were next crossed with the lysozyme-Cre transgenic mouse, which expresses the Cre recombinase specifically in myeloid lineage cells. The resulting *Cfp$^{fl/fl}$Lys-Cre$^+$* mice had diminished properdin mRNA expression in CD11b$^+$ cells.[56] Significant decrease in LPS-induced alternative pathway complement activity in the sera of *Cfpfl/flLys-Cre$^+$* mice was observed and they develop significantly attenuated arthritis disease as compared to wild-type mice.

REFERENCES

1. Nolan KF, Schwaeble W, Kaluz S, Dierich MP, Reid KB. Molecular cloning of the cDNA coding for properdin, a positive regulator of the alternative pathway of human complement. *Eur J Immunol* 1991;**21**(3):771–6.
2. Nolan KF, Reid KBM. Complete primary structure of human properdin – a positive regulator of the alternative pathway of the serum complement-system. *Biochem Soc Trans* 1990;**18**(6):1161–2.
3. Sun Z, Reid KBM, Perkins SJ. The dimeric and trimeric solution structures of the multidomain complement protein properdin by X-ray scattering, analytical ultracentrifugation and constrained modelling. *J Mol Biol* 2004;**343**(5):1327–43.
4. Fearon DT, Austen KF. Properdin – binding to C3b and stabilization of C3b-dependent C3 convertase. *J Exp Med* 1975;**142**(4):856–63.
5. Hartmann S, Hofsteenge J. Properdin, the positive regulator of complement, is highly C-mannosylated. *J Biol Chem* 2000;**275**(37):28569–74.
6. Higgins JMG, Wiedemann H, Timpl R, Reid KBM. Characterization of mutant forms of recombinant human properdin lacking single thrombospondin type-I repeats – identification of modules important for function. *J Immunol* 1995;**155**(12):5777–85.
7. Smith CA, Pangburn MK, Vogel CW, Muller-Eberhard HJ. Molecular architecture of human properdin, a positive regulator of the alternative pathway of complement. *J Biol Chem* 1984;**259**(7):4582–8.
8. Pangburn MK. Analysis of the natural polymeric forms of human properdin and their functions in complement activation. *J Immunol* 1989;**142**(1):202–7.
9. Farries TC, Finch JT, Lachmann PJ, Harrison RA. Resolution and analysis of native and activated properdin. *Biochem J* 1987;**243**(2):507–17.
10. Ferreira VP, Cortes C, Pangburn MK. Native polymeric forms of properdin selectively bind to targets and promote activation of the alternative pathway of complement. *Immunobiology* 2010;**215**(11):932–40.

11. Smith KF, Nolan KF, Reid KB, Perkins SJ. Neutron and X-ray scattering studies on the human complement protein properdin provide an analysis of the thrombospondin repeat. *Biochemistry* 1991;**30**(32):8000–8.

12. Tan KM, Duquette M, Liu JH, Dong YC, Zhang RG, Joachimiak A, et al. Crystal structure of the TSP-1 type 1 repeats: a novel layered fold and its biological implication. *J Cell Biol* 2002;**159**(2):373–82.

13. Klenotic PA, Page RC, Misra S, Silverstein RL. Expression, purification and structural characterization of functionally replete thrombospondin-1 type 1 repeats in a bacterial expression system. *Protein Expr Purif* 2011;**80**(2):253–9.

14. Alcorlo M, Tortajada A, Rodriguez de Cordoba S, Llorca O. Structural basis for the stabilization of the complement alternative pathway C3 convertase by properdin. *Proc Natl Acad Sci USA* 2013;**110**(33):13504–9.

15. Farries TC, Lachmann PJ, Harrison RA. Analysis of the interactions between properdin, the 3rd component of complement (C-3), and its physiological activation products. *Biochem J* 1988;**252**(1):47–54.

16. Farries TC, Lachmann PJ, Harrison RA. Analysis of the interaction between properdin and Factor-B, components of the alternative-pathway C-3 convertase of complement. *Biochem J* 1988;**253**(3):667–75.

17. Pangburn MK, Muller-Eberhard HJ. The C3 convertase of the alternative pathway of human complement. Enzymic properties of the bimolecular proteinase. *Biochem J* 1986;**235**(3):723–30.

18. Blatt AZ, Pathan S, Ferreira VP. Properdin: a tightly regulated critical inflammatory modulator. *Immunol Rev* 2016;**274**:172–190.

19. Kemper C, Atkinson JP, Hourcade DE. Properdin: emerging roles of a pattern-recognition molecule. *Annu Rev Immunol* 2010;**28**:131–55.

20. Pauly D, Nagel BM, Reinders J, Killian T, Wulf M, Ackermann S, et al. A novel antibody against human properdin inhibits the alternative complement system and specifically detects properdin from blood samples. *PLoS One* 2014;**9**(5):e96371.

21. Whiteman LY, Purkall DB, Ruddy S. Covalent linkage of C3 to properdin during complement activation. *Eur J Immunol* 1995;**25**(5):1481–4.

22. Holt GD, Pangburn MK, Ginsburg V. Properdin binds to sulfatide [Gal(3-SO4)beta 1-1 Cer] and has a sequence homology with other proteins that bind sulfated glycoconjugates. *J Biol Chem* 1990;**265**(5):2852–5.

23. Harboe M, Garred P, Lindstad JK, Pharo A, Muller F, Stahl GL, et al. The role of properdin in zymosan- and *Escherichia coli*-induced complement activation. *J Immunol* 2012;**189**(5):2606–13.

24. Saggu G, Cortes C, Emch HN, Ramirez G, Worth RG, Ferreira VP. Identification of a novel mode of complement activation on stimulated platelets mediated by properdin and C-3(H2O). *J Immunol* 2013;**190**(12):6457–67.

25. Wirthmueller U, Dewald B, Thelen M, Schafer MKH, Stover C, Whaley K, et al. Properdin, a positive regulator of complement activation, is released from secondary granules of stimulated peripheral blood neutrophils. *J Immunol* 1997;**158**(9):4444–51.

26. Cortes C, Ohtola JA, Saggu G, Ferreira VP. Local release of properdin in the cellular microenvironment: role in pattern recognition and amplification of the alternative pathway of complement. *Front Immunol* 2013;**3**:412–8.

27. Camous L, Roumenina L, Bigot S, Brachemi S, Fremeaux-Bacchi V, Lesavre P, et al. Complement alternative pathway acts as a positive feedback amplification of neutrophil activation. *Blood* 2011;**117**(4):1340–9.

28. Bongrazio M, Pries AR, Zakrzewicz A. The endothelium as physiological source of properdin: role of wall shear stress. *Mol Immunol* 2003;**39**(11):669–75.

29. NCBI. Properdin precursor [*Homo sapiens*]. https://www.ncbi.nlm.nih.gov/protein/NP_002612.1; 2015.

30. de Castro E, Sigrist CJA, Gattiker A, Bulliard V, Langendijk-Genevaux PS, Gasteiger E, et al. ScanProsite: detection of PROSITE signature matches and ProRule-associated functional and structural residues in proteins. *Nucleic Acids Res* 2006;**34**:W362–5.

31. Coleman MP, Murray JC, Willard HF, Nolan KF, Reid KBM, Blake DJ, et al. Genetic and physical mapping around the properdin-P gene. *Genomics* 1991;**11**(4):991–6.

32. NCBI. *Homo sapiens* complement factor properdin (CFP), RefSeqGene (LRG_129) on chromosome X. http://www.ncbi.nlm.nih.gov/nuccore/224451044; 2014.

33. Evans EP, Burtenshaw MD, Laval SH, Goundis D, Reid KBM, Boyd Y. Localization of the properdin factor complement locus Pfc to band-A3 on the mouse X-chromosome. *Genet Res* 1990;**56**(2–3):153–5.

34. NCBI. *Homo sapiens* complement factor properdin (CFP), transcript variant 1, mRNA. https://www.ncbi.nlm.nih.gov/nuccore/NM_002621.2; 2015.

35. Maves KK, Weiler JM. Detection of properdin mRNA in human peripheral blood monocytes and spleen. *J Lab Clin Med* 1992;**120**(5):762–6.

36. Nolan KF, Kaluz S, Higgins JM, Goundis D, Reid KB. Characterization of the human properdin gene. *Biochem J* 1992;**287**(Pt 1):291–7.

37. NCBI. Complement factor properdin [*Homo sapiens* (human)]. https://www.ncbi.nlm.nih.gov/gene/5199; 2016.

38. Fredrikson GN, Westberg J, Kuijper EJ, Tijssen CC, Sjoholm AG, Uhlen M, et al. Molecular characterization of properdin deficiency type III: dysfunction produced by a single point mutation in exon 9 of the structural gene causing a tyrosine to aspartic acid interchange. *J Immunol* 1996;**157**(8):3666–71.

39. NCBI. Complement factor properdin [*Mus musculus* (house mouse)]. http://www.ncbi.nlm.nih.gov/gene/18636; 2016.

40. NCBI. Properdin precursor [*Mus musculus*]. http://www.ncbi.nlm.nih.gov/protein/NP_032849.2; 2015.

41. Goundis D, Reid KB. Properdin, the terminal complement components, thrombospondin and the circumsporozoite protein of malaria parasites contain similar sequence motifs. *Nature* 1988;**335**(6185):82–5.

42. Maves KK, Guenthner ST, Densen P, Moser DR, Weiler JM. Cloning and characterization of the cDNA encoding guinea-pig properdin: a comparison of properdin from three species. *Immunology* 1995;**86**(3):475–9.

43. NCBI. *Rattus norvegicus* complement factor properdin (Cfp), mRNA. http://www.ncbi.nlm.nih.gov/nuccore/166157770; 2016.

44. NCBI. *Danio rerio* complement factor properdin (cfp), mRNA. http://www.ncbi.nlm.nih.gov/nuccore/NM_001160126.1; 2015.

45. NCBI. Properdin: SNP. http://www.ncbi.nlm.nih.gov/snp/?term=properdin; 2016.

46. Fijen CA, van den Bogaard R, Schipper M, Mannens M, Schlesinger M, Nordin FG, et al. Properdin deficiency: molecular basis and disease association. *Mol Immunol* 1999;**36**(13–14):863–7.

47. Agarwal S, Ferreira VP, Cortes C, Pangburn MK, Rice PA, Ram S. An evaluation of the role of properdin in alternative pathway activation on *Neisseria meningitidis* and *Neisseria gonorrhoeae*. *J Immunol* 2010;**185**(1):507–16.

48. Tsao N, Tsai WH, Lin YS, Chuang WJ, Wang CH, Kuo CF. Streptococcal pyrogenic exotoxin B cleaves properdin and inhibits complement-mediated opsonophagocytosis. *Biochem Biophys Res Commun* 2006;**339**(3):779–84.

49. van den Bogaard R, Fijen CA, Schipper MG, de Galan L, Kuijper EJ, Mannens MM. Molecular characterisation of 10 Dutch properdin type I deficient families: mutation analysis and X-inactivation studies. *Eur J Hum Genet* 2000;**8**(7):513–8.

50. Helminen M, Seitsonen S, Jarva H, Meri S, Jarvela IE. A novel mutation W388X underlying properdin deficiency in a Finnish family. *Scand J Immunol* 2012;**75**(4):445–8.

51. Westberg J, Fredrikson GN, Truedsson L, Sjoholm AG, Uhlen M. Sequence-based analysis of properdin deficiency: identification of point mutations in two phenotypic forms of an X-linked immunodeficiency. *Genomics* 1995;**29**(1):1–8.

52. Fredrikson GN, Gullstrand B, Westberg J, Sjoholm AG, Uhlen M, Truedsson L. Expression of properdin in complete and incomplete deficiency: normal in vitro synthesis by monocytes in two cases with properdin deficiency type II due to distinct mutations. *J Clin Immunol* 1998;**18**(4):272–82.

53. Holmberg V, Onkamo P, Lahtela E, Lahermo P, Bedu-Addo G, Mockenhaupt FP, et al. Mutations of complement lectin pathway genes MBL2 and MASP2 associated with placental malaria. *Malar J* 2012;**11**:61–9.

54. Stover CM, Luckett JC, Echtenacher B, Dupont A, Figgitt SE, Brown J, et al. Properdin plays a protective role in polymicrobial septic peritonitis. *J Immunol* 2008;**180**(5):3313–8.

55. Kimura Y, Miwa T, Zhou L, Song WC. Activator-specific requirement of properdin in the initiation and amplification of the alternative pathway complement. *Blood* 2008;**111**(2):732–40.

56. Kimura Y, Zhou L, Miwa T, Song WC. Genetic and therapeutic targeting of properdin in mice prevents complement-mediated tissue injury. *J Clin Invest* 2010;**120**(10):3545–54.

Chapter 28

CR1

Ionita Ghiran, Anne Nicholson-Weller

Beth Israel Deaconess Medical Center, Boston, MA, United States

OTHER NAMES

Complement receptor 1, CD35, immune adherence receptor, C3b/C4b receptor.

PHYSICOCHEMICAL PROPERTIES

CR1 is a type I transmembrane glycoprotein of 2039 amino acids; the leader sequence is between amino acids 1 and 41.[1,2] The mature peptide is located between position 42 and 2039, with a 25 amino acid transmembrane domain, between position 1972–1996, and with a 43 amino acid C-terminal domain, containing two PDZ motifs.[3] There are four major structural allotypes recognised in humans: 1 (F, fast or A), 2 (S, slow or B), 3 (Cor F) and 4 (D). The most abundant is allotype 1, CR1*1, which will be used for all the subsequent descriptions. The isoelectric point of CR1 is 6.57. The relative molecular mass of CR1 depends on the cellular source of CR1, type of electrophoresis system and buffers used. CR1 obtained from nucleated blood cells run higher than CR1 from red cells, due to altered N-glycosylation.[2]

Allotype	Relative Molecular Mass (M_r, kDa, Reduced)	Relative Molecular Mass (M_r, kDa, Unreduced)
CR1*1	220–250	190–210
CR1*2	250–280	220–250
CR1*3	190–220	160–190
CR1*4	Over 280	Over 250

There are 21 glycosylation sites on mature CR1 molecule, with the following locations: Asn 56, Asn 156, Asn 252, Asn 410, Asn 447, Asn 509, Asn 578, Asn 702, Asn 860, Asn 897, Asn 959, Asn 1028, Asn 1152, Asn 1310, Asn 1481, Asn 1504, Asn 1534, Asn 1540, Asn 1605, Asn 1763, Asn 1908. Glycosylation contributes with 20–25 kDa to the molecular mass of CR1. Radiolabelled experiments using tagged glucosamine indicate lack of O-linked oligosaccharide on CR1.[4]

The Complement FactsBook. http://dx.doi.org/10.1016/B978-0-12-810420-0.00028-6
295

STRUCTURE

The extracellular domain of CR1 is comprised of 30 short consensus repeats (SCRs), SCRs are also known as complement control protein repeats or Sushi domains (Fig. 28.1). Extracellular SCRs are further grouped in four, tandem large homologous repeats, LHR-A, LHR-B, LHR-C and LHR-D, each 45 kDa in size, and composed of seven SCRs.[5] There are two unassigned SCRs near the plasma membrane, SCR 29 and 30. The homology between the four LHRs varies between 60% and 99%.[5] The S variant of CR1, CR1*2, has an additional LHR-S inserted between LHR-B and LHR-C. The transmembrane domain, higher primates, baboons and macaques have GPI-anchored version of CR1 with identical extracellular portions. The crystal structure is shown for the first two domains (Fig. 28.2).

FUNCTION

In humans, the main function of CR1 is binding of complement fragments C3b, C4b, C1q and MBL (mannan-binding protein).[6] The interaction between CR1 and iC3b is weak. CR1 on circulating red blood cells (RBCs) binds complement-opsonised immune complexes and microbes present in blood and delivers them to tissue-resident macrophages in liver and spleen through a process called immune adhesion clearance. By preventing immune complexes to persist in circulation, and directly interact with nucleated blood cells, RBC CR1 contributes to maintaining an anti-inflammatory environment in blood. Following the transfer of immune complexes, RBCs along with CR1 are returned to circulation. EBV (Epstein–Barr virus) binds to CR1 by activating classical complement pathway in the absence of antibody,[7] as well as by direct protein–protein interaction.[8] CR1 also served as an adhesion for *Plasmodium falciparum*.[9] In rodents, immune clearance is performed by platelets and soluble factor H.[10] Complement-opsonised particles are phagocytosed by neutrophils and circulating monocytes, but only after priming. On T cells and dendritic cells, CR1 participates in antigen presentation. In RBCs, CR1 promotes Ca^{++} influx and increases membrane deformability.[11] CR1 modulates B cell function and is involved in polarisation of Tregs. The complement regulatory function of CR1 depends on soluble factor I, which acts as a cofactor during degradation of complement fragment C3b to iC3b and C3f, and of iC3b to C3c and C3d,g.[12] In the presence of factor I, CR1 also cleaves C4b to C4c and C4d and accelerates the degradation of the C3 and C5 convertases of the classical and alternative complement activation pathways.[13]

FIGURE 28.1 Diagram of protein domains for CR1.

(A)

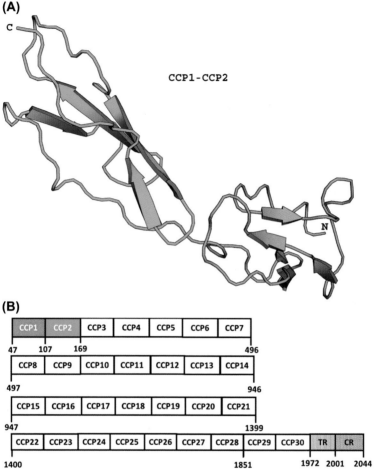

CCP1-CCP2

(B)

FIGURE 28.2   Diagram of crystal structure for CR1. (A) Ribbon diagram of the solution NMR structure of complement receptor type 1 (CR1) CCP1-CCP2 domains (PDB ID: 2MCZ). (B) Schematic representation of the single-chain CR1 domains.

TISSUE DISTRIBUTION

In blood, CR1 is expressed on RBCs, neutrophils, monocytes, B cells, sub-population of T and activated NK cells.[14–16] In tissue, CR1 is found on podocytes, activated endothelial cells, neurons, astrocytes and follicular dendritic cells.[17] A soluble form of CR1, with concentration ranging 13–81 ng/mL, exists in plasma that could be either actively secreted by the cells, likely circulating neutrophils, or the product of enzymatic cleavage.[18,19] The soluble form of CR1 may play a protective role during inflammatory conditions.[20]

REGULATION OF EXPRESSION

While the expression levels of CR1 on circulating RBCs is genetically deter-mined, during the 120 days life cycle in circulation, RBCs lose about 40% of their original CR1 numbers. During acute and chronic inflammatory conditions, such as systemic lupus erythematosus (SLE), lepromatous leprosy, AIDS, auto-immune haemolytic anaemias, circulating RBCs lose CR1 at an accelerated rate due to abnormally high immune clearance process. The low CR1 expression level on circulating RBCs is a consequence of the disease, and not a predisposi-tion. In blood neutrophils, most of the CR1 is stored in secretory vesicles and becomes exposed at the plasma membrane only upon activation.[21]

PROTEIN SEQUENCE

```
MGASSPRSPE PVGPPAPGLP FCCGGSLLAV VVLLALPVAW GQCNAPEWLP   50
FARPTNLTDE FEFPIGTYLN YECRPGYSGR PFSIICLKNS VWTGAKDRCR  100
RKSCRNPPDP VNGMVHVIKG IQFGSQIKYS CTKGYRLIGS SSATCIISGD  150
TVIWDNETPI CDRIPCGLPP TITNGDFIST NRENFHYGSV VTYRCNPGSG  200
GRKVFELVGE PSIYCTSNDD QVGIWSGPAP QCIIPNKCTP PNVENGILVS  250
DNRSLFSLNE VVEFRCQPGF VMKGPRRVKC QALNKWEPEL PSCSRVCQPP  300
PDVLHAERTQ RDKDNFSPGQ EVFYSCEPGY DLRGAASMRC TPQGDWSPAA  350
PTCEVKSCDD FMGQLLNGRV LFPVNLQLGA KVDFVCDEGF QLKGSSASYC  400
VLAGMESLWN SSVPVCEQIF CPSPPVIPNG RHTGKPLEVF PFGKTVNYTC  450
DPHPDRGTSF DLIGESTIRC TSDPQGNGVW SSPAPRCGIL GHCQAPDHFL  500
FAKLKTQTNA SDFPIGTSLK YECRPEYYGR PFSITCLDNL VWSSPKDVCK  550
RKSCKTPPDP VNGMVHVITD IQVGSRINYS CTTGHRLIGH SSAECILSGN  600
AAHWSTKPPI CQRIPCGLPP TIANGDFIST NRENFHYGSV VTYRCNPGSG  650
GRKVFELVGE PSIYCTSNDD QVGIWSGPAP QCIIPNKCTP PNVENGILVS  700
DNRSLFSLNE VVEFRCQPGF VMKGPRRVKC QALNKWEPEL PSCSRVCQPP  750
PDVLHAERTQ RDKDNFSPGQ EVFYSCEPGY DLRGAASMRC TPQGDWSPAA  800
PTCEVKSCDD FMGQLLNGRV LFPVNLQLGA KVDFVCDEGF QLKGSSASYC  850
VLAGMESLWN SSVPVCEQIF CPSPPVIPNG RHTGKPLEVF PFGKAVNYTC  900
DPHPDRGTSF DLIGESTIRC TSDPQGNGVW SSPAPRCGIL GHCQAPDHFL  950
FAKLKTQTNA SDFPIGTSLK YECRPEYYGR PFSITCLDNL VWSSPKDVCK 1000
RKSCKTPPDP VNGMVHVITD IQVGSRINYS CTTGHRLIGH SSAECILSGN 1050
TAHWSTKPPI CQRIPCGLPP TIANGDFIST NRENFHYGSV VTYRCNLGSR 1100
GRKVFELVGE PSIYCTSNDD QVGIWSGPAP QCIIPNKCTP PNVENGILVS 1150
DNRSLFSLNE VVEFRCQPGF VMKGPRRVKC QALNKWEPEL PSCSRVCQPP 1200
PEILHGEHTP SHQDNFSPGQ EVFYSCEPGY DLRGAASLHC TPQGDWSPEA 1250
PRCAVKSCDD FLGQLPHGRV LFPLNLQLGA KVSFVCDEGF RLKGSSVSHC 1300
VLVGMRSLWN NSVPVCEHIF CPNPPAILNG RHTGTPSGDI PYGKEISYTC 1350
DPHPDRGMTF NLIGESTIRC TSDPHGNGVW SSPAPRCELS VRAGHCKTPE 1400
QFPFASPTIP INDFEFPVGT SLNYECRPGY FGKMFSISCL ENLVWSSVED 1450
NCRRKSCGPP PEPFNGMVHI NTDTQFGSTV NYSCNEGFRL IGSPSTTCLV 1500
SGNNVTWDKK APICEIISCE PPPTISNGDF YSNNRTSFHN GTVVTYQCHT 1550
```

PROTEIN SEQUENCE—*Continued*

```
GPDGEQLFEL VGERSIYCTS KDDQVGVWSS PPPRCISTNK CTAPEVENAI 1600
RVPGNRSFFT LTEIIRFRCQ PGFVMVGSHT VQCQTNGRWG PKLPHCSRVC 1650
QPPPEILHGE HTLSHQDNFS PGQEVFYSCE PSYDLRGAAS LHCTPQGDWS 1700
PEAPRCTVKS CDDFLGQLPH GRVLLPLNLQ LGAKVSFVCD EGFRLKGRSA 1750
SHCVLAGMKA LWNSSVPVCE QIFCPNPPAI LNGRHTGTPF GDIPYGKEIS 1800
YACDTHPDRG MTFNLIGESS IRCTSDPQGN GVWSSPAPRC ELSVPAACPH 1850
PPKIQNGHYI GGHVSLYLPG MTISYICDPG YLLVGKGFIF CTDQGIWSQL 1900
DHYCKEVNCS FPLFMNGISK ELEMKKVYHY GDYVTLKCED GYTLEGSPWS 1950
QCQADDRWDP PLAKCTSRTH DALIVGTLSG TIFFILLIIF LSWIILKHRK 2000
GNNAHENPKE VAIHLHSQGG SSVHPRTLQT NEENSRVLP
```

The leader sequence (residues 1–46) is underlined and the potential *N*-linked glycosylation sites.

PROTEIN MODULES

1 or 6–46	Leader peptide	exon 1
47–106	CCP1, start LHR-A	exon 2
107–168	CCP2	exons 3/4
169–238	CCP3	exons 5
239–300	CCP4	exon 5
301–360	CCP5	exon 6
361–423	CCP6	exons 7/8
424–496	CCP7, end LHR-A	exon 9
497–556	CCP8, start LHR-B	exon 10
557–618	CCP9	exons 11/12
619–688	CCP10	exon 13
669–750	CCP11	exon 13
751–810	CCP12	exon 14
811–873	CCP13	exons 15/16
874–946	CCP14, end LHR-B	exon 17
947–1006	CCP15, begin LHR-C	exon 18
1007–1068	CCP16	exons 19/20
1069–1138	CCP17	exon 21
1139–1200	CCP18	exon 21
1201–1260	CCP19	exon 22
1261–1323	CCP20	exons 23/24
1324–1399	CCP21, end LHR-C	exon 25
1400–1459	CCP22, begin LHR-D	exon 26
1460–1521	CCP23	exons 27/28
1522–1591	CCP24	exon 29
1592–1653	CCP25	exon 29
1654–1713	CCP26	exon 30
1714–1776	CCP27	exons 31/32
1777–1851	CCP28, end of LHR-D	exon 33
1852–1911	CCP29	exon 34
1912–1972	CCP30	exon 35
1977–2001	Transmembrane region	exons 36/37
2002–2044	Cytoplasmic domain	exon 38

LIGAND BINDING SITES

47–300	CCPs 1–4	LHR-A	C4b-binding site; C3b, low affinity
497–750	CCPs 8–11	LHR-B	C3b-binding site, C4b low affinity
947–1200	CCPs 15–18	LHR-C	C3b-binding site, C4b low affinity

PROTEIN–PROTEIN INTERACTION SITES

| 2040–2043 | | Cytoplasmic domain type-1 PDZ-motif, interaction with FAP-1. |

CHROMOSOMAL LOCATION

Human: chromosome 1q, 32.2 (Ensembl).
Telomere … CD46 … **CR1** … CR2 … CD55 … C4bp … Centromere.
Mouse analogue (CR1/CR2): chromosome 1q, 40.
Telomere … Crry … CR1/CR2 … Cfh … C4bp … Centromere.

HUMAN cDNA SEQUENCE

```
TTTTGTCCCG GAACCCCGCA GCCCTCCCCA CACTCTGGGC GCGGAGCACA ATGATTGGTC    60

ACTCCTATTT TCGCTGAGCT TTTCCTCTTA TTTCAGTTTT CTTCGAGATC AAATCTGGTT   120

TGTAGATGTG CTTGGGGAGA ATGGGGGCCT CTTCTCCAAG AAGCCCGGAG CCTGTCGGGC   180

CGCCGGCGCC CGGTCTCCCC TTCTGCTGCG GAGGATCCCT GCTGGCGGTT GTGGTGCTGC   240

TTGCGCTGCC GGTGGCCTGG GGTCAATGCA ATGCCCCAGA ATGGCTTCCA TTTGCCAGGC   300

CTACCAACCT AACTGATGAA TTTGAGTTTC CCATTGGGAC ATATCTGAAC TATGAATGCC   360

GCCCTGGTTA TTCCGGAAGA CCGTTTTCTA TCATCTGCCT AAAAAACTCA GTCTGGACTG   420

GTGCTAAGGA CAGGTGCAGA CGTAAATCAT GTCGTAATCC TCCAGATCCT GTGAATGGCA   480

TGGTGCATGT GATCAAAGGC ATCCAGTTCG GATCCCAAAT TAAATATTCT TGTACTAAAG   540

GATACCGACT CATTGGTTCC TCGTCTGCCA CATGCATCAT CTCAGGTGAT ACTGTCATTT   600

GGGATAATGA AACACCTATT TGTGACAGAA TTCCTTGTGG GCTACCCCCC ACCATCACCA   660

ATGGAGATTT CATTAGCACC AACAGAGAGA ATTTTCACTA TGGATCAGTG GTGACCTACC   720

GCTGCAATCC TGGAAGCGGA GGGAGAAAGG TGTTTGAGCT TGTGGGTGAG CCCTCCATAT   780

ACTGCACCAG CAATGACGAT CAAGTGGGCA TCTGGAGCGG CCCCGCCCCT CAGTGCATTA   840

TACCTAACAA ATGCACGCCT CCAAATGTGG AAAATGGAAT ATTGGTATCT GACAACAGAA   900

GCTTATTTTC CTTAAATGAA GTTGTGGAGT TTAGGTGTCA GCCTGGCTTT GTCATGAAAG   960
```

HUMAN cDNA SEQUENCE—*Continued*

```
GACCCCGCCG TGTGAAGTGC CAGGCCCTGA ACAAATGGGA GCCGGAGCTA CCAAGCTGCT 1020
CCAGGGTATG TCAGCCACCT CCAGATGTCC TGCATGCTGA GCGTACCCAA AGGGACAAGG 1080
ACAACTTTTC ACCTGGGCAG GAAGTGTTCT ACAGCTGTGA GCCCGGCTAC GACCTCAGAG 1140
GGGCTGCGTC TATGCGCTGC ACACCCCAGG GAGACTGGAG CCCTGCAGCC CCCACATGTG 1200
AAGTGAAATC CTGTGATGAC TTCATGGGCC AACTTCTTAA TGGCCGTGTG CTATTTCCAG 1260
TAAATCTCCA GCTTGGAGCA AAAGTGGATT TTGTTTGTGA TGAAGGATTT CAATTAAAAG 1320
GCAGCTCTGC TAGTTACTGT GTCTTGGCTG GAATGGAAAG CCTTTGGAAT AGCAGTGTTC 1380
CAGTGTGTGA ACAAATCTTT TGTCCAAGTC CTCCAGTTAT TCCTAATGGG AGACACACAG 1440
GAAAACCTCT GGAAGTCTTT CCCTTTGGGA AAACAGTAAA TTACACATGC GACCCCCACC 1500
CAGACAGAGG GACGAGCTTC GACCTCATTG GAGAGAGCAC CATCCGCTGC ACAAGTGACC 1560
CTCAAGGGAA TGGGGTTTGG AGCAGCCCTG CCCCTCGCTG TGGAATTCTG GGTCACTGTC 1620
AAGCCCCAGA TCATTTTCTG TTTGCCAAGT TGAAAACCCA AACCAATGCA TCTGACTTTC 1680
CCATTGGGAC ATCTTTAAAG TACGAATGCC GTCCTGAGTA CTACGGGAGG CCATTCTCTA 1740
TCACATGTCT AGATAACCTG GTCTGGTCAA GTCCCAAAGA TGTCTGTAAA CGTAAATCAT 1800
GTAAAACTCC TCCAGATCCA GTGAATGGCA TGGTGCATGT GATCACAGAC ATCCAGGTTG 1860
GATCCAGAAT CAACTATTCT TGTACTACAG GGCACCGACT CATTGGTCAC TCATCTGCTG 1920
AATGTATCCT CTCGGGCAAT GCTGCCCATT GGAGCACGAA GCCGCCAATT TGTCAACGAA 1980
TTCCTTGTGG GCTACCCCCC ACCATCGCCA ATGGAGATTT CATTAGCACC AACAGAGAGA 2040
ATTTTCACTA TGGATCAGTG GTGACCTACC GCTGCAATCC TGGAAGCGGA GGGAGAAAGG 2100
TGTTTGAGCT TGTGGGTGAG CCCTCCATAT ACTGCACCAG CAATGACGAT CAAGTGGGCA 2160
TCTGGAGCGG CCCGGCCCCT CAGTGCATTA TACCTAACAA ATGCACGCCT CCAAATGTGG 2220
AAAATGGAAT ATTGGTATCT GACAACAGAA GCTTATTTTC CTTAAATGAA GTTGTGGAGT 2280
TTAGGTGTCA GCCTGGCTTT GTCATGAAAG GACCCCGCCG TGTGAAGTGC CAGGCCCTGA 2340
ACAAATGGGA GCCGGAGCTA CCAAGCTGCT CCAGGGTATG TCAGCCACCT CCAGATGTCC 2400
TGCATGCTGA GCGTACCCAA AGGGACAAGG ACAACTTTTC ACCCGGGCAG GAAGTGTTCT 2460
ACAGCTGTGA GCCCGGCTAT GACCTCAGAG GGGCTGCGTC TATGCGCTGC ACACCCCAGG 2520
GAGACTGGAG CCCTGCAGCC CCCACATGTG AAGTGAAATC CTGTGATGAC TTCATGGGCC 2580
AACTTCTTAA TGGCCGTGTG CTATTTCCAG TAAATCTCCA GCTTGGAGCA AAAGTGGATT 2640
TTGTTTGTGA TGAAGGATTT CAATTAAAAG GCAGCTCTGC TAGTTATTGT GTCTTGGCTG 2700
GAATGGAAAG CCTTTGGAAT AGCAGTGTTC CAGTGTGTGA ACAAATCTTT TGTCCAAGTC 2760
CTCCAGTTAT TCCTAATGGG AGACACACAG GAAAACCTCT GGAAGTCTTT CCCTTTGGAA 2820
AAGCAGTAAA TTACACATGC GACCCCCACC CAGACAGAGG GACGAGCTTC GACCTCATTG 2880
GAGAGAGCAC CATCCGCTGC ACAAGTGACC CTCAAGGGAA TGGGGTTTGG AGCAGCCCTG 2940
CCCCTCGCTG TGGAATTCTG GGTCACTGTC AAGCCCCAGA TCATTTTCTG TTTGCCAAGT 3000
TGAAAACCCA AACCAATGCA TCTGACTTTC CCATTGGGAC ATCTTTAAAG TACGAATGCC 3060
```

Continued

HUMAN cDNA SEQUENCE—*Continued*

```
GTCCTGAGTA CTACGGGAGG CCATTCTCTA TCACATGTCT AGATAACCTG GTCTGGTCAA 3120
GTCCCAAAGA TGTCTGTAAA CGTAAATCAT GTAAAACTCC TCCAGATCCA GTGAATGGCA 3180
TGGTGCATGT GATCACAGAC ATCCAGGTTG GATCCAGAAT CAACTATTCT TGTACTACAG 3240
GGCACCGACT CATTGGTCAC TCATCTGCTG AATGTATCCT CTCAGGCAAT ACTGCCCATT 3300
GGAGCACGAA GCCGCCAATT TGTCAACGAA TTCCTTGTGG GCTACCCCCA ACCATCGCCA 3360
ATGGAGATTT CATTAGCACC AACAGAGAGA ATTTTCACTA TGGATCAGTG GTGACCTACC 3420
GCTGCAATCT TGGAAGCAGA GGGAGAAAGG TGTTTGAGCT TGTGGGTGAG CCCTCCATAT 3480
ACTGCACCAG CAATGACGAT CAAGTGGGCA TCTGGAGCGG CCCCGCCCCT CAGTGCATTA 3540
TACCTAACAA ATGCACGCCT CCAAATGTGG AAAATGGAAT ATTGGTATCT GACAACAGAA 3600
GCTTATTTTC CTTAAATGAA GTTGTGGAGT TTAGGTGTCA GCCTGGCTTT GTCATGAAAG 3660
GACCCCGCCG TGTGAAGTGC CAGGCCCTGA ACAAATGGGA GCCAGAGTTA CCAAGCTGCT 3720
CCAGGGTGTG TCAGCCGCCT CCAGAAATCC TGCATGGTGA GCATACCCCA AGCCATCAGG 3780
ACAACTTTTC ACCTGGGCAG GAAGTGTTCT ACAGCTGTGA GCCTGGCTAT GACCTCAGAG 3840
GGGCTGCGTC TCTGCACTGC ACACCCCAGG GAGACTGGAG CCCTGAAGCC CCGAGATGTG 3900
CAGTGAAATC CTGTGATGAC TTCTTGGGTC AACTCCCTCA TGGCCGTGTG CTATTTCCAC 3960
TTAATCTCCA GCTTGGGGCA AAGGTGTCCT TTGTCTGTGA TGAAGGGTTT CGCTTAAAGG 4020
GCAGTTCCGT TAGTCATTGT GTCTTGGTTG GAATGAGAAG CCTTTGGAAT AACAGTGTTC 4080
CTGTGTGTGA ACATATCTTT TGTCCAAATC CTCCAGCTAT CCTTAATGGG AGACACACAG 4140
GAACTCCCTC TGGAGATATT CCCTATGGAA AAGAAATATC TTACACATGT GACCCCCACC 4200
CAGACAGAGG GATGACCTTC AACCTCATTG GGGAGAGCAC CATCCGCTGC ACAAGTGACC 4260
CTCATGGGAA TGGGGTTTGG AGCAGCCCTG CCCCTCGCTG TGAACTTTCT GTTCGTGCTG 4320
GTCACTGTAA AACCCCAGAG CAGTTTCCAT TTGCCAGTCC TACGATCCCA ATTAATGACT 4380
TTGAGTTTCC AGTCGGGACA TCTTTGAATT ATGAATGCCG TCCTGGGTAT TTTGGGAAAA 4440
TGTTCTCTAT CTCCTGCCTA GAAAACTTGG TCTGGTCAAG TGTTGAAGAC AACTGTAGAC 4500
GAAAATCATG TGGACCTCCA CCAGAACCCT TCAATGGAAT GGTGCATATA AACACAGATA 4560
CACAGTTTGG ATCAACAGTT AATTATTCTT GTAATGAAGG GTTTCGACTC ATTGGTTCCC 4620
CATCTACTAC TTGTCTCGTC TCAGGCAATA ATGTCACATG GGATAAGAAG GCACCTATTT 4680
GTGAGATCAT ATCTTGTGAG CCACCTCCAA CCATATCCAA TGGAGACTTC TACAGCAACA 4740
ATAGAACATC TTTTCACAAT GGAACGGTGG TAACTTACCA GTGCCACACT GGACCAGATG 4800
GAGAACAGCT GTTTGAGCTT GTGGGAGAAC GGTCAATATA TTGCACCAGC AAAGATGATC 4860
AAGTTGGTGT TTGGAGCAGC CCTCCCCCTC GGTGTATTTC TACTAATAAA TGCACAGCTC 4920
CAGAAGTTGA AAATGCAATT AGAGTACCAG GAAACAGGAG TTTCTTTACC CTCACTGAGA 4980
TCATCAGATT TAGATGTCAG CCCGGGTTTG TCATGGTAGG GTCCCACACT GTGCAGTGCC 5040
AGACCAATGG CAGATGGGGG CCCAAGCTGC CACACTGCTC CAGGGTGTGT CAGCCGCCTC 5100
```

HUMAN cDNA SEQUENCE—*Continued*

```
CAGAAATCCT GCATGGTGAG CATACCCTAA GCCATCAGGA CAACTTTTCA CCTGGGCAGG 5160

AAGTGTTCTA CAGCTGTGAG CCCAGCTATG ACCTCAGAGG GGCTGCGTCT CTGCACTGCA 5220

CGCCCCAGGG AGACTGGAGC CCTGAAGCCC CTAGATGTAC AGTGAAATCC TGTGATGACT 5280

TCCTGGGCCA ACTCCCTCAT GGCCGTGTGC TACTTCCACT TAATCTCCAG CTTGGGGCAA 5340

AGGTGTCCTT TGTTTGCGAT GAAGGGTTCC GATTAAAAGG CAGGTCTGCT AGTCATTGTG 5400

TCTTGGCTGG AATGAAAGCC CTTTGGAATA GCAGTGTTCC AGTGTGTGAA CAAATCTTTT 5460

GTCCAAATCC TCCAGCTATC CTTAATGGGA GACACACAGG AACTCCCTTT GGAGATATTC 5520

CCTATGGAAA AGAAATATCT TACGCATGCG ACACCCACCC AGACAGAGGG ATGACCTTCA 5580

ACCTCATTGG GGAGAGCTCC ATCCGCTGCA CAAGTGACCC TCAAGGGAAT GGGGTTTGGA 5640

GCAGCCCTGC CCCTCGCTGT GAACTTTCTG TTCCTGCTGC CTGCCCACAT CCACCCAAGA 5700

TCCAAAACGG GCATTACATT GGAGGACACG TATCTCTATA TCTTCCTGGG ATGACAATCA 5760

GCTACATTTG TGACCCCGGC TACCTGTTAG TGGGAAAGGG CTTCATTTTC TGTACAGACC 5820

AGGGAATCTG GAGCCAATTG GATCATTATT GCAAAGAAGT AAATTGTAGC TTCCCACTGT 5880

TTATGAATGG AATCTCGAAG GAGTTAGAAA TGAAAAAGT ATATCACTAT GGAGATTATG 5940

TGACTTTGAA GTGTGAAGAT GGGTATACTC TGGAAGGCAG TCCCTGGAGC CAGTGCCAGG 6000

CGGATGACAG ATGGGACCCT CCTCTGGCCA AATGTACCTC TCGTACACAT GATGCTCTCA 6060

TAGTTGGCAC TTTATCTGGT ACGATCTTCT TTATTTTACT CATCATTTTC CTCTCTTGGA 6120

TAATTCTAAA GCACAGAAAA GGCAATAATG CACATGAAAA CCCTAAAGAA GTGGCTATCC 6180

ATTTACATTC TCAAGGAGGC AGCAGCGTTC ATCCCCGAAC TCTGCAAACA AATGAAGAAA 6240

ATAGCAGGGT CCTTCCTTGA CAAAGTACTA TACAGCTGAA GAACATCTCG AATACAATTT 6300

TGGTGGGAAA GGAGCCAATT GATTTCAACA GAATCAGATC TGAGCTTCAT AAAGTCTTTG 6360

AAGTGACTTC ACAGAGACGC AGACATGTGC ACTTGAAGAT GCTGCCCCTT CCCTGGTACC 6420

TAGCAAAGCT CCTGCCTCTT TGTGTGCGTC ACTGTGAAAC CCCCACCCTT CTGCCTCGTG 6480

CTAAACGCAC ACAGTATCTA GTCAGGGGAA AAGACTGCAT TTAGGAGATA GAAAATAGTT 6540

TGGATTACTT AAAGGAATAA GGTGTTGCCT GGAATTTCTG GTTTGTAAGG TGGTCACTGT 6600

TCTTTTTTAA AATATTTGTA ATATGGAATG GGCTCAGTAA GAAGAGCTTG GAAAATGCAG 6660

AAAGTTATGA AAAATAAGTC ACTTATAATT ATGCTACCTA CTGATAACCA CTCCTAATAT 6720

TTTGATTCAT TTTCTGCCTA TCTTCTTTCA CATATGTGTT TTTTTACATA CGTACTTTTC 6780

CCCCTTAGTT TGTTTCCTTT TATTTTATAG AGCAGAACCC TAGTCTTTTA AACAGTTTAG 6840

AGTGAAATAT ATGCTATATC AGTTTTTACT TTCTCTAGGG AGAAAAATTA ATTTACTAGA 6900

AAGGCATGAA ATGATCATGG GAAGAGTGGT TAAGACTACT GAAGAGAAAT ATTTGGAAAA 6960

TAAGATTTCG ATATCTTCTT TTTTTTTGAG ATGGAGTCTG GCTCTGTCTC CCAGGCTGGA 7020

GTGCAGTGGC GTAATCTCGG CTCACTGCAA GCTCCGCCTC CCGGGTTGAC ACCATTTTCC 7080

TGCCTCAGCC TCCTGAGTAG TTGGGATTAC CAGTAGATGG GACTACAGGC ACCTGCCAAC 7140

ACGCCCGGCT AATTTTTTTG TATTTTTAGT AGAGACGGGG TTTCACCATG TTAGCCAGGA 7200
```

Continued

HUMAN cDNA SEQUENCE—*Continued*

```
TGGTCTGGAT CTCCTGACCT CGTGATCCAC CTGCCTCGGC CTCCCAAAGT GCTGCGATTA 7260
CAGGCATGAG CCACCGCGCC TGGCCGCTTT CGATATTTTC TAAACTTTAA TTCAAAAGCA 7320
CTTTGTGCTG TGTTCTATAT AAAAAACATA ATAAAAATTG AAATGAAAGA ATAATTGTTA 7380
TTATAAAGT ACTAGCTTAC TTTTGTATGG ATTCAGAATA TACTAAATTA ACTTTTTAAA 7440
ACACAACTTT TAAAAAATGT ATCAAAAT̲A̲ T̲A̲A̲ACGTGTT CTGATATTTT TAAAATAAGT 7500
GACCTTGTGT TCTTTAACCA GTCCACATCT TTAGAGAACA AAAATGTGTT ATGATATTAT 7560
GGGCCATGCT AATGACCTCT AGAAAACATC AGAATATTTC TGGATATTTA ATAATAGCTT 7620
TATATATGAC TAATGCTCAT TTCTATGTAA TTCTGTTTAA TAGTTGCTTT AAAGGTGAAT 7680
TTTGCCACAT TTACTTTGAC AGCAGTATAA GGAGTGAGAT AGACATGAAC CTGAATTTCA 7740
ATTTAAAATC ATGGAAGAGA GGGAAAAAAA ACCAGCTTAA GAAAAATCAA CTGATAAACT 7800
GCAAGAAAAA AATGCAACTT ACATCACAAA AGCTAATTGC TTTATTATTT AGAGAGTACT 7860
TAAAAATTAA AGACCAAACT TCTCTCCACC CAACAAAAAT GGGCAAAGGA CATACAGCTA 7920
GGTCACCAAG AAAGAAGGGC AAATAGGTGG TGAGTATATG TAAAGATACT TGATAGGACT 7980
TTTGCTTAGT TGAATCTTTA GCAAATCTCT TTTATTTCTT GGGATTTTGA AGAAGTAATT 8040
TTTAAAGGAG GACTAGAAAC TAAGTGATTG GGAATTGGCC TTTTTAGAAT TAAAATTTCC 8100
CATTACAAGA AAAAAAAATC CTGTGTTCTT TTTTTTTTCC AGAATGGAGT AGGTCAGTGA 8160
GCAATGTGAT TAATAAATAT TTCAATGTCT GTGACTTTTG ATTTATTTTG GAGACAGGGT 8220
CTTGCTCTGT TACCCAGGCT GGAGTGCAGT GGTGCTATCT AGGCTTACTG CAACCTCACC 8380
TGTCACTTTT TAATTGCAAG AAAGCTGAAA GGTTTTTTTC TATTATATCA GTTATAATGA 8440
TAAATACTGT ATATACTAAC TATGAGTAAA ATACTATATT GCCTAACTTG TATTATTAAG 8500
CAATTCTGCT AACCTGTGAC CTTACATTTT CATCTGAAAA GCAGGGGCTG GACACCAATT 8460
GCCCTATGAA GCTATTGCTA GTCCTAACAT TCTTTGTTTT GTTTGCTTTT TTGGCACACT 8520
TAAGTGTGTA CTATGAAGTT TATGATGCTT TAATGAAATT TTCTGTCTCT ACCATTGTAA 8580
TGAGAAAGGA ATAAAATACT TTATTTTGCA AATCTAAAAA AAAAAAAA
```

The first nucleotides in each exon are underlined. The two possible initiation codons (ATG, methionine) are underlined and so are the termination codon (TGA) and the polyadenylation site (AATAAA).

GENOMIC STRUCTURE

The gene for CR1 is found on chromosome 1, in the cluster of RCA (regulators of complement activation) region. The gene presented here (allotype F or 1) is the most common transcript, extends 133 kb and encodes 39 exons. The second most abundant form, allotype S or 2, is larger, 150–160 kb and spans 47 exons. The additional eight exons form an additional long homologue repeat (LHR-S) (Fig. 28.3).

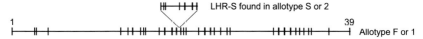

FIGURE 28.3   Diagram of the genomic structure for CR1. Allotype F or 1 has 39 exons, while allotype Sor 2, has an additional LHR containing 8 exons.

ACCESSION NUMBERS

		cDNA	Genomic
Human	CR1	Y000816	L17390–L17430
Chimpanzee	CR1	L24920–L24922	
Baboon	CR1	L39791	
Mouse	CR1/CR2	M61132	
		M36470	
		M29281	
		M35684	
		J04153	
		M33527	
		U17123–U17128	
		X98171	
Mouse	Crry	M23529	
		M34164–M34173	
Rat	Crry	L36532	
		D42115	

DEFICIENCY

To date, there are no known individuals with innate CR1 deficiency. Acquired CR1 deficiency, seen as low expression levels of CR1 on circulating RBCs, is seen during pregnancy[22] and in chronic inflammatory conditions such as HIV, SLE, sepsis and acute respiratory syndrome due to generation and clearance of abnormally large numbers of circulating immune complexes.[23] Innate low expression of CR1 on circulating RBCs is seen in individuals expressing Knops group antigens (Knops, McCoy, Swain-Langley and York) on CR1 that also have circulating anti-CR1 antibodies.

POLYMORPHIC VARIANTS

CR1 polymorphism due to insertion, duplication and deletion in CR1 gene has triple consequences: (1) structural, detected as variation in molecular mass of CR1 variants, (2) functional, with altered binding avidity for C3b due to changes in LHR-B segments, and changes in the cofactor activity effectiveness,[24] and (3) quantitative, with H allotype having expression levels of CR1 on circulating RBCs over ten times larger compared to the L allotype. In general population, RBC CR1 levels vary between 100 in L allotype expresser to

600 in mixed HL allotype and 1200 in H allotype expressers.[18] The expression levels of CR1 in circulating neutrophils do not display an H/L allotype variation. Allotype S of CR1 has been found to be associated with increased risk of Alzheimer disease.[25]

POLYMORPHISM FREQUENCIES[26]

Structural Alleles	White Population	African Americans	Mexican Americans	Chinese Taiwanese
CR1*1	0.86–0.93	0.82–0.84	0.89	0.96
CR1*2	0.07–0.26	0.11–0.12	0.11	0.03
CR1*3	0–0.02	0.04–0.06	0	0.01
CR1*4	<0.01	<0.01	<0.01	0

Quantitative Allele	White Population	African Americans	Mexican Americans	Chinese Taiwanese
H	0.75–0.78	0.85	0.80	0.71
L	0.25–0.22	0.15	0.20	0.28

Knops Phenotype	White Population	African Americans	Mexican Americans	Chinese Taiwanese
Kn(a+)	0.98	0.99	Unknown	0.99
McC(a+)	0.98	0.94	Unknown	1.00
McC(b+)	0.01	0.51	Unknown	0.02
SL(a+)	0.99	0.65	Unknown	0.97
Yk(a+)	0.98	0.92	Unknown	Unknown

MUTANT ANIMALS

In mice, CR1/CR2 knockout animals were generated by mutation of *Cr2* genes in embryonic stem cells by homologous recombination.[27] Functional studies of the mice have demonstrated the role of CR1 and CR2 in B cell activation, regulation of B cell-specific humoral immune response cell defects and increased risk to develop cardiomyopathy, altered IgG3 antibody production and expression of significant defects in T cell functions.[27–30]

REFERENCES

1. Fearon DT. Identification of the membrane glycoprotein that is the C3b receptor of the human erythrocyte, polymorphonuclear leukocyte, B lymphocyte and monocyte. *J Exp Med* 1980;**152**:20–30.
2. Klickstein LB, Moulds JM. CR1. In: Morley BJ, Walport MJ, editors. *The complement facts book*. New York: Academic Press; 2000. p. 136–45.
3. Ghiran I, Glodek AM, Weaver G, Klickstein LB, Nicholson-Weller A. Ligation of erythrocyte CR1 induces its clustering in complex with scaffolding protein FAP-1. *Blood* 2008;**112**(8):3465–73.

4. Lublin DM, Griffith RC, Atkinson JP. Influence of glycosylation on allelic and cell-specific Mr variation, receptor processing, and ligand binding of the human complement C3b/C4b receptor. *J Biol Chem* 1986;**261**(13):5736–44.

5. Klickstein LB, Wong WW, Smith JA, Weis JH, Wilson JG, Fearon DT. Human C3b/C4b receptor (CR1). Demonstration of long homologous repeating domains that are composed of the short consensus repeats characteristics of C3/C4 binding proteins. *J Exp Med* 1987;**165**(4):1095–112.

6. Birmingham DJ, Hebert LA. CR1 and CR1-like: the primate immune adherence receptors. *Immunol Rev* 2001;**180**:100–11.

7. Martin H, McConnell I, Gorick B, Hughes-Jones NC. Antibody-independent activation of the classical pathway of complement by Epstein-Barr virus. *Clin Exp Immunol* 1987;**67**(3):531–6.

8. Ogembo JG, Kannan L, Ghiran I, Nicholson-Weller A, Finberg RW, Tsokos GC, et al. Human complement receptor type 1/CD35 is an Epstein–Barr Virus receptor. *Cell Rep* 2013;**3**(2):371–85.

9. Rowe JA, Moulds JM, Newbold CI, Miller LH. *P. Falciparum rosetting* mediated by a parasite-variant erythrocyte membrane protein and complement-receptor 1. *Nature* 1997;**388**(6639):292–5.

10. Alexander JJ, Hack BK, Cunningham PN, Quigg RJ. A protein with characteristics of factor H is present on rodent platelets and functions as the immune adherence receptor. *J Biol Chem* 2001;**276**(34):32129–35.

11. Glodek AM, Mirchev R, Golan DE, Khoory JA, Burns JM, Shevkoplyas SS, et al. Ligation of complement receptor 1 increases erythrocyte membrane deformability. *Blood* 2010;**116**(26):6063–71.

12. Ross GD, Lambris JD, Cain JA, Newman SL. Generation of three different fragments of bound C3 with purified factor I or serum. I. Requirements for factor H vs. CR1 cofactor activity. *J Immunol* 1982;**129**(5):2051–60.

13. Daha MR, Fearon DT, Austen KF. C3 requirements for formation of alternative pathway C5 convertase. *J Immunol* 1976;**117**(2):630–4.

14. Collard CD, Bukusoglu C, Agah A, Colgan SP, Reenstra WR, Morgan BP, et al. Hypoxia-induced expression of complement receptor type 1 (CR1, CD35) in human vascular endothelial cells. *Am J Physiol* 1999;**276**:C450–8.

15. Min X, Liu C, Wei Y, Wang N, Yuan G, Liu D, et al. Expression and regulation of complement receptors by human natural killer cells. *Immunobiology* 2014;**219**(9):671–9.

16. Cohen JH, Atkinson JP, Klickstein LB, Oudin S, Subramanian VB, Moulds JM. The C3b/C4b receptor (CR1, CD35) on erythrocytes: methods for study of the polymorphisms. *Mol Immunol* 1999;**36**(13–14):819–25.

17. Sim RB, Malhotra V, Day AJ, Erdei A. Structure and specificity of complement receptors. *Immunol Lett* 1987;**14**(3):183–90.

18. Yoon SH, Fearon DT. Characterization of a soluble form of the C3b/C4b receptor (CR1) in human plasma. *J Immunol* 1985;**134**(5):3332–8.

19. Pascual M, Duchosal MA, Steiger G, Giostra E, Pechere A, Paccaud JP, et al. Circulating soluble CR1 (CD35). Serum levels in diseases and evidence for its release by human leukocytes. *J Immunol* 1993;**151**(3):1702–11.

20. Mulligan MS, Yeh CG, Rudolph AR, Ward PA. Protective effects of soluble CR1 in complement- and neutrophil-mediated tissue injury. *J Immunol* 1992;**148**:1479–85.

21. Olesen B, Thomsen BS, Kristensen HS. Loss of erythrocyte complement receptors (CR1; CD35) in patients with acute episodes of septicaemia or bacterial meningitis. *Scand J Infect Dis* 1992;**24**(2):189–95.

22. Imrie HJ, McGonigle TP, Liu DT, Jones DR. Reduction in erythrocyte complement receptor 1 (CR1, CD35) and decay accelerating factor (DAF, CD55) during normal pregnancy. *J Reprod Immunol* 1996;**31**(3):221–7.

23. Larcher C, Schulz TF, Hofbauer J, Hengster P, Romani N, Wachter H, et al. Expression of the C3d/EBV receptor and of other cell membrane surface markers is altered upon HIV-1 infection of myeloid, T, and B cells. *J Acquir Immune Defic Syndr* 1990;**3**(2):103–8.

24. Wong WW, Farrell SA. Proposed structure of the F' allotype of human CR1. Loss of a C3b binding site may be associated with altered function. *J Immunol* 1991;**146**:656–62.

25. Mahmoudi R, Kisserli A, Novella JL, Donvito B, Drame M, Reveil B, et al. Alzheimer's disease is associated with low density of the long CR1 isoform. *Neurobiol Aging* 2015;**36**(4). 1766 e5–1766 e12.

26. Moulds JM, Brai M, Cohen J, Cortelazzo A, Cuccia M, Lin M, et al. Reference typing report for complement receptor 1 (CR1). *Exp Clin Immunogenet* 1998;**15**(4):291–4.

27. Molina H, Holers VM, Li B, Fung Y, Mariathasan S, Goellner J, et al. Markedly impaired humoral immune response in mice deficient in complement receptors 1 and 2. *Proc Natl Acad Sci USA* 1996;**93**(8):3357–61.

28. Boackle SA, Culhane KK, Brown JM, Haas M, Bao L, Quigg RJ, et al. CR1/CR2 deficiency alters IgG3 autoantibody production and IgA glomerular deposition in the MRL/lpr model of SLE. *Autoimmunity* 2004;**37**(2):111–23.

29. Fang Y, Xu C, Fu YX, Holers VM, Molina H. Expression of complement receptors 1 and 2 on follicular dendritic cells is necessary for the generation of a strong antigen-specific IgG response. *J Immunol* 1998;**160**(11):5273–9.

30. Fairweather D, Frisancho-Kiss S, Njoku DB, Nyland JF, Kaya Z, Yusung SA, et al. Complement receptor 1 and 2 deficiency increases coxsackievirus B3-induced myocarditis, dilated cardiomyopathy, and heart failure by increasing macrophages, IL-1beta, and immune complex deposition in the heart. *J Immunol* 2006;**176**(6):3516–24.

Chapter 29

CRIg

Menno van Lookeren Campagne, Luz D. Orozco

Genentech Inc., South San Francisco, CA, United States

OTHER NAMES

Z39Ig, V-set and Ig domain-containing 4, VSIG4.

PHYSICOCHEMICAL PROPERTIES

Complement receptor of the immunoglobulin superfamily (CRIg) is a type 1 transmembrane protein that is composed of 399 amino acids, with a 19 amino acid leader sequence, 2 Ig-like domains, a 21 amino acid transmembrane region and 95 amino acid cytoplasmic domain.

Mature Protein:
pI	6.02
M_r (K) Predicted	~42
N-linked glycosylation sites	None

STRUCTURE

CRIg is a type 1 transmembrane Ig superfamily member that is found in two alternatively spliced forms: CRIg(L) that encodes both V and C_2-type terminal Ig domains and CRIg(S) which encodes only a single V-type domain. The cytoplasmic tail contains the consensus AP-2 internalisation motif YARL. There are three putative phosphorylation sites on the intracellular cytoplasmic tail (Fig. 29.1).

A crystal structure of CRIg N-terminal IgV domain in complex with C3b is presented in Fig. 29.2. The binding interface between CRIg and C3b is large and buries 2670 Å^2 of solvent-accessible surface. CRIg binds primarily to the β-chain of C3b.[1] Four macroglobulin (MG) domains and the linker (LNK) domain in C3b all interact with CRIg. The largest contributions are

The Complement FactsBook. http://dx.doi.org/10.1016/B978-0-12-810420-0.00029-8

FIGURE 29.1   Protein structure of complement receptor of the immunoglobulin superfamily.

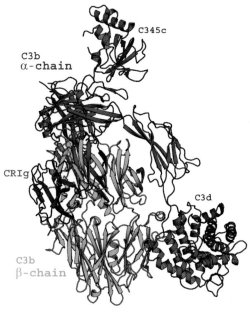

FIGURE 29.2   Crystal structure of human complement receptor of the immunoglobulin superfamily (*blue ribbon*) bound to C3b (pink and green) (PDB accession number 2ICF).

made by MG3 and MG6 at 30% and 40%, respectively, of the total buried interface. The C3b:CRIg structure shows that the reorientation of MG3, along with the movement of a helical section in the LNK region, is necessary to form the CRIg binding site, thus explaining why CRIg cannot bind to C3.

FUNCTION

CRIg was first described as a novel gene located on the human X chromosome[2] and expressed in human lung, placenta and synovium.[3] In 2006, CRIg was identified as a novel member of the complement receptor family critical for pathogen clearance by Kupffer cells[4] and as a novel negative regulator of T cell activation.[5] Over the past decade, CRIg has been implicated in various functions including the clearance of systemic pathogens,[6–8] inhibition of alternative pathway complement activation in disease models[9–14] and in the regulation of adaptive immune response.[15–18] CRIg is the only complement receptor and inhibitor described so far that has Ig domains. No mutations in CRIg have been linked to disease.

CRIg binds to C3b and iC3b molecules that are covalently attached to particle surfaces, including pathogens.[1,4] CRIg binds its C3 ligands with its N-terminal IgV domain and subsequently internalises together with its phagocytised cargo. Following intravenous infection with *Listeria monocytogenes*, CRIg plays an important role in clearing C3-dependent platelet–pathogen complexes by liver Kupffer cells.[7] CRIg has been described to directly bind lipoteichoic acid expressed in the cell wall of Gram-positive bacteria independent of C3[6] indicative of a broader role for CRIg as a pathogen recognition receptor. While CRIg has been described to have T cell modulatory properties,[5,15,17,19,20] a binding partner of CRIg on T lymphocytes has not yet been identified.

Next to functioning as a C3b/iC3b receptor, CRIg's extracellular domain inhibits complement activation through the alternative, but not the classical or lectin, pathway, providing a reagent to selectively block the alternative complement pathway in mouse models of disease. Murine CRIg IgV domain fused to the Fc fragment of an immunoglobulin has been shown to protect in preclinical rodent models of arthritis,[13,14] uveitis,[12] intestinal ischaemia-reperfusion injury,[11] type 1 diabetes[17] and lupus nephritis.[10] A targeted complement inhibitor, in which the extracellular domain of murine CRIg was fused to the MAC inhibitor CD59, improved recovery in a rodent model of traumatic brain injury.[9]

TISSUE DISTRIBUTION

Murine CRIg is expressed on tissue-resident macrophages with particular high expression on liver Kupffer cells[4] and a subpopulation of resident peritoneal macrophages.[21] Expression is low or undetectable in splenic or lymph node macrophages, dendritic cells, Langerhans cells of the skin (Immgen. org database) and microglia (unpublished results). In humans, CRIg is widely expressed in the lung, adipose tissue, spleen, adrenal gland, small intestine, bladder, colon, breast and macrophages associated with blood vessels. In these tissues, CRIg expression correlates with expression of macrophage genes, with a particular strong association with CD163 expressing macrophages.

REGULATION OF EXPRESSION

Dexamethasone and IL-10 induced upregulation of CRIg on human monocyte-derived macrophages.[22] Downregulation of CRIg was promoted by TNF-α and PKCα.[23] Promoter elements that control CRIg expression have not yet been described.

HUMAN PROTEIN SEQUENCE

```
MGILLGLLLL GHLTVDTYGR PILEVPESVT GPWKGDVNLP CTYDPLQGYT        50

QVLVKWLVQR GSDPVTIFLR DSSGDHIQQA KYQGRLHVSH KVPGDVSLQL       100

STLEMDDRSH YTCEVTWQTP DGNQVVRDKI TELRVQKLSV SKPTVTTGSG       150

YGFTVPQGMR ISLQCQARGS PPISYIWYKQ QTNNQEPIKV ATLSTLLFKP       200

AVIADSGSYF CTAKGQVGSE QHSDIVKFVV KDSSKLLKTK TEAPTTMTYP       250

LKATSTVKQS WDWTTDMDGY LGETSAGPGK SLPVFAIILI ISLCCMVVFT       300

MAYIMLCRKT SQQEHVYEAA RAHAREANDS GETMRVAIFA SGCSSDEPTS       350

QNLGNNYSDE PCIGQEYQII AQINGNYARL LDTVPLDYEF LATEGKSVC
```

The signal sequence is underlined. Putative phosphorylation sites are denoted in bold.

PROTEIN MODULES

Sequence	Domain	Exon
1–21	Leader peptide	1
22–84	IgV	2
85–142	IgC2	3
145–203	Extracellular	4/5
206–264	Transmembrane	6
273–329	Cytoplasmic	7/8

CHROMOSOMAL LOCALISATION

CRIg is localised to the distal arm of human chromosome X, at 66 Mb. The CRIg gene locus is flanked by moesin (433 Kb upstream) and hephaestin (140 Kb downstream) in mouse and human. No other complement genes are located in the vicinity of CRIg.

Genome Reference Consortium human build 38 (GRCh38) positions:

Human	Chromosome X: 66,021,738–66,040,125, reverse strand
Mouse	Chromosome X: 96,247,203–96,293,438, reverse strand
Rat	Chromosome X: 65,370,851–65,400,298, reverse strand
Chimp	Chromosome X: 65,623,709–65,642,545, reverse strand
Cynomolgus	Chromosome X: 62,554,891–62,573,499, reverse strand

cDNA SEQUENCE

```
GGTAGCACCA CGCTGGAAGA AAGGACAGAA GTAGCTCTGG CTGTGATGGG GATCTTACTG      60

GGCCTGCTAC TCCTGGGGCA CCTAACAGTG GACACTTATG GCCGTCCCAT CCTGGAAGTG     120

CCAGAGAGTG TAACAGGACC TTGGAAAGGG GATGTGAATC TTCCCTGCAC CTATGACCCC     180

CTGCAAGGCT ACACCCAAGT CTTGGTGAAG TGGCTGGTAC AACGTGGCTC AGACCCTGTC     240

ACCATCTTTC TACGTGACTC TTCTGGAGAC CATATCCAGC AGGCAAAGTA CCAGGGCCGC     300

CTGCATGTGA GCCACAAGGT TCCAGGAGAT GTATCCCTCC AATTGAGCAC CCTGGAGATG     360

GATGACCGGA GCCACTACAC GTGTGAAGTC ACCTGGCAGA CTCCTGATGG CAACCAAGTC     420

GTGAGAGATA AGATTACTGA GCTCCGTGTC CAGAAACTCT CTGTCTCCAA GCCCACAGTG     480

ACAACTGGCA GCGGTTATGG CTTCACGGTG CCCCAGGGAA TGAGGATTAG CCTTCAATGC     540

CAGGCTCGGG GTTCTCCTCC CATCAGTTAT ATTTGGTATA AGCAACAGAC TAATAACCAG     600

GAACCCATCA AAGTAGCAAC CCTAAGTACC TTACTCTTCA AGCCTGCGGT GATAGCCGAC     660

TCAGGCTCCT ATTTCTGCAC TGCCAAGGGC CAGGTTGGCT CTGAGCAGCA CAGCGACATT     720

GTGAAGTTTG TGGTCAAAGA CTCCTCAAAG CTACTCAAGA CCAAGACTGA GGCACCTACA     780

ACCATGACAT ACCCCTTGAA AGCAACATCT ACAGTGAAGC AGTCCTGGGA CTGGACCACT     840

GACATGGATG GCTACCTTGG AGAGACCAGT GCTGGGCCAG GAAAGAGCCT GCCTGTCTTT     900

GCCATCATCC TCATCATCTC CTTGTGCTGT ATGGTGGTTT TTACCATGGC CTATATCATG     960

CTCTGTCGGA AGACATCCCA ACAAGAGCAT GTCTACGAAG CAGCCAGGGC ACATGCCAGA    1020

GAGGCCAACG ACTCTGGAGA AACCATGAGG GTGGCCATCT TCGCAAGTGG CTGCTCCAGT    1080

GATGAGCCAA CTTCCCAGAA TCTGGGCAAC AACTACTCTG ATGAGCCCTG CATAGGACAG    1140

GAGTACCAGA TCATCGCCCA GATCAATGGC AACTACGCCC GCCTGCTGGA CACAGTTCCT    1200

CTGGATTATG AGTTTCTGGC CACTGAGGGC AAAAGTGTCT GTTAAAAATG CCCCATTAGG    1260

CCAGGATCTG CTGACATAAT TGCCTAGTCA GTCCTTGCCT TCTGCATGGC CTTCTTCCCT    1320

GCTACCTCTC TTCCTGGATA GCCCAAAGTG TCCGCCTACC AACACTGGAG CCGCTGGGAG    1380

TCACTGGCTT TGCCCTGGAA TTTGCCAGAT GCATCTCAAG TAAGCCAGCT GCTGGATTTG    1440

GCTCTGGGCC CTTCTAGTAT CTCTGCCGGG GGCTTCTGGT ACTCCTCTCT AAATACCAGA    1500

GGGAAGATGC CCATAGCACT AGGACTTGGT CATCATGCCT ACAGACACTA TTCAACTTTG    1560

GCATCTTGCC ACCAGAAGAC CCGAGGGAGG CTCAGCTCTG CCAGCTCAGA GGACCAGCTA    1620

TATCCAGGAT CATTTCTCTT TCTTCAGGGC CAGACAGCTT TTAATTGAAA TTGTTATTTC    1680

ACAGGCCAGG GTTCAGTTCT GCTCCTCCAC TATAAGTCTA ATGTTCTGAC TCTCTCCTGG    1740

TGCTCAATAA ATATCTAATC ATAACAGCAA
```

The first five nucleotides in each exon are underlined. The initiation and termination codons (ATG and TAA, respectively) and putative polyadenylation signals are double underlined.

GENOMIC STRUCTURE

Human CRIg is encoded by eight exons. Exons 1, 2 and 3 encode the signal peptides, IgV and IgC2 domains, respectively. Exon 6 encodes the transmembrane domains and exons 7 and 8, the cytoplasmic domain (Fig. 29.3).

ACCESSION NUMBERS

Human: 11,326, NM007268, Q9Y279, CCDS14383.
Mouse: 278,180, NM177789, F6TUL9, CCDS30288.

DEFICIENCY AND POLYMORPHIC VARIANTS

There are no known GWAS hits in the CRIg (VSIG4) locus, and no familial mutations or phenotypes have been described so far. Linkage analysis in rats suggests that VSIG4 is linked to collagen-induced arthritis, blood pressure, stress response and memory (Gene Editing Rat Resource Center). Known coding variants (SNPs) in human CRIg are rs34581041, rs34222730 and rs17315645 (Fig. 29.4).

FIGURE 29.3 Genomic structure of human complement receptor of the immunoglobulin superfamily. *Open boxes* represent 5' and 3' UTR regions.

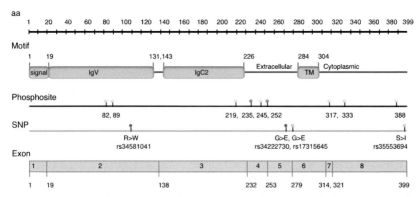

FIGURE 29.4 Protein structure of human complement receptor of the immunoglobulin superfamily.

DEFICIENT MICE

Vsig4^{tm1Gne}/Vsig4^{tm1Gne} mice were made in C57BL/6NCrl genetic background by intragenic deletion, where exon 1 was replaced with a neomycin resistance cassette.[4] Mice homozygous for a gene-targeted knockout allele fail to exhibit complement-dependent clearance of pathogens from the circulation.[4,8] CRIg-deficient mice succumb to intravenous *L. monocytogenes* infection.[4]

REFERENCES

1. Wiesmann C, Katschke KJ, Yin J, et al. Structure of C3b in complex with CRIg gives insights into regulation of complement activation. *Nature* 2006;**444**:217–20.
2. Langnaese K, Colleaux L, Kloos DU, Fontes M, Wieacker P. Cloning of Z39Ig, a novel gene with immunoglobulin-like domains located on human chromosome X. *Biochimica Biophys Acta* 2000;**1492**:522–5.
3. Walker MG. Z39Ig is co-expressed with activated macrophage genes. *Biochim Biophys Acta* 2002;**1574**:387–90.
4. Helmy KY, Katschke Jr KJ, Gorgani NN, et al. CRIg: a macrophage complement receptor required for phagocytosis of circulating pathogens. *Cell* 2006;**124**:915–27.
5. Vogt L, Schmitz N, Kurrer MO, et al. VSIG4, a B7 family-related protein, is a negative regulator of T cell activation. *J Clin Invest* 2006;**116**:2817–26.
6. Zeng Z, Surewaard BG, Wong CH, Geoghegan JA, Jenne CN, Kubes P. CRIg functions as a macrophage pattern recognition receptor to directly bind and capture blood-borne gram-positive bacteria. *Cell Host Microbe* 2016;**20**:99–106.
7. Broadley SP, Plaumann A, Coletti R, et al. Dual-track clearance of circulating bacteria balances rapid restoration of blood sterility with induction of adaptive immunity. *Cell Host Microbe* 2016;**20**:36–48.
8. He JQ, Katschke Jr KJ, Gribling P, et al. CRIg mediates early Kupffer cell responses to adenovirus. *J Leukoc Biol* 2013;**93**:301–6.
9. Ruseva MM, Ramaglia V, Morgan BP, Harris CL. An anticomplement agent that homes to the damaged brain and promotes recovery after traumatic brain injury in mice. *Proc Natl Acad Sci USA* 2015;**112**:14319–24.
10. Lieberman LA, Mizui M, Nalbandian A, Bosse R, Crispin JC, Tsokos GC. Complement receptor of the immunoglobulin superfamily reduces murine lupus nephritis and cutaneous disease. *Clin Immunol* 2015;**160**:286–91.
11. Chen J, Crispin JC, Dalle Lucca J, Tsokos GC. A novel inhibitor of the alternative pathway of complement attenuates intestinal ischemia/reperfusion-induced injury. *J Surg Res* 2011;**167**:e131–136.
12. Chen M, Muckersie E, Luo C, Forrester JV, Xu H. Inhibition of the alternative pathway of complement activation reduces inflammation in experimental autoimmune uveoretinitis. *Eur J Immunol* 2010;**40**:2870–81.
13. Li B, Xi H, Diehl L, et al. Improving therapeutic efficacy of a complement receptor by structure-based affinity maturation. *J Biol Chem* 2009;**284**:35605–11.
14. Katschke Jr KJ, Helmy KY, Steffek M, et al. A novel inhibitor of the alternative pathway of complement reverses inflammation and bone destruction in experimental arthritis. *J Exp Med* 2007;**204**:1319–25.
15. Li Y, Wang YQ, Wang DH, et al. Costimulatory molecule VSIG4 exclusively expressed on macrophages alleviates renal tubulointerstitial injury in VSIG4 KO mice. *J Nephrol* 2014;**27**:29–36.

16. Jung K, Kang M, Park C, et al. Protective role of V-set and immunoglobulin domain-containing 4 expressed on kupffer cells during immune-mediated liver injury by inducing tolerance of liver T- and natural killer T-cells. *Hepatology* 2012;**56**:1838–48.

17. Fu W, Wojtkiewicz G, Weissleder R, Benoist C, Mathis D. Early window of diabetes determinism in NOD mice, dependent on the complement receptor CRIg, identified by noninvasive imaging. *Nat Immunol* 2012;**13**:361–8.

18. He JQ, Wiesmann C, van Lookeren Campagne M. A role of macrophage complement receptor CRIg in immune clearance and inflammation. *Mol Immunol* 2008;**45**:4041–7.

19. Jung K, Seo SK, Choi I. Endogenous VSIG4 negatively regulates the helper T cell-mediated antibody response. *Immunol Lett* 2015;**165**:78–83.

20. Xu S, Sun Z, Li L, et al. Induction of T cells suppression by dendritic cells transfected with VSIG4 recombinant adenovirus. *Immunol Lett* 2010;**128**:46–50.

21. Gorgani NN, He JQ, Katschke Jr KJ, et al. Complement receptor of the Ig superfamily enhances complement-mediated phagocytosis in a subpopulation of tissue resident macrophages. *J Immunol* 2008;**181**:7902–8.

22. Gorgani NN, Thathaisong U, Mukaro VR, et al. Regulation of CRIg expression and phagocytosis in human macrophages by arachidonate, dexamethasone, and cytokines. *Am J Pathol* 2011;**179**:1310–8.

23. Ma Y, Usuwanthim K, Munawara U, et al. Protein kinase calpha regulates the expression of complement receptor Ig in human monocyte-derived macrophages. *J Immunol* 2015;**194**:2855–61.

Chapter 30

Factor H and Factor H-like Protein 1

Paul N. Barlow

University of Edinburgh, Edinburgh, United Kingdom

OTHER NAMES

Factor H isoform a: adrenomedullin-binding protein 1 (rarely used); beta-1H globulin (historical, no longer used).

Factor H-like 1: factor H isoform b; reconectin (rarely used).

PHYSICOCHEMICAL PROPERTIES

Factor H[1,2] (FH) is synthesised as a single-chain polypeptide of 1231 amino acid residues, including an 18-residue leader sequence.

pI mature glycoprotein	5.4–6
M_r	155,000
Disulphide bonds	40
Potential N-glycosylation sites	511, 700, 784, 814, 864, 893, 1011 and 1077 (mature protein numbering)

These are occupied primarily by diantennary disialylated glycans of $M_r = 2204$.[3] After cleavage of the leader sequence, splice variant, FH-like 1 (FHL1) has 431 residues, $M_r = 49,000$, predicted pI = 6.45 and no N-glycans.

STRUCTURE

Factor H is predominantly monomeric, but weakly self-associates ($K_D = 28\,\mu M$) and may oligomerise in the presence of glycosaminoglycans or high concentrations of metal ions.[4] It is composed from 20 homologous units called complement control protein modules (CCPs) (SCRs or sushi domains)[5] (Fig. 30.1).

The splice variant, FHL-1, consists of the first seven CCPs followed by the C-terminal sequence, Ser-Pro-Leu-Thr. Each CCP contains ~60 residues including four invariant cysteines forming Cys[I]-Cys[III],Cys[II]-Cys[IV] disulphides. Neighbouring modules are linked by sequences of three to eight residues. No

The Complement FactsBook. http://dx.doi.org/10.1016/B978-0-12-810420-0.00030-4

FIGURE 30.1   Diagram of protein domains for factor H.

(A)

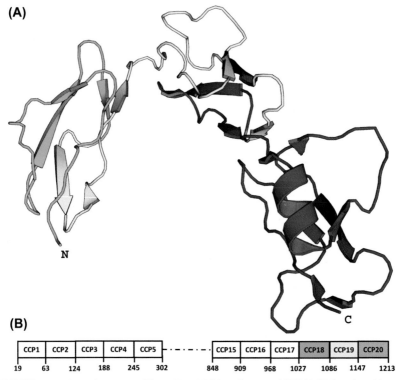

(B)

CCP1	CCP2	CCP3	CCP4	CCP5	· - · - · -	CCP15	CCP16	CCP17	CCP18	CCP19	CCP20
19	63	124	188	245	302	848	909	968	1027	1086	1147	1213

FIGURE 30.2   Crystal structure of factor H. (A) Ribbon diagram of CCP18-20 domains of human complement factor H crystal structure (PDB ID: 3SW0). (B) Schematic representation of factor H CCP domains.

three-dimensional structure of FH has been determined, probably because its flexibility hinders crystallisation, but partial crystal structure is shown in Fig. 30.2. In negatively stained electron micrographs, FH molecules[6] adopted multiple conformations, but were predominantly doubled over; analytical ultracentrifugation, small-angle X-ray scattering[7] (models deposited in PDB, e.g., 3GAV) and chemical cross-linking[8] also suggested a chain of modules bent back on itself. High-resolution structures were determined for several FH segments, either alone or in complex with other molecules (PDB identifiers provided): CCPs 1–2 (2RLP), 2–3 (2RLQ), 1–4 (2WII), 5, 6–7 (e.g., 2W80, 2YBY), 7 (2JGW and 2JGX), 6–8 (2UWN and 2V8E), 9 (4K12), 10–11

(4B2R), 11–12 (4B2S), 12–13 (2KMS), 15 (1HFI), 16 (1HCC), 15–16 (1HFH), 18–20 (3SWO), 19–20 (e.g., 2BZM, 2G7I and 4ONT). Each CCP resembles a prolate spheroid, with a long axis of ~4 nm and short axes of ~2 nm, and contains β-strands in antiparallel sheets aligned approximately with the long axis (Fig. 30.2). Its N and C termini lie towards either extremity of its long axis, promoting end-to-end arrangements of tandem CCPs with variable intermodular contacts, tilts and twists.[9]

FUNCTION

Factor H inhibits C3b amplification. It is a cofactor for factor I-catalysed cleavage of C3b to iC3b, an opsonin and ligand for complement receptors 2 and 3. It accelerates irreversible dissociation of alternative pathway C3 convertase, C3bBb, and might also compete with factor B for binding to C3b during proconvertase formation. Being soluble, FH is important for protecting surfaces, including extracellular matrix (ECM), lacking membrane-bound, regulators. These functions are performed via binding sites[10] for C3b (CCPs 1–4 and 19–20) and C3b/iC3b/C3d(g) (CCPs 19–20), as well as recognition sites for self-associated molecular patterns containing glycosaminoglycans and sialic acids (CCPs 7 and 20).[11,12] Factor H binds adrenomedullin, a peptide hormone, and may prevent its degradation. A role for FH in the management of cellular senescence, stress or injury is suggested by interactions reported with C-reactive protein,[13] pentraxin 3,[14] DNA, histones, annexin II,[15] malondialdehyde acetaldehyde adducts of proteins[16] and oxidised lipids.[17] FHL1 also has factor I-cofactor activity and C3bBb decay-accelerating activity. FHL1 retains a self-recognition site in CCP 7 but lacks CCP 20. CCP 4, common to FH and FHL1, contains an Arg-Gly-Asp motif.

TISSUE DISTRIBUTION

FH is present at 135–250 mg/L in serum.[18] Its primary site of synthesis is in hepatocytes. Secondary sites include Kupffer cells, lung cells, retinal pigment epithelium, peripheral blood lymphocytes, myoblasts, rhabdomyosarcoma cells, fibroblasts, umbilical vein endothelial cells, glomerular mesangial cells, neurons, astrocytes and glia cells.[19] FHL1 is present in serum in 10- to 50-fold lower quantities (by weight) than FH,[20] but appears to be the preeminent complement regulator in Bruch's membrane, a layer of specialised ECM within the macula.[21]

REGULATION OF EXPRESSION

The cytokine, IFN-γ, was shown to upregulate expression of the FH gene, *CFH*, in hepatocytes, astrocytes, lung cells and astroglioma cells, consistent with the presence of two IFN-γ activation sites within the *CFH* proximal promoter.[22] Interleukin-27 upregulates the expression of *CFH* in retinal cells through activation of STAT1 and upregulation of STAT1 target genes.[23]

HUMAN PROTEIN SEQUENCE

```
MRLLAKIICL MLWAICVAED CNELPPRRNT EILTGSWSDQ TYPEGTQAIY    50
KCRPGYRSLG NIIMVCRKGE WVALNPLRKC QKRPCGHPGD TPFGTFTLTG   100
GNVFEYGVKA VYTCNEGYQL LGEINYRECD TDGWTNDIPI CEVVKCLPVT   150
APENGKIVSS AMEPDREYHF GQAVRFVCNS GYKIEGDEEM HCSDDGFWSK   200
EKPKCVEISC KSPDVINGSP ISQKIIYKEN ERFQYKCNMG YEYSERGDAV   250
CTESGWRPLP SCEEKSCDNP YIPNGDYSPL RIKHRTGDEI TYQCRNGFYP   300
ATRGNTAKCT STGWIPAPRC TLKPCDYPDI KHGGLYHENM RRPYFPVAVG   350
KYYSYYCDEH FETPSGSYWD HIHCTQDGWS PAVPCLRKCY FPYLENGYNQ   400
NYGRKFVQGK SIDVACHPGY ALPKAQTTVT CMENGWSPTP RCIRVKTCSK   450
SSIDIENGFI SESQYTYALK EKAKYQCKLG YVTADGETSG SITCGKDGWS   500
AQPTCIKSCD IPVFMNARTK NDFTWFKL**N**D TLDYECHDGY ESNTGSTTGS   550
IVCGYNGWSD LPICYERECE LPKIDVHLVP DRKKDQYKVG EVLKFSCKPG   600
FTIVGPNSVQ CYHFGLSPDL PICKEQVQSC GPPPELLNGN VKEKTKEEYG   650
HSEVVEYYCN PRFLMKGPNK IQCVDGEWTT LPVCIVEEST CGDIPELEHG   700
WAQLSSPPYY YGDSVEF**N**CS ESFTMIGHRS ITCIHGVWTQ LPQCVAIDKL   750
KKCKSSNLII LEEHLKNKKE FDHNSNIRYR CRGKEGWIHT VCINGRWDPE   800
V**N**CSMAQIQL CPPPPQIPNS H**N**MTTTLNYR DGEKVSVLCQ ENYLIQEGEE   850
ITCKDGRWQS IPLCVEKIPC SQPPQIEHGT I**N**SSRSSQES YAHGTKLSYT   900
CEGGFRISEE **N**ETTCYMGKW SSPPQCEGLP CKSPPEISHG VVAHMSDSYQ   950
YGEEVTYKCF EGFGIDGPAI AKCLGEKWSH PPSCIKTDCL SLPSFENAIP  1000
MGEKKDVYKA GEQVTYTCAT YYKMDGAS**N**V TCINSRWTGR PTCRDTSCVN  1050
PPTVQNAYIV SRQMSKYPSG ERVRYQCRSP YEMFGDEEVM CLNG**N**WTEPP  1100
QCKDSTGKCG PPPPIDNGDI TSFPLSVYAP ASSVEYQCQN LYQLEGNKRI  1150
TCRNGQWSEP PKCLHPCVIS REIMENYNIA LRWTAKQKLY SRTGESVEFV  1200
CKRGYRLSSR SHTLRTTCWD GKLEYPTCAK R
```

Signal peptide is underlined. Utilised glycosylation sites are in bold. Numbering is for the immature protein (plus 18 with respect to the mature protein). The FHL1 sequence is identical until residue 445 and then terminates in the sequence '…SFTL'.

PROTEIN MODULES

−18 to −1	Leader peptide	exon 1
3–62	CCP1	exon 2
67–123	CCP2	exons 3 and 4
128–187	CCP3	exon 5
192–244	CCP4	exon 6
249–302	CCP5	exon 7
307–367	CCP6	exon 8
371–424	CCP7	exon 9
430–487	CCP8	exon 10
491–546	CCP9	exon 11
551–605	CCP10	exon 12
612–666	CCP11	exon 13
673–726	CCP12	exon 14
735–785	CCP13	exon 15
793–846	CCP14	exon 16
852–908	CCP15	exon 17
913–966	CCP16	exon 18
971–1025	CCP17	exon 19
1030–1084	CCP18	exon 20
1091–1145	CCP19	exon 21
1149–1210	CCP20	exon 22

The numbering used is for the mature protein. Module boundaries were predicted in the SMART database.

CHROMOSOMAL LOCATION

The gene for human FH (*CFH*) lies within the regulator of complement activation (RCA) gene cluster in 1q32. The human RCA gene cluster spans 21·45 cM and includes 15 complement-related genes. *CFH* lies within a centromeric 650 kb-long DNA segment that contains *CFH*, *CFHR3*, *CFHR1*, *CFHR4*, *CFHR2* and *CFHR5*.

HUMAN cDNA SEQUENCE

```
1     AATTCTTGGA AGAGGAGAAC TGGACGTTGT GAACAGAGTT AGCTGGTAAA TGTCCTCTTA
61    AAAGATCCAA AAAATGAGAC TTCTAGCAAA GATTATTTGC CTTATGTTAT GGGCTATTTG
121   TGTAGCAGAA GATTGCAATG AACTTCCTCC AAGAAGAAAT ACAGAAATTC TGACAGGTTC
181   CTGGTCTGAC CAAACATATC CAGAAGGCAC CCAGGCTATC TATAAATGCC GCCCTGGATA
241   TAGATCTCTT GGAAATGTAA TAATGGTATG CAGGAAGGGA GAATGGGTTG CTCTTAATCC
301   ATTAAGGAAA TGTCAGAAAA GGCCCTGTGG ACATCCTGGA GATACTCCTT TTGGTACTTT
361   TACCCTTACA GGAGGAAATG TGTTTGAATA TGGTGTAAAA GCTGTGTATA CATGTAATGA
421   GGGGTATCAA TTGCTAGGTG AGATTAATTA CCGTGAATGT GACACAGATG GATGGACCAA
481   TGATATTCCT ATATGTGAAG TTGTGAAGTG TTTACCAGTG ACAGCACCAG AGAATGGAAA
541   AATTGTCAGT AGTGCAATGG AACCAGATCG GGAATACCAT TTTGGACAAG CAGTACGGTT
601   TGTATGTAAC TCAGGCTACA AGATTGAAGG AGATGAAGAA ATGCATTGTT CAGACGATGG
661   TTTTTGGAGT AAAGAGAAAC CAAAGTGTGT GGAAATTTCA TGCAAATCCC CAGATGTTAT
721   AAATGGATCT CCTATATCTC AGAAGATTAT TTATAAGGAG AATGAACGAT TTCAATATAA
781   ATGTAACATG GGTTATGAAT ACAGTGAAAG AGGAGATGCT GTATGCACTG AATCTGGATG
841   GCGTCCGTTG CCTTCATGTG AAGAAAAATC ATGTGATAAT CCTTATATTC CAAATGGTGA
901   CTACTCACCT TTAAGGATTA AACACAGAAC TGGAGATGAA ATCACGTACC AGTGTAGAAA
961   TGGTTTTTAT CCTGCAACCC GGGGGAAATAC AGCCAAATGC ACAAGTACTG GCTGGATACC
1021  TGCTCCGAGA TGTACCTTGA AACCTTGTGA TTATCCAGAC ATTAAACATG GAGGTCTATA
1081  TCATGAGAAT ATGCGTAGAC CATACTTTCC AGTAGCTGTA GGAAAATATT ACTCCTATTA
1141  CTGTGATGAA CATTTTGAGA CTCCGTCAGG AAGTTACTGG GATCACATTC ATTGCACACA
1201  AGATGGATGG TCGCCAGCAG TACCATGCCT CAGAAAATGT TATTTTCCTT ATTTGGAAAA
1261  TGGATATAAT CAAAATCATG GAAGAAAGTT TGTACAGGGT AAATCTATAG ACGTTGCCTG
1321  CCATCCTGGC TACGCTCTTC CAAAAGCGCA GACCACAGTT ACATGTATGG AGAATGGCTG
1381  GTCTCCTACT CCCAGATGCA TCCGTGTCAA AACATGTTCC AAATCAAGTA TAGATATTGA
1441  GAATGGGTTT ATTTCTGAAT CTCAGTATAC ATATGCCTTA AAAGAAAAAG CGAAATATCA
1501  ATGCAAACTA GGATATGTAA CAGCAGATGG TGAAACATCA GGATCAATTA GATGTGGGAA
1561  AGATGGATGG TCAGCTCAAC CCACGTGCAT TAAATCTTGT GATATCCCAG TATTTATGAA
1621  TGCCAGAACT AAAAATGACT TCACATGGTT TAAGCTGAAT GACACATTGG ACTATGAATG
1681  CCATGATGGT TATGAAAGCA ATACTGGAAG CACCACTGGT TCCATAGTGT GTGGTTACAA
1741  TGGTTGGTCT GATTACCCCA TATGTTATGA AAGAGAATGC GAACTTCCTA AAATAGATGT
1801  ACACTTAGTT CCTGATCGCA AGAAAGACCA GTATAAAGTT GGAGAGGTGT TGAAATTCTC
1861  CTGCAAACCA GGATTTACAA TAGTTGGACC TAATTCCGTT CAGTGCTACC ACTTTGGATT
```

HUMAN cDNA SEQUENCE—*Continued*

```
1921 GTCTCCTGAC CTCCCAATAT GTAAAGAGCA AGTACAATCA TGTGGTCCAC CTCCTCAACT

1981 CCTCAATGGG AATGTTAAGG AAAAAACGAA AGAAGAATAT GGACACAGTG AAGTGGTGGA

2041 ATATTATTGC AATCCTAGAT TTCTAATGAA GGGACCTAAT AAAATTCAAT GTGTTGATGG

2101 AGAGTGGACA ACTTTACCAG TGTGTATTGT GGAGGAGAGT ACCTGTGGAG ATATACCTGA

2161 ACTTGAACAT GGCTGGGCCC AGCTTTCTTC CCCTCCTTAT TACTATGGAG ATTCAGTGGA

2221 ATTCAATTGC TCAGAATCAT TTACAATGAT TGGACACAGA TCAATTACGT GTATTCATGG

2281 AGTATGGACC CAACTTCCCC AGTGTGTGGC AATAGATAAA CTTAAGAAGT GCAAATCATC

2341 AAATTTAATT ATACTTGAGG AACATTTAAA AAACAAGAAG GAATTCGATC ATAATTCTAA

2401 CATAAGGTAC AGATGTAGAG GAAAAGAAGG ATGGATACAC ACAGTCTGCA TAAATGGAAG

2461 ATGGGATCCA GAAGTGAACT GCTCAATGGC ACAAATACAA TTATGCCCAC CTCCACCTCA

2521 GATTCCCAAT TCTCACAATA TGACAACCAC ACTGAATTAT CGGGATGGAG AAAAAGTATC

2581 TGTTCTTTGC CAAGAAAATT ATCTAATTCA GGAAGGAGAA GAAATTACAT GCAAAGATGG

2641 AAGATGGCAG TCAATACCAC TCTGTGTTGA AAAAATTCCA TGTTCACAAC CACCTCAGAT

2701 AGAACACGGA ACCATTAATT CATCCAGGTC TTCACAAGAA AGTTATGCAC ATGGGACTAA

2761 ATTGAGTTAT ACTTGTGAGG GTGGTTTCAG GATATCTGAA GAAAATGAAA CAACATGCTA

2821 CATGGGAAAA TGGAGTTCTC CACCTCAGTG TGAAGGCCTT CCTTGTAAAT CTCCACCTGA

2881 GATTTCTCAT GGTGTTGTAG CTCACATGTC AGACAGTTAT CAGTATGGAG AAGAAGTTAC

2941 GTACAAATGT TTTGAAGGTT TTGGAATTGA TGGGCCTGCA ATTGCAAAAT GCTTAGGAGA

3001 AAAATGGTCT CACCCTCCAT CATGCATAAA AACAGATTGT CTCAGTTTAC CTAGCTTTGA

3061 AAATGCCATA CCCATGGGAG AGAAGAAGGA TGTGTATAAG GCGGGTGAGC AAGTGACTTA

3121 CACTTGTGCA ACATATTACA AAATGGATGG AGCCAGTAAT GTAACATGCA TTAATAGCAG

3181 ATGGACAGGA AGGCCAACAT GCAGAGACAC CTCCTGTGTG AATCCGCCCA CAGTACAAAA

3241 TGCTTATATA GTGTCGAGAC AGATGAGTAA ATATCCATCT GGTGAGAGAG TACGTTATCA

3301 ATGTAGGAGC CCTTATGAAA TGTTTGGGGA TGAAGAAGTG ATGTGTTTAA ATGGAAACTG

3361 GACGGAACCA CCTCAATGCA AAGATTCTAC AGGAAAATGT GGGCCCCCTC CACCTATTGA

3421 CAATGGGGAC ATTACTTCAT TCCCGTTGTC AGTATATGCT CCAGCTTCAT CAGTTGAGTA

3481 CCAATGCCAG AACTTGTATC AACTTGAGGG TAACAAGCGA ATAACATGTA GAAATGGACA

3541 ATGGTCAGAA CCACCAAAAT GCTTACATCC GTGTGTAATA TCCCGAGAAA TTATGGAAAA

3601 TTATAACATA GCATTAAGGT GGACAGCCAA ACAGAAGCTT TATTCGAGAA CAGGTGAATC

3661 AGTTGAATTT GTGTGTAAAC GGGGATATCG TCTTTCATCA CGTTCTCACA CATTGCGAAC

3721 AACATGTTGG GATGGGAAAC TGGAGTATCC AACTTGTGCA AAAAGA**TAGA** ATCAATCATA

3781 AAGTGCACAC CTTTATTCAG AACTTTAGTA TTAAATCAGT CTCAATTTC ATTTTTTATG

3841 TATTGTTTTA CTCCTTTTTA TTCATACGTA AAATTTTGGA TTAATTTGTG AAAATGTAAT

3901 TATAAGCTGA GACCGGTGGC TCTCTT
```

Start and stop codons in bold. The first five bases of each intron (except intron 1) are underlined. To infer the FHL-1 DNA sequence, see Fig. 30.3.

GENOMIC STRUCTURE

The positions of introns (numbered) within human *CFH* are shown and the locations within chromosome 1 of the first base of intron 1 and the last base of intron 22 are indicated. Alternative splicing gives rise to FHL1 (10a is the last of 10 introns encoding FHL1). Exon 10a encodes the last four amino acids of FHL1. Mice do not produce an FHL1 orthologue.

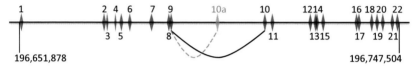

FIGURE 30.3   Diagram of genomic structure for factor H.

ACCESSION NUMBERS

Human factor H:	HGNC:HGNC:4883
	Ensembl:ENSG00000000971
	HPRD:00601
	MIM:134370
	Vega:OTTHUMG00000035607
Mouse factor H:	MGI:MGI:88385
	Ensembl:ENSMUSG00000026365
	Vega:OTTMUSG00000024648
Rat factor H:	RGD:620428
	Ensembl:ENSRNOG00000030715

DEFICIENCY

Deficiencies in plasma FH are linked to dense deposit disease (DDD)[24] and characterised by unrestricted activation of the alternative pathway of complement causing depletion of serum C3, damage to glomerular cells and deposition of complement product in the glomerular basement membrane.

POLYMORPHIC VARIANTS

Numerous single-nucleotide polymorphism and mutations have been identified.[25] Some are linked to diseases including DDD, atypical haemolytic uraemic syndrome (aHUS), and age-related macular degeneration (AMD).[19] Notable examples include (mature protein numbering) H402Y linked to increased risk of AMD,[26] V62 L thought to be protective against AMD and R1210C, a rare mutation, strongly linked to AMD and aHUS.[27]

MUTANT ANIMALS

A lineage of Norwegian Yorkshire pigs was described in which DDD-like symptoms were linked to a hereditary FH deficiency. This trait was subsequently bred out of the commercial pig population.[28]

In *cfh*[-/-] mice, created by gene targeting of embryonic stem cells, continuous uncontrolled plasma C3 cleavage occurred along with symptoms resembling DDD, such as glomerular deposits of C3 cleavage products.[29] Two-year-old *cfh*[-/-] mice had impaired visual acuity accompanied by damage to rod photoreceptors and restricted retinal blood vessels, but lacked the subretinal pigment epithelial (RPE) deposits that are a hallmark of AMD.[30] Crossing *cfh*[-/-] mice with a human FH transgenic strain (created using bacterial artificial chromosomes expressing full-length human *CFH*) ameliorated some of the kidney and retinal damage observed in the knockout. Interestingly, aged *cfh*[-/-] and *cfh*[+/-] mice fed a high-fat diet did exhibit sub-RPE deposits while *cfh*[+/-] mice also showed a significant visual function decline.[31]

Transgenic mice expressing mouse FH lacking CCPs 16–20 (FHΔ16–20) were generated by injecting a transgene into fertilised mouse eggs. These animals were intercrossed with *cfh*[-/-] mice. In the resultant *cfh*[-/-]FHΔ16–20 animals, plasma C3 was near-normal and aHUS-like, rather than DDD-like, symptoms were reported.[32]

Chimeric mice were generated by replacing CCPs 6–8 of mouse FH with human FH CCPs 6–8 (incorporating the H402Y SNP), using constructs driven by an apolipoprotein (ApoE) promoter.[33] In these mice, C3 plasma levels were maintained and AMD-reminiscent subretinal drusen-like deposits were observed. Immunohistochemistry showed a thicker sub-RPE band of C3d staining. These mice were more susceptible, upon aging, to retinal/RPE oxidative stress.

REFERENCES

1. Makou E, Herbert AP, Barlow PN. Functional anatomy of complement factor H. *Biochemistry* 2013;**52**:3949–62.
2. Ferreira VP, Pangburn MK, Cortes C. Complement control protein factor H: the good, the bad, and the inadequate. *Mol Immunol* 2010;**47**:2187–97.
3. Fenaille F, Le Mignon M, Groseil C, Ramon C, Riande S, Siret L, et al. Site-specific N-glycan characterization of human complement factor H. *Glycobiology* 2007;**17**:932–44.
4. Perkins SJ, Nan R, Li K, Khan S, Miller A. Complement Factor H-ligand interactions: self-association, multivalency and dissociation constants. *Immunobiology* 2012;**217**:281–97.
5. Soares DC, Barlow PN. Complement control protein modules in the regulators of complement activation. In: Morikis D, Lambris JD, editors. *Structural biology of the complement system.* Boca Raton: CRC Press, Taylor & Francis Group; 2005. p. 19–62.
6. DiScipio RG. Ultrastructures and interactions of complement factors H and I. *J Immunol* 1992;**149**:2592–9.
7. Aslam M, Perkins SJ. Folded-back solution structure of monomeric factor H of human complement by synchrotron X-ray and neutron scattering, analytical ultracentrifugation and constrained molecular modelling. *J Biol Chem* 2001;**309**:1117–38.

8. Herbert AP, Makou E, Chen ZA, Kerr H, Richards A, Rappsilber J, et al. Complement evasion mediated by enhancement of captured factor H: implications for protection of self-surfaces from complement. *J Immunol* 2015;**195**:4986–98.

9. Barlow PN, Steinkasserer A, Norman DG, Kieffer B, Wiles AP, Sim RB, et al. Solution structure of a pair of complement modules by nuclear magnetic resonance. *J Biol Chem* 1993;**232**:268–84.

10. Schmidt CQ, Herbert AP, Kavanagh D, Gandy C, Fenton CJ, Blaum BS, et al. A new map of glycosaminoglycan and C3b binding sites on factor H. *J Immunol* 2008;**181**:2610–9.

11. Giannakis E, Male DA, Ormsby RJ, Mold C, Jokiranta TS, Ranganathan S, et al. Multiple ligand binding sites on domain seven of human complement factor H. *Int Immunopharmacol* 2001;**1**:433–43.

12. Pangburn MK. Cutting edge: localization of the host recognition functions of complement factor H at the carboxyl-terminal: implications for hemolytic uremic syndrome. *J Immunol* 2002;**169**:4702–6.

13. Jarva H, Jokiranta TS, Hellwage J, Zipfel PF, Meri S. Regulation of complement activation by C-reactive protein: targeting the complement inhibitory activity of factor H by an interaction with short consensus repeat domains 7 and 8-11. *J Immunol* 1999;**163**:3957–62.

14. Deban L, Jarva H, Lehtinen MJ, Bottazzi B, Bastone A, Doni A, et al. Binding of the long pentraxin PTX3 to factor H: interacting domains and function in the regulation of complement activation. *J Immunol* 2008;**181**:8433–40.

15. Leffler J, Herbert AP, Norstrom E, Schmidt CQ, Barlow PN, Blom AM, et al. Annexin-II, DNA, and histones serve as factor H ligands on the surface of apoptotic cells. *J Biol Chem* 2010;**285**:3766–76.

16. Weismann D, Hartvigsen K, Lauer N, Bennett KL, Scholl HP, Charbel Issa P, et al. Complement factor H binds malondialdehyde epitopes and protects from oxidative stress. *Nature* 2011;**478**:76–81.

17. Shaw PX, Zhang L, Zhang M, Du H, Zhao L, Lee C, et al. Complement factor H genotypes impact risk of age-related macular degeneration by interaction with oxidized phospholipids. *Proc Natl Acad Sci USA* 2012;**109**:13757–62.

18. Hakobyan S, Harris CL, Tortajada A, Goicochea de Jorge E, García-Layana A, et al. Measurement of factor H variants in plasma using variant-specific monoclonal antibodies: application to assessing risk of age-related macular degeneration. *IOVS* 2008;**49**:1983–90.

19. de Cordoba SR, de Jorge EG. Translational mini-review series on complement factor H: genetics and disease associations of human complement factor H. *Clin Exp Immunol* 2008;**151**:1–13.

20. Schwaeble W, Zwirner J, Schulz TF, Linke RP, Dierich MP, Weiss EH. Human complement factor H: expression of an additional truncated gene product of 43 kDa in human liver. *Eur J Immunol* 1987;**17**:1485–9.

21. Clark SJ, Schmidt CQ, White AM, Hakobyan S, Morgan BP, Bishop PN. Identification of factor H-like protein 1 as the predominant complement regulator in Bruch's membrane: implications for age-related macular degeneration. *J Immunol* 2014;**193**:4962–70.

22. Fraczek LA, Martin BK. Transcriptional control of genes for soluble complement cascade regulatory proteins. *Mol Immunol* 2010;**48**:9–13.

23. Amadi-Obi A, Yu CR, Dambuza I, Kim SH, Marrero B, Egwuagu CE. Interleukin 27 induces the expression of complement factor H (CFH) in the retina. *PLoS One* 2012;**7**:e45801.

24. Levy M, Halbwachs-Mecarelli L, Gubler MC, Kohout G, Bensenouci A, Niaudet P, et al. H deficiency in two brothers with atypical dense intramembranous deposit disease. *Kindey Int* 1986;**30**:949–56.

25. Rodriguez E, Rallapalli PM, Osborne AJ, Perkins SJ. New functional and structural insights from updated mutational databases for complement factor H, Factor I, membrane cofactor protein and C3. *Biosci Rep* 2014;**34**. pii:e00146.

26. Hageman GS, Anderson DH, Johnson LV, Hancox LS, Taiber AJ, Hardisty LI, et al. A common haplotype in the complement regulatory gene factor H (HF1/CFH) predisposes individuals to age-related macular degeneration. *Proc Natl Acad Sci USA* 2005;**102**:7227–32.

27. Raychaudhuri S, Iartchouk O, Chin K, Tan PL, Tai AK, Ripke S, et al. A rare penetrant mutation in CFH confers high risk of age-related macular degeneration. *Nat Genet* 2011;**43**:1232–6.

28. Hogasen K, Jansen JH, Harboe M. Eradication of porcine factor H deficiency in Norway. *Vet Rec* 1997;**140**:392–5.

29. Pickering MC, Cook HT, Warren J, Bygrave AE, Moss J, Walport MJ, et al. Uncontrolled C3 activation causes membranoproliferative glomerulonephritis in mice deficient in complement factor H. *Nat Genet* 2002;**31**:424–8.

30. Coffey PJ, Gias C, McDermott CJ, Lundh P, Pickering MC, Sethi C, et al. Complement factor H deficiency in aged mice causes retinal abnormalities and visual dysfunction. *Proc Natl Acad Sci USA* 2007;**104**:16651–6.

31. Toomey CB, Kelly U, Saban DR, Bowes Rickman C. Regulation of age-related macular degeneration-like pathology by complement factor H. *Proc Natl Acad Sci USA* 2015;**112**:E3040–9.

32. Pickering MC, de Jorge EG, Martinez-Barricarte R, Recalde S, Garcia-Layana A, Rose KL, et al. Spontaneous hemolytic uremic syndrome triggered by complement factor H lacking surface recognition domains. *J Exp Med* 2007;**204**:1249–56.

33. Ufret-Vincenty RL, Aredo B, Liu X, McMahon A, Chen PW, Sun H, et al. Transgenic mice expressing variants of complement factor H develop AMD-like retinal findings. *IOVS* 2010;**51**:5878–87.

Chapter 31

Factor H-Related Proteins 1–5

Scott R. Barnum[1], Paul N. Barlow[2]

[1]*University of Alabama at Birmingham, Birmingham, AL, United States;* [2]*University of Edinburgh, Edinburgh, United Kingdom*

OTHER NAMES

Complement factor H-related proteins 1–5 (CFHR1, CFHR2, CFHR3, CFHR4, CFHR5)

PHYSICOCHEMICAL PROPERTIES

The five complement factor H-related proteins (CFHRs) are encoded by separate genes and synthesised as single-chain precursors of lengths between 243 and 578 amino acid residues.[1]

	Precursor	Leader	Mature Protein	pI Calc. (Av.)	M_r/1000	N-linked Glycosylation
CFHR1	330	18	312	6.4 (6.6)	35.7	2 (**126**, **194**)
CFHR2	243	18	225	5.6 (5.5)	25.9	1 (**126**)
CFHR3	331	18	313	6.7 (6.9)	35.5	4 (108, 185, 205, 309)
CFHR4A	578	18	560	5.8 (5.9)	63.4	6 (127, 186, 206, 374, 433 453)
CFHR4B	331	18	313	6.1 (6.1)	35.3	unknown
CFHR5	569	18	551	6.4 (6.3)	62.5	2 (126, 400)

CFHR4A and CFHR4B are splice variants.

*p*I values were calculated at the Website http://isoelectric.ovh.org/calculate.php that also provides a mean of *p*I values (shown in parentheses) calculated using multiple algorithms.[2]

The total number and locations (residue numbers) of predicted N-linked glycosylation sites are provided with the use of bold to indicate those sites for which glycosylation has been confirmed.

The Complement FactsBook. http://dx.doi.org/10.1016/B978-0-12-810420-0.00031-6

STRUCTURE

The CFHRs are composed exclusively from complement control protein modules (CCPs) (also known as SCRs or sushi domains)[3] each containing about 60 amino acid residues: five in CFHR1; four in CFHR2; five in CFHR3; nine and five in the splice variants CFHR4A and CFHR4B, respectively; and nine in CFHR5 (Fig. 31.1). Two allotypic variants of CFHR1, which differ by three residues in CCP3, are referred to as CFHR1-A (with sequence HLE) and CFHR1-B (with sequence YVQ).[4] The CCPs are joined by short linking sequences. The CFHRs may be divided into two groups according to whether they possess a shared dimerisation motif. In group I, CFHRs 1, 2 and 5 share an N-terminal pair of CCPs (with >85% sequence identity between proteins) that mediate homodimer formation.[5] Heterodimers CFHR1:CFHR2, CFHR1:CFHR5 and CFHR2:CFHR5 have also been reported.[5] Notably, the C-terminal two CCPs of CFHR1 differ by just two residues from the C-terminal two CCPs (CCPs 19 and 20) of complement factor H (CFH) that are important for the recognition of self-surfaces by CFH.[6] In group II, CFHRs 3, 4A and 4B lack this dimerisation motif. CFHR3 and CFHR4, in group II, share nearly identical pairs of C-terminal CCPs that are about 65% and 35% identical in sequence to CFH CCPs 19 and 20, respectively. Of note, the CFHR3 CCPs 1–2 are highly similar in sequence (>85%) to CFH CCPs 6–7 that (like CCPs 19–20) mediate recognition of sialic acids and glycosaminoglycans by CFH.[6]

There are no high-resolution structures of any of the CFHRs but crystal structures of the following have been deposited at the PDB: CFHR1 CCPs 1–2 (3ZD2), CFHR1 CCPs 4–5 (4MUC) and CFHR2 CCPs 3–4 (3ZD1). As expected, the structure of CFHR1 CCPs 4–5[7] is almost identical to that of CFH CCPs 19–20 (see chapter on factor H). The structure of CFHR1 CCPs 1–2 revealed a tight head-to-tail dimer in which Tyr34, Ser36 and Tyr39 are key interface residues[5]; these residues are conserved in CFHR1, CFHR2 and CFHR5.

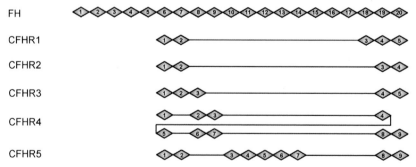

FIGURE 31.1 Protein domain/module structures of human CFHR 1–5. In each CFHR, the numbered CCP modules (that in all cases are joined by short linkers) are aligned with equivalent (based on sequence similarities) modules of the full-length CFH protein. The CFHR proteins share high sequence identity with CFH CCPs 6–9, CCPs 10–14 and the C-terminal CCPs 18–20.

FUNCTION

With respect to binding partners (reviewed in Ref. 1), all CFHRs have been reported to interact with C3b, C3d(g) and iC3b. All but CFHR4 were reported to bind to heparin (a soluble surrogate for glycosaminoglycans). All of them, bar CFHR3, have been found in particles containing high-density lipoproteins.[8] CFHRs 1, 2, 4 and 5 have been reported to bind to C-reactive protein (CRP). Various CFHRs are binding partners for several microbial proteins.

CFHR1

CFHR1 was reported to bind C5, as well as to C3b and cell surfaces; in these studies CFHR1 inhibited the C5 convertase and consequently hindered membrane attack complex formation, but it did not inhibit C3 convertase.[9] CFHR1 was also reported to bind to complement receptor-type 3.[10] Other reports[5,11] suggested that CFHR1 does not bind significantly to C5 at physiological concentrations of CFHR1. Instead these reports claimed that the principal role of CFHR1 is to compete with the complement regulator CFH for binding to at least some of the self-surface markers recognised by CFH [including C3d(g)]. This property is a consequence of the 98% sequence identity between CFHR1 CCPs 4–5 and CFH CCPs 19–20 (a key self-surface recognition domain of CFH), and also of the avidity arising from the dimerisation of CFHR1 (mediated by its CCPs 1–2) and hence the presence of two binding regions per molecule. CFHR1 lacks any CCPs with sequence similarity to those in CFH (namely, CCPs 1–4) that are responsible for disruption, via decay accelerating activity and cofactor activity for the protease factor I, of the alternative pathway C3 convertase, C3bBb. Consequently it is hypothesised that CFHR1 not only competes with but also blocks, or antagonises, the actions of CFH in specific surface contexts (as was demonstrated on guinea pig erythrocytes[5]). CFHR1, which might be present in plasma at roughly equivalent molar concentrations to CFH, may therefore *de*regulate the C3b-amplification loop of the complement system. CFHR1 may even act as a decoy,[12] being captured by bacterial proteins instead of CFH, and thereby frustrating the ability of the bacterial cells to recruit CFH for the purposes of complement evasion. Heterodimers of CFHR1 with CFHR2 and CFHR5 are likewise expected to be potential CFH antagonists although this depends upon their concentrations, both in plasma and locally to sites of infection or complement activation. Thus CFHRs 1, 2 and 5 (if not also CFHRs in group II) apparently have the potential to exhibit a surface context-dependent combinatorial range of complement deregulatory efficacies.[12]

CFHR2

CFHR2 was initially reported to inhibit alternative pathway C3 convertase activity.[13] But there is uncertainty over levels of CFHR2 in plasma (see below) and the physiological relevance of this activity has been questioned.[12] CFHR2 possesses a dimerisation domain in its CCPs 1 and 2. In addition to forming homodimers, it has been reported to form heterodimers with CFHR1 and CFHR5. Because, unlike CFHR1, CFHR2 lacks a CCP that mimics (in terms of having a high level of sequence identity) the key self-surface-recognition CCP 20 of CFH, the CFHR1:CFHR2 heterodimer is expected to be a less potent antagonist of CFH compared to the CFHR1 homodimer. It is also possible that the C-terminal CCPs of CFHR2 have evolved to recognise specific molecular patterns and thereby antagonise CFH action in specific contexts.[12]

CFHR3

Several studies reported that CFHR3, at higher concentrations than are likely to be present in plasma (see below), has factor I-cofactor activity for C3b.[14] It was also reported to act synergistically with CFH.[15] The binding of CFHR3 to C3d was reported to block the interaction of C3d with CD21[16]; thus CFHR3 could perturb the adjuvant properties of C3d. The presence in CFHR3 of CCPs 1–2 that closely resemble CCPs 6–7 of CFH implies that it could compete with CFH for binding to certain self-surface markers. Indeed it was shown that *Neisseria meningitidis* cannot distinguish between CFHR3 and CFH when the microbe attempts to hijack CFH for self-protection.[17] Doubt about competition between CFH and CFHR3 has been raised in a report that shows CFHR3 plasma levels to be about 130-fold lower in molar terms than CFH plasma levels.[18]

CFHR4

Although two splice variants exist, with five or nine CCP modules, there are few reports of how their functional properties differ. Reports of CFHR4 possessing factor I-cofactor activity towards C3b suggested synergism with CFH[15] but there exist uncertainties over plasma levels of CFHR4 and the physiological relevance of these findings. Like the other CFHRs, CFHR4 might compete with CFH for binding to similar regions of C3b and C3d(g) and, as a consequence of being relatively inefficient at inactivation of C3b, could antagonise CFH regulation of C3b amplification. But CFHR4 lacks a dimerisation domain (and hence an avidity effect). Furthermore it may not be present at sufficient levels in plasma to compete with the much more abundant CFH. Nonetheless there is evidence that CFHR4 (like CFHR5, see below) serves as a platform for C3bBb assembly following the initial binding of C3b.[19] This could explain a report of the enhanced opsonisation of dying cells in the presence of CFHR4.[20] An enhancement of opsonisation could also be attributed to the recruitment by CFHR4, when resident on necrotic cells, of the pentameric versions of CRP to which it binds.[20]

CFHR5

Higher concentrations of CFHR5 that are likely to be present in plasma were reported to inhibit the alternative pathway C3 convertase[21] in a similar way to CFH. Another view[12] is that CFHR5 homodimers (or heterodimers with CFHR1 or CFHR2) may, thanks to an avidity effect, outcompete the more abundant CFH for binding to some of its surface-associated ligands. In this manner, and because of its relatively low convertase–inhibition activity, CFHR5 antagonises CFH and promotes rather than inhibits complement activation. The degree of competition with CFH and hence deregulation likely depends on the relative levels of CFHR5 and CFH (and indeed of CFHR1 and CFHR2), as well as the nature of the surface target and its molecular context. In a further report, evidence was presented that CFHR5 supports assembly of the alternative pathway C3 convertase but also binds to monomeric CRP, to pentraxin 3 and to extracellular matrix.[22]

TISSUE DISTRIBUTION

The CFHR proteins are all plasma proteins produced primarily by hepatocytes. The CFHR genes appear not to be expressed in the retina.[23]

Some estimates of plasma concentration of CFHRs are listed below.

CFHR1 – 70–100 μg/mL[9]
CFHR2 – 50 μg/mL[13]
CFHR3 – 70–100 μg/mL[14] (but see below)
CHFR4 – 6–54 μg/mL[19]
CHFR5 – 3–6 μg/m[21]

A subsequent study[24] based on identification and quantification of all five CFHRs in plasma using mass spectrometry proposed much smaller values. A further, rigorous estimate of CFHR3 levels, using five anti-CFHR3 antibodies, yielded 0.7 μg/mL, in line with the mass spectrometric estimate.[18]

REGULATION OF EXPRESSION

There is limited evidence suggesting that plasma levels of CFHRs may be modulated under various circumstances, for example, during infections.[25] On the other hand, only small increases in CFHR3 levels were detected in a group of sepsis patients.[18] More study is important since the adjustment of CFHR levels, and in particular of their ratios to CFH levels, could allow modulation of the complement system to ensure it is tailored to meet any of a range of microbial and other hazards and challenges.

PROTEIN MODULES
CFHR1

Sequence	Domain	Exon
1–18	Leader peptide	exon 1
22–84	CCP 1	exon 2
85–142	CCP 2	exon 3
145–203	CCP 3	exon 4
206–264	CCP 4	exon 5
273–329	CCP 5	exon 6

CFHR2

Sequence	Domain	Exon
1–18	Leader peptide	exon 1
22–84	CCP 1	exon 2
85–142	CCP 2	exon 3
147–205	CCP 3	exon 4
206–268	CCP 4	exon 5

CFHR3

Sequence	Domain	Exon
1–18	Leader peptide	exon 1
22–84	CCP 1	exon 2
85–142	CCP 2	exon 3
144–205	CCP 3	exon 4
208–266	CCP 4	exon 5
267–330	CCP 5	exon 6

CFHR4

Sequence	Domain	Exon
1–18	Leader peptide	exon 1
20–85	CCP 1	exon 2
86–147	CCP 2	exon 3
148–206	CCP 3	exon 4
209–267	CCP 4	exon 5
268–332	CCP 5	exon 6
333–394	CCP 6	exon 7
395–453	CCP 7	exon 8
456–514	CCP 8	exon 9
515–578	CCP 9	exon 10

CFHR5

Sequence	Domain	Exon
1–18	Leader peptide	exon 1
23–83	CCP 1	exon 2
85–142	CCP 2	exon 3
145–203	CCP 3	exon 4
206–264	CCP 4	exon 5
267–324	CCP 5	exon 6
329–383	CCP 6	exon 7
387–444	CCP 7	exon 8
447–505	CCP 8	exon 9
507–569	CCP 9	exon 10

CHROMOSOMAL LOCATION

The five *CFHR* genes lie within the regulators of complement activation cluster at chromosome 1q32. They occupy a centromeric 360-kb segment along with the gene for CFH . The region shows evidence of large genomic duplications as discussed below.[26,27]

GENOMIC STRUCTURE

ACCESSION NUMBERS

Human

CHFR1: Ensembl – ENST00000320493
CHFR2: Ensembl – ENST00000367415
CHFR3: Ensembl – ENST00000367425
CHFR4: Ensembl – ENST00000251424
CHFR5: Ensembl – ENST00000256785

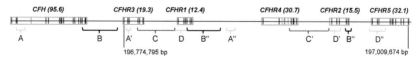

FIGURE 31.2 The exon/intron structures of the human *CFHR* gene cluster and part of the adjacent *CFH* gene. Genes are represented by *rectangles* (sizes indicated in kilobases) and approximate exon positions by *diamonds*. In CFH, the *grey diamond* represents an alternative 10th exon that codes for the C-terminus of the short, seven-CCP module, CFH-like 1 splice variant; in *CFHR4*, the four grey exons are expressed in addition to the other six in the full-length CFHR4A splice variant, but not in CFHR4B. Areas of repetitive sequence that are potential sites of homologous recombination are indicated beneath, e.g., A, A′ and A″ are regions of similar sequence. *Adapted from Jozsi M, Tortajada A, Uzonyi B, Goicoechea de Jorge E, Rodriguez de Cordoba S. Factor H-related proteins determine complement-activating surfaces. Trends Immunol June 2015;36(6):374–84. PubMed PMID: 25979655.*

DEFICIENCY

The *CFH* (coding for both CFH and CFH-like 1) and *CFHRs 1–5* gene cluster includes extensive regions of genomic duplications (see Fig. 31.2). These increase the likelihood of genomic rearrangements through gene conversion and nonallelic homologous recombination. Such events are associated, at varying confidence levels, with a list of diseases that include systemic lupus erythematosus (SLE); age-related macular degeneration (AMD); and the kidney diseases, IgA nephropathy (IgAN), C3 glomerulopathy (C3G) and atypical haemolytic uraemic syndrome (aHUS) (reviewed in Ref. 12).

The *CFH*, *CFHR3* and *CFHR1* genes lie adjacent to one another (Fig. 31.2). Several haplotypes extend over all three genes, and are linked to plasma levels of these proteins and their relative concentrations. Haplotype 3 increases the ratio of CFHR3 to CFH levels and, interestingly, decreases susceptibility to meningococcal disease.[28] At the other extreme, a nonallelic homologous recombination event involving duplicated regions lying downstream of *CFH* and of *CFHR1* leads to deletions of *CFHR1* and *CFHR3* with an allelic frequency of up to 0.55 in some populations.[29] Such deletions are protective against AMD and IgAN[30,31] consistent with the role of CFHR1 and CFHR3 as deregulators yet, strikingly, they are linked to an increased risk of aHUS and SLE[32,33]; in the case of aHUS this might be as a result of an associated occurrence of CFH autoantibodies.[34]

POLYMORPHIC VARIANTS

Genomic rearrangements that create genes encoding hybrids of CFH and CFHR1 are strongly associated with aHUS. In one example, the C-terminal CCP (CCP 5) of CFHR1 replaces the C-terminal CCP (CCP 20) of CFH. The swapped modules differ by just two residues but these might be critical for recognising sialic acids.[35] The resulting protein is functionally defective as it does not efficiently recognise and bind self-surfaces. Two aHUS-associated CFH:CFHR3 hybrids are similarly functionally deficient.[36,37] An inverse version of dimeric CFHR1, in which CCP 5 is replaced by CCP 20 of CFH, is probably a pathogenic gain-of-function variant of CFHR1 that outcompetes CFH for binding to self-surface markers and therefore overly deregulates complement.[38] A CFHR5:CFHR2 hybrid (CCPs 1–2 of CFHR5 followed by the four CCPs of CFHR2) was reported to cause a case of C3G.[39] A further set of genomic rearrangements leads to duplication of the dimerisation motif at the N-termini of CFHRs 1, 2 and 5. These rearrangements induce multimerisation of the gene products that may (like the CFHR1:CFH hybrid) compete too strongly with CFH for binding to its targets, and thereby impair the regulatory efficacy of CFH. These rearrangements are also strongly linked to C3G.[11,40]

MUTANT ANIMALS

No mutant CFHR animals have been reported.

REFERENCES

1. Skerka C, Chen Q, Fremeaux-Bacchi V, Roumenina LT. Complement factor H related proteins (CFHRs). *Mol Immunol* December 15, 2013;**56**(3):170–80. PubMed PMID: 23830046.
2. Kozlowski LP. IPC – isoelectric point calculator. *Biol Direct* October 21, 2016;**11**(1):55. PubMed PMID: 27769290.
3. Soares DC, Barlow PN. Complement control protein modules in the regulators of complement activation. In: Morikis D, Lambris JD, editors. *Structural biology of the complement system*. Boca Raton: CRC Press, Taylor & Francis Group; 2005. p. 19–62.
4. Abarrategui-Garrido C, Martinez-Barricarte R, Lopez-Trascasa M, de Cordoba SR, Sanchez-Corral P. Characterization of complement factor H-related (CFHR) proteins in plasma reveals novel genetic variations of CFHR1 associated with atypical hemolytic uremic syndrome. *Blood* November 5, 2009;**114**(19):4261–71. PubMed PMID: 19745068.
5. Goicoechea de Jorge E, Caesar JJ, Malik TH, Patel M, Colledge M, Johnson S, et al. Dimerization of complement factor H-related proteins modulates complement activation in vivo. *Proc Natl Acad Sci USA* March 19, 2013;**110**(12):4685–90. PubMed PMID: 23487775. Pubmed Central PMCID: 3606973.
6. Schmidt CQ, Herbert AP, Kavanagh D, Gandy C, Fenton CJ, Blaum BS, et al. A new map of glycosaminoglycan and C3b binding sites on factor H. *J Immunol* August 15, 2008;**181**(4):2610–9. PubMed PMID: 18684951.
7. Bhattacharjee A, Reuter S, Trojnar E, Kolodziejczyk R, Seeberger H, Hyvarinen S, et al. The major autoantibody epitope on factor H in atypical hemolytic uremic syndrome is structurally different from its homologous site in factor H-related protein 1, supporting a novel model for induction of autoimmunity in this disease. *J Biol Chem* April 10, 2015;**290**(15):9500–10. PubMed PMID: 25659429. Pubmed Central PMCID: 4392255.
8. Park CT, Wright SD. Plasma lipopolysaccharide-binding protein is found associated with a particle containing apolipoprotein A-I, phospholipid, and factor H-related proteins. *J Biol Chem* July 26, 1996;**271**(30):18054–60. PubMed PMID: 8663389.
9. Heinen S, Hartmann A, Lauer N, Wiehl U, Dahse HM, Schirmer S, et al. Factor H-related protein 1 (CFHR-1) inhibits complement C5 convertase activity and terminal complex formation. *Blood* September 17, 2009;**114**(12):2439–47. PubMed PMID: 19528535. Epub 2009/06/17. eng.
10. Losse J, Zipfel PF, Jozsi M. Factor H and factor H-related protein 1 bind to human neutrophils via complement receptor 3, mediate attachment to *Candida albicans*, and enhance neutrophil antimicrobial activity. *J Immunol* January 15, 2010;**184**(2):912–21. PubMed PMID: 20008295. Epub 2009/12/17. eng.
11. Tortajada A, Yebenes H, Abarrategui-Garrido C, Anter J, Garcia-Fernandez JM, Martinez-Barricarte R, et al. C3 glomerulopathy-associated CFHR1 mutation alters FHR oligomerization and complement regulation. *J Clin Invest* June 2013;**123**(6):2434–46. PubMed PMID: 23728178. Pubmed Central PMCID: 3668852.
12. Jozsi M, Tortajada A, Uzonyi B, Goicoechea de Jorge E, Rodriguez de Cordoba S. Factor H-related proteins determine complement-activating surfaces. *Trends Immunol* June 2015;**36**(6):374–84. PubMed PMID: 25979655.

13. Eberhardt HU, Buhlmann D, Hortschansky P, Chen Q, Bohm S, Kemper MJ, et al. Human factor H-related protein 2 (CFHR2) regulates complement activation. *PLoS One* 2013;**8**(11):e78617. PubMed PMID: 24260121. Pubmed Central PMCID: 3832495.

14. Fritsche LG, Lauer N, Hartmann A, Stippa S, Keilhauer CN, Oppermann M, et al. An imbalance of human complement regulatory proteins CFHR1, CFHR3 and factor H influences risk for age-related macular degeneration (AMD). *Hum Mol Genet* December 1, 2010;**19**(23):4694–704. PubMed PMID: 20843825.

15. Hellwage J, Jokiranta TS, Koistinen V, Vaarala O, Meri S, Zipfel PF. Functional properties of complement factor H-related proteins FHR-3 and FHR-4: binding to the C3d region of C3b and differential regulation by heparin. *FEBS Lett* December 3, 1999;**462**(3):345–52. PubMed PMID: 10622723.

16. Buhlmann D, Eberhardt HU, Medyukhina A, Prodinger WM, Figge MT, Zipfel PF, et al. FHR3 blocks C3d-mediated coactivation of human B cells. *J Immunol* July 15, 2016;**197**(2):620–9. PubMed PMID: 27279373.

17. Caesar JJ, Lavender H, Ward PN, Exley RM, Eaton J, Chittock E, et al. Competition between antagonistic complement factors for a single protein on *N. meningitidis* rules disease susceptibility. *eLife* December 23, 2014;**3**. PubMed PMID: 25534642. Pubmed Central PMCID: 4273445.

18. Pouw RB, Brouwer MC, Geissler J, van Herpen LV, Zeerleder SS, Wuillemin WA, et al. Complement factor H-related protein 3 serum levels are low compared to factor H and mainly determined by gene copy number variation in CFHR3. *PLoS One* 2016;**11**(3):e0152164. PubMed PMID: 27007437. Pubmed Central PMCID: 4805260.

19. Hebecker M, Jozsi M. Factor H-related protein 4 activates complement by serving as a platform for the assembly of alternative pathway C3 convertase via its interaction with C3b protein. *J Biol Chem* June 1, 2012;**287**(23):19528–36. PubMed PMID: 22518841. Pubmed Central PMCID: 3365989.

20. Mihlan M, Hebecker M, Dahse HM, Halbich S, Huber-Lang M, Dahse R, et al. Human complement factor H-related protein 4 binds and recruits native pentameric C-reactive protein to necrotic cells. *Mol Immunol* January 2009;**46**(3):335–44. PubMed PMID: 19084272. Epub 2008/12/17. eng.

21. McRae JL, Duthy TG, Griggs KM, Ormsby RJ, Cowan PJ, Cromer BA, et al. Human factor H-related protein 5 has cofactor activity, inhibits C3 convertase activity, binds heparin and C-reactive protein, and associates with lipoprotein. *J Immunol* May 15, 2005;**174**(10):6250–6. PubMed PMID: 15879123.

22. Csincsi AI, Kopp A, Zoldi M, Banlaki Z, Uzonyi B, Hebecker M, et al. Factor H-related protein 5 interacts with pentraxin 3 and the extracellular matrix and modulates complement activation. *J Immunol* May 15, 2015;**194**(10):4963–73. PubMed PMID: 25855355. Pubmed Central PMCID: 4416742.

23. Hughes AE, Bridgett S, Meng W, Li M, Curcio CA, Stambolian D, et al. Sequence and expression of complement factor H gene cluster variants and their roles in age-related macular degeneration risk. *Invest Ophthalmol Vis Sci* May 1, 2016;**57**(6):2763–9. PubMed PMID: 27196323. Pubmed Central PMCID: 4884056.

24. Zhang P, Zhu M, Geng-Spyropoulos M, Shardell M, Gonzalez-Freire M, Gudnason V, et al. A novel, multiplexed targeted mass spectrometry assay for quantification of complement factor H (CFH) variants and CFH-related proteins 1–5 in human plasma. *Proteomics* March 27, 2017. PubMed PMID: 27647805.

25. Narkio-Makela M, Hellwage J, Tahkokallio O, Meri S. Complement-regulator factor H and related proteins in otitis media with effusion. *Clin Immunol* July 2001;**100**(1):118–26. PubMed PMID: 11414752.

26. Diaz-Guillen MA, Rodriguez de Cordoba S, Heine-Suner D. A radiation hybrid map of complement factor H and factor H-related genes. *Immunogenetics* June 1999;**49**(6):549–52. PubMed PMID: 10380701.

27. Male DA, Ormsby RJ, Ranganathan S, Giannakis E, Gordon DL. Complement factor H: sequence analysis of 221 kb of human genomic DNA containing the entire fH, fHR-1 and fHR-3 genes. *Mol Immunol* January–Febraury 2000;**37**(1–2):41–52. PubMed PMID: 10781834.

28. Davila S, Wright VJ, Khor CC, Sim KS, Binder A, Breunis WB, et al. Genome-wide association study identifies variants in the CFH region associated with host susceptibility to meningococcal disease. *Nat Genet* September 2010;**42**(9):772–6. PubMed PMID: 20694013.

29. Holmes LV, Strain L, Staniforth SJ, Moore I, Marchbank K, Kavanagh D, et al. Determining the population frequency of the CFHR3/CFHR1 deletion at 1q32. *PLoS One* 2013;**8**(4):e60352. PubMed PMID: 23613724. Pubmed Central PMCID: 3629053.

30. Hughes AE, Orr N, Esfandiary H, Diaz-Torres M, Goodship T, Chakravarthy U. A common CFH haplotype, with deletion of CFHR1 and CFHR3, is associated with lower risk of age-related macular degeneration. *Nat Genet* October 2006;**38**(10):1173–7. PubMed PMID: 16998489.

31. Gharavi AG, Kiryluk K, Choi M, Li Y, Hou P, Xie J, et al. Genome-wide association study identifies susceptibility loci for IgA nephropathy. *Nat Genet* March 13, 2011;**43**(4):321–7. PubMed PMID: 21399633. Pubmed Central PMCID: 3412515.

32. Zhao J, Wu H, Khosravi M, Cui H, Qian X, Kelly JA, et al. Association of genetic variants in complement factor H and factor H-related genes with systemic lupus erythematosus susceptibility. *PLoS Genet* May 2011;**7**(5):e1002079. PubMed PMID: 21637784. Pubmed Central PMCID: 3102741.

33. Zipfel PF, Edey M, Heinen S, Jozsi M, Richter H, Misselwitz J, et al. Deletion of complement factor H-related genes CFHR1 and CFHR3 is associated with atypical hemolytic uremic syndrome. *PLoS Genet* March 16, 2007;**3**(3):e41. PubMed PMID: 17367211. Pubmed Central PMCID: 1828695. Epub 2007/03/21. eng.

34. Moore I, Strain L, Pappworth I, Kavanagh D, Barlow PN, Herbert AP, et al. Association of factor H autoantibodies with deletions of CFHR1, CFHR3, CFHR4, and with mutations in CFH, CFI, CD46, and C3 in patients with atypical hemolytic uremic syndrome. *Blood* January 14, 2010;**115**(2):379–87. PubMed PMID: 19861685. Pubmed Central PMCID: 2829859. Epub 2009/10/29. eng.

35. Blaum BS, Hannan JP, Herbert AP, Kavanagh D, Uhrin D, Stehle T. Structural basis for sialic acid-mediated self-recognition by complement factor H.. *Nat Chem Biol* November 24, 2014;**11**:77–82. PubMed PMID: 25402769.

36. Francis NJ, McNicholas B, Awan A, Waldron M, Reddan D, Sadlier D, et al. A novel hybrid CFH/CFHR3 gene generated by a microhomology-mediated deletion in familial atypical hemolytic uremic syndrome. *Blood* January 12, 2012;**119**(2):591–601. PubMed PMID: 22058112.

37. Challis RC, Araujo GS, Wong EK, Anderson HE, Awan A, Dorman AM, et al. A de novo deletion in the regulators of complement activation cluster producing a hybrid complement factor H/Complement factor H-Related 3 gene in atypical hemolytic uremic syndrome. *J Am Soc Nephrol* June 2016;**27**(6):1617–24. PubMed PMID: 26490391. Pubmed Central PMCID: 4884102.

38. Valoti E, Alberti M, Tortajada A, Garcia-Fernandez J, Gastoldi S, Besso L, et al. A novel atypical hemolytic uremic syndrome-associated hybrid CFHR1/CFH gene encoding a fusion protein that antagonizes factor H-dependent complement regulation. *J Am Soc Nephrol* January 2015;**26**(1):209–19. PubMed PMID: 24904082. Pubmed Central PMCID: 4279739.

39. Xiao X, Ghossein C, Tortajada A, Zhang Y, Meyer N, Jones M, et al. Familial C3 glomerulonephritis caused by a novel CFHR5-CFHR2 fusion gene. *Mol Immunol* September 2016;**77**:89–96. PubMed PMID: 27490940.

40. Medjeral-Thomas NR, O'Shaughnessy MM, O'Regan JA, Traynor C, Flanagan M, Wong L, et al. C3 glomerulopathy: clinicopathologic features and predictors of outcome. *Clin J Am Soc Nephrol* January 2014;**9**(1):46–53. PubMed PMID: 24178974. Pubmed Central PMCID: 3878702.

Chapter 32

Clusterin

Valeria Naponelli, Saverio Bettuzzi

University of Parma, Parma, Italy

OTHER NAMES

Clusterin, CLU; sulphated glycoprotein-2, SGP-2; testosterone-repressed prostate message 2, TRPM-2; apolipoprotein J, ApoJ; complement cytolysis inhibitor, CLI; complement-associated protein SP-40, SP-40; NA1/NA2: ionising radiation (IR)-induced protein-8, XIP-8; Ku70-binding protein 1; aging-associated gene 4 protein, AAG4.

PHYSICOCHEMICAL PROPERTIES

The biosynthesis of human secreted clusterin (CLU) is canonical. As other secreted proteins, translation of CLU mRNA results in a preprotein composed of 449 amino acids (psCLU). The first 22 amino acids constitute a signal sequence for cotranslational translocation into the endoplasmic reticulum. sCLU, detectable as a 60–65 kDa band by SDS-PAGE, is transported into the Golgi and then heavily glycosylated. A proteolysis process removes the leader signal polypeptide and cleaves the preprotein into an α-chain and a β-chain. The two chains are linked in an antiparallel fashion through five disulphide bonds.[1] The mature form of CLU (sCLU) is a secreted heterodimeric glycoprotein of 75–80 kDa (theoretical molecular weight and pI: 52.5; 5.88).

Subunit		α-chain	β-chain
Amino acid number		228–449	23–227
M_r (K)	Predicted	26.7	24.4
	Observed	35–40	35–40
N-linked glycosylation sites		4	3
	Position	291, 317, 354, 374	86, 103, 145
Interchain disulphide bonds (5)		313, 305, 302, 295, 285	102, 113, 116, 121, 129

STRUCTURE

Although CLU has been a matter of research for the past 30 years, its structural properties are still not completely understood. Obtaining sCLU X-ray structure,

The Complement FactsBook. http://dx.doi.org/10.1016/B978-0-12-810420-0.00032-8

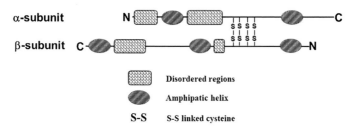

FIGURE 32.1   Location of ordered and disordered regions in the clusterin protein.

or reliable NMR spectra, is still a challenge nowadays probably because of its aggregating nature and specific protein features. In fact, sCLU forms oligomers: in addition it can interact with distinct ligands forming complexes with different molecular mass and diameter.[2,3]

The primary structure of sCLU is highly conserved between different species: about 30% of the mass of the mature protein is made of carbohydrates.[2] The secondary structure of CLU has been predicted through circular dichroism and infrared spectroscopy analyses.[3]

Structural predictions performed in rat, human, and bovine CLU showed highly conserved amphipathic α-helices regions with hydrophobic and hydrophilic features.[4] Intrinsically disordered regions have been also predicted, such as coil-like and molten globule-like regions.[5] These domains are predominantly located at the N- and C- termini of the α- and β-chains. Predicted ordered regions have been found around the conserved cysteine residues implicated in the formation of the five disulphide bonds. A short disordered region comprises the posttranslational cleavage site generating α- and β-chains. The structure of CLU is supposed to be highly flexible, due to the presence of amphipatic α-helix ordered structures and disordered regions (Fig. 32.1). This feature may account for its strong binding activity of unfolded proteins, and/or other putative partners. It has been suggested to act as an intra- and extracellular chaperone.[6]

FUNCTION

Although CLU was identified more than 30 years ago, an understanding of its biological functions is still elusive. It was first described in 1983 as a secreted glycoprotein in ram rete testis fluid and able to elicit aggregation of a variety of cells in vitro.[2] Later, the human homologue of the rat protein was described as an integral component of the soluble C5b-9 complement complex, which is assembled in plasma on activation of the complement cascade.[3,7] sCLU tissue distribution was similar to that of the terminal complement components. A further study showed that CLU inhibits the activity of the complement system.[8] sCLU binds to the terminal C complexes and prevents their insertion into cell membranes. The resulting complexes are soluble and unable to induce complement lysis.[9] sCLU specifically binds to C7, the β-subunit of C8 and C9.[10] The conformational changes occurring during the formation of MAC expose the interaction sites for sCLU that binds to a structural motif common to C7, C8 and C9β inhibiting the correct complex assembly.[11,12]

It has been speculated that one of sCLU in vivo function is to control terminal complement-mediated damage, preventing uncontrolled membrane attack complex (MAC) activity.[13]

sCLU has been found on MAC bound to circulating immune complexes (CIC) in systemic lupus erythematosus patients.[14] The authors speculated that the presence of sCLU in MAC-CIC might be necessary to keep these complexes in soluble phase. The presence of CLU has been associated to a protective function from complement-mediated injury[15] or from potentially damaging agents of the extracellular environment.[11,16]

The association between sCLU and complement proteins has been reported in different human pathological conditions suggesting a protective effect of sCLU under stress conditions.[17–21]

However, the inhibitory function of sCLU on complement activation was not supported by the experimental evidences by Hochgrebe et al.[22]: their data suggested that, under physiological conditions, sCLU is not a relevant regulator of complement activation.

Data showed that a protein produced by *Pseudomonas aeruginosa* binds sCLU, blocking C5b-9 deposition and protecting the bacterium from complement damage.[23] The evasion strategy adopted by *P. aeruginosa* supports an important role played by sCLU in the modulation of complement activity. The interaction between sCLU and other pathogenic microbes has been demonstrated.[24–27]

Sabatte et al. reported that only seminal plasma CLU bears highly fucosilated glycans necessary to bind DC-SIGN, a C-type lectin receptor selectively expressed on dendritic cells (DCs).[28] They also demonstrated that seminal plasma CLU interacts with stress-damaged proteins targeting them to DCs via DC-SIGN.[29] Therefore, seminal plasma CLU may contribute to mediate female tolerance to seminal antigens.

A protective effect of sCLU has been also described independently from complement activation in kidney and heart.[11,30]

There is evidence of sCLU in regulating chronic inflammation and autoimmunity.[31–34] CLU has been found to interact with immunoglobulins by a multivalent mechanism, binding to both the Fc and Fab regions.[35]

TISSUE DISTRIBUTION

sCLU is ubiquitously expressed in almost all mammalian tissues and has been found in all human body fluids and analysed.[36] sCLU circulates in human plasma at a concentration of 150–540 μg/mL. It is about 10 times higher in human seminal plasma.[37,38]

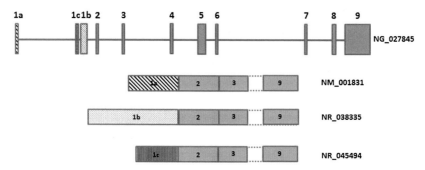

FIGURE 32.2 Structure of clusterin gene and transcription products.

In plasma CLU is a soluble protein, or a component of a lipid-poor subclass of high-density lipoproteins of 70–200 kDa. Proteomic analyses revealed that CLU is also bound to LDLs and VLDLs.[39] DCs express high level of CLU: CLU expression increased by more than 30-fold during DC maturation.[11]

Under stress conditions CLU may escape the canonical pathway of secretory proteins, and the different forms that are originated from altered biogenesis may localise in various intracellular compartments such as cytoplasm, mitochondria, microsomes, nuclei and cell membranes.[40]

REGULATION OF EXPRESSION

CLU is considered an acute-phase protein[11] because it is secreted after endotoxin (LPS), tumour necrosis factor and interleukin IL-1 stimulation in liver and in serum.[20] The main transcriptional product of *CLU* gene is transcript variant which produces a secreted protein of 449 aa (sCLU). Other transcript isoforms have been described. All variants are transcribed as pre-mRNAs and comprise nine exons and eight introns. Exon 1 sequences (called 1a, 1b and 1c) are unique to each of the three mRNAs while they share the same sequence from exon 2 to exon 9.[41] The CLU promoter (called P1) has been sequenced and found to be rather unique. Bonacini et al. discovered and characterised a novel promoter, called P2, directly responsible for sCLU mRNA variant 2 transcription (Fig. 32.2).[40]

PROTEIN SEQUENCE

```
MMKTLLLFVG  LLLTWESGQV  LGDQTVSDNE  LQEMSNQGSK  YVNKEIQNAV   50
NGVKQIKTLI  EKTNEERKTL  LSNLEEAKKK  KEDALNETRE  SETKLKELPG  100
VCNETMMALW  EECKPCLKQT  CMKFYARVCR  SGSGLVGRQL  EEFLNQSSPF  150
YFWMNGDRID  SLLENDRQQT  HMLDVMQDHF  SRASSIIDEL  FQDRFFTREP  200
QDTYHYLPFS  LPHRRPHFFF  PKSRIVR//SLM  PFSPYEPLNF  HAMFQPFLEM  250
IHEAQQAMDI  HFHSPAFQHP  PTEFIREGDD  DRTVCREIRH  NSTGCLRMKD  300
QCDKCREILS  VDCSTNNPSQ  AKLRRELDES  LQVAERLTRK  YNELLKSYQW  350
KMLNTSSLLE  QLNEQFNWVS  RLANLTQGED  QYYLRVTTVA  SHTSDSDVPS  400
GVTEVVVKLF  DSDPITVTVP  VEVSRKNPKF  METVAEKALQ  EYRKKHREE
```

The leader sequence is underlined. The // symbol denotes break between the β- and α-subunits. The *N*-linked glycosylation sites are indicated in bold (**N**).

PROTEIN MODULES

1–22	Signal peptide	exon 2
67–82	Heparin-binding domain	exon 3
104–131	Cysteine domain	exon 4
176–189	Amphipathic helix	exon 5
212–218	Heparin-binding domain	exon 5
243–259	Amphipathic helix	exon 5
285–313	Cysteine domain	exon 6/7
423–430	Heparin-binding domain	exon 8
424–440	Amphipathic helix	exon 8
442–449	Heparin-binding domain	exon 8/9

CHROMOSOMAL LOCATION

The gene coding for sCLU is well conserved through evolution and located in chromosome 8 in humans, in the region 8p21-p12. The *CLU* gene is organised into nine exons and eight introns, spanning a region of about 18 kbp. The *CLU* gene structure is well conserved in different species.

Mouse *CLU* gene is located on chromosome 14.[42]

cDNA SEQUENCE[43]

```
GCTTTCCGCG GCATTCTTTG GGCGTGAGTC ATGCAGGTTT GCAGCCAGCC CCAAAGGGGG    60
TGTGTGCGCG AGCAGAGCGC TATAAATACG GCGCCTCCCA GTGCCCACAA CGCGGCGTCG   120
CCAGGAGGAG CGCGCGGGCA CAGGGTGCCG CTGACCGAGG CGTGCAAAGA CTCCAGAATT   180
GGAGGCATGA TGAAGACTCT GCTGCTGTTT GTGGGGCTGC TGCTGACCTG GGAGAGTGGG   240
CAGGTCCTGG GGGACCAGAC GGTCTCAGAC AATGAGCTCC AGGAAATGTC CAATCAGGGA   300
AGTAAGTACG TCAATAAGGA AATTCAAAAT GCTGTCAACG GGGTGAAACA GATAAAGACT   360
CTCATAGAAA AAACAAACGA AGAGCGCAAG ACACTGCTCA GCAACCTAGA AGAAGCCAAG   420
AAGAAGAAAG AGGATGCCCT AAATGAGACC AGGGAATCAG AGACAAAGCT GAAGGAGCTC   480
CCAGGAGTGT GCAATGAGAC CATGATGGCC CTCTGGGAAG AGTGTAAGCC CTGCCTGAAA   540
CAGACCTGCA TGAAGTTCTA CGCACGCGTC TGCAGAAGTG GCTCAGGCCT GGTTGGCCGC   600
CAGCTTGAGG AGTTCCTGAA CCAGAGCTCG CCCTTCTACT TCTGGATGAA TGGTGACCGC   660
ATCGACTCCC TGCTGGAGAA CGACCGGCAG CAGACGCACA TGCTGGATGT CATGCAGGAC   720
CACTTCAGCC GCGCGTCCAG CATCATAGAC GAGCTCTTCC AGGACAGGTT CTTCACCCGG   780
GAGCCCCAGG ATACCTACCA CTACCTGCCC TTCAGCCTGC CCCACCGGAG GCCTCACTTC   840
TTCTTTCCCA AGTCCCGCAT CGTCCGCAGC TTGATGCCCT TCTCTCCGTA CGAGCCCCTG   900
AACTTCCACG CCATGTTCCA GCCCTTCCTT GAGATGATAC ACGAGGCTCA GCAGGCCATG   960
GACATCCACT TCCATAGCCC GGCCTTCCAG CACCCGCCAA CAGAATTCAT ACGAGAAGGC  1020
GACGATGACC GGACTGTGTG CCGGGAGATC CGCCACAACT CCACGGGCTG CCTGCGGATG  1080
AAGGACCAGT GTGACAAGTG CCGGGAGATC TTGTCTGTGG ACTGTTCCAC CAACAACCCC  1140
TCCCAGGCTA AGCTGCGGCG GGAGCTCGAC GAATCCCTCC AGGTCGCTGA GAGGTTGACC  1200
AGGAAATACA ACGAGCTGCT AAAGTCCTAC CAGTGGAAGA TGCTCAACAC CTCCTCCTTG  1260
CTGGAGCAGC TGAACGAGCA GTTTAACTGG GTGTCCCGGC TGGCAAACCT CACGCAAGGC  1320
GAAGACCAGT ACTATCTGCG GGTCACCACG GTGGCTTCCC ACACTTCTGA CTCGGACGTT  1380
CCTTCCGGTG TCACTGAGGT GGTCGTGAAG CTCTTTGACT CTGATCCCAT CACTGTGACG  1440
GTCCCTGTAG AAGTCTCCAG GAAGAACCCT AAATTTATGG AGACCGTGGC GGAGAAAGCG  1500
CTGCAGGAAT ACCGCAAAAA GCACCGGGAG GAGTGAGATG TGGATGTTGC TTTTGCACCT  1560
ACGGGGGCAT CTGAGTCCAG CTCCCCCCAA GATGAGCTGC AGCCCCCCAG AGAGAGCTCT  1620
GCACGTCACC AAGTAACCAG GCCCCAGCCT CCAGGCCCCC AACTCCGCCC AGCCTCTCCC  1680
CGCTCTGGAT CCTGCACTCT AACACTCGAC TCTGCTGCTC ATGGGAAGAA CAGAATTGCT  1740
CCTGCATGCA ACTAATTCAA TAAAACTGTC TTGTGAGCTG ATCGCTTGGA GGGTCCTCTT  1800
TTTATGTTGA GTTGCTGCTT CCCGGCATGC CTTCATTTTG CTATGGGGGG CAGGCAGGGG  1860
GGATGGAAAA TAAGTAGAAA CAAAAAAGCA GTGGCTAAGA TGGTATAGGG ACTGTCATAC  1920
CAGTGAAGAA TAAAAGGGTG AAGAATAAAA GGGATATGAT GACAAGGTTG ATCCACTTCA  1980
```

cDNA SEQUENCE—*Continued*

```
AGAATTGCTT GCTTTCAGGA AGAGAGATGT GTTTCAACAA GCCAACTAAA ATATATTGCT  2040
GCAAATGGAA GCTTTTCTGT TCTATTATAA AACTGTCGAT GTATTCTGAC CAAGGTGCGA  2100
CAATCTCCTA AAGGAATACA CTGAAAGTTA AGGAGAAGAA TCAGTAAGTG TAAGGTGTAC  2160
TTGGTATTAT AATGCATAAT TGATGTTTTC GTTATGAAAA CATTTGGTGC CCAGAAGTCC  2220
AAATTATCAG TTTTATTTGT AAGAGCTATT GCTTTTGCAG CGGTTTTATT TGTAAAAGCT  2280
GTTGATTTCG AGTTGTAAGA GCTCAGCATC CCAGGGGCAT CTTCTTGACT GTGGCATTTC  2340
CTGTCCACCG CCGGTTTATA TGATCTTCAT ACCTTTCCCT GGACCACAGG CGTTTCTCGG  2044
CTTTTAGTCT GAACCATAGC TGGGCTGCAG TACCCTACGC TGCCAGCAGG TGGCCATGAC  2460
TACCCGTGGT ACCAATCTCA GTCTTAAAGC TCAGGCTTTT CGTTCATTAA CATTCTCTGA  2520
TAGAATTCTG GTCATCAGAT GTACTGCAAT GGAACAAAAC TCATCTGGCT GCATCCCAGG  2580
TGTGTAGCAA AGTCCACATG TAAATTTATA GCTTAGAATA TTCTTAAGTC ACTGTCCCTT  2640
GTCTCTCTTT GAAGTTATAA ACAACAAACT TAAAGCTTAG CTTATGTCCA AGGTAAGTAT  2700
TTTAGCATGG CTGTCAAGGA AATTCAGAGT AAAGTCAGTG TGATTCACTT AATGATATAC  2760
ATTAATTAGA ATTATGGGGT CAGAGGTATT TGCTTAAGTG ATCATAATTG TAAAGTATAT  2821
GTCACATTGT CACATTAATG TCACACTGTT TCAAAGTTA AAAAAAAAAA AAAAAA
```

Human CLU cDNA. The methionine initiation codon (<u>ATG</u>), the termination codon (<u>TGA</u>) and the polyadenylation signal (<u>ATTAAA</u>)[43] are indicated. The first five nucleotides in each exon are underlined to indicate intron–exon boundaries.

GENOMIC STRUCTURE

The *CLU* gene is organised into nine exons and eight introns, spanning a region of about 18 kbp (Fig. 32.3). The CLU gene structure is well conserved in different species.

ACCESSION NUMBERS

Human	RefSeq Gene NG_027845.1, gene
	GenBank NM_001831.2, transcript variant 1
	GenBank NR_038335, transcript variant 2
	GenBank NR_045494, transcript variant 3
	UniProt ID P10909, protein
Mouse	RefSeq NP_038520.2, protein
	RefSeq NM_006518567, mRNA
	RefSeq XP_006518567.1, isoform X2
	RefSeq XM_006518504.2, transcript variant X2

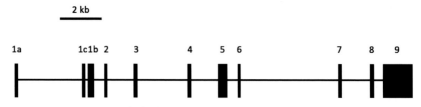

FIGURE 32.3 Human clusterin gene.

DEFICIENCY

There are no known CLU deficiencies in the population.

POLYMORPHIC VARIANTS

Several single nucleotide polymorphisms (SNPs) have been identified in CLU gene in coding regions, UTRs and introns, but a strong association to a pathologic condition is yet to be found.[44,45] 6316delT, an insertion (I)/deletion (D) polymorphism, has been studied in a Japanese population.[46] D/D genotype was shown to associate with significantly higher total cholesterol levels and low-LDL levels in hypertensive females.

MUTANT ANIMALS

CLU knockout mice have been generated by McLaughlin[47] and are available in The Jackson Laboratory. A targeting vector was constructed in which exons 1 and 2 of the endogenous gene were replaced with a mouse phosphoglycerol kinase promoter driven hypoxanthine phosphoribosyl transferase gene.

Bettuzzi et al. generated TRAMP/CLU KO mice to investigate the role of CLU in prostate cancer progression.[48] CLU knockout (CluKO) mice were crossed with TRAMP mice. TRAMP/CLU KO mice (available in Prof. Bettuzzi's lab) are strongly prostate cancer prone and highly metastatic.

REFERENCES

1. Choi-Miura NH, Takahashi Y, Nakano Y, Tobe T, Tomita M. Identification of the disulfide bonds in human plasma protein SP-40,40 (apolipoprotein-J). *J Biochem* 1992;**112**:557–61.
2. Blaschuk O, Burdzy K, Fritz IB. Purification and characterization of a cell-aggregating factor (clusterin), the major glycoprotein in ram rete testis fluid. *J Biol Chem* 1983;**258**:7714–20.
3. Rohne P, Prochnow H, Koch Brandt C. The CLU-files: disentanglement of a mystery. *Biomol Concepts* 2016;**7**:1–15.
4. Bailey RW, Dunker AK, Brown CJ, Garner EC, Griswold MD. Clusterin, a binding protein with a molten globule-like region. *Biochemistry* 2001;**40**:11828–40.
5. Dunker AK, Lawson JD, Brown CJ, Williams RM, Romero P, Oh JS, et al. Intrinsically disordered protein. *J Mol Graph Model* 2001;**19**:26–59.
6. Wyatt A, Yerbury J, Poon S, Dabbs R, Wilson M. Chapter 6: the chaperone action of clusterin and its putative role in quality control of extracellular protein folding. *Adv Cancer Res* 2009;**104**:89–114.
7. Murphy BF, Kirszbaum L, Walker ID, d'Apice AJ. SP-40,40, a newly identified normal human serum protein found in the SC5b-9 complex of complement and in the immune deposits in glomerulonephritis. *J Clin Invest* 1988;**81**:1858–64.
8. Jenne DE, Tschopp J. Molecular structure and functional characterization of a human complement cytolysis inhibitor found in blood and seminal plasma: identity to sulfated glycoprotein 2, a constituent of rat testis fluid. *Proc Natl Acad Sci USA* 1989;**86**:7123–7.
9. Wilson MR, Roeth PJ, Easterbrook-Smith SB. Clusterin enhances the formation of insoluble immune complexes. *Biochem Biophys Res Commun* 1991;**177**:985–90.

10. Tschopp J, Chonn A, Hertig S, French LE. Clusterin, the human apolipoprotein and complement inhibitor, binds to complement C7, C8 beta, and the b domain of C9. *J Immunol* 1993;**151**:2159–65.

11. Falgarone G, Chiocchia G. Chapter 8: clusterin: a multifacet protein at the crossroad of inflammation and autoimmunity. *Adv Cancer Res* 2009;**104**:139–70.

12. McDonald JF, Nelsestuen GL. Potent inhibition of terminal complement assembly by clusterin: characterization of its impact on C9 polymerization. *Biochemistry* 1997;**36**:7464–73.

13. Choi NH, Mazda T, Tomita M. A serum protein sp40,40 modulates the formation of membrane attack complex of complement on erythrocytes. *Mol Immunol* 1989;**26**:835–40.

14. Chauhan AK, Moore TL. Presence of plasma complement regulatory proteins clusterin (Apo J) and vitronectin (S40) on circulating immune complexes (CIC). *Clin Exp Immunol* 2006;**145**:398–406.

15. Cuida M, Legler DW, Eidsheim M, Jonsson R. Complement regulatory proteins in the salivary glands and saliva of Sjogren's syndrome patients and healthy subjects. *Clin Exp Rheumatol* 1997;**15**:615–23.

16. Aronow BJ, Lund SD, Brown TL, Harmony JA, Witte DP. Apolipoprotein J expression at fluid-tissue interfaces: potential role in barrier cytoprotection. *Proc Natl Acad Sci USA* 1993;**90**:725–9.

17. D'Cruz OJ, Wild RA. Evaluation of endometrial tissue specific complement activation in women with endometriosis. *Fertil Steril* 1992;**57**:787–95.

18. Balslev E, Thomsen HK, Danielsen L, Sheller J, Garred P. The terminal complement complex is generated in chronic leg ulcers in the absence of protectin (cd59). *APMIS* 1999;**107**:997–1004.

19. Bykov I, Junnikkala S, Pekna M, Lindros KO, Meri S. Effect of chronic ethanol consumption on the expression of complement components and acute-phase proteins in liver. *Clin Immunol* 2007;**124**:213–20.

20. Hardardottir I, Kunitake ST, Moser AH, Doerrler WT, Rapp JH, Grunfeld C, et al. Endotoxin and cytokines increase hepatic messenger rna levels and serum concentrations of apolipoprotein J (clusterin) in Syrian hamsters. *J Clin Invest* 1994;**94**:1304–9.

21. Chiang KC, Goto S, Chen CL, Lin CL, Lin YC, Pan TL, et al. Clusterin may be involved in rat liver allograft tolerance. *Transpl Immunol* 2000;**8**:95–9.

22. Hochgrebe TT, Humphreys D, Wilson MR, Easterbrook-Smith SB. A reexamination of the role of clusterin as a complement regulator. *Exp Cell Res* 1999;**249**:13–21.

23. Hallstrom T, Uhde M, Singh B, Skerka C, Riesbeck K, Zipfel PF. *Pseudomonas aeruginosa* uses dihydrolipoamide dehydrogenase (lpd) to bind to the human terminal pathway regulators vitronectin and clusterin to inhibit terminal pathway complement attack. *PLoS One* 2015;**10**:e0137630.

24. Akesson P, Sjoholm AG, Bjorck L. Protein sic, a novel extracellular protein of *Streptococcus pyogenes* interfering with complement function. *J Biol Chem* 1996;**271**:1081–8.

25. Kurosu T, Chaichana P, Yamate M, Anantapreecha S, Ikuta K. Secreted complement regulatory protein clusterin interacts with dengue virus nonstructural protein 1. *Biochem Biophys Res Commun* 2007;**362**:1051–6.

26. Li DQ, Lundberg F, Ljungh A. Binding of vitronectin and clusterin by coagulase-negative staphylococci interfering with complement function. *J Mater Sci Mater Med* 2001;**12**:979–82.

27. Partridge SR, Baker MS, Walker MJ, Wilson MR. Clusterin, a putative complement regulator, binds to the cell surface of *Staphylococcus aureus* clinical isolates. *Infect Immun* 1996;**64**:4324–9.

28. Sabatte J, Faigle W, Ceballos A, Morelle W, Rodriguez Rodrigues C, Remes Lenicov F, et al. Semen clusterin is a novel DC-SIGN ligand. *J Immunol* 2011;**187**:5299–309.

29. Merlotti A, Dantas E, Remes Lenicov F, Ceballos A, Jancic C, Varese A, et al. Fucosylated clusterin in semen promotes the uptake of stress-damaged proteins by dendritic cells via DC-SIGN. *Hum Reprod* 2015;**30**:1545–56.

30. Van Dijk A, Vermond RA, Krijnen PA, Juffermans LJ, Hahn NE, Makker SP, et al. Intravenous clusterin administration reduces myocardial infarct size in rats. *Eur J Clin Invest* 2010;**40**:893–902.

31. Sonn CH, Yu YB, Hong YJ, Shim YJ, Bluestone JA, Min BH, et al. Clusterin synergizes with IL-2 for the expansion and IFN-gamma production of natural killer cells. *J Leukoc Biol* 2010;**88**:955–63.

32. Afanasyeva MA, Britanova LV, Korneev KV, Mitkin NA, Kuchmiy AA, Kuprash DV. Clusterin is a potential lymphotoxin beta receptor target that is upregulated and accumulates in germinal centers of mouse spleen during immune response. *PLoS One* 2014;**9**:e98349.

33. Hong GH, Kwon HS, Moon KA, Park SY, Park S, Lee KY, et al. Clusterin modulates allergic airway inflammation by attenuating CCL20-mediated dendritic cell recruitment. *J Immunol* 2016;**196**:2021–30.

34. Shim YJ, Kang BH, Choi BK, Park IS, Min BH. Clusterin induces the secretion of TNF-alpha and the chemotactic migration of macrophages. *Biochem Biophys Res Commun* 2012;**422**:200–5.

35. Wilson MR, Easterbrook-Smith SB. Clusterin binds by a multivalent mechanism to the Fc and Fab regions of IgG. *Biochim Biophys Acta* 1992;**1159**:319–26.

36. Bettuzzi S. Conclusions and perspectives. *Adv Cancer Res* 2009;**105**:133–50.

37. Guo W, Ma X, Xue C, Luo JF, Zhu XL, Xiang JQ, et al. Serum clusterin as a tumor marker and prognostic factor for patients with esophageal cancer. *Dis Markers* 2014. http://dx.doi.org/10.1155/2014/168960.

38. Trougakos IP, Gonos ES. Clusterin/apolipoprotein J in human aging and cancer. *Int J Biochem Cell Biol* 2002;**34**:1430–48.

39. Baralla A, Sotgiu E, Deiana M, Pasella S, Pinna S, Mannu A, et al. Plasma clusterin and lipid profile: a link with aging and cardiovascular diseases in a population with a consistent number of centenarians. *PLoS One* 2015;**10**:e0128029.

40. Bonacini M, Coletta M, Ramazzina I, Naponelli V, Modernelli A, Davalli P, et al. Distinct promoters, subjected to epigenetic regulation, drive the expression of two clusterin mRNAs in prostate cancer cells. *Biochim Biophys Acta* 2015;**1849**:44–54.

41. Rizzi F, Bettuzzi S. The clusterin paradigm in prostate and breast carcinogenesis. *Endocr Relat Cancer* 2010;**17**:R1–17.

42. Birkenmeier EH, Letts VA, Frankel WN, Magenheimer BS, Calvet JP. Sulfated glycoprotein-2 (Sgp-2) maps to mouse chromosome 14. *Mamm Genome* 1993;**4**:131–2.

43. Kirszbaum L, Sharpe JA, Murphy B, d'Apice AJ, Classon B, Hudson P, et al. Molecular cloning and characterization of the novel, human complement-associated protein, sp-40,40: a link between the complement and reproductive systems. *EMBO J* 1989;**8**:711–8.

44. Savas S, Schmidt S, Jarjanazi H, Ozcelik H. Functional nssnps from carcinogenesis-related genes expressed in breast tissue: potential breast cancer risk alleles and their distribution across human populations. *Hum Genomics* 2006;**2**:287–96.

45. Bettens K, Brouwers N, Engelborghs S, Lambert JC, Rogaeva E, Vandenberghe R, et al. Both common variations and rare non-synonymous substitutions and small insertion/deletions in CLU are associated with increased Alzheimer risk. *Mol Neurodegener* 2012;**7**:3.

46. Miwa Y, Takiuchi S, Kamide K, Yoshii M, Horio T, Tanaka C, et al. Insertion/deletion polymorphism in clusterin gene influences serum lipid levels and carotid intima-media thickness in hypertensive Japanese females. *Biochem Biophys Res Commun* 2005;**331**:1587–93.

47. McLaughlin L, Zhu G, Mistry M, Ley-Ebert C, Stuart WD, Florio CJ, et al. Apolipoprotein J/clusterin limits the severity of murine autoimmune myocarditis. *J Clin Invest* 2000;**106**:1105–1113.

48. Bettuzzi S, Davalli P, Davoli S, Chayka O, Rizzi F, Belloni L, et al. Genetic inactivation of ApoJ/clusterin: effects on prostate tumourigenesis and metastatic spread. *Oncogene* 2009;**28**:4344–52.

Chapter 33

Vitronectin

Yu-Ching Su, Kristian Riesbeck

Lund University, Malmö, Sweden

OTHER NAMES

Serum spreading factor, S protein (site-specific protein), plasminogen activator inhibitor-1 (PAI-1) binding protein and epibolin.

PHYSICOCHEMICAL PROPERTIES

Vitronectin (Vn) is synthesised as a precursor polypeptide consisting of 478 amino acids (aa).[1,2] The precursor protein sequence consists of a 19 aa signal peptide and 459 aa mature protein.[3–9]

Mature protein:	
pI	4.75–5.25
M_r (K)	
Predicted	52.5
Observed	~75
N-linked glycosylation sites	3 (86, 169, 242)
Phosphorylation sites	3 (Thr-50, Thr-57, Ser-362)
Sulphation sites	2 (Tyr-56, Tyr-59)

STRUCTURE

After the signal peptide, the first 43 aa on the Vn molecule is identical to the somatomedin-B (SMB) consensus sequence[10] (Fig. 33.1). This region serves also as a binding domain for plasminogen activator inhibitor-1 (PAI-1) and urokinase-type plasminogen activator receptor (uPAR).[11] This is followed by a triplet residue Arg-Gly-Asp (RGD) (aa residues 45–47) that functions as an anchoring site for cell integrin receptors (αvβ1, αvβ3, αvβ5, αvβ6, αvβ8 and αIIbβ3).[12–17] Downstream of the RGD sequence is a stretch of highly acidic amino acids (residues 53–64) that has binding affinity for the thrombin–antithrombin III complex (TAT) and collagen.[18,19] This polyanionic acidic

The Complement FactsBook. http://dx.doi.org/10.1016/B978-0-12-810420-0.00033-X
Copyright © 2018 Elsevier Ltd. All rights reserved.

FIGURE 33.1 Schematic representation of protein domains of the vitronectin molecule.

region is also important for neutralising the C-terminal polycationic basic region (residue 348–379) to stabilise the folded structure of either monomeric Vn or the intermolecule assembly of the Vn multimeric complex. Vn has three heparin-binding sites, namely HBD-1, HBD-2 and HBD-3 that are located at residues 82–137, 175–219 and 346–361, respectively.[20–22] In addition to heparin, the HBD is also responsible for the binding capacity of Vn to other glycosaminoglycans such as dextran blue, fucoidan or cellular heparan sulphate proteoglycans. The highly basic HBD-3 overlaps with domains involved in the binding of Vn to plasminogen (residues 332–348), PAI-1 (residues 348–370, secondary binding site), collagen type I and osteonectin.[18,23–25] Vn harbours four hemopexin repeats (Hpx; heme-binding protein in plasma) distributed along the central region (Hpx-1, Hpx-2 and Hpx-3 at residues 142–285) and C-terminal edge (Hpx-4 at residues 406–453).[26,27] As a complement regulator, more than half of the Vn molecule (residues 51–310) accommodates interaction sites for complement proteins C5b-C7 and C9.[28,29]

Vn circulates in two forms in human blood: (1) a 75 kDa-full length single-chain fragment and (2) a disulphide-linked dimer consisting of a two-chain fragment (65 and 10 kDa). The two-chain fragment is a product of the 75-kDa mature protein cleaved by an unknown protease between Arg-379 and Ala-380. The proteolytic products are linked by a single disulphide bridge at residues Cys-274 and Cys-453 of N-terminal of 65 kDa and C-terminal 10 kDa fragments, respectively.[9]

Vn has a high degree of conformational flexible structure. In circulating blood, Vn exists as a heterogeneous mixture consisting of different forms that serve different biological roles, either as monomeric or multimeric Vn. Notably, the native monomeric Vn (also known as the folded form) is mainly found in human plasma.[30,31] Vn isolated from serum, platelets releasates, extracellular matrices (ECM) and upon interaction with cell receptors (integrins) is mostly in the multimeric conformation (unfolded or active form). The active form of total Vn in serum is higher (7%) than in human plasma (2%).[30] The multimeric Vn can be generated via exposure to heparin, chaotropes (urea), detergents, low pH (<6.0) and heat (56°C), as well as through complex formation under physiological conditions with PAI-1, thrombin–antithrombin (TAT) and complement C5b-C9.[31–35] The molecule undergoes conformation changes into an unfolded or denatured structure, and further self-associates into a large complex with the surrounding unfolded Vn molecules to form multimers. The intermolecule binding occurs via disulphide linkage and noncovalent complementary charge interactions between the N-terminal acidic region (residues 53-64) of one Vn molecule with the C-terminal basic heparin-binding domain (residues

348-379) of another neighbouring Vn molecule. The event of Vn multimeri-sation is concentration dependent in that it requires a critical number of free molecules. Formation of 3 to 16-mers can generate multimeric Vn that has a molecular weight ranging from 200 to 1200 kD. When observed in electron microscopy, monomeric Vn is visualised as a 6–8 nm wide and 11 nm 'peanut' shape whereas the multimeric form appears as a globular structure with a diameter of 15–28 nm.[31,36]

FUNCTION

Key biological functions of Vn, mainly in its activated form, include (1) regulation of the innate immune system, (2) maintenance of vascular hae-mostasis (thrombosis and fibrinolysis), and, finally, (3) promotion of cell adhesion and migration in tissue repair and regeneration. To protect host cells from innocent bystander cell lysis (self-killing), Vn functions as a regulator at the terminal lytic step of the (1) complement pathway and (2) perforin-mediated cell lysis.[28,29] In the terminal complement pathway, Vn interacts with the premembrane attack complex (pre-MAC) of C5b-7 that in turn occu-pies the metastable membrane binding site of the complex. This eventually hinders the insertion of the C5b-7 into the cell membrane and thus precludes the downstream completion with C9 to form the MAC lytic pore (C5b-9). Another regulatory mechanism is via interaction of free (nonmembrane inserted) C5b-7, C5b-8 and C5b-9 with Vn into water-soluble complexes (sC5b-7 etc.) that are haemolytically nonactive. The sC5b-7 will also further attract free C8 and C9 to form the sC5b-8 and sC5b-9, respectively.[28,37,38] These nonlytic complexes are eventually scavenged by Vn and clusterin upon removal from the host. In perforin-mediated cell lysis, free polyanionic antimicrobial peptides degranulated from cytolytic T-lymphocytes are cap-tured by Vn through its polycationic HBD-3 site. This causes the Vn-perforin complex to be sequestered and stored within provisional extracellular matrix (ECM), where Vn is interacting via its collagen- and glycosaminoglycans/heparin-binding domains. This consequently traps and inhibits the lytic activ-ity of free perforin.[29] The regulatory role of Vn in controlling/preventing a self-antigen attack by autologous complement and perforin is important yet limited to anatomical locations that are rich in Vn such as the vessel wall.

DEGRADATION PATHWAY

The C-terminal basic aa cluster of Vn is susceptible to proteolytic degradation by the proteases thrombin, elastase and plasmin at wound healing site. This reduces its ability to stabilise PAI-1 and consequently promotes further activa-tion of plasminogen. The Vn degradation pathway regulates the role of Vn by switching between antifibrinolytic (bind PAI and stabilise its inhibitory role) and profibrinolytic proteins.

TISSUE DISTRIBUTION

Vn is mainly synthesised in the liver as a single-chain 75 kDa polypeptide precursor and is released to the circulation and interstitial space or ECM through receptor-mediated trans-, endo- or exocytosis.[31,39–41] Vn is also produced by extrahepatic tissues but less compared to hepatocytes. In addition to the liver as the major organ of Vn biosynthesis, appreciable amounts of Vn deposits or/and its mRNA are also detected in various normal tissues and organs. This includes brain, male genital tract, lung and bronchoalveolar lavage fluid (BAL), smooth and skeletal muscles, photoreceptors of inner retina, heart, renal tissue in kidney, dermal elastic fibre in skin, uterus, thymus, ovarian surface and vascular wall.[12,42–47] Vn is also found in seminal plasma, urine, amniotic fluid and cerebrospinal fluid.[42,43,48] Vn is present in normal human plasma at a concentration between 200 and 400 µg/mL (constitutes 0.2%–0.5% of total plasma proteins), mainly as a monomeric form and is predominantly derived from the liver.[30,49,50] Vn deposited within platelet (mainly as multimeric aggregates) is the second main circulating pool of Vn and accounts for 0.8%–1% of the total circulating Vn.[31] Tissues experiencing stress or trauma had upregulated levels of Vn mRNA as a proinflammatory response with increased Vn accumulated in the extravascular space of the traumatised tissues.[12] In addition, Vn deposition is also found colocalised with the immune complex C5b-C9 in diseased tissues associated with chronic inflammatory disease such as rheumatoid arthritis, chronic fibrotic diseases (lung, liver and kidney fibrosis), elastotic skin lesions, glomerulonephritis, Alzheimer disease (in the plaques), atherosclerosis and degenerative central nervous system disorder (drusen).[12,51–53] Increased levels of Vn at 10-fold higher concentrations is present in BAL from patients with interstitial lung diseases.[47,54]

REGULATION OF EXPRESSION

It has been suggested that Vn is an acute-phase protein since its expression in liver is upregulated during inflammation or acute injury. Increased plasma level of about twofold was conserved up to 5 days following surgery.[55] Studies with animal models and cell lines revealed that the expression of Vn gene is induced upon acute-phase reactions and chronic inflammation and is primarily mediated by interleukin-6.[51,55]

HUMAN PROTEIN SEQUENCE[1]

```
MAPLRPLLIL ALLAWVALAD QESCKGRCTE GFNVDKKCQC DELCSYYQSC   50
CTDYTAECKP QVTRGDVFTM PEDEYTVYDD GEEKNNATVH EQVGGPSLTS  100
DLQAQSKGNP EQTPVLKPEE EAPAPEVGAS KPEGIDSRPE TLHPGRPQPP  150
AEEELCSGKP FDAFTDLKNG SLFAFRGQYC YELDEKAVRP GYPKLIRDVW  200
GIEGPIDAAF TRINCQGKTY LFKGSQYWRF EDGVLDPDYP RNISDGFDGI  250
PDNVDAALAL PAHSYSGRER VYFFKGKQYW EYQFQHQPSQ EECEGSSLSA  300
VFEHFAMMQR DSWEDIFELL FWGRTSAGTR QPQFISRDWH GVPGQVDAAM  350
AGRIYISGMA PRPSLAKKQR FRHRNRKGYR SQRGHSRGRN QNSRRPSRAT  400
WLSLFSSEES NLGANNYDDY RMDWLVPATC EPIQSVFFFS GDKYYRVNLR  450
TRRVDTVDPP YPRSIAQYWL GCPAPGHL                          478
```

The signal peptide is underlined.
Residues with bold font indicate glycosylation sites.

PROTEIN MODULES[1,51]

1–19	Leader sequence	exon 1
20–63	SMB	exon 2
64–156	Hpx-1	exons 3/4
158–202	Hpx-1	exons 3/4
203–250	HBD-2, Hpx-2	exons 4/5
251–305	Hpx-3	exons 5/6
306–418	Plg, HBD-3	exons 6/7
419–478	Hpx-4	exons 7/8

CHROMOSOMAL LOCATION[51]

Human: 17q11.2
Mouse: Chromosome 11, 46.74 cM
Rat: 11q25

cDNA SEQUENCE[51]

```
GAGCAAACAG AGCAGCAGAA AAGGCAGTTC CTCTTCTCCA GTGCCCTCCT TCCCTGTCTC    60
TGCCTCTCCC TCCCTTCCTC AGGCATCAGA GCGGAGACTT CAGGGAGACC AGAGCCCAGC   120
TTGCCAGGCA CTGAGCTAGA AGCCCTGCCA TGGCACCCCT GAGACCCCTT CTCATACTGG   180
CCCTGCTGGC ATGGGTTGCT CTGGCTGACC AAGAGTCATG CAAGGGCCGC TGCACTGAGG   240
GCTTCAACGT GGACAAGAAG TGCCAGTGTG ACGAGCTCTG CTCTTACTAC CAGAGCTGCT   300
GCACAGACTA TACGGCTGAG TGCAAGCCCC AAGTGACTCG CGGGGATGTG TTCACTATGC   360
CGGAGGATGA GTACACGGTC TATGACGATG GCGAGGAGAA AAACAATGCC ACTGTCCATG   420
AACAGGTGGG GGGCCCCTCC CTGACCTCTG ACCTCCAGGC CCAGTCCAAA GGGAATCCTG   480
AGCAGACACC TGTTCTGAAA CCTGAGGAAG AGGCCCCTGC GCCTGAGGTG GGCGCCTCTA   540
AGCCTGAGGG GATAGACTCA AGGCCTGAGA CCCTTCATCC AGGGAGACCT CAGCCCCCAG   600
CAGAGGAGGA GCTGTGCAGT GGGAAGCCCT TCGACGCCTT CACCGACCTC AAGAACGGTT   660
CCCTCTTTGC CTTCCGAGGG CAGTACTGCT ATGAACTGGA CGAAAAGGCA GTGAGGCCTG   720
GGTACCCCAA GCTCATCCGA GATGTCTGGG GCATCGAGGG CCCCATCGAT GCCGCCTTCA   780
CCCGCATCAA CTGTCAGGGG AAGACCTACC TCTTCAAGGG TAGTCAGTAC TGGCGCTTTG   840
AGGATGGTGT CCTGGACCCT GATTACCCCC GAAATATCTC TGACGGCTTC GATGGCATCC   900
CGGACAACGT GGATGCAGCC TTGGCCCTCC CTGCCCATAG CTACAGTGGC CGGGAGCGGG   960
TCTACTTCTT CAAGGGGAAA CAGTACTGGG AGTACCAGTT CCAGCACCAG CCCAGTCAGG  1020
AGGAGTGTGA AGGCAGCTCC CTGTCGGCTG TGTTTGAACA CTTTGCCATG ATGCAGCGGG  1080
ACAGCTGGGA GGACATCTTC GAGCTTCTCT TCTGGGGCAG AACCTCTGCT GGTACCAGAC  1140
AGCCCCAGTT CATTAGCCGG GACTGGCACG GTGTGCCAGG GCAAGTGGAC GCAGCCATGG  1200
CTGGCCGCAT CTACATCTCA GGCATGGCAC CCCGCCCCTC CTTGGCCAAG AAACAAAGGT  1260
TTAGGCATCG CAACCGCAAA GGCTACCGTT CACAACGAGG CCACAGCCGT GGCCGCAACC  1320
AGAACTCCCG CCGGCCATCC CGCGCCACGT GGCTGTCCTT GTTCTCCAGT GAGGAGAGCA  1380
ACTTGGGAGC CAACAACTAT GATGACTACA GGATGGACTG GCTTGTGCCT GCCACCTGTG  1440
AACCCATCCA GAGTGTCTTC TTCTTCTCTG GAGACAAGTA CTACCGAGTC AATCTTCGCA  1500
CACGGCGAGT GGACACTGTG GACCCTCCCT ACCCACGCTC CATCGCTCAG TACTGGCTGG  1560
GCTGCCCAGC TCCTGGCCAT CTGTAGGAGT CAGAGCCCAC ATGGCCGGGC CCTCTGTAGC  1620
TCCCTCCTCC CATCTCCTTC CCCCAGCCCA ATAAAGGTCC CTTAGCCCCG AGTTTAAA    1678
```

The first five nucleotides in each exon are underlined to indicate the intron–exon boundaries. The methionine translation initiation codon (ATG) and the stop codon (TAG) are indicated as bold text, whereas the putative polyadenylation signal (AATAA) is shown as shaded nucleotides.

FIGURE 33.2   Diagram indicates the relative position of each exon (numbered as 1–8) on the vitronectin gene. The scale bar represents 500 bp of nucleotide sequence.

GENOMIC STRUCTURE[51]

The human Vn gene spans 3076 bp on chromosome 17 from nucleotide 28,367,277 and end at 28,370,352. The Vn gene consists of eight exons and seven introns (Fig. 33.2). Exon sizes vary between 64 and 345 bp, whereas introns have size between 93–461 bp.

ACCESSION NUMBERS

Human: NM_000638
Mouse: NM_011707

DEFICIENCY

Genetic deficiency of Vn has not been reported in humans; however, patients with liver failure have low plasma level of Vn.[56]

POLYMORPHIC VARIANTS

Polymorphism in the Vn protein sequence has been reported at three amino acid residues, namely A103S, R249Q and T381M. Thr and Met at the position 381 affects the distribution of a single 75 kDa or dimer-form 65/10 kDa Vn. Presence of Thr rather than Met increases the susceptibility of Vn to cleavage by an unknown protease located between Arg-379 and Ala-380.[9]

MUTANT ANIMALS

Mice with complete Vn deficiency demonstrated normal development, fertility and survival,[57] suggesting the glycoprotein to be dispensable in organ development and that its roles can be replaced by other ECM proteins. However, dermal wound healing was slightly delayed with increased wound fibrinolysis and decreased microvascular angiogenesis in these mice suggesting a role for Vn in tissue injury and repair.[58] Unstable thrombi were observed in these mice, suggesting Vn is important in thrombus stability.[59] Wild-type vascular smooth muscles migrated slower in naïve mice as compared to Vn-deficient mice.[44]

REFERENCES

1. Schvartz I, Seger D, Shaltiel S. Vitronectin. *Int J Biochem Cell Biol* 1999;**31**(5).539–44.
2. Preissner KT, May AE, Wohn KD, Germer M, Kanse SM. Molecular crosstalk between adhesion receptors and proteolytic cascades in vascular remodelling. *Thromb Haemost* 1997;**78**(1):88–95.
3. Ogawa H, Yoneda A, Seno N, Hayashi M, Ishizuka I, Hase S, et al. Structures of the N-linked oligosaccharides on human plasma vitronectin. *Eur J Biochem* 1995;**230**(3):994–1000.
4. Shaltiel S, Schvartz I, Korc-Grodzicki B, Kreizman T. Evidence for an extra-cellular function for protein kinase A. *Mol Cell Biochem* 1993;**127–128**:283–91.
5. Seger D, Gechtman Z, Shaltiel S. Phosphorylation of vitronectin by casein kinase II. Identification of the sites and their promotion of cell adhesion and spreading. *J Biol Chem* 1998;**273**(38):24805–13.
6. Korc-Grodzicki B, Tauber-Finkelstein M, Chain D, Shaltiel S. Vitronectin is phosphorylated by a cAMP-dependent protein kinase released by activation of human platelets with thrombin. *Biochem Biophys Res Commun* 1988;**157**(3):1131–8.
7. Hwang H, Lee JY, Lee HK, Park GW, Jeong HK, Moon MH, et al. In-depth analysis of site-specific N-glycosylation in vitronectin from human plasma by tandem mass spectrometry with immunoprecipitation. *Anal Bioanal Chem* 2014;**406**(30):7999–8011.
8. Clerc F, Reiding KR, Jansen BC, Kammeijer GS, Bondt A, Wuhrer M. Human plasma protein N-glycosylation. *Glycoconj J* 2016;**33**(3):309–43.
9. Tollefsen DM, Weigel CJ, Kabeer MH. The presence of methionine or threonine at position 381 in vitronectin is correlated with proteolytic cleavage at arginine 379. *J Biol Chem* 1990;**265**(17):9778–81.
10. Zhou A. Functional structure of the somatomedin B domain of vitronectin. *Protein Sci* 2007;**16**(7):1502–8.
11. Deng G, Royle G, Wang S, Crain K, Loskutoff DJ. Structural and functional analysis of the plasminogen activator inhibitor-1 binding motif in the somatomedin B domain of vitronectin. *J Biol Chem* 1996;**271**(22):12716–23.
12. Preissner KT, Seiffert D. Role of vitronectin and its receptors in haemostasis and vascular remodeling. *Thromb Res* 1998;**89**(1):1–21.
13. Felding-Habermann B, Cheresh DA. Vitronectin and its receptors. *Curr Opin Cell Biol* 1993;**5**(5):864–8.
14. Seiffert D, Smith JW. The cell adhesion domain in plasma vitronectin is cryptic. *J Biol Chem* 1997;**272**(21):13705–10.
15. Cherny RC, Honan MA, Thiagarajan P. Site-directed mutagenesis of the arginine-glycine-aspartic acid in vitronectin abolishes cell adhesion. *J Biol Chem* 1993;**268**(13):9725–9.
16. Chillakuri CR, Jones C, Mardon HJ. Heparin binding domain in vitronectin is required for oligomerization and thus enhances integrin mediated cell adhesion and spreading. *FEBS Lett* 2010;**584**(15):3287–91.
17. Choi Y, Kim E, Lee Y, Han MH, Kang IC. Site-specific inhibition of integrin alpha v beta 3-vitronectin association by a ser-asp-val sequence through an Arg-Gly-Asp-binding site of the integrin. *Proteomics* 2010;**10**(1):72–80.
18. Gebb C, Hayman EG, Engvall E, Ruoslahti E. Interaction of vitronectin with collagen. *J Biol Chem* 1986;**261**(35):16698–703.
19. Gechtman Z, Belleli A, Lechpammer S, Shaltiel S. The cluster of basic amino acids in vitronectin contributes to its binding of plasminogen activator inhibitor-1: evidence from thrombin-, elastase- and plasmin-cleaved vitronectins and anti-peptide antibodies. *Biochem J* 1997;**325**(Pt 2):339–49.

20. Gibson AD, Lamerdin JA, Zhuang P, Baburaj K, Serpersu EH, Peterson CB. Orientation of heparin-binding sites in native vitronectin. Analyses of ligand binding to the primary glycosaminoglycan-binding site indicate that putative secondary sites are not functional. *J Biol Chem* 1999;**274**(10):6432–42.

21. Liang OD, Rosenblatt S, Chhatwal GS, Preissner KT. Identification of novel heparin-binding domains of vitronectin. *FEBS Lett* 1997;**407**(2):169–72.

22. Kost C, Stuber W, Ehrlich HJ, Pannekoek H, Preissner KT. Mapping of binding sites for heparin, plasminogen activator inhibitor-1, and plasminogen to vitronectin's heparin-binding region reveals a novel vitronectin-dependent feedback mechanism for the control of plasmin formation. *J Biol Chem* 1992;**267**(17):12098–105.

23. Schar CR, Blouse GE, Minor KH, Peterson CB. A deletion mutant of vitronectin lacking the somatomedin B domain exhibits residual plasminogen activator inhibitor-1-binding activity. *J Biol Chem* 2008;**283**(16):10297–309.

24. Schar CR, Jensen JK, Christensen A, Blouse GE, Andreasen PA, Peterson CB. Characterization of a site on PAI-1 that binds to vitronectin outside of the somatomedin B domain. *J Biol Chem* 2008;**283**(42):28487–96.

25. Blouse GE, Dupont DM, Schar CR, Jensen JK, Minor KH, Anagli JY, et al. Interactions of plasminogen activator inhibitor-1 with vitronectin involve an extensive binding surface and induce mutual conformational rearrangements. *Biochemistry* 2009;**48**(8):1723–35.

26. Tolosano E, Altruda F. Hemopexin: structure, function, and regulation. *DNA Cell Biol* 2002;**21**(4):297–306.

27. Piccard H, Van den Steen PE, Opdenakker G. Hemopexin domains as multifunctional liganding modules in matrix metalloproteinases and other proteins. *J Leukoc Biol* 2007;**81**(4):870–92.

28. Sheehan M, Morris CA, Pussell BA, Charlesworth JA. Complement inhibition by human vitronectin involves non-heparin binding domains. *Clin Exp Immunol* 1995;**101**(1):136–41.

29. Tschopp J, Masson D, Schafer S, Peitsch M, Preissner KT. The heparin binding domain of S-protein/vitronectin binds to complement components C7, C8, and C9 and perforin from cytolytic T-cells and inhibits their lytic activities. *Biochemistry* 1988;**27**(11):4103–9.

30. Izumi M, Yamada KM, Hayashi M. Vitronectin exists in two structurally and functionally distinct forms in human plasma. *Biochim Biophys Acta* 1989;**990**(2):101–8.

31. Stockmann A, Hess S, Declerck P, Timpl R, Preissner KT. Multimeric vitronectin. Identification and characterization of conformation-dependent self-association of the adhesive protein. *J Biol Chem* 1993;**268**(30):22874–82.

32. Bittorf SV, Williams EC, Mosher DF. Alteration of vitronectin. Characterization of changes induced by treatment with urea. *J Biol Chem* 1993;**268**(33):24838–46.

33. Tomasini BR, Mosher DF. Conformational states of vitronectin: preferential expression of an antigenic epitope when vitronectin is covalently and noncovalently complexed with thrombin-antithrombin III or treated with urea. *Blood* 1988;**72**(3):903–12.

34. Tomasini BR, Owen MC, Fenton 2nd JW, Mosher DF. Conformational lability of vitronectin: induction of an antigenic change by alpha-thrombin-serpin complexes and by proteolytically modified thrombin. *Biochemistry* 1989;**28**(19):7617–23.

35. Seiffert D. Evidence that conformational changes upon the transition of the native to the modified form of vitronectin are not limited to the heparin binding domain. *FEBS Lett* 1995;**368**(1):155–9.

36. Lynn GW, Heller WT, Mayasundari A, Minor KH, Peterson CB. A model for the three-dimensional structure of human plasma vitronectin from small-angle scattering measurements. *Biochemistry* 2005;**44**(2):565–74.

37. Milis L, Morris CA, Sheehan MC, Charlesworth JA, Pussell BA. Vitronectin-mediated inhibition of complement: evidence for different binding sites for C5b-7 and C9. *Clin Exp Immunol* 1993;**92**(1):114–9.

38. Podack ER, Preissner KT, Muller-Eberhard HJ. Inhibition of C9 polymerization within the SC5b-9 complex of complement by S-protein. *Acta Pathol Microbiol Immunol Scand Suppl* 1984;**284**:89–96.

39. Kobayashi J, Yamada S, Kawasaki H. Distribution of vitronectin in plasma and liver tissue: relationship to chronic liver disease. *Hepatology* 1994;**20**(6):1412–7.

40. Kemkes-Matthes B, Preissner KT, Langenscheidt F, Matthes KJ, Muller-Berghaus G. S protein/vitronectin in chronic liver diseases: correlations with serum cholinesterase, coagulation factor X and complement component C3. *Eur J Haematol* 1987;**39**(2):161–5.

41. Seiffert D, Schleef RR. Two functionally distinct pools of vitronectin (Vn) in the blood circulation: identification of a heparin-binding competent population of Vn within platelet alphagranules. *Blood* 1996;**88**(2):552–60.

42. Seiffert D, Iruela-Arispe ML, Sage EH, Loskutoff DJ. Distribution of vitronectin mRNA during murine development. *Dev Dyn* 1995;**203**(1):71–9.

43. Seiffert D, Crain K, Wagner NV, Loskutoff DJ. Vitronectin gene expression in vivo. Evidence for extrahepatic synthesis and acute phase regulation. *J Biol Chem* 1994;**269**(31):19836–42.

44. Garg N, Goyal N, Strawn TL, Wu J, Mann KM, Lawrence DA, et al. Plasminogen activator inhibitor-1 and vitronectin expression level and stoichiometry regulate vascular smooth muscle cell migration through physiological collagen matrices. *J Thromb Haemost* 2010;**8**(8):1847–54.

45. Anderson DH, Hageman GS, Mullins RF, Neitz M, Neitz J, Ozaki S, et al. Vitronectin gene expression in the adult human retina. *Invest Ophthalmol Vis Sci* 1999;**40**(13):3305–15.

46. Dahlback K, Wulf HC, Dahlback B. Vitronectin in mouse skin: immunohistochemical demonstration of its association with cutaneous amyloid. *J Invest Dermatol* 1993;**100**(2):166–70.

47. Salazar-Pelaez LM, Abraham T, Herrera AM, Correa MA, Ortega JE, Pare PD, et al. Vitronectin expression in the airways of subjects with asthma and chronic obstructive pulmonary disease. *PLoS One* 2015;**10**(3):e0119717.

48. Bronson RA, Preissner KT. Measurement of vitronectin content of human spermatozoa and vitronectin concentration within seminal fluid. *Fertil Steril* 1997;**68**(4):709–13.

49. Sano K, Miyamoto Y, Kawasaki N, Hashii N, Itoh S, Murase M, et al. Survival signals of hepatic stellate cells in liver regeneration are regulated by glycosylation changes in rat vitronectin, especially decreased sialylation. *J Biol Chem* 2010;**285**(23):17301–9.

50. Barnes DW, Silnutzer J. Isolation of human serum spreading factor. *J Biol Chem* 1983;**258**(20):12548–52.

51. Seiffert D. Constitutive and regulated expression of vitronectin. *Histol Histopathol* 1997; **12**(3):787–97.

52. Shin TM, Isas JM, Hsieh CL, Kayed R, Glabe CG, Langen R, et al. Formation of soluble amyloid oligomers and amyloid fibrils by the multifunctional protein vitronectin. *Mol Neurodegener* 2008;**3**:16.

53. Dufourcq P, Louis H, Moreau C, Daret D, Boisseau MR, Lamaziere JM, et al. Vitronectin expression and interaction with receptors in smooth muscle cells from human atheromatous plaque. *Arterioscler Thromb Vasc Biol* 1998;**18**(2):168–76.

54. Singh B, Janardhan KS, Kanthan R. Expression of angiostatin, integrin alphavbeta3, and vitronectin in human lungs in sepsis. *Exp Lung Res* 2005;**31**(8):771–82.

55. Seiffert D, Geisterfer M, Gauldie J, Young E, Podor TJ. IL-6 stimulates vitronectin gene expression in vivo. *J Immunol* 1995;**155**(6):3180–5.

56. Conlan MG, Tomasini BR, Schultz RL, Mosher DF. Plasma vitronectin polymorphism in normal subjects and patients with disseminated intravascular coagulation. *Blood* 1988;**72**(1):185–90.
57. Zheng X, Saunders TL, Camper SA, Samuelson LC, Ginsburg D. Vitronectin is not essential for normal mammalian development and fertility. *Proc Natl Acad Sci USA* 1995;**92**(26):12426–30.
58. Jang YC, Tsou R, Gibran NS, Isik FF. Vitronectin deficiency is associated with increased wound fibrinolysis and decreased microvascular angiogenesis in mice. *Surgery* 2000;**127**(6):696–704.
59. Koschnick S, Konstantinides S, Schafer K, Crain K, Loskutoff DJ. Thrombotic phenotype of mice with a combined deficiency in plasminogen activator inhibitor 1 and vitronectin. *J Thromb Haemost* 2005;**3**(10):2290–5.

Chapter 34

CD59

Paul Morgan

Cardiff University, Cardiff, United Kingdom

PHYSICOCHEMICAL PROPERTIES

The immature protein of CD59 is synthesised as a single-chain precursor of 128 amino acids, including a leader peptide of 25 amino acids and a 26 amino acid C-terminal sequence that includes the motif for glycosyl phosphatidylinositol (GPI) anchor addition. CD59 becomes GPI-anchored in the act of secretion into the endoplasmic reticulum (ER), the anchor attaching to residue N102 coincident with removal of the C-terminus. The mature protein comprises 77 amino acids. For all amino acid numbering the translation initiator methionine is numbered as +1.

pI (calculated from mature)	5.06
M_r (kDa) (mature, predicted)	8.96 (nonglycosylated)
M_r (kDa) (observed)	18–23 (glycosylated)
N-Glycosylation sites	N43 (occupied; complex glycan)[1]
O-Glycosylation sites	T76, T77 (predicted)
GPI-anchor site	N102
Intrachain disulphides[2]	28↔51, 31↔38, 44↔64, 70↔88, 89↔94

STRUCTURE

CD59 is a heavily glycosylated single-chain GPI-anchored membrane protein. It is related to the Ly6/uPAR superfamily of GPI-anchored (usually), cysteine-rich proteins characterised by a conserved disulphide bond pattern that creates the three-fingered Ly6/uPAR domain.[3] The single, large (4–6 kDa) N-linked carbohydrate group is placed lateral to the protein core, and in erythrocyte-derived CD59 the carbohydrate is complex and highly variable.[1] The protein is also O-glycosylated, likely on T76/T77. The seven-amino acid sequence between C94 and N102 provides a flexible linker.

Both NMR and crystal structures of CD59 have been published.[4–6] In all published structures, the protein is comprised of three β-sheets and an α-helix. The highest resolution structure reveals an additional small α-helix.[6] The amino

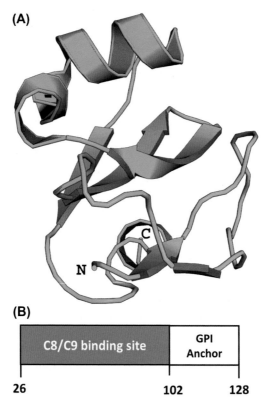

FIGURE 34.1 Crystal structure for human complement component CD59. (A) Ribbon diagram of human CD59 crystal structure (PDB code: 2UWR). (B) Schematic representation of CD59 domains.

acids involved in binding C8/C9 to inhibit membrane attack complex (MAC) formation have been identified by mutagenesis; these cluster around W65 and map to a surface-exposed hydrophobic groove on the CD59 crystal (Fig. 34.1).[7]

FUNCTION

CD59 inhibits the formation of MAC pores in the membranes of expressing cells: it is a 'suicide inhibitor', locking onto C8 in the forming MAC to block the recruitment of C9 into the complex.[8] In common with other GPI-linked proteins, cross-linking of CD59 can cause cell activation.[9] The biological relevance of this is uncertain. CD59 has been described as an alternative ligand for CD2 and implicated in T cell activation.[10,11] CD59 has been 'pirated' for use as an essential receptor for the streptococcal cytolytic toxin, intermedilysin.[12]

The regions in C8β and C9 that bind CD59 have been defined by mutagenesis and peptide-binding studies.[13,14] The complementary site on CD59 is described above. CD59 has been cocrystallised with intermedilysin, identifying the binding sites on both partners.[15]

TISSUE DISTRIBUTION

CD59 is broadly expressed, present on all circulating cells, vascular endothelium and on most cells examined in tissues. Cell surface distribution is usually patchy, reflecting its propensity to reside in lipid rafts. Expression on neurons and glia is comparatively low.[16] A soluble form of CD59, likely derived from epithelia, is found in urine and was the source of CD59 for the first structural studies.[4] Soluble CD59 is also found in trace amounts in plasma and other biological fluids.

REGULATION OF EXPRESSION

There is limited information. A poorly defined enhancer element in the first intron was shown to be required for high-level expression in cell lines.[16] The transcription factor Sp1 has been implicated as a regulator of constitutive expression of CD59, and NF-κB and CREB-dependent pathways implicated in inflammation-induced upregulation of CD59 expression.[17] The microRNA miR-224 was reported to downregulate CD59 expression by binding to its 3'-untranslated region.[18]

HUMAN PROTEIN SEQUENCE

The leader peptide is highlighted. The single *N*-glycosylation site (N43) is in bold. The site of GPI-anchor addition (N102) is in bold and double-underlined.

```
MGIQGGSVLF GLLLVLAVFC HSGHSLQCYN CPNPTADCKT AVNCSSDFDA   50

CLITKAGLQV YNKCWKFEHC NFNDVTTRLR ENELTYYCCK KDLCNFNEQL  100

ENGGTSLSEK TVLLLVTPFL AAAWSLHP
```

Protein Modules

CD59 is essentially a single-domain protein, the five intrachain disulphide bonds creating the rigid, three-fingered Ly6/uPAR domain.

CHROMOSOMAL LOCATION

The gene encoding human CD59 is at Chr11p14-p13.
    The gene encoding mouse Cd59 is at Chr2(E2-E4).[19] The mouse *Cd59* gene is duplicated; *Cd59a* is broadly expressed, while expression of *Cd59b* is restricted to testis.[20]

HUMAN cDNA SEQUENCE

```
CGCAGAAGCG GCTCGAGGCT GGAAGAGGAT CCTGGGCGCC GCCAGGTTCT GTGGACAATC    60

ACAATGGGAA TCCAAGGAGG GTCTGTCCTG TTCGGGCTGC TGCTCGTCCT GGCTGTCTTC   120

TGCCATTCAG GTCATAGCCT GCAGTGCTAC AACTGTCCTA ACCCAACTGC TGACTGCAAA   180

ACAGCCGTCA ATTGTTCATC TGATTTTGAT GCGTGTCTCA TTACCAAAGC TGGGTTACAA   240

GTGTATAACA AGTGTTGGAA GTTTGAGCAT TGCAATTTCA ACGACGTCAC AACCCGCTTG   300

AGGGAAAATG AGCTAACGTA CTACTGCTGC AAGAAGGACC TGTGTAACTT TAACGAACAG   360

CTTGAAAATG GTGGGACATC CTTATCAGAG AAAACAGTTC TTCTGCTGGT GACTCCATTT   420

CTGGCAGCAG CCTGGAGCCT TCATCCCTAA GTCAACACCA GGAGAGCTTC TCCCAAACTC   480

CCCGTTCCTG CGTAGTCCGC TTTCTCTTGC TGCCACATTC TAAAGGCTTG ATATTTTCCA   540

AATGGATCCT GTTGGGAAAG AATAAAATTA GCTTGAGCAA CCTGGCTAAG ATAGAGGGGT   600

CTGGGAGACT TTGAAGACCA GTCCTGCCCG CAGGGAAGCC CCACTTGAAG GAAGAAGTCT   660

AAGAGTGAAG TAGGTGTGAC TTGAACTAGA TTGCATGCTT CCTCCTTTGC TCTTGGGAAG   720

ACCAGCTTTG CAGTGACAGC TTGAGTGGGT TCTCTGCAGC CCTCAGATTA TTTTTCCTCT   780

GGCTCCTTGG ATGTAGTCAG TTAGCATCAT TAGTACATCT TTGGAGGGTG GGGCAGGAGT   840

ATATGAGCAT CCTCTCTCAC ATGGAACGCT TTCATAAACT TCAGGGATCC CGTGTTGCCA   900

TGGAGGCATG CCAAATGTTC CATATGTGGG TGTCAGTCAG GGACAACAAG ATCCTTAATG   960

CAGAGCTAGA GGACTTCTGG CAGGGAAGTG GGGAAGTGTT CCAGATTCCA GATAGCAGGG  1020

CATGAAAACT TAGAGAGGTA CAAGTGGCTG AAAATCGAGT TTTTCCTCTG TCTTTAAATT  1080

TTATATGGGC TTTGTTATCT TCCACTGGAA AAGTGTAATA GCATACATCA ATGGTGTGTT  1140
```

The sequence shown lacks the alternatively spliced exon 2, which is present in a minority (10%–20%) of the expressed mRNA. Multiple species of mRNA are found, differing in degree of polyadenylation. Intron/exon boundaries are shown by underlining the first five nucleotides in each exon; the initiation (ATG) and termination (TAA) sequences are double-underlined.

GENOMIC STRUCTURE

In early reports, the human *CD59* gene is described as spanning ~26 kb and containing five exons, including the alternatively spliced exon 2.[21,22]

A report states that the *CD59* gene consists of seven exons and six introns spanning ~33.5 kb including 5'-/3'- noncoding regions (NCBI Reference Sequence: NG_008057.1). Exon 1 occupies bases 1–98; exon 2, bases 2870–2985; exon 3, bases 5012–5056; exon 4, bases 13528–13563, exon 5, bases 13753–14101; exon 6, bases 19009–19110; exon 7, bases 26137–33470. The protein coding sequence is in exons 5, 6 and 7 (bases 14035–26354) (Fig. 34.2).

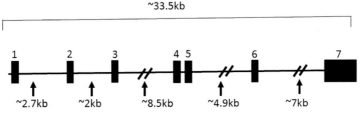

FIGURE 34.2 Diagram of genomic structure for CD59.

ACCESSION NUMBERS

Human CD59   UniProtKB: P13987
Mouse CD59a   UniProtKB: A2BI31; MGI: 109177; NCBI: 12509
Rat CD59   UniProtKB: P27274

DEFICIENCY

CD59 deficiency is rare; until now only a single case of complete deficiency presenting with a paroxysmal nocturnal haemoglobinuria (PNH)-like syndrome and multiple strokes has been reported.[23] In 2013, a cluster of cases was reported in Israel, affecting Jewish immigrants of North African origin.[24] Patients presented in infancy with Coombs-negative haemolysis and relapsing inflammatory demyelinating polyneuropathy. In all cases, a homozygous mutation, C89Y, in CD59 was present and no CD59 expression was detected on blood cells. Carrier frequency in the affected community was 1:66. An unrelated family with an identical clinical presentation had a homozygous D49V mutation in affected individuals and no CD59 expression.[25]

In PNH, affected circulating cells lack all GPI-anchored proteins, including CD59; absence of CD59 on erythrocytes results in intravascular haemolysis and anaemia, and absence on platelets and leucocytes predisposes to stroke.

POLYMORPHIC VARIANTS

No coding region polymorphisms in CD59 have been described. A promoter SNP that correlated with CD59 expression levels influenced donor lung survival in a transplant study.[26] A dinucleotide repeat polymorphism upstream of exon 1 (impact uncertain) has been described.[27] Nine CD59 SNPs were tested for association with AMD; negative results.[28]

MUTANT ANIMALS

CD59a knockout mice were described in 2001.[29] The phenotype was mild (compensated haemolysis), but when tested in models of complement-driven disease the mice were much more severely affected.[30] Mice deleted for the gene encoding CD59b presented a severe haemolytic anaemia and male reproductive deficits.[31]

REFERENCES

1. Rudd PM, Morgan BP, Wormald MR, Harvey DJ, van den Berg CW, Davis SJ, et al. The glycosylation of the complement regulatory protein, human erythrocyte CD59. *J Biol Chem* 1997;**272**:7229–44.

2. Sugita Y, Nakano Y, Oda E, Noda K, Tobe T, Miura NH, et al. Determination of carboxyl-terminal residue and disulfide bonds of MACIF (CD59), a glycosyl-phosphatidylinositol-anchored membrane protein. *J Biochem* 1993;**114**:473–7.

3. Galat A. The three-fingered protein domain of the human genome. *Cell Mol Life Sci* November 2008;**65**(21):3481–93.

4. Fletcher CM, Harrison RA, Lachmann PJ, Neuhaus D. Structure of a soluble, glycosylated form of the human complement regulatory protein CD59. *Structure* 1994;**2**:185–99.

5. Huang Y, Fedarovich A, Tomlinson S, Davies C. Crystal structure of CD59: implications for molecular recognition of the complement proteins C8 and C9 in the membrane-attack complex. *Acta Crystallogr D Biol Crystallogr* 2007;**63**:714–21.

6. Leath KJ, Johnson S, Roversi P, Hughes TR, Smith RA, Mackenzie L, et al. High-resolution structures of bacterially expressed soluble human CD59. *Acta Crystallogr Sect F Struct Biol Cryst Commun* 2007;**63**:648–52.

7. Bodian DL, Davis SJ, Morgan BP, Rushmere NK. Mutational analysis of the active site and anti-body epitopes of the complement-inhibitory glycoprotein, CD59. *J Exp Med* 1997;**185**:507–16.

8. Meri S, Morgan BP, Davies A, Daniels RH, Olavesen MG, Waldmann H, et al. Human protectin (CD59), an 18,000-20,000 MW complement lysis restricting factor, inhibits C5b-8 catalysed insertion of C9 into lipid bilayers. *Immunology* 1990;**71**:1–9.

9. van den Berg CW, Cinek T, Hallett MB, Horejsi V, Morgan BP. Exogenous glycosyl phospha-tidylinositol-anchored CD59 associates with kinases in membrane clusters on U937 cells and becomes Ca(2+)-signaling competent. *J Cell Biol* 1995;**131**:669–77.

10. Deckert M, Kubar J, Zoccola D, Bernard-Pomier G, Angelisova P, Horejsi V, et al. CD59 mol-ecule: a second ligand for CD2 in T cell adhesion. *Eur J Immunol* 1992;**22**:2943–7.

11. Longhi MP, Harris CL, Morgan BP, Gallimore A. Holding T cells in check–a new role for complement regulators? *Trends Immunol* 2006;**27**:102–8.

12. Giddings KS, Zhao J, Sims PJ, Tweten RK. Human CD59 is a receptor for the cholesterol-dependent cytolysin intermedilysin. *Nat Struct Mol Biol* 2004;**11**:1173–8.

13. Huang Y, Qiao F, Abagyan R, Hazard S, Tomlinson S. Defining the CD59-C9 binding interac-tion. *J Biol Chem* 2006;**281**:27398–404.

14. Husler T, Lockert DH, Sims PJ. Role of a disulfide-bonded peptide loop within human complement C9 in the species-selectivity of complement inhibitor CD59. *Biochemistry* 1996;**35**:3263–9.

15. Johnson S, Brooks NJ, Smith RA, Lea SM, Bubeck D. Structural basis for recognition of the pore-forming toxin intermedilysin by human complement receptor CD59. *Cell Rep* 2013;**3**:1369–77.

16. Tone M, Diamond LE, Walsh LA, Tone Y, Thompson SA, Shanahan EM, et al. High level tran-scription of the complement regulatory protein CD59 requires an enhancer located in intron 1. *J Biol Chem* 1999;**274**:710–6.

17. Du Y, Teng X, Wang N, Zhang X, Chen J, Ding P, et al. NF-κB and enhancer-binding CREB protein scaffolded by CREB-binding protein (CBP)/p300 proteins regulate CD59 protein expression to protect cells from complement attack. *J Biol Chem* 2014;**289**:2711–24.

18. Song G, Song G, Ni H, Gu L, Liu H, Chen B, et al. Deregulated expression of miR-224 and its target gene: CD59 predicts outcome of diffuse large B-cell lymphoma patients treated with R-CHOP. *Curr Cancer Drug Targets* 2014;**14**:659–70.

19. Holt DS, Powell MB, Rushmere NK, Morgan BP. Genomic structure and chromosome location of the gene encoding mouse CD59. *Cytogenet Cell Genet* 2000;**89**:264–7.

20. Baalasubramanian S, Harris CL, Donev RM, Mizuno M, Omidvar N, Song WC, et al. CD59a is the primary regulator of membrane attack complex assembly in the mouse. *J Immunol* 2004;**173**:3684–92.

21. Petranka JG, Fleenor DE, Sykes K, Kaufman RE, Rosse WF. Structure of the CD59-encoding gene: further evidence of a relationship to murine lymphocyte antigen Ly-6 protein. *Proc Natl Acad Sci USA* 1992;**89**:7876–9.

22. Holguin MH, Martin CB, Eggett T, Parker CJ. Analysis of the gene that encodes the complement regulatory protein, membrane inhibitor of reactive lysis (CD59). Identification of an alternatively spliced exon and characterization of the transcriptional regulatory regions of the promoter. *J Immunol* 1996;**157**:1659–68.

23. Motoyama N, Okada N, Yamashina M, Okada H. Paroxysmal nocturnal hemoglobinuria due to hereditary nucleotide deletion in the HRF20 (CD59) gene. *Eur J Immunol* 1992;**22**:2669–73.

24. Nevo Y, Ben-Zeev B, Tabib A, Straussberg R, Anikster Y, Shorer Z, et al. CD59 deficiency is associated with chronic hemolysis and childhood relapsing immune-mediated polyneuropathy. *Blood* 2013;**121**:129–35.

25. Haliloglu G, Maluenda J, Sayinbatur B, Aumont C, Temucin C, Tavil B, Cetin M, Oguz KK, Gut I, Picard V, Melki J, Topaloglu H. Early-onset chronic axonal neuropathy, strokes, and hemolysis: inherited CD59 deficiency. *Neurology* 2015;**84**:1220–4.

26. Budding K, van de Graaf EA, Kardol-Hoefnagel T, Broen JC, Kwakkel-van Erp JM, Oudijk EJ, et al. A promoter polymorphism in the CD59 complement regulatory protein gene in donor lungs correlates with a higher risk for chronic rejection after lung transplantation. *Am J Transpl* 2016;**16**:987–98.

27. Dewald G, Nöthen MM. Dinucleotide repeat polymorphism at the human CD59 locus. *Clin Genet* 1995;**47**:165–6.

28. Cipriani V, Matharu BK, Khan JC, Shahid H, Stanton CM, Hayward C, et al. Genetic variation in complement regulators and susceptibility to age-related macular degeneration. *Immunobiology* 2012;**217**:158–61.

29. Holt DS, Botto M, Bygrave AE, Hanna SM, Walport MJ, Morgan BP. Targeted deletion of the CD59 gene causes spontaneous intravascular hemolysis and hemoglobinuria. *Blood* 2001;**98**:442–9.

30. Mead RJ, Neal JW, Griffiths MR, Linington C, Botto M, Lassmann H, Morgan BP. Deficiency of the complement regulator CD59a enhances disease severity, demyelination and axonal injury in murine acute experimental allergic encephalomyelitis. *Lab Invest* 2004;**84**:21–8.

31. Qin X, Krumrei N, Grubissich L, Dobarro M, Aktas H, Perez G, et al. Deficiency of the mouse complement regulatory protein mCd59b results in spontaneous hemolytic anemia with platelet activation and progressive male infertility. *Immunity* 2003;**18**:217–27.

Chapter 35

CSMD1

Scott R. Barnum

University of Alabama at Birmingham, Birmingham, AL, United States

OTHER NAMES

CUB and sushi domains protein 1

PHYSICOCHEMICAL PROPERTIES

CSMD1 is synthesised as a single-chain precursor of 3565 amino acids including a 26 amino acid leader peptide.[1]

pI	5.3–5.7
M_r (K) predicted	~389
N-linked glycosylation sites	39 possible

STRUCTURE

CSMD1 is composed of 14 CUB domains and 28 complement control protein domains (CCP; a.k.a., sushi domains). The overall secondary and tertiary structures of the molecule have not been examined; however, the C-terminal 15 CCP domains likely project the rest of the molecule away from the cell membrane. There are 72 disulphide bridges throughout the extracellular portion of the molecule. The transmembrane region is composed of 21 predominately hydrophobic amino acids followed by a 56 amino acid cytoplasmic tail that contains multiple predicted phosphorylation sites on serine, threonine and tyrosine residues.[1,2] The crystal structure of CSMD1 is not yet available. The human sequence is highly conserved with mouse and rat (93% sequence identity) (Fig 35.1).[1,2]

FUNCTION

Studies using a membrane-bound, truncated form of human CSMD1 composed of the 15 CCP domains demonstrated that it effectively inhibited the

The Complement FactsBook. http://dx.doi.org/10.1016/B978-0-12-810420-0.00035-3

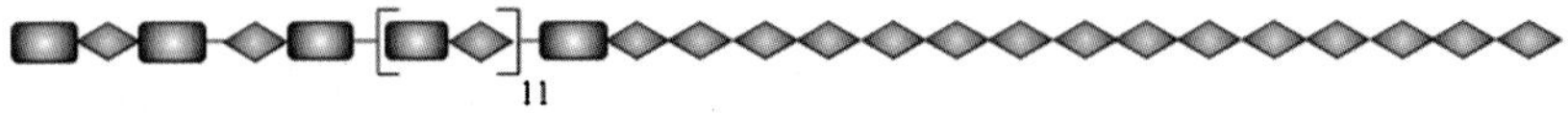

FIGURE 35.1 Domain structure of CSMD1. CSMD1 is composed of 14 CUB and 28 sushi domains.

classical pathway and served as a cofactor in the factor I-mediated cleavage of C3b.[3] Transfection of the human full-length molecule into a cancer cell line demonstrated reduced C3b deposition on the cell surface compared to cells in which CSMD1 had been knocked down.[3] Two additional family members, CSMD2 and CSMD3, have not been evaluated for their function. Given their high degree of structural homology to CSMD1, it is likely they inhibit complement in a fashion similar to CSMD1.[4] CSMD1 was originally described as a putative tumour suppressor based on its deletion in cancers of the head, neck, prostate and bladder with gene deletions on chromosome 8p23.[1] Studies in melanoma cells suggested that CSMD1 induces apoptosis in a Smad-dependent mechanism.[5] Studies link CSMD1 as one of a subset of genes that contribute to attention deficit hyperactivity disorder, autism, bipolar disorder, anxiety disorders, major depressive disorder and schizophrenia.[6]

DEGRADATION PATHWAY

No known degradation pathway or proteolytic fragments.

TISSUE DISTRIBUTION

CSMD1 is a type 1 transmembrane protein expressed in multiple tissues as assessed by qtPCR. In humans, expression is highest in cerebral cortex, brain white matter, cerebellum, foetal brain and testis. There is also low-level expression throughout the gastrointestinal system, the placenta and thyroid gland.[3] A similar expression pattern has been reported for rat CSMD1.[2] Studies in the rat demonstrated expression of CSMD1 on neuronal growth cones and in neuronal cell bodies.[2] There are no reports of a soluble form of CSMD1 in blood or other body fluids.

REGULATION OF EXPRESSION

There are no reports of CSMD1 being an acute-phase protein or its expression being regulated by cytokines, chemokines or growth factors.

HUMAN PROTEIN SEQUENCE

```
MTAWRRFQSL LLLLGLLVLC ARLLTAAKGQ NCGGLVQGPN GTIESFGFDI   50

GYPNYANCTW IIITGERNRI QLSFHTFALE ENFDILSVYD GQPQQGNLKV  100

RLSGFQLPSS IVSTGSILTL WFTTDFAVSA QGFKALYEVL PSHTCGNPGE  150

ILKGVLHGTR FNIGDKIRYS CLPGYILEGH AILTCIVSPG NGASWDFPAP  200

FCRAEGACGG TLRGTSSSIS SPHFPSEYEN NADCTWTILA EPGDTIALVF  250

TDFQLEEGYD FLEISGTEAP SIWLTGMNLP SPVISSKNWL RLHFTSDSNH  300

RRKGFNAQFQ VKKAIELKSR GVKMLPSKDG SHKNSVLSQG GVALVSHMCL  350

DPGIPENGRR AGSDFSRVGA NVQFSCEDNY VLQGSKSITC QRVTETLAAW  400

SDHRPICRAR TCGSNLRGPS GVITSPNYPV QYEDNAHCVW VITTTDPDKV  450

IKLAFEEFEL ERGYDTLTVG DAGKVGDTRS VLYVLTGSSV PDLIVSMSNQ  500

MWLHLQSDDS IGSPGFKAVY QEIEKGGCGD PGIPAYGKRT GSSFLHGDTL  550

TFECPAAFEL VGERVITCQQ NNQWSGNKPS CVFSCFFNFT ASSGIILSPN  600

YPEEYGNNMN CVWLIISEPG SRIHLIFNDF DVEPQFDFLA VKDDGISDIT  650

VLGTFSGNEV PSQLASSGHI VRLEFQSDHS TTGRGFNITY TTFGQNECHD  700

PGIPINGRRF GDRFLLGSSV SFHCDDGFVK TQGSESITCI LQDGNVVWSS  750

TVPRCEAPCG GHLTASSGVI LPPGWPGYYK DSLHCEWIIE AKPGHSIKMT  800

FDRFQTEVNY DTLEVRDGPA SSSPLIGEYH GTQAPQFLIS TGNFMYLLFT  850

TDNSRSSIGF LIHYESVTLE SDSCLDPGIP VNGHRHGGDF GIRSTVTFSC  900

DPGYTLSDDE PLVCERNHQW NHALPSCDAL CGGYIQGKSG TVLSPGFPDF  950

YPNSLNCTWT IEVSHGKGVQ MIFHTFHLES SHDYLLITED GSFSEPVARL 1000

TGSVLPHTIK AGLFGNFTAQ LRFISDFSIS YEGFNITFSE YDLEPCDDPG 1050

VPAFSRRIGF HFGVGDSLTF SCFLGYRLEG ATKLTCLGGG RRVWSAPLPR 1100

CVAECGASVK GNEGTLLSPN FPSNYDNTHE CIYKIETEAG KGIHLRTRSF 1150

QLFEGDTLKV YDGKDSSSRP LGTFTKNELL GLILNSTSNH LWLEFNTNGS 1200

DTDQGFQLTY TSFDLVKCED PGIPNYGYRI RDEGHFTDTV VLYSCNPGYA 1250

MHGSNTLTCL SGDRRVWDKP LPSCIAECGG QIHAATSGRI LSPGYPAPYD 1300

NNLHCTWIIE ADPGKTISLH FIVFDTEMAH DILKVWDGPV DSDILLKEWS 1350
```

HUMAN PROTEIN SEQUENCE – *Continued*

```
GSALPEDIHS TFNSLTLQFD SDFFISKSGF SIQFSTSIAA TCNDPGMPQN  1400

GTRYGDSREA GDTVTFQCDP GYQLQGQAKI TCVQLNNRFF WQPDPPTCIA  1450

ACGGNLTGPA GVILSPNYPQ PYPPGKECDW RVKVNPDFVI ALIFKSFNME  1500

PSYDFLHIYE GEDSNSPLIG SYQGSQAPER IESSGNSLFL AFRSDASVGL  1550

SGFAIEFKEK PREACFDPGN IMNGTRVGTD FKLGSTITYQ CDSGYKILDP  1600

SSITCVIGAD GKPSWDQVLP SCNAPCGGQY TGSEGVVLSP NYPHNYTAGQ  1650

ICLYSITVPK EFVVFGQFAY FQTALNDLAE LFDGTHAQAR LLSSLSGSHS  1700

GETLPLATSN QILLRFSAKS GASARGFHFV YQAVPRTSDT QCSSVPEPRY  1750

GRRIGSEFSA GSIVRFECNP GYLLQGSTAL HCQSVPNALA QWNDTIPSCV  1800

VPCSGNFTQR RGTILSPGYP EPYGNNLNCI WKIIVTEGSG IQIQVISFAT  1850

EQNWDSLEIH DGGDVTAPRL GSFSGTTVPA LLNSTSNQLY LHFQSDISVA  1900

AAGFHLEYKT VGLAACQEPA LPSNSIKIGD RYMVNDVLSF QCEPGYTLQG  1950

RSHISCMPGT VRRWNYPSPL CIATCGGTLS TLGGVILSPG FPGSYPNNLD  2000

CTWRISLPIG YGAHIQFLNF STEANHDFLE IQNGPYHTSP MIGQFSGTDL  2050

PAALLSTTHE TLIHFYSDHS QNRQGFKLAY QAYELQNCPD PPPFQNGYMI  2100

NSDYSVGQSV SFECYPGYIL IGHPVLTCQH GINRNWNYPF PRCDAPCGYN  2150

VTSQNGTIYS PGFPDEYPIL KDCIWLITVP PGHGVYINFT LLQTEAVNDY  2200

IAVWDGPDQN SPQLGVFSGN TALETAYSST NQVLLKFHSD FSNGGFFVLN  2250

FHAFQLKKCQ PPPAVPQAEM LTEDDDFEIG DFVKYQCHPG YTLVGTDILT  2300

CKLSSQLQFE GSLPTCEAQC PANEVRTGSS GVILSPGYPG NYFNSQTCSW  2350

SIKVEPNYNI TIFVDTFQSE KQFDALEVFD GSSGQSPLLV VLSGNHTEQS  2400

NFTSRSNQLY LRWSTDHATS KKGFKIRYAA PYCSLTHPLK NGGILNRTAG  2450

AVGSKVHYFC KPGYRMVGHS NATCRRNPLG MYQWDSLTPL CQAVSCGIPE  2500

SPGNGSFTGN EFTLDSKVVY ECHEGFKLES SQQATAVCQE DGLWSNKGKP  2550

PMCKPVACPS IEAQLSEHVI WRLVSGSLNE YGAQVLLSCS PGYYLEGWRL  2600

LRCQANGTWN IGDERPSCRV ISCGSLSFPP NGNKIGTLTV YGATAIFTCN  2650

TGYTLVGSHV RECLANGLWS GSETRCLAGH CGSPDPIVNG HISGDGFSYR  2700

DTVVYQCNPG FRLVGTSVRI CLQDHKWSGQ TPVCVPITCG HPGNPAHGFT  2750

NGSEFNLNDV VNFTCNTGYL LQGVSRAQCR SNGQWSSPLP TCRVVNCSDP  2800
```

HUMAN PROTEIN SEQUENCE—*Continued*

```
GFVENAIRHG QQNFPESFEY GMSILYHCKK GFHLLGSSAL TCMANGLWDR 2850

SLPKCLAISC GHPGVPANAV LTGELFTYGA VVHYSCRGSE SLIGNDTRVC 2900

QEDSHWSGAL PHCTGNNPGF CGDPGTPAHG SRLGDDFKTK SLLRFSCEMG 2950

HQLRGSPERT CLLNGSWSGL QPVCEAVSCG NPGTPTNGMI VSSDGILFSS 3000

SVIYACWEGY KTSGLMTRHC TANGTWTGTA PDCTIISCGD PGTLANGIQF 3050

GTDFTFNKTV SYQCNPGYVM EAVTSATIRC TKDGRWNPSK PVCKAVLCPQ 3100

PPPVQNGTVE GSDFRWGSSI SYSCMDGYQL SHSAILSCEG RGVWKGEIPQ 3150

CLPVFCGDPG IPAEGRLSGK SFTYKSEVFF QCKSPFILVG SSRRVCQADG 3200

TWSGIQPTCI DPAHNTCPDP GTPHFGIQNS SRGYEVGSTV FFRCRKGYHI 3250

QGSTTRTCLA NLTWSGIQTE CIPHACRQPE TPAHADVRAI DLPTFGYTLV 3300

YTCHPGFFLA GGSEHRTCKA DMKWTGKSPV CKSKGVREVN ETVTKTPVPS 3350

DVFFVNSLWK GYYEYLGKRQ PATLTVDWFN ATSSKVNATF SEASPVELKL 3400

TGIYKKEEAH LLLKAFQIKG QADIFVSKFE NDNWGLDGYV SSGLERGGFT 3450

FQGDIHGKDF GKFKLERQDP LNPDQDSSSH YHGTSSGSVA AAILVPFFAL 3500

ILSGFAFYLY KHRTRPKVQY NGYAGHENSN GQASFENPMY DTNLKPTEAK 3550

AVRFDTTLNT VCTVV
```

The leader sequence is underlined. Putative *N*-linked glycosylation sites are denoted by **N**, while putative phosphorylation sites are denoted by **S**, **T** and **Y**.

PROTEIN MODULES

Amino acid sequences for each relevant section of the protein from leader to the carboxy-terminal end of the protein. The list shows associated exons for each module.

	Domain	Exon
1–26	Leader peptide	exon 1
32–140	CUB 1	exons 2/3
143–204	Sushi 1	exon 4
208–312	CUB 2	exons 5–8
347–409	Sushi 2	exon 9
412–523	CUB 3	exons 10–12
526–583	Sushi 3	exon 13
585–693	CUB 4	exon 14
696–757	Sushi 4	exon 15
759–867	CUB 5	exons 16/17
872–929	Sushi 5	exon 18
931–1041	CUB 6	exons 19/20

	Domain	Exon
1044–1103	Sushi 6	exon 21
1105–1213	CUB 7	exons 22/23
1216–1276	Sushi 7	exon 24
1278–1387	CUB 8	exons 25/26
1390–1450	Sushi 8	exon 27
1452–1560	CUB 9	exon 28/29
1563–1624	Sushi 9	exon 30
1626–1734	CUB 10	exons 31–33
1740–1801	Sushi 10	exon 34
1803–1911	CUB 11	exons 35–37
1914–1973	Sushi 11	exons 38/39
1975–2083	CUB 12	exons 40/41
2086–2145	Sushi 12	exon 42
2147–2258	CUB 13	exons 43/44
2257–2318	Sushi 13	exons 45/46
2320–2431	CUB 14	exons 47–48
2431–2493	Sushi 14	exon 49
2494–2555	Sushi 15	exon 50
2556–2620	Sushi 16	exon 51
2621–2678	Sushi 17	exon 52
2679–2736	Sushi 18	exon 53
2737–2794	Sushi 19	exon 54
2795–2857	Sushi 20	exon 55
2858–2915	Sushi 21	exon 56
2919–2976	Sushi 22	exon 57
2977–3035	Sushi 23	exon 58
3036–3095	Sushi 24	exon 59
3096–3153	Sushi 25	exon 60
3154–3211	Sushi 26	exon 61
3215–3273	Sushi 27	exons 62/63
3274–3333	Sushi 28	exon 64
3489–3509	Transmembrane	exon 67
3510–3565	Cytoplasmic tail	exons 68–69

CHROMOSOMAL LOCATION

Human: 8p23
Mouse: 8
Rat: 16

HUMAN cDNA SEQUENCE

```
GCGAGGCTCCTCACTGCAGCGAAGGGTCAGAACTGTGGAGGCTTAGTCCAGGGTCCCAAT     120

GGCACTATTGAGAGCCCAGGGTTTCCTCACGGGTATCCGAACTATGCCAACTGCACCTGG     180

ATCATCATCACGGGCGAGCGCAATAGGATACAGTTGTCCTTCCATACCTTTGCTCTTGAA     240

GAAGATTTTGATATTTTATCAGTTTACGATGGACAGCCTCAACAAGGGAATTTAAAAGTG     300

AGATTATCGGGATTTCAGCTGCCCTCCTCTATAGTGAGTACAGGATCTATCCTCACTCTG     360

TGGTTCACGACAGACTTCGCTGTGAGTGCCCAAGGTTTCAAAGCATTATATGAAGTTTTA     420

CCTAGCCACACTTGTGGAAATCCTGGAGAAATCCTGAAAGGAGTTCTGCATGGAACGAGA     480
```

HUMAN cDNA SEQUENCE—*Continued*

```
TTCAACATAGGAGACAAAATCCGGTACAGCTGCCTCCCTGGCTACATCTTGGAAGGCCAC     540

GCCATCCTGACCTGCATCGTCAGCCCAGGAAATGGTGCATCGTGGGACTTCCCAGCTCCC     600

TTTTGCAGAGCTGAGGGAGCCTGCGGAGGAACCTTACGCGGGACCAGCAGCTCCATCTCC     660

AGCCCGCACTTCCCTTCAGAGTACGAGAACAACGCGGACTGCACCTGGACCATTCTGGCT     720

GAGCCCGGGGACACCATTGCGCTGGTCTTCACTGACTTTCAGCTAGAAGAAGGATATGAT     780

TTCTTAGAGATCAGTGGCACGGAAGCTCCATCCATATGGCTAACTGGCATGAACCTCCCC     840

TCTCCAGTTATCAGTAGCAAGAATTGGCTACGACTCCATTTCACCTCTGACAGCAACCAC     900

CGACGCAAAGGATTTAACGCTCAGTTCCAAGTGAAAAAGGCGATTGAGTTGAAGTCAAGA     960

GGAGTCAAGATGCTGCCCAGCAAGGATGGAAGCCATAAAAACTCTGTCTTGAGCCAAGGA    1020

GGTGTTGCATTGGTCTCTGACATGTGTCCAGATCCTGGGATTCCAGAAAATGGTAGAAGA    1080

GCAGGTTCCGACTTCAGGGTTGGTGCAAATGTACAGTTTTCATGTGAGGACAATTACGTG    1140

CTCCAGGGATCTAAAAGCATCACCTGTCAGAGAGTTACAGAGACGCTCGCTGCTTGGAGT    1200

GACCACAGGCCCATCTGCCGAGCGAGAACATGTGGATCCAATCTGCGTGGGCCCAGCGGC    1260

GTCATTACCTCCCCTAATTATCCGGTTCAGTATGAAGATAATGCACACTGTGTGTGGGTC    1320

ATCACCACCACCGACCCGGACAAGGTCATCAAGCTTGCCTTTGAAGAGTTTGAGCTGGAG    1380

CGAGGCTATGACACCCTGACGGTTGGTGATGCTGGGAAGGTGGGAGACACCAGATCGGTC    1440

TTGTACGTGCTCACGGGATCCAGTGTTCCTGACCTCATTGTGAGCATGAGCAACCAGATG    1500

TGGCTACATCTGCAGTCGGATGATAGCATTGGCTCACCTGGGTTTAAAGCTGTTTACCAA    1560

GAAATTGAAAAGGGAGGGTGTGGGGATCCTGGAATCCCCGCCTATGGGAAGCGGACGGGC    1620

AGCAGTTTCCTCCATGGAGATACACTCACCTTTGAATGCCCGGCGGCCTTTGAGCTGGTG    1680

GGGGAGAGAGTTATCACCTGTCAGCAGAACAATCAGTGGTCTGGCAACAAGCCCAGCTGT    1740

GTATTTTTCATGTTTCTTCAACTTTACGGCATCATCTGGGATTATTCTGTCACCAAATTAT    1800

CCAGAGGAATATGGGAACAACATGAACTGTGTCTGGTTGATTATCTCGGAGCCAGGAAGT    1860

CGAATTCACCTAATCTTTAATGATTTTGATGTTGAGCCTCAGTTTGACTTTCTCGCGGTC    1920

AAGGATGATGGCATTTCTGACATAACTGTCCTGGGTACTTTTTCTGGCAATGAAGTGCCT    1980

TCCCAGCTGGCCAGCAGTGGGCATATAGTTCGCTTGGAATTTCAGTCTGACCATTCCACT    2040

ACTGGCAGAGGGTTCAACATCACTTACACCACATTTGGTCAGAATGAGTGCCATGATCCT    2100

GGCATTCCTATAAACGGACGACGTTTTGGTGACAGGTTTCTACTCGGGAGCTCGGTTTCT    2160

TTCCACTGTGATGATGGCTTTGTCAAGACCCAGGGATCCGAGTCCATTACCTGCATACTG    2220

CAAGACGGGAACGTGGTCTGGAGCTCCACCGTGCCCCGCTGTGAAGCTCCATGTGGTGGA    2280

CATCTGACAGCGTCCAGCGGAGTCATTTTGCCTCCTGGATGGCCAGGATATTATAAGGATT   2340

CTTTACATTGTGAATGGATAATTGAAGCAAAACCAGGCCACTCTATCAAAATAACTTTTG    2400

ACAGATTTCAGACAGAGGTCAATTATGACACCTTGGAGGTCAGAGATGGGCCAGCCAGTT    2460
```

HUMAN cDNA SEQUENCE – *Continued*

```
CGTCCCCACTGATCGGCGAGTACCACGGCACCCAGGCACCCCAGTTCCTCATCAGCACCG   2520

GGAACTTCATGTACCTGCTGTTCACCACTGACAACAGCCGCTCCAGCATCGGCTTCCTCA   2580

TCCACTATGAGAGTGTGACGCTTGAGTCGGATTCCTGCCTGGACCCGGGCATCCCTGTGA   2640

ACGGCCATCGCCACGGTGGAGACTTTGGCATCAGGTCCACAGTGACTTTCAGCTGTGACC   2700

CGGGGTACACACTAAGTGACGACGAGCCCCTCGTCTGTGAGAGGAACCACCAGTGGAACC   2760

ACGCCTTGCCCAGCTGCGACGCTCTATGTGGAGGCTACATCCAAGGGAAGAGTGGAACAG   2820

TCCTTTCTCCTGGGTTTCCAGATTTTTATCCAAACTCTCTAAACTGCACGTGGACCATTG   2880

AAGTGTCTCATGGGAAAGGAGTTCAAATGATCTTTCACACCTTTCATCTTGAGAGTTCCC   2940

ACGACTATTTACTGATCACAGAGGATGGAAGTTTTTCCGAGCCCGTTGCCAGGCTCACCG   3000

GGTCGGTGTTGCCTCATACGATCAAGGCAGGCCTGTTTGGAAACTTCACTGCCCAGCTTC   3060

GGTTTATATCAGACTTCTCAATTTCGTACGAGGGCTTCAATATCACATTTTCAGAATATG   3120

ACCTGGAGCCATGTGATGATCCTGGAGTCCCTGCCTTCAGCCGAAGAATTGGTTTTCACT   3180

TTGGTGTGGGGAGACTCTCTGACGTTTTCCTGCTTCCTGGGATATCGTTTAGAAGGTGCCA   3240

CCAAGCTTACCTGCCTGGGTGGGGGCCGCCGTGTGTGGAGTGCACCTCTGCCAAGGTGTG   3300

TGGCCGAATGTGGAGCAAGTGTCAAAGGAAATGAAGGAACATTACTGTCTCCAAATTTTC   3360

CATCCAATTATGATAATAACCATGAGTGTATCTATAAAATAGAAACAGAAGCCGGCAAGG   3420

GCATCCACCTTAGAACACGAAGCTTCCAGCTGTTTGAAGGAGATACTCTAAAGGTATATG   3480

ATGGAAAAGACAGTTCCTCACGTCCACTGGGCACGTTCACTAAAAATGAACTTCTGGGGC   3540

TGATCCTAAACAGCACATCCAATCACCTGTGGCTAGAGTTCAACACCAATGGATCTGACA   3600

CCGACCAAGGTTTTCAACTCACCTATACCAGTTTTGATCTGGTAAAATGTGAGGATCCGG   3660

GCATCCCTAACTACGGCTATAGGATCCGTGATGAAGGCCACTTTACCGACACTGTAGTTC   3720

TGTACAGTTGCAACCCGGGGTACGCCATGCATGGCAGCAACACCCTGACCTGTTTGAGTG   3780

GAGACAGGAGAGTGTGGGACAAACCACTACCTTCGTGCATAGCGGAATGTGGTGGTCAGA   3840

TCCATGCAGCCACATCAGGACGAATATTGTCCCCTGGCTATCCAGCTCCGTATGACAACA   3900

ACCTCCACTGCACCTGGATTATAGAGGCAGACCCAGGAAAGACCATTAGCCTCCATTTCA   3960

TTGTTTTCGACACGGAGATGGCTCACGACATCCTCAAGGTCTGGGACGGGCCGGTGGACA   4020

GTGACATCCTGCTGAAGGAGTGGAGTGGCTCCGCCCTTCCGGAGGACATCCACAGCACCT   4080

TCAACTCACTCACCCTGCAGTTCGACAGCGACTTCTTCATCAGCAAGTCTGGCTTCTCCA   4140

TCCAGTTCTCCACCTCAATTGCAGCCACCTGTAACGATCCAGGTATGCCCCAAAATGGCA   4200

CCCGCTATGGAGACAGCAGAGAGGCTGGAGACACCGTCACATTCCAGTGTGACCCTGGCT   4260

ATCAGCTCCAAGGACAAGCCAAAATCACCTGTGTGCAGCTGAATAACCGGTTCTTTTGGC   4320

AACCAGACCCTCCTACATGCATAGCTGCTTGTGGAGGGAATCTGACGGGCCCAGCAGGTG   4380

TTATTTTGTCACCCAACTACCCACAGCCGTATCCTCCTGGGAAGGAATGTGACTGGAGAG   4440

TAAAAGTGAACCCGGACTTTGTCATCGCCTTGATATTCAAAAGTTTCAACATGGAGCCCA   4500
```

HUMAN cDNA SEQUENCE—*Continued*

```
GCTATGACTTCCTACACATCTATGAAGGGGAAGATTCCAACAGCCCCCTCATTGGGAGTT  4560

ACCAGGGCTCTCAGGCCCCAGAAAGAATAGAGAGTAGCGGAAACAGCCTGTTTCTGGCAT  4620

TTCGGAGTGATGCCTCCGTGGGCCTTTCAGGGTTCGCCATTGAATTTAAAGAGAAACCAC  4680

GGGAAGCTTGTTTTGACCCAGGAAATATAATGAATGGGACAAGAGTTGGAACAGACTTCA  4740

AGCTTGGCTCCACCATCACCTACCAGTGTGACTCTGGCTATAAGATTCTTGACCCCTCAT  4800

CCATCACCTGTGTGATTGGGGCTGATGGGAAACCCTCCTGGGACCAAGTGCTGCCCTCCT  4860

GCAATGCTCCCTGTGGAGGCCAGTACACGGGATCAGAAGGGGTAGTTTTATCACCAAACT  4920

ACCCCCATAATTACACAGCTGGTCAAATATGCCTCTATTCCATCACGGTACCAAAGGAAT  4980

TCGTGGTCTTTGGACAGTTTGCCTATTTCCAGACAGCCCTGAATGATTTGGCAGAATTAT  5040

TTGATGGAACCCATGCACAGGCCAGACTTCTCAGCTCACTCTCGGGGTCTCACTCAGGGG  5100

AAACATTGCCCTTGGCTACGTCAAATCAAATTCTGCTCCGATTCAGTGCAAAGAGCGGTG  5160

CCTCTGCCCGCGGCTTCCACTTCGTGTATCAAGCTGTTCCTCGTACCAGTGACACCCAAT  5220

GCAGCTCTGTCCCCGAGCCCAGATACGGAAGGAGAATTGGTTCTGAGTTTTCTGCCGGCT  5280

CCATCGTCCGATTCGAGTGCAACCCGGGATACCTGCTTCAGGGTTCCACGGCGCTCCACT  5340

GCCAGTCCGTGCCCAACGCCTTGGCACAGTGGAACGACACGATCCCCAGCTGTGTGGTAC  5400

CCTGCAGTGGCAATTTCACTCAACGAAGAGGTACAATCCTGTCCCCCGGCTACCCTGAGC  5460

CATACGGAAACAACTTGAACTGTATATGGAAGATCATAGTTACGGAGGGCTCGGGAATTC  5520

AGATCCAAGTGATCAGTTTTGCCACGGAGCAGAACTGGGACTCCCTTGAGATCCACGATG  5580

GTGGGGATGTGACCGCACCCAGACTGGGAAGCTTCTCAGGCACCACAGTACCGGCACTGC  5640

TGAACAGTACTTCCAACCAACTCTACCTGCATTTCCAGTCTGACATTAGTGTGGCAGCTG  5700

CTGGTTTCCACCTGGAATACAAAACTGTAGGTCTTGCTGCATGCCAAGAACCAGCCCTCC  5760

CCAGCAACAGCATCAAAATCGGAGATCGGTACATGGTGAACGACGTGCTCTCCTTCCAGT  5820

GCGAGCCCGGGTACACCCTGCAGGGCCGTTCCCACATTTCCTGTATGCCAGGGACCGTTC  5880

GCCGTTGGAACTATCCGTCTCCCCTGTGCATTGCAACCTGTGGAGGGACGCTGAGCACCT  5940

TGGGTGGTGTGATCCTGAGCCCCGGCTTCCCAGGTTCTTACCCCAACAACTTAGACTGCA  6000

CCTGGAGGATCTCATTACCCATCGGCTATGGTGCACATATTCAGTTTCTGAATTTTTCTA  6060

CCGAAGCTAATCATGACTTCCTTGAAATTCAAAATGGACCTTACCACACCAGCCCCATGA  6120

TTGGACAATTTAGCGGCACGGATCTCCCCGCGGCCCTGCTGAGCACAACGCATGAAACCC  6180

TCATCCACTTTTATAGTGACCATTCGCAAAACCGGCAAGGATTTAAACTTGCTTACCAAG  6240

CCTATGAATTACAGAACTGTCCAGATCCACCCCCATTTCAGAATGGGTACATGATCAACT  6300

CGGATTACAGCGTGGGGCAATCAGTATCTTTCGAGTGTTATCCTGGGTACATTCTAATAG  6360

GCCATCCTGTCCTCACTTGTCAGCATGGGATCAACAGAAACTGGAACTACCCTTTTCCAA  6420

GATGTGATGCCCCTTGTGGGTACAACGTAACTTCTCAGAACGGCACCATCTACTCCCCTG  6480

GCTTTCCTGATGAGTATCCGATCCTGAAGGACTGCATTTGGCTCATCACGGTGCCTCCAG  6540

GGCACGGAGTTTACATCAACTTCACCCTGTTACAGACGGAAGCTGTCAACGATTACATTG  6600
```

HUMAN cDNA SEQUENCE—*Continued*

```
CTGTTTGGGACGGTCCCGATCAGAACTCACCCCAGCTGGGAGTTTTCAGTGGCAACACAG   6660

CCCTCGAAACGGCGTATAGCTCCACCAACCAAGTCCTGCTCAAGTTCCACAGCGACTTTT   6720

CAAATGGAGGCTTCTTTGTCCTCAATTTCCACGCATTTCAGCTCAAGAAATGTCAACCTC   6780

CCCCAGCGGTTCCACAGGCAGAAATGCTTACTGAGGATGATGATTTCGAAATAGGAGATT   6840

TTGTGAAGTACCAGTGCCACCCCGGGTACACCTTGGTGGGGACCGACATTCTGACTTGCA   6900

AGCTCAGTTCCCAGTTGCAGTTTGAGGGTTCTCTCCCAACATGTGAAGCACAATGCCCAG   6960

CAAATGAAGTCCGGACTGGATCATCGGGAGTCATTCTCAGTCCAGGGTATCCGGGTAATT   7020

ATTTTAACTCCCAGACTTGCTCTTGGAGTATTAAAGTGGAACCAAACTACAACATTACCA   7080

TCTTTGTGGACACATTTCAAAGTGAAAAGCAGTTTGATGCACTGGAAGTGTTTGATGGTT   7140

CTTCTGGGCAAAGTCCTCTGCTAGTAGTCTTAAGTGGGAATCATACTGAACAATCAAATT   7200

TTACAAGCAGGAGTAATCAGTTATATCTCCGCTGGTCCACTGACCATGCCACCAGTAAGA   7260

AAGGATTCAAGATTCGCTATGCAGCACCTTACTGCAGTTTGACCCACCCCCTGAAGAATG   7320

GGGGTATTCTAAACAGGACTGCAGGAGCGGTTGGAAGCAAAGTGCATTATTTTTGCAAGC   7380

CTGGATACCGAATGGTCGGCCACAGCAATGCAACCTGTAGACGAAACCCACTTGGCATGT   7440

ACCAGTGGGACTCCCTCACGCCACTCTGCCAGGCTGTGTCCTGTGGAATCCCAGAATCCC   7500

CAGGAAACGGTTCATTTACCGGGAACGAGTTCACTTTGGACAGTAAAGTGGTCTATGAAT   7560

GTCATGAGGGCTTCAAGCTTGAATCCAGCCAGCAAGCAACAGCCGTGTGTCAAGAAGATG   7620

GGTTGTGGAGTAACAAGGGGAAGCCGCCCACGTGTAAGCCGGTCGCTTGCCCCAGCATTG   7680

AAGCTCAGCTCTCAGAACATGTCATCTGGAGGCTGGTTTCAGGATCCTTGAATGAGTACG   7740

GTGCTCAAGTATTGCTGAGCTGCAGTCCTGGTTACTACTTAGAAGGCTGGAGGCTCCTGC   7800

GGTGCCAGGCCAATGGGACGTGGAACATAGGAGATGAGAGGCCAAGCTGTCGAGTTATCT   7860

CGTGTGTGGAAGCCTTTCCTTTCCCCCAAATGGCAACAAGATTGGAACGTTGACAGTTTATG   7920

GGGCCACAGCTATATTTACGTGCAACACCGGCTACACGCTTGTGGGGTCTCATGTCAGAG   7980

AGTGCTTGGCAAATGGGCTCTGGAGCGGCAGCGAAACTCGATGTCTGGCTGGCCACTGCG   8040

GTTCCCCAGACCCGATTGTGAACGGTCACATTAGTGGAGATGGCTTCAGTTACAGAGACA   8100

CGGTGGTTTACCAGTGCAATCCTGGTTTCCGGCTTGTGGGAACTTCCGTGAGGATATGCC   8160

TGCAAGACCACAAGTGGTCTGGACAAACGCCTGTCTGTGTCCCCATCACATGTGGTCACC   8220

CTGGAAACCCTGCCCACGGATTCACTAATGGCAGTGAGTTCAACCTGAATGATGTCGTGA   8280

ATTTCACCTGCAACACGGGCTATTTGCTGCAGGGCGTGTCTCGAGCCCAGTGTCGGAGCA   8340

ACGGCCAGTGGAGTAGCCCTCTGCCCACGTGTCGAGTGGTGAACTGTTCTGATCCAGGCT   8400

TTGTGGAAAATGCCATTCGTCACGGGCAACAGAACTTCCCTGAGAGTTTTGAGTATGGAA   8460

TGAGTATCCTGTACCATTGCAAGAAGGGATTTTACTTGCTGGGATCTTCAGCCTTGACCT   8520

GTATGGCAAATGGCTTATGGGACCGATCCCTGCCCAAGTGTTTGGCTATATCGTGTGGAC   8580

ACCCAGGGGTCCCTGCCAACGCCGTCCTCACTGGAGAGCTGTTTACCTATGGCGCCGTCG   8640

TGCACTACTCCTGCAGAGGGAGCGAGAGCCTCATAGGCAACGACACGAGAGTGTGCCAGG   8700
```

HUMAN cDNA SEQUENCE—*Continued*

```
AAGACAGTCACTGGAGCGGGGCACTGCCCCACTGCACAGGAAATAATCCTGGATTCTGTG   8760
GTGATCCGGGGACCCCAGCACATGGGTCTCGGCTTGGTGATGACTTTAAGACAAAGAGTC   8820
TTCTCCGCTTCTCCTGTGAAATGGGGCACCAGCTGAGGGGCTCCCCTGAACGCACGTGTT   8880
TGCTCAATGGGTCATGGTCAGGACTGCAGCCGGTGTGTGAGGCCGTGTCCTGTGGCAACC   8940
CTGGCACACCCACCAACGGAATGATTGTCAGTAGTGATGGCATTCTGTTCTCCAGCTCGG   9000
TCATCTATGCCTGCTGGGAAGGCTACAAGACCTCAGGGCTCATGACACGGCATTGCACAG   9060
CCAATGGGACCTGGACAGGCACTGCTCCCGACTGCACAATTATAAGTTGTGGGGATCCAG   9120
GCACACTAGCAAATGGCATCCAGTTTGGGACCGACTTCACCTTCAACAAGACTGTGAGCT   9180
ATCAGTGTAACCCAGGCTATGTCATGGAAGCAGTCACATCCGCCACTATTCGCTGTACCA   9240
AAGACGGCAGGTGGAATCCGAGCAAACCTGTCTGCAAAGCCGTGCTGTGTCCTCAGCCGC   9300
CGCCGGTGCAGAATGGAACAGTGGAGGGAAGTGATTTCCGCTGGGGCTCCAGCATAAGTT   9360
ACAGCTGCATGGACGGTTACCAGCTCTCTCACTCCGCCATCCTCTCCTGTGAAGGTCGCG   9420
GGGTGTGGAAAGGAGAGATCCCCCAGTGTCTCCCTGTGTTCTGCGGGAGACCCTGGCATCC   9480
CCGCAGAAGGGCGACTTAGTGGGAAAAGTTTCACCTATAAGTCCGAAGTCTTCTTCCAGT   9540
GCAAATCTCCATTTATACTCGTGGGATCCTCCAGAAGAGTCTGCCAAGCTGACGGCACGT   9600
GGAGCGGCATACAACCCACCTGCATTGATCCTGCTCATAACACCTGCCCAGACCCTGGTA   9660
CGCCACACTTTGGAATACAGAATAGCTCCAGAGGCTATGAGGTTGGAAGCACGGTTTTTT   9720
TCAGGTGCAGAAAAGGCTACCATATTCAAGGTTCCACGACTCGCACCTGCCTTGCCAATT   9780
TAACATGGAGTGGGATACAGACCGAATGTATACCTCATGCCTGCAGACAGCCAGAAACCC   9840
CGGCACACGCGGATGTGAGAGCCATCGATCTTCCTACTTTCGGCTACACCTTAGTGTACA   9900
CCTGCCATCCAGGCTTTTTCCTCGCAGGGGGATCTGAGCACAGAACATGTAAAGCAGACA   9960
TGAAATGGACAGGAAAGTCGCCTGTGTGTAAAAGTAAAGGAGTGAGAGAAGTTAATGAAA  10040
CAGTTACTAAAACTCCAGTTCCTTCAGATGTCTTTTTCGTCAATTCACTGTGGAAGGGGT  10100
ATTATGAATATTTAGGGAAAAGACAACCCGCCACTCTAACTGTTGACTGGTTCAATGCAA  10160
CAAGCAGTAAGGTGAATGCCACCTTCAGCGAAGCCTCGCCAGTGGAGCTGAAGTTGACAG  10220
GCATTTACAAGAAGGAGGAGGCCCACTTACTCCTGAAAGCTTTTCAAATTAAAGGCCAGG  10280
CAGATATTTTTGTAAGCAAGTTCGAAAATGACAACTGGGGACTAGATGGTTATGTGTCAT  10340
CTGGACTTGAAAGAGGAGGATTTACTTTTCAAGGTGACATTCATGGAAAAGACTTTGGAA  10400
AATTTAAGCTAGAAAGGCAAGATCCTTTAAACCCAGATCAAGACTCTTCCAGTCATTACC  10460
ACGGCACCAGCAGTGGCTCTGTGGCGGCTGCCATTCTGGTTCCTTTCTTTGCTCTAATTT  10520
TATCAGGGTTTGCATTTTACCTCTACAAACACAGAACGAGACCAAAAGTTCAATACAATG  10580
GCTATGCTGGGCATGAAAACAGCAATGGACAAGCATCGTTTGAAAACCCCATGTATGATA  10640
CAAACTTAAAACCCACAGAAGCCAAGGCTGTGAGGTTTGACACAACTCTGAACACAGTCT  10700
GTACAGTGGTATAG
```

GENOMIC STRUCTURE

The CSMD1 gene is estimated to span more than 2 Mb and contains 69 exons (Fig 35.2).[1]

ACCESSION NUMBERS

Human	AF333704
	NP150094
Mouse	AY017475
	NP444401
Rat	DQ124115

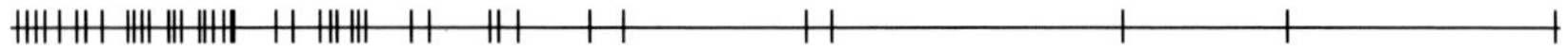

FIGURE 35.2 The human exon/intron structure of CSMD1. CSMD1 is composed of 69 exons (denoted as *vertical black bars*) interspersed over approximately a 2 MB region on chromosome 8.

DEFICIENCY

No total CSMD1 deficiency has been reported. Loss of expression of CSMD1 due to partial deletion is associated with many cancers including head and neck squamous cell cancer; lung squamous cell carcinoma; and cancers of the breast, colon, prostate and liver.[1,7–10]

POLYMORPHIC VARIANTS

There are a total of four human CSMD isoforms. CSMD1 is the canonical full-length isoform. Isoform 2 ('short') is generated by alternative slicing, is missing sushi domains 17 and 18 and has a predicted molecular weight of 370K. Isoform 3 has an alternate sequence for sushi domain 12 and is truncated at the end of that domain with a predicted molecular weight of 230K. Isoform 4 has an amino acid change of A–K at sequence 2013 and is truncated at that point with an estimated molecular weight of 219K. CSMD2 and CSMD3 are two additional family members based on the high degree of sequence homology to CSMD1. Both proteins contain 14 CUB domains and either 26 (CSMD2) or 27 sushi domains (CSMD3).[4] Neither of the latter two proteins has been assessed for complement inhibitory activity.

MUTANT ANIMALS

No mutant CSMD1 mutant animal has been reported.

REFERENCES

1. Sun PC, Uppaluri R, Schmidt AP, et al. Transcript map of the 8p23 putative tumor suppressor region. *Genomics* 2001;**75**:17–25.

2. Kraus DM, Elliott GS, Chute H, et al. CSMD1 is a novel multiple domain complement-regulatory protein highly expressed in the central nervous system and epithelial tissues. *J Immunol* 2006;**176**:4419–30.

3. Escudero-Esparza A, Kalchishkova N, Kurbasic E, Jiang WG, Blom AM. The novel complement inhibitor human CUB and Sushi multiple domains 1 (CSMD1) protein promotes factor I-mediated degradation of C4b and C3b and inhibits the membrane attack complex assembly. *FASEB J* 2013;**27**:5083–93.

4. Lau WL, Scholnick SB. Identification of two new members of the CSMD gene family. *Genomics* 2003;**82**:412–5.

5. Tang MR, Wang YX, Guo S, Han SY, Wang D. CSMD1 exhibits antitumor activity in A375 melanoma cells through activation of the Smad pathway. *Apoptosis* 2012;**17**:927–37.

6. Lotan A, Fenckova M, Bralten J, et al. Neuroinformatic analyses of common and distinct genetic components associated with major neuropsychiatric disorders. *Front Neurosci* 2014;**8**:331.

7. Farrell C, Crimm H, Meeh P, et al. Somatic mutations to CSMD1 in colorectal adenocarcinomas. *Cancer Biol Ther* 2008;**7**:609–13.

8. Midorikawa Y, Yamamoto S, Tsuji S, et al. Allelic imbalances and homozygous deletion on 8p23.2 for stepwise progression of hepatocarcinogenesis. *Hepatology* 2009;**49**:513–22.

9. Sigbjornsdottir BI, Ragnarsson G, Agnarsson BA, et al. Chromosome 8p alterations in sporadic and BRCA2 999del5 linked breast cancer. *J Med Genet* 2000;**37**:342–7.

10. Kamal M, Shaaban AM, Zhang L, et al. Loss of CSMD1 expression is associated with high tumour grade and poor survival in invasive ductal breast carcinoma. *Breast Cancer Res Treat* 2010;**121**:555–63.

Part VI

Anaphylatoxin and Leucocyte Receptors

Chapter 36

C3aR1

Liam G. Coulthard[1,2], Owen A. Hawksworth[3], Trent M. Woodruff[3]

[1]Royal Brisbane and Women's Hospital, Herston, QLD, Australia; [2]University of Queensland, Herston, QLD, Australia; [3]University of Queensland, St. Lucia, QLD, Australia

OTHER NAMES

Complement factor 3a receptor, C3a-R, anaphylatoxin C3a receptor.

PHYSIOCHEMICAL PROPERTIES

C3aR is produced as a single-polypeptide chain product of 482 residues. The human receptor has a predicted molecular weight of 53.9 kDa and isoelectric point of 6.2. The mature receptor contains a disulphide bond between the cysteines at positions 95 and 172, which are contained within extracellular loops 1 and 2, respectively. Several other posttranslational modifications have been reported, including two N-linked glycosylation sites (residues 9 and 194), one O-linked glycosylation site (266), three sulphotyrosines (174, 184, 318) and two potential phosphorylation sites on the intracellular tail (459, 463).

STRUCTURE

C3aR is a G-protein coupled receptor with seven transmembrane α-helices and an extracellular N-terminus (Fig. 36.1). The outer extracellular loops participate in ligand binding and activation, with these functions occurring at spatial separate regions of the receptor, similar to the other anaphylatoxin receptors.[1] Interestingly, C3aR exhibits a large second extracellular loop of 172 amino acids.[2] The size of this extracellular loop is largest in mammals, with cloning of rainbow trout and Xenopus C3aR demonstrating smaller loops that are not as remarkable, in terms of size, as the mammalian form.[3] The loop is of interest because the residues that are important for C3a binding appear to be well conserved, rising questions why the mammalian version is so large.[3–5] It is unlikely that this evolutionarily sequential extension of extracellular loop 2 is without benefits, and it is hypothesised that the loop may contribute to the binding of other, as yet unidentified, ligands.[4]

The Complement FactsBook. http://dx.doi.org/10.1016/B978-0-12-810420-0.00036-5
385

FIGURE 36.1 Diagram of the domain structure of C3aR.

The intracellular C-terminus contains several residues that can undergo phosphorylation. C3a-induced desensitisation of C3aR involves phosphorylation of the receptor by G-protein-coupled receptor kinases (GRK) in a rapid, dose-dependent and reversible fashion.[6] This desensitisation is augmented by receptor internalisation in response to C3a. Internalisation of the C3a receptor has been demonstrated in granulocytes to be dependent on the activity of protein kinase C and is insensitive to pertussis toxin inhibition of C3aR signalling.[7]

FUNCTION

C3aR has traditionally been grouped collectively with C5aR1 as a proin-flammatory mediator, although now this description has been challenged.[8] The function of C3aR is dependent on cell type and, within the same cell type, other environmental cues. In the immune response, C3aR has separate and paradoxical functions on granulocytes, which are likely dependent on cell state and C3a concentration. In the bone marrow, C3aR signalling acts to prevent mobilisation of neutrophils, whereas at the site of infection C3aR causes degranulation of basophils, eosinophils and mast cells.[9-12] C3aR has several proinflammatory properties overlapping with C5aR; however, the anaphylatoxin C3a is much less potent than C5a and also mediates anti-inflammatory functions.[1] In vitro investigations into the effects of C3a binding to C3aR on LPS-induced cytokine release by peripheral blood mononuclear cells (PBMCs) demonstrate different effects depending on the activation state of the cell. C3a augments proinflammatory cytokine production of adherent cells, but suppresses that of nonadherent cells.[13] Depending on the adhesive state of cells, C3a suppresses TNF-α, IL-1β and IL-6 synthesis in lipopolysaccharide-primed nonadherent PBMCs and B lymphocytes and enhances their synthesis in adherent cell systems.[13-15]

In human eosinophils, C3a binding to C3aR induces calcium mobilisation, oxidative burst and degranulation.[11,16] In mast cells, C3a can induce histamine secretion in a dose-related fashion.[17-19] C3a can also increase intracellular calcium levels in human monocytes, and induce production of IL-1β, TNF-α, IL-6 and PGE$_2$.[13,14,20-23] This eosinophil and mast cell-specificity has highlighted C3aR as a drug-target candidate for allergic conditions. Such inferences have been strengthened through observations of *C3aR*$^{-/-}$ animals that exhibit reduced responses in atopic diseases such as allergic asthma and allergic dermatitis.[24] Although the C3a/C3aR axis has been demonstrated to cause chemotaxis of eosinophils and basophils, interestingly and in stark contrast to C5a/C5aR activation, neutrophils do not undergo chemotaxis, granular release or oxidative burst in the presence of C3a.[25] Earlier studies had reported a functional role for C3aR on neutrophils;[16] however, these were observations of indirect mechanisms. For instance, C3a has been shown to cause chemotaxis of neutrophils in cultures that contain >5% eosinophils, and eosinophils release a soluble factor in response to C3a that causes neutrophil chemotaxis.[25]

The C3a/C3aR axis has a role in modulating the response of lymphocytes to a pathogenic insult via the actions of an intermediate cell type.[26–28] Several studies have demonstrated functional roles for C3aR within the lymphoid lineage itself. C3aR-expressing, T lymphocytes were demonstrated in patients with severe inflammatory skin diseases, and C3aR expression could be induced by administration of type I interferons.[29] Alternatively, an emerging paradigm is the presence of intracellular anaphylatoxin receptors,[30] which may have eluded earlier attempts at detection. Human T cells express both C3aR and the C3a ligand within intracellular vesicles. In the resting state, production of C3a contributes to T cell survival. Once activated, both the receptor and the ligand move to the surface to promote a Th1 response.[30]

Administration of C3a has been demonstrated to augment the T cell response, promote T cell proliferation/survival, and prolong the inflammatory response through suppressing regulatory T cells (T_{reg}) production.[31–33] In support of this, T cell populations are reduced in $C3aR^{-/-}$ mice in several models of disease.[31,33,34] T cell receptor stimulation upregulates C3aR mRNA expression in isolated T cell populations, and the presence of C3aR appears to promote a T_h1 response, as measured by IFNγ.[32] This is supported by in vivo observations of increased IFNγ production in mice lacking the complement suppressive factor, *Daf*.[35] In addition, C3a is required for the inflammasome-mediated production of IL-β within monocytes, promoting Th$_{17}$ activity.[36] C3aR signalling also acts in concert with C5aR1 signalling to suppress the production of TGF-β1 from dendritic cells, reducing the stimulus for differentiation to T_{reg} cells.[33] The induced T_{reg} cells heavily suppress the proinflammatory CD4⁺ T cell function, leading to abrogation of the inflammatory process. In addition, C3aR signalling in tissue antigen presenting cells suppresses IL-4 production, inhibiting a T_H2 polarised response.[37] However, this effect on T cell biology is not entirely APC-mediated, as altered T_{reg} responses also occur with adoptive transfer of $C5aR1^{-/-}/C3aR^{-/-}$ T cells into a wild-type animal.[33] $C5aR1^{-/-}/C3aR^{-/-}$ T_{reg} cells also demonstrate prolonged survival and enhanced function, suggesting that C3aR/C5aR1 stimulus of T cells is important in regulating inflammatory responses.[34]

In an experimental model of allograft rejection mediated by CD4⁺ priming of cytotoxic T cells, CD4⁺ T cell-depleted mice have only a weak CD8⁺

response to heart transplants. Local complement dysregulation by removal of *Daf* prevented the protection of CD4$^+$ T cell depletion after heart transplant. Although not confirmed by pharmacological blockade of any complement receptor, authors postulated that there is a direct activation effect of C3a/C5a on CD8$^+$ T cells.[38]

In development, chemotactic signalling of C3aR controls morphogenetic processes including radial intercalation during epiboly; collective migration of neural crest cells; and migration of cerebellar granule neurons, cardiac progenitor cells and mesenchymal stem cells.[39–42] In adults, C3aR contributes to the regulation of metabolism, regulation of Chematopoietic stem/progenitors cells mobilisation and the survival of hepatocytes following liver injury.[43–45]

TISSUE DISTRIBUTION

C3aR was initially shown to be expressed on leucocytes of myeloid lineage including neutrophils, basophils, eosinophils, mast cells and monocyte/macrophages.[46] C3aR expression has also been demonstrated intracellularly within T cells, and the receptor is shuttled to the surface upon cell activation.[30] It is now well accepted that C3aR is expressed on endothelial cells; smooth muscle cells and cells of brain origin, such as astrocytes, microglia and neurons.[22,39,47,48] C3aR expression has also been demonstrated in mesenchymal, neural and haematopoietic stem cells.[39,42,49]

REGULATION OF EXPRESSION

The C3aR gene, *C3AR*, has a conserved region 70 to 35 bases upstream of the transcriptional start site, which is critical to cell type-specific regulation.[50] This region contains a GATA, ETS-like and AP-1 site, but the relative contribution of these sites to gene transcription has been demonstrated to be dependent on the cell type. For instance, transcriptional control in myeloid-derived cells, which are reliant on the functionality of both the ETS and AP-1 elements, differs to that reported in astrocytes, where there is heavy reliance on the AP-1 site. The deletion of the ETS-like site does not reduce transcription significantly.[51,52] Additionally, although transcription from the AP-1 site in macrophages was initiated by c-Jun binding, there was no indication of this interaction occurring in primary astrocytes or the Ast2.1 cell line.[52]

HUMAN PROTEIN SEQUENCE

The human peptide sequence for C3aR is presented below. Underlined are the putative glycosylation sites (N9, N194, S266), sulphation sites (Y174, Y184, Y318) and disulphide bond (C95-C172, *additionally highlighted*).

```
MASFSAETNS TDLLSQPWNE PPVILSMVIL SLTFLLGLPG NGLVLWVAGL  50

KMQRTVNTIW FLHLTLADLL CCLSLPFSLA HLALQGQWPY GRFLCKLIPS 100

IIVLNMFASV FLLTAISLDR CLVVFKPIWC QNHRNVGMAC SICGCIWVVA 150

FVMCIPVFVY REIFTTDNHN RCGYKFGLSS SLDYPDFYGD PLENRSLENI 200

VQPPGEMNDR LDPSSFQTND HPWTVPTVFQ PQTFQRPSAD SLPRGSARLT 250

SQNLYSNVFK PADVVSPKIP SGFPIEDHET SPLDNSDAFL STHLKLFPSA 300

SSNSFYESEL PQGFQDYYNL GQFTDDDQVP TPLVAITITR LVVGFLLPSV 350

IMIACYSFIV FRMQRGRFAK SQSKTFRVAV VVVAVFLVCW TPYHIFGVLS 400

LLTDPETPLG KTLMSWDHVC IALASANSCF NPFLYALLGK DFRKKARQSI 450

QGILEAAFSE ELTRSTHCPS NNVISERNST TV                   482
```

PROTEIN MODULES

AA1-23	N-Terminus	exon 2
AA24-46	Transmembrane domain 1	exon 2
AA47-57	Intracellular loop 1	exon 2
AA58-80	Transmembrane domain 2	exon 2
AA81-96	Extracellular loop 1	exon 2
AA97-118	Transmembrane domain 3	exon 2
AA119-139	Intracellular loop 2	exon 2
AA140-160	Transmembrane domain 4	exon 2
AA161-340	Extracellular loop 2	exon 2
AA341-360	Transmembrane domain 5	exon 2
AA361-377	Intracellular loop 3	exon 2
AA378-400	Transmembrane domain 6	exon 2
AA401-417	Extracellular loop 3	exon 2
AA418-438	Transmembrane domain 7	exon 2
AA439-482	C-Terminus	exon 2

CHROMOSOMAL LOCATION

The C3aR gene (*C3AR*) has the chromosomal location 12p13.31 in humans and 6F1 in mice.

HUMAN cDNA SEQUENCE

Human cDNA sequence is presented below. Three cDNA variants have been identified, which differ only in the length of the 3′ untranslated regions. For simplicity only variant 1 has been presented. The first five nucleotides of exon 2 are underlined (position 195–199). Initiation and termination signals are highlighted and underlined (positions 205–207 and 1650–1652, respectively). Putative poly-adenylation signal is underlined (position 3496–3501).

```
CTGTGAGGTC AGATAGTGGT CTAGAGCATA AGACTTAACT TATTGCCGGA AACAGAGAGA    60

GAACAGAAGA AGAGAAAGCT CAGCAAATTT TCTTGCCATA CTTCATGACT TCACTGTGGC   120

TAAGTGTGGG GACCAGACAG GACTCGTGGA GACATCCAGG TGCTGAAGCC TTCAGCTACT   180

GTCTCAGTTT TTTGAAGTTT AGCAATGGCG TCTTTCTCTG CTGAGACCAA TTCAACTGAC   240

CTACTCTCAC AGCCATGGAA TGAGCCCCCA GTAATTCTCT CCATGGTCAT TCTCAGCCTT   300

ACTTTTTTAC TGGGATTGCC AGGCAATGGG CTGGTGCTGT GGGTGGCTGG CCTGAAGATG   360

CAGCGGACAG TGAACACAAT TTGGTTCCTC CACCTCACCT TGGCGGACCT CCTCTGCTGC   420

CTCTCCTTGC CCTTCTCGCT GGCTCACTTG GCTCTCCAGG GACAGTGGCC CTACGGCAGG   480

TTCCTATGCA AGCTCATCCC CTCCATCATT GTCCTCAACA TGTTTGCCAG TGTCTTCCTG   540

CTTACTGCCA TTAGCCTGGA TCGCTGTCTT GTGGTATTCA AGCCAATCTG GTGTCAGAAT   600

CATCGCAATG TAGGGATGGC CTGCTCTATC TGTGGATGTA TCTGGGTGGT GGCTTTTGTG   660

ATGTGCATTC CTGTGTTCGT GTACCGGGAA ATCTTCACTA CAGACAACCA TAATAGATGT   720

GGCTACAAAT TTGGTCTCTC CAGCTCATTA GATTATCCAG ACTTTTATGG AGATCCACTA   780

GAAAACAGGT CTCTTGAAAA CATTGTTCAG CCGCCTGGAG AAATGAATGA TAGGTTAGAT   840

CCTTCCTCTT TCCAAACAAA TGATCATCCT TGGACAGTCC CCACTGTCTT CCAACCTCAA   900

ACATTTCAAA GACCTTCTGC AGATTCACTC CCTAGGGGTT CTGCTAGGTT AACAAGTCAA   960

AATCTGTATT CTAATGTATT TAAACCTGCT GATGTGGTCT CACCTAAAAT CCCCAGTGGG  1020

TTTCCTATTG AAGATCACGA AACCAGCCCA CTGGATAACT CTGATGCTTT TCTCTCTACT  1080

CATTTAAAGC TGTTCCCTAG CGCTTCTAGC AATTCCTTCT ACGAGTCTGA GCTACCACAA  1140

GGTTTCCAGG ATTATTACAA TTTAGGCCAA TTCACAGATG ACGATCAAGT GCCAACACCC  1200

CTCGTGGCAA TAACGATCAC TAGGCTAGTG GTGGGTTTCC TGCTGCCCTC TGTTATCATG  1260

ATAGCCTGTT ACAGCTTCAT TGTCTTCCGA ATGCAAAGGG GCCGCTTCGC CAAGTCTCAG  1320

AGCAAAACCT TTCGAGTGGC CGTGGTGGTG GTGGCTGTCT TTCTTGTCTG CTGGACTCCA  1380

TACCACATTT TTGGAGTCCT GTCATTGCTT ACTGACCCAG AAACTCCCTT GGGGAAAACT  1440

CTGATGTCCT GGGATCATGT ATGCATTGCT CTAGCATCTG CCAATAGTTG CTTTAATCCC  1500

TTCCTTTATG CCCTCTTGGG GAAAGATTTT AGGAAGAAAG CAAGGCAGTC CATTCAGGGA  1560

ATTCTGGAGG CAGCCTTCAG TGAGGAGCTC ACACGTTCCA CCCACTGTCC CTCAAACAAT  1620

GTCATTTCAG AAAGAAATAG TACAACTGTG TGAAAATGTG GAGCAGCCAA CAAGCAGGGG  1680

CTCTTAGGCA ATCACATAGT GAAAGTTTAT AAGAGGATGA AGTGATATGG TGAGCAGCGG  1740

ACTTCAAAAA CTGTCAAAGA ATCAATCCAG CGGTTCTCAA ACGGTACACA GACTATTGAC  1800
```

HUMAN cDNA SEQUENCE – *Continued*

```
ATCA GCATCA CCTAGAAACT TGTTAGAAAT GCAAATTCTC AAGCCGCATC CCAGACTTGC    1860

TGAATCGGAA TCTCTGGGGG TTGGGACCCA GCAAGGGCAC TTAACAAACC CTCGTTTCTG    1920

ATTAATGCTA AATGTAAGAA TCATTGTAAA CATTAGTTCT ATTTCTATCC CAAACTAAGC    1980

TATGTGAAAT AAGAGAAGCT ACTTTGTTTT TAAATGATGT TGAATATTTG TCGATATTTC    2040

CATCATTAAA TTTTTCCTTA GCATTGTCTA AGTCTTCCAG ATTCCATTTA AAACCATTTC    2100

TTGTTCTCCT ACGTGAGTGA AAGATGATCA TATATCCTAA TGCTTTGTTG TCGTGTGGTG    2160

TTGATGGTTT TAAACGAAAA GAAAGTGCAA AAAGAAAATG CCTGTGAAGA CAAGAAGCCA    2220

TGAGACTGAG TCTGGAGCAT AGGGTTATGC AATGATGCCT GTCCCTGGGA CACCCCTGG     2280

GTACAGGATA TAGAAATTTC CACTATTACA TAGAGTTTCC ACTATTACAA CTAAATAAGC    2340

ATCTATTGTG TGAAAACTGA CTCATGAAAT GTTATGAAAG CTGTGGTTTG GGGAGTTCTG    2400

TTTCTTCTAA CTGCCTACCG GTTGGGCACC TATTTTCCAC TCCTCTTCCT AAGCTCCTTA    2460

ATTTCCTTAT TACTCCCCAG CCTCCAAATC TTCCACATCA GACTTTGTGC CTCAAACAAC    2520

CTCTAATTTC GTAAGATTCT AGTTACTCCC TTCCTCTTGC TCCAAATGAA TACTTTCTAA    2580

GAAAGTATTT CAAGTGGAAG GAGAAAGAGG GTGGAGGATG GAGCAGCAAT TCTTCTACTC    2640

TCTGCAACTG AGTACCCTAC CAGGCTTGCC ATCACATTTT AAAACATGAC GACAGGCAAC    2700

TTACATGCCA AAATTACCAA ATATATCTTC TGGGTTTTTT AAATCCTTTT CTTTGCCAAA    2760

GTAATACATG CACATAGTTT TAAAATAATT TAATAAGGTA TATAATGAAA TATGAGGTCT    2820

CCTACCTCAC TGTGCCCAAA AGTTCCCTCC TCCCACTCTC ATTTCCCAGA GATAATCCTT    2880

GCACAATTTT AGATGTTTCC TTTGATAATT ATCATGATGT TTCTAAATCA TGTGCTTATG    2940

CTGCTCTTTT CTGGAGGCAT GATAAAACGA CTTCTTGTTT TGAAAGATGA AGATGTTTAT    3000

CCAAGCACCC CATATTTTTA ATTTGTTTAT CCAGCATCCC AACATTCATT AATAACCATA    3060

TTTTAATTCA TTCATGACCA CATATTTTTC TTCTACTTTG TCTATACACT CCAACCATTT    3120

ATATAGCTTT CCTTCTGTCC CTTTTTCATT TAAAACAAAA TTACCTAACT CCCTACCACC    3180

TTCTCATTTT TCTGTATATA TAAATGTTTG TGTCAAACGT CTGAAATTTC TGGCTTGTTT    3240

GTATCACAAC GTGGCCTCAT CTAAACCAAA TACAATGATG TAGTCTAAAA ACAGAAAATG    3300

ACATGTGTTT TAGACCTGCA AGACACTATC TGTTCAATGG CTGAGGTGAG GGTCTGGACT    3360

ACAGATTTTT TATAAAGTAT ATGCAGAAAA ATTACAAATC ACTAGGAATT CTTTCAGTTG    3420

TGAAGAATGT CTGACATAAG ATTTGAAGTG CTACCTTTCC AGCTTATATA TTAATTTGCT    3480

TATATATTTG ATATG<u>AATAA A</u>TGCTTTTTT TCTCATGGGT CCTTGCGAGG CTCAGAGATT    3540

TATGAAAAAA AAAAAAAA                                                  3558
```

GENOMIC STRUCTURE

The gene encoding C3aR (*C3AR1*) consists of two exons separated by a large (~6 kB) intron. The entire coding sequence for C3aR is contained within exon 2.

5,001   14,628

Exon 1   Exon 2

FIGURE 36.2   The structure of the human gene on chromosome 12. Numbers at each end of the locus represent the nucleotide position within the chromosome. The darker grey colour represents the coding sequence within exon 2.

C3aR shares similarity in genomic structure with the other complement anaphylatoxins. In addition, this structure is conserved between the mammalian species with nonhuman primates and rodents sharing the 'two exon' structure (Fig. 36.2).

ACCESSION NUMBERS

Homo sapiens
 Variant 1   NM_004054.3
 Variant 2   NM_001326475.1
 Variant 3   NM_001326477.1
Mus musculus   NM_009779.2
Rattus norvegicus   NM_032060.1
Macaca fascicularis   NM_001319375.1
Gallus gallus   NM_001030769.2
Xenopus tropicalis (predicted) XM_002941206.3
Danio rerio (predicted)   XM_009305062.1

DEFICIENCY

No cases of human C3aR deficiency have been reported.

POLYMORPHIC VARIANTS

V136A – No change in function has been reported for this polymorphic variant.

MUTANT ANIMALS

C3aR knockout mice are available from the JAX laboratory (C.129S4-C3ar1[tm1Cge]/J).[53] These mice were generated using a neomycin cassette to knockout a 736-bp fragment including the 3′ end of the C3aR intron and the 5′ end of exon 2. The knockout construct was created using 129S4/SvJae embryonic stem cells, and the resulting litters were backcrossed to BALB/cAnNCrl mice for 15 generations. Several labs have also generated C3aR knockout mice backcrossed to other strains, including the C57BL6/J strain.[54]

 C3aR[-/-] mice exhibit no gross morphological deficits or impaired development. In infectious diseases, the phenotype of *C3aR*[-/-] animals is complex. They have increased mortality in models of sepsis, potentially due to the role of C3aR in suppressing neutrophil mobilisation from the bone marrow.[9,54] An increased rate of renal damage is seen in models of lupus nephritis in *C3aR*[−/−] mice. However, knockout of C3aR is protective in models of *Pseudomonas aeruginosa* pneumonia and the development of allergic asthma.[55,56] In the animal model for multiple sclerosis, experimental autoimmune encephalomyelitis, *C3aR*[−/−] mice presented with attenuated disease and reduced inflammatory infiltrate into the central nervous system (CNS), while transgenic mice expression of C3a in the CNS had exacerbated disease, massive meningeal infiltration of macrophages and significantly increased mortality.[57]

REFERENCES

1. Ember JA, Hugli TE. Complement factors and their receptors. *Immunopharmacology* 1997;**38**:3–15.
2. Ames RS, Nuthulaganti P, Kumar C. In *Xenopus* oocytes the human C3a and C5a receptors elicit a promiscuous response to the anaphylatoxins. *FEBS Lett* 1996;**395**:157–9.
3. Boshra H, et al. Characterization of a C3a receptor in rainbow trout and *Xenopus*: the first identification of C3a receptors in nonmammalian species. *J Immunol* 2005;**175**:2427–37.
4. Gao J, et al. Sulfation of tyrosine 174 in the human C3a receptor is essential for binding of C3a anaphylatoxin. *J Biol Chem* 2003;**278**:37902–8.
5. Chao TH, et al. Role of the second extracellular loop of human C3a receptor in agonist binding and receptor function. *J Biol Chem* 1999;**274**:9721–8.
6. Langkabel P, Zwirner J, Oppermann M. Ligand-induced phosphorylation of anaphylatoxin receptors C3aR and C5aR is mediated by G protein-coupled receptor kinases. *Eur J Immunol* 1999;**29**:3035–46.
7. Settmacher B, et al. Modulation of C3a activity: internalization of the human C3a receptor and its inhibition by C5a. *J Immunol* 1999;**162**:7409–16.
8. Coulthard LG, Woodruff TM. Is the complement activation product C3a a proinflammatory molecule? Re-evaluating the evidence and the myth. *J Immunol* 2015;**194**:3542–8.
9. Wu MCL, et al. The receptor for complement component C3a mediates protection from intestinal ischemia-reperfusion injuries by inhibiting neutrophil mobilization. *Proc Natl Acad Sci USA* 2013;**110**:9439–44.
10. Vibhuti A, Gupta K, Subramanian H, Guo Q, Ali H. Distinct and shared roles of β-arrestin-1 and β-arrestin-2 on the regulation of C3a receptor signaling in human mast cells. *PLoS One* 2011;**6**:e19585.
11. Takafuji S, Tadokoro K, Ito K, Dahinden CA. Degranulation from human eosinophils stimulated with C3a and C5a. *Int Arch Allergy Immunol* 1994;**104**(Suppl. 1):27–9.
12. Ahamed J, Haribabu B, Ali H. Cutting edge: differential regulation of chemoattractant receptor-induced degranulation and chemokine production by receptor phosphorylation. *J Immunol* 2001;**167**:3559–63.
13. Takabayashi T, et al. A new biologic role for C3a and C3a desArg: regulation of TNF-alpha and IL-1 beta synthesis. *J Immunol* 1996;**156**:3455–60.
14. Takabayashi T, et al. Both C3a and C3a(desArg) regulate interleukin-6 synthesis in human peripheral blood mononuclear cells. *J Infect Dis* 1998;**177**:1622–8.
15. Fischer WH, Hugli TE. Regulation of B cell functions by C3a and C3a(desArg): suppression of TNF-alpha, IL-6, and the polyclonal immune response. *J Immunol* 1997;**159**:4279–86.

16. Elsner J, Oppermann M, Czech W, Kapp A. C3a activates the respiratory burst in human polymorphonuclear neutrophilic leukocytes via pertussis toxin-sensitive G-proteins. *Blood* 1994;**83**:3324–31.

17. Kubota Y. The effect of human anaphylatoxins and neutrophils on histamine release from isolated human skin mast cells. *J Dermatol* 1992;**19**:19–26.

18. Mousli M, Hugli TE, Landry Y, Bronner C. A mechanism of action for anaphylatoxin C3a stimulation of mast cells. *J Immunol* 1992;**148**:2456–61.

19. Woolhiser MR, Brockow K, Metcalfe DD. Activation of human mast cells by aggregated IgG through FcgammaRI: additive effects of C3a. *Clin Immunol* 2004;**110**:172–80.

20. Haeffner-Cavaillon N, Cavaillon JM, Laude M, Kazatchkine MD. C3a(C3adesArg) induces production and release of interleukin 1 by cultured human monocytes. *J Immunol* 1987;**139**:794–9.

21. Monsinjon T, Gasque P, Ischenko A, Fontaine M. C3A binds to the seven transmembrane anaphylatoxin receptor expressed by epithelial cells and triggers the production of IL-8. *FEBS Lett* 2001;**487**:339–46.

22. Monsinjon T, et al. Regulation by complement C3a and C5a anaphylatoxins of cytokine production in human umbilical vein endothelial cells. *FASEB J* 2003;**17**:1003–14.

23. Fischer WH, Jagels MA, Hugli TE. Regulation of IL-6 synthesis in human peripheral blood mononuclear cells by C3a and C3a(desArg). *J Immunol* 1999;**162**:453–9.

24. Bautsch W, et al. Cutting edge: guinea pigs with a natural C3a-receptor defect exhibit decreased bronchoconstriction in allergic airway disease: evidence for an involvement of the C3a anaphylatoxin in the pathogenesis of asthma. *J Immunol* 2000;**165**:5401–5.

25. Daffern PJ, Pfeifer PH, Ember JA, Hugli TE. C3a is a chemotaxin for human eosinophils but not for neutrophils. I. C3a stimulation of neutrophils is secondary to eosinophil activation. *J Exp Med* 1995;**181**:2119–27.

26. Morgan EL, Thoman ML, Hobbs MV, Weigle WO, Hugli TE. Human C3a-mediated suppression of the immune response. II. Suppression of human in vitro polyclonal antibody responses occurs through the generation of nonspecific OKT8+ suppressor T cells. *Clin Immunol Immunopathol* 1985;**37**:114–23.

27. Francis K, et al. Complement C3a receptors in the pituitary gland: a novel pathway by which an innate immune molecule releases hormones involved in the control of inflammation. *FASEB J* 2003;**17**:2266–8.

28. Francis K, Lewis BM, Monk PN, Ham J. Complement C5a receptors in the pituitary gland: expression and function. *J Endocrinol* 2008;**199**:417–24.

29. Werfel T, et al. Activated human T lymphocytes express a functional C3a receptor. *J Immunol* 2000;**165**:6599–605.

30. Liszewski MK, et al. Intracellular complement activation sustains T cell homeostasis and mediates effector differentiation. *Immunity* 2013;**39**:1143–57.

31. Lim H, et al. Negative regulation of pulmonary Th17 responses by C3a anaphylatoxin during allergic inflammation in mice. *PLoS One* 2012;**7**:e52666.

32. Strainic MG, et al. Locally produced complement fragments C5a and C3a provide both costimulatory and survival signals to naive CD4+ T cells. *Immunity* 2008;**28**:425–35.

33. Strainic MG, Shevach EM, An F, Lin F, Medof ME. Absence of signaling into CD4(+) cells via C3aR and C5aR enables autoinductive TGF-β1 signaling and induction of Foxp3(+) regulatory T cells. *Nat Immunol* 2012. http://dx.doi.org/10.1038/ni.2499.

34. Kwan W-H, van der Touw W, Paz-Artal E, Li MO, Heeger PS. Signaling through C5a receptor and C3a receptor diminishes function of murine natural regulatory T cells. *J Exp Med* 2013;**210**:257–68.

35. Liu J, et al. IFN-gamma and IL-17 production in experimental autoimmune encephalomyelitis depends on local APC-T cell complement production. *J Immunol* 2008;**180**:5882–9.

36. Asgari E, et al. C3a modulates IL-1β secretion in human monocytes by regulating ATP efflux and subsequent NLRP3 inflammasome activation. *Blood* 2013;**122**:3473–81.

37. Kawamoto S, et al. The anaphylatoxin C3a downregulates the Th2 response to epicutaneously introduced antigen. *J Clin Invest* 2004;**114**:399–407.

38. Vieyra M, et al. Complement regulates CD4 T-cell help to CD8 T cells required for murine allograft rejection. *Am J Pathol* 2011;**179**:766–74.

39. Rahpeymai Y, et al. Complement: a novel factor in basal and ischemia-induced neurogenesis. *EMBO J* 2006;**25**:1364–74.

40. Carmona-Fontaine C, et al. Complement fragment C3a controls mutual cell attraction during collective cell migration. *Dev Cell* 2011;**21**:1026–37.

41. Szabó A, et al. The molecular basis of radial intercalation during tissue spreading in early development. *Dev Cell* 2016;**37**:213–25.

42. Schraufstatter IU, Discipio RG, Zhao M, Khaldoyanidi SK. C3a and C5a are chemotactic factors for human mesenchymal stem cells, which cause prolonged ERK1/2 phosphorylation. *J Immunol* 2009;**182**:3827–36.

43. Lim J, et al. C5aR and C3aR antagonists each inhibit diet-induced obesity, metabolic dysfunction, and adipocyte and macrophage signaling. *FASEB J* 2013;**27**:822–31.

44. Hsieh C-C, et al. The role of complement component 3 (C3) in differentiation of myeloid-derived suppressor cells. *Blood* 2013. http://dx.doi.org/10.1182/blood-2012-06-440214.

45. Markiewski MM, et al. C3a and C3b activation products of the third component of complement (C3) are critical for normal liver recovery after toxic injury. *J Immunol* 2004;**173**:747–54.

46. Martin U, et al. The human C3a receptor is expressed on neutrophils and monocytes, but not on B or T lymphocytes. *J Exp Med* 1997;**186**:199–207.

47. Gasque P, et al. The receptor for complement anaphylatoxin C3a is expressed by myeloid cells and nonmyeloid cells in inflamed human central nervous system: analysis in multiple sclerosis and bacterial meningitis. *J Immunol* 1998;**160**:3543–54.

48. Schraufstatter IU, Trieu K, Sikora L, Sriramarao P, DiScipio R. Complement C3a and C5a induce different signal transduction cascades in endothelial cells. *J Immunol* 2002;**169**:2102–10.

49. Reca R, et al. Functional receptor for C3a anaphylatoxin is expressed by normal hematopoietic stem/progenitor cells, and C3a enhances their homing-related responses to SDF-1. *Blood* 2003;**101**:3784–93.

50. Martin CB, Ingersoll SA, Martin BK. Transcriptional control of the C3a receptor gene in glial cells: dependence upon AP-1 but not Ets. *Mol Immunol* 2007;**44**:703–12.

51. Schaefer M, et al. The transcription factors AP-1 and Ets are regulators of C3a receptor expression. *J Biol Chem* 2005;**280**:42113–23.

52. Martin CB, Martin BK. Characterization of the murine C3a receptor enhancer-promoter: expression control by an activator protein 1 sequence and an Ets-like site. *J Immunol* 2005;**175**:3123–32.

53. Humbles AA, et al. A role for the C3a anaphylatoxin receptor in the effector phase of asthma. *Nature* 2000;**406**:998–1001.

54. Kildsgaard J, et al. Cutting edge: targeted disruption of the C3a receptor gene demonstrates a novel protective anti-inflammatory role for C3a in endotoxin-shock. *J Immunol* 2000;**165**:5406–9.

55. Wenderfer SE, Wang H, Ke B, Wetsel RA, Braun MC. C3a receptor deficiency accelerates the onset of renal injury in the MRL/lpr mouse. *Mol Immunol* 2009;**46**:1397–404.

56. Mueller-Ortiz SL, Hollmann TJ, Haviland DL, Wetsel RA. Ablation of the complement C3a anaphylatoxin receptor causes enhanced killing of *Pseudomonas aeruginosa* in a mouse model of pneumonia. *Am J Physiol Lung Cell Mol Physiol* 2006;**291**:L157–65.

57. Boos L, Campbell IC, Ames R, Wetsel RA, Barnum SR. Deletion of the complement anaphylatoxin C3a receptor attenuates, whereas ectopic expression of C3a in the brain exacerbates, experimental autoimmune encephalomyelitis. *J Immunol* 2004;**173**:4708–14.

Chapter 37

C5aR1

Liam G. Coulthard[1,2], Owen A. Hawksworth[2], Trent M. Woodruff[2]

[1]Royal Brisbane and Women's Hospital, Herston, QLD, Australia; [2]University of Queensland, Herston, QLD, Australia

OTHER NAMES

Complement factor C5a receptor, complement factor C5a receptor-1, C5aR, CD88, C5a anaphylatoxin chemotactic receptor.

PHYSIOCHEMICAL PROPERTIES

C5aR1 is produced as a single-polypeptide chain product of 350 residues. The receptor has a predicted molecular weight of 39.3 kDa and isoelectric point of 9.2. There is a single sulphide bond between the cysteine residues (Cys109, Cys188) of the first and second extracellular loops.[1] Several other posttranslational modifications are noted on this receptor, including an *N*-linked glycosylation site (position 5) and two sulphation sites (position 11 and 14) within the N-terminus. Several phosphorylation sites also exist on the intracellular C-terminus (positions 314, 317, 327, 332, 334 and 338), which likely contribute to signal transduction.

STRUCTURE

C5aR1 is a G-protein coupled receptor with seven transmembrane α-helices and an extracellular N-terminus. The N-terminus and the three extracellular loops form a complex that binds C5a.[2] The C-terminus exists intracellularly and contains a cluster of serine residues that contribute to signalling and desensitisation.[3]

There have been several point mutagenesis studies demonstrating the relative contributions of various residues that are well reviewed in Monk et al.[1] Briefly, the aspartic acid residues of the N-terminus; the sulphated Tyr11 and 14; and residues Arg175, Glu199 and Asp282 of the extracellular loops are most likely to form the C5a binding site. Activation of the receptor is also dependent on the presence of a Try-X-Phe-Gly motif in the first extracellular loop[4] (Fig. 37.1).

The Complement FactsBook. http://dx.doi.org/10.1016/B978-0-12-810420-0.00037-7

FIGURE 37.1   Domain structure of C5aR1.

FUNCTION

The functions of C5aR1 fall into two broad categories; functions relevant to the immune response and newly discovered functions relating to development and regeneration.

The classical functions of innate immunity have been well described in the literature. In this arena, C5aR1 functions to detect its ligand, C5a, and direct leucocytes towards sites of injury or pathogens. Receptor activation induces chemotaxis of all myeloid leucocytes including neutrophils, monocytes, eosinophils, mast cells and basophils, as well as tissue-resident macrophages, including brain microglia. Additionally, C5aR1 controls effector functions of these cells, inducing degranulation and proinflammatory mediator production/release and enhancing phagocytosis. It is interesting to note that in subsets of leucocytes, C5aR1 displays bimodal signalling dependent on extracellular C5a concentrations. This has been best demonstrated in neutrophils where, at low concentrations, C5aR1 acts as a sensing receptor for chemotaxis,[1] and as concentrations increase C5aR1 signalling causes degranulation.[5] Additionally, C5aR1 signalling inhibits neutrophil apoptosis, which further enhances pathogen destruction at the sites of infection.

In the surrounding tissue, C5aR1 is expressed on endothelial cells of blood vessel walls, with receptor activation causing vasodilation, increased expression of adhesion molecules and angiogenesis,[6] all of which promote extravasation of circulating immune cells to sites of insult.

Although the presence of C5aR1 in the lymphocyte lineage has been controversial, there is building evidence that the actions of C5aR1 also extend to adaptive immunity. C5aR1 exhibits varied signalling responses in T cell subsets regulating T cell biology and function. Signalling via C5aR1 has been shown to promote CD4+ T cell survival and to inhibit CD4+ Foxp3+ Treg cell functioning.[7] Additionally, it has been shown that intracellularly localised C5aR1 signals, in activated T cells, induce generation of reactive oxygen species and promote NLRP3 inflammasome assembly, augmenting positively regulating T helper 1 responses to infection.[8]

Outside of the immune system, C5aR1 signalling has also been shown to control mobilisation of supportive cell niches; inducing chemotaxis of mesenchymal stem cells, hematopoietic stem cells and cardiac progenitor cells.[9,10] Furthermore, C5aR1 promotes repair and regeneration responses following injury through induction of proliferation in hepatic, cardiac and endothelial cells.[6,10] In the developing brain, C5aR1 signalling has been shown to increase proliferation and survival of immature cerebellar granule neurons[11] and to prevent folate-deficiency-induced neural tube defects.[12] In the adult brain, mixed functions have been observed with C5aR1 acting as a prosurvival and proapoptotic signalling receptor depending on the subset of neurons and mechanism of injury.[13,14] Additionally, C5aR1 has been reported to induce astrocyte proliferation which, in the setting of spinal cord injury, increases neuronal support, reducing pathology and improving recovery of motor functions.[15]

TISSUE DISTRIBUTION

C5aR1 is expressed throughout a wide variety of tissues and cell types.[16] The receptor is expressed on all cells of a myeloid lineage in high abundance. However, notably, its expression in the lymphoid lineage is much more restricted. There has been considerable debate as to the status of C5aR1 expression in this lineage. Two separate reports have used GFP reporter mice to demonstrate C5aR1 expression, and both strains were unable to demonstrate GFP expression in T cell populations.[17,18] By contrast, studies in human CD4+ T cells demonstrate that C5aR1 is in fact expressed intracellularly.[8] Outside of the immune system, C5aR1 is expressed at lower levels in numerous nonmyeloid cells where it performs organ-specific functions. Expression of C5aR1 also occurs within the central nervous system; cells of the neuroectodermal lineage (neurons, astrocytes, oligodendrocytes) and microglia express C5aR1 at detectable levels.[19–21] In the adult brain, neuronal expression of C5aR1 is restricted to specific areas, namely the pyramidal neurons of the neocortex, neurons of the hippocampus and Purkinje cells of the cerebellum.[20,22] In development, neural progenitor cells of both the embryo and adult mice have also been demonstrated to express C5aR1.[12,23] At the earliest point of development, C5aR1 is expressed and functional on embryonic stem cells.[24]

REGULATION OF EXPRESSION

The important promoter regions of *C5ar1* occur within 500 bp of exon 1. This region contains a CCAAT enhancer box sequence (C/EB) and AP-4 and NF-κB binding sites.[25] The C/EB region is the recognition site for transcriptional activity in cells of the myeloid lineage, such as macrophages and microglia. In the human genome, these regions are conserved and important for expression of human C5aR1.[26] However, the wider promoter region (~2 kb 5′ to exon 1) of C5aR1 is not responsible for upregulation of C5aR1 with differentiation of the U937 monocyte cell line, suggesting more complex regulation of the gene yet to be discovered.[26] The expression of C5aR1 in nonmyeloid lineages is dependent on other motifs within the C5aR1 promoter. Astrocytic *C5ar1* expression has been shown to be independent of the C/EB motif used for myeloid expression and employs regulatory elements either upstream of this site, or perhaps unknown elements within the large intronic sequence.[21] Highlighting this lineage-specific difference, the NF-κB site within the C5aR1 promoter is a suppressor site in astroglia, but not microglia.[27] While both cell types are located within the same tissue and perform a supportive role in brain function, microglia are of the myeloid lineage, whereas astroglia are of the neuroectodermal lineage.

There is a paucity of reports within the available literature on the functions of the intron and 3′ untranslated region (UTR) in C5aR1 expression and function. The length and phylogenic conservation of the intronic sequence suggest that it may be involved in the transcriptional regulation of the gene.[28] Interestingly, a knock-in mouse generated to insert an IRES-EGFP sequence after the second exon of C5ar1 led to increased protein production and impaired trafficking.[18] In the homozygous knock-in animal, C5aR1 protein was confined to the intracellular compartment and was not detectable within the cellular membrane. Reports suggest the disruption of the 3′ UTR caused the dysregulated production and impaired trafficking of the protein product, although this was not demonstrated.[18]

HUMAN PROTEIN SEQUENCE

The human peptide sequence for C5aR1 is presented below. Underlined are the glycosylation site (N5), sulphation sites (Y11, Y14) and the phosphoserines of the C-terminus (S314, S317, S327, S332, S334 and S338). The cysteine residues (C109, C188) that form the disulphide bond between extracellular loops 1 and 2 are highlighted and underlined.

```
MNSFNYTTPD  YGHYDDKDTL  DLNTPVDKTS  NTLRVPDILA  LVIFAVVFLV   50
GVLGNALVVW  VTAFEAKRTI  NAIWFLNLAV  ADFLSCLALP  ILFTSIVQHH  100
HWPFGGAACS  ILPSLILLNM  YASILLLATI  SADRFLLVFK  PIWCQNFRGA  150
GLAWIACAVA  WGLALLLTIP  SFLYRVVREE  YFPPKVLCGV  DYSHDKRRER  200
AVAIVRLVLG  FLWPLLTLTI  CYTFILLRTW  SRRATRSTKT  LKVVVAVVAS  250
FFIFWLPYQV  TGIMMSFLEP  SSPTFLLLNK  LDSLCVSFAY  INCCINPIIY  300
VVAGQGFQGR  LRKSLPSLLR  NVLTEESVVR  ESKSFTRSTV  DTMAQKTQAV  350
```

PROTEIN MODULES

AA1-37	N-Terminus	exons 1/2
AA38-60	Transmembrane domain 1	exon 2
AA61-71	Intracellular loop 1	exon 2
AA72-94	Transmembrane domain 2	exon 2
AA95-110	Extracellular loop 1	exon 2
AA111-132	Transmembrane domain 3	exon 2
AA133-153	Intracellular loop 2	exon 2
AA154-174	Transmembrane domain 4	exon 2

AA175-200	Extracellular loop 2	exon 2
AA201-226	Transmembrane domain 5	exon 2
AA227-242	Intracellular loop 3	exon 2
AA243-265	Transmembrane domain 6	exon 2
AA266-282	Extracellular loop 3	exon 2
AA283-303	Transmembrane domain 7	exon 2
AA304-350	C-Terminus	exon 2

CHROMOSOMAL LOCATION

The genetic locus for C5aR1 is located on chromosome 19q13.3–13.4 in humans and chromosome 7 A1 in the mouse.

HUMAN cDNA SEQUENCE

Human cDNA sequence is presented below. One predicted variant exists, which differs only in the length of the 3′ UTR. For simplicity only the confirmed sequence has been presented. The first five nucleotides of exon 2 are underlined (position 53–57). Initiation and termination signals are highlighted and underlined (positions 50–52 and 1100–1102, respectively). Putative polyadenylation signal is underlined (position 3496–3501).

```
CTTGGGCAGG AGGGACCTTC GATCCTCGGG GAGCCCAGGA GACCAGAACA TGAACTCCTT    60
CAATTATACC ACCCCTGATT ATGGGCACTA TGATGACAAG GATACCCTGG ACCTCAACAC   120
CCCTGTGGAT AAAACTTCTA ACACGCTGCG TGTTCCAGAC ATCCTGGCCT TGGTCATCTT   180
TGCAGTCGTC TTCCTGGTGG GAGTGCTGGG CAATGCCCTG GTGGTCTGGG TGACGGCATT   240
CGAGGCCAAG CGGACCATCA ATGCCATCTG GTTCCTCAAC TTGGCGGTAG CCGACTTCCT   300
CTCCTGCCTG GCGCTGCCCA TCTTGTTCAC GTCCATTGTA CAGCATCACC ACTGGCCCTT   360
TGGCGGGGCC GCCTGCAGCA TCCTGCCCTC CCTCATCCTG CTCAACATGT ACGCCAGCAT   420
CCTGCTCCTG GCCACCATCA GCGCCGACCG CTTTCTGCTG GTGTTTAAAC CCATCTGGTG   480
CCAGAACTTC CGAGGGGCCG GCTTGGCCTG GATCGCCTGT GCCGTGGCTT GGGGTTTAGC   540
CCTGCTGCTG ACCATACCCT CCTTCCTGTA CCGGGTGGTC CGGGAGGAGT ACTTTCCACC   600
AAAGGTGTTG TGTGGCGTGG ACTACAGCCA CGACAAACGG CGGGAGCGAG CCGTGGCCAT   660
CGTCCGGCTG GTCCTGGGCT TCCTGTGGCC TCTACTCACG CTCACGATTT GTTACACTTT   720
CATCCTGCTC CGGACGTGGA GCCGCAGGGC CACGCGGTCC ACCAAGACAC TCAAGGTGGT   780
GGTGGCAGTG GTGGCCAGTT TCTTTATCTT CTGGTTGCCC TACCAGGTGA CGGGGATAAT   840
GATGTCCTTC CTGGAGCCAT CGTCACCCAC CTTCCTGCTG CTGAATAAGC TGGACTCCCT   900
GTGTGTCTCC TTTGCCTACA TCAACTGCTG CATCAACCCC ATCATCTACG TGGTGGCCGG   960
CCAGGGCTTC CAGGGCCGAC TGCGGAAATC CCTCCCCAGC CTCCTCCGGA ACGTGTTGAC  1020
TGAAGAGTCC GTGGTTAGGG AGAGCAAGTC ATTCACGCGC TCCACAGTGG ACACTATGGC  1080
CCAGAAGACC CAGGCAGTGT AGGCGACAGC CTCATGGGCC ACTGTGGCCC GATGTCCCCT  1140
TCCTTCCCGG CCATTCTCCC TCTTGTTTTC ACTTCACTTT TCGTGGGATG GTGTTACCTT  1200
AGCTAACTAA CTCTCCTCCA TGTTGCCTGT CTTTCCCAGA CTTGTCCCTC CTTTTCCAGC  1260
GGGACTCTTC TCATCCTTCC TCATTTGCAA GGTGAACACT TCCTTCTAGG GAGCACCCTC  1320
CCACCCCCCA CCCCCCCCAC ACACACCATC TTTCCATCCC AGGCTTTTGA AAAACAAACA  1380
GAAACCCGTG TATCTGGGAT ATTTCCATAT GGCAATAGGT GTGAACAGGG AACTCAGAAT  1440
ACAGACAAGT AGAAAGATTC TCGCTTAAAA AAAATGTATT TATTTTATGG CAAGTTGGAA  1500
```

Continued

HUMAN cDNA SEQUENCE—*Continued*

```
AATATGTAAC TGGAATCTCA AAAGTTCTTT GGGACAAAAC AGAAGTCCAT GGAGTTATCT 1560
AAGCTCTTGT AAGTGAGTTA ATTTAAAAAA GAAAATTAGG CTGAGAGCAG TGGCTCACGC 1620
CTGTAATCCC AGAACTTTGG GAGGCTAAGG TGGGTGGATC ACCTGAGGTC AAGAGTTCCA 1680
GACCAGGCTG GCCAGCATGG TGAAACCCCG TCTGTACTAA AAATACAAAA AATTAACTGG 1740
GCATGGTAGT GGGTGCCTGT AATCCCAGCT ACTTGGGAGG CTGAGGTGGG AGAATTGCTC 1800
GAACTTGGAG GTGGAGGTTG TGGTGAGCCA TGATCGCACC ACTGCACTCT AGCCTGGGTG 1860
ACCGAGGGAG GCTCTGTCTC AAAAGCAAAG CAAAAACAAA AACAAAAACA CCTAAAAAAC 1920
CTGCAGTTTT GTTTGTACTT TGTTTTTAAA TTATGCTTTC TATTTTGAGA TCATTGCAAA 1980
CTCAACACAA TTGTAAGTAA TGATACAGAG GGATCTTGTG TACCCTTCAC CCAGCCTCCC 2040
CCAATGGCAA CATCTTGCAA AACTACAATG TAGTCTCATA ACCAGGATAT TGACATTGAT 2100
ACAGTGAAGA TACAGGACAT TCTCATCACC ACAGGGATCC CCAGGATGCC CACTTCCCTC 2160
CACCCCCACA CCCCAGCCGT GTCCCTAACC CCTGGCAACC AGGAATCCAC TCTCCATTTC 2220
TATAATGTTG TCATTTCAAG AATGTTATTC AATGGAATCA TATAGTATGT AACCTGTTTT 2280
GAGCTTAAAA AAAAAGTATA CATGACTTTA ATGAGGAAAA TAAAAATGAA TATTGAAATG 2340
TT                                                                2342
```

GENOMIC STRUCTURE

The C5aR1 locus consists of two exons separated by a large (~10 kb) intron. The first exon, the shorter of the two, contains the 131 bp 5′ UTR and the start codon (ATG) of the C5aR1 sequence. Exon 2 contains the remaining coding sequence and a large, 1.5 kb 3′ UTR region. C5aR1 shares similarity in genomic structure with the other complement anaphylatoxins (see Chapters 36 and 38) (Fig. 37.2).

ACCESSION NUMBERS

Homo sapiens	NM_001736.3
Predicted variant	XM_005259190.4
Mus musculus	
Variant 1	NM_007577.4
Variant 2	NM_001173550.1
Rattus norvegicus	NM_053619.1
Macaca fascicularis (predicted)	XM_005589697.2
Gallus gallus (predicted)	XM_015273054.1
Xenopus tropicalis (predicted)	
Variant 1	XM_002941282.3
Variant 2	XM_012969371.1
Danio rerio (predicted)	XM_005159274.3

FIGURE 37.2 The structure of the human gene on chromosome 19. Numbers at each end of the locus represent the nucleotide position within the chromosome. The darker grey colour represents the coding sequence.

DEFICIENCY

There are no reports of C5aR1 deficiency in humans. However, there has been reported a subgroup of Han Chinese individuals with polymorphism of C5aR1 who are linked with a predisposition to coronary artery disease.[29]

POLYMORPHIC VARIANTS

D2N

The D2N polymorphism has no known association with human disease.

N279K

The N279K polymorphism has been associated with familial Mediterranean fever; however, this mutation does not affect the binding of C5a nor the degranulation response of neutrophils.[30] Therefore, it is not yet understood whether this polymorphism has any role in the pathophysiology of familial Mediterranean fever, or if it is merely a coincidental polymorphism within these families.

MUTANT ANIMALS

Knockout mice for C5aR1 are available through JAX laboratories on the BALB/cJ background. These were created through targeted knockout of the entire C5aR1 coding sequence in 129S4/SvJae embryonic stem cells, with a neomycin resistance cassette. C57BL6/J blastocysts were used as a vector for the creation of chimeras and the resultant litters were backcrossed onto a BALB/c background.[31] The BALB/cJ background C5aR1 knockout mice are available through the Jackson laboratories; however, other groups have generated separate knockout mice that have been backcrossed to the C57BL6/J background.[32]

Neither gross morphological defects nor behavioural abnormalities have been reported for the $C5aR^{-/-}$ mouse. This strain is more susceptible to bacterial infection and carries a higher bacterial load compared to wild-type counterparts in models of sepsis.[33]

REFERENCES

1. Monk PN, Scola A-M, Madala P, Fairlie DP. Function, structure and therapeutic potential of complement C5a receptors. *Br J Pharmacol* 2007;**152**:429–48.
2. Klos A, et al. The role of the anaphylatoxins in health and disease. *Mol Immunol* 2009;**46**: 2753–66.
3. Giannini E, Brouchon L, Boulay F. Identification of the major phosphorylation sites in human C5a anaphylatoxin receptor in vivo. *J Biol Chem* 1995;**270**:19166–72.
4. Sarma JV, Ward PA. New developments in C5a receptor signaling. *Cell Health Cytoskelet* 2012;**4**:73–82.
5. Huber-Lang MS, et al. Structure-function relationships of human C5a and C5aR. *J Immunol* 2003;**170**:6115–24.

6. Kurihara R, et al. C5a promotes migration, proliferation, and vessel formation in endothelial cells. *Inflamm Res* 2010;**59**:659–66.

7. Strainic MG, Shevach EM, An F, Lin F, Medof ME. Absence of signaling into CD4(+) cells via C3aR and C5aR enables autoinductive TGF-β1 signaling and induction of Foxp3(+) regulatory T cells. *Nat Immunol* 2012. http://dx.doi.org/10.1038/ni.2499.

8. Arbore G, et al. T helper 1 immunity requires complement-driven NLRP3 inflammasome activity in CD4+T cells. *Science* 2016;**352**. aad1210–aad1210.

9. Schraufstatter IU, Discipio RG, Zhao M, Khaldoyanidi SK. C3a and C5a are chemotactic factors for human mesenchymal stem cells, which cause prolonged ERK1/2 phosphorylation. *J Immunol* 2009;**182**:3827–36.

10. Lara-Astiaso D, et al. Complement anaphylatoxins C3a and C5a induce a failing regenerative program in cardiac resident cells. Evidence of a role for cardiac resident stem cells other than cardiomyocyte renewal. *Springerplus* 2012;**1**:63.

11. Bénard M, et al. Role of complement anaphylatoxin receptors (C3aR, C5aR) in the development of the rat cerebellum. *Mol Immunol* 2008;**45**:3767–74.

12. Denny KJ, et al. C5a receptor signaling prevents folate deficiency-induced neural tube defects in mice. *J Immunol* 2013;**190**:3493–9.

13. Pavlovski D, et al. Generation of complement component C5a by ischemic neurons promotes neuronal apoptosis. *FASEB J* 2012. http://dx.doi.org/10.1096/fj.11-202382.

14. Bénard M, et al. Characterization of C3a and C5a Receptors in Rat Cerebellar Granule neurons during maturation neuroprotective effect of C5a against apoptotic cell death. *J Biol Chem* 2004;**279**:43487–96.

15. Brennan FH, et al. The complement receptor C5aR controls acute inflammation and astrogliosis following spinal cord injury. *J Neurosci* 2015;**35**:6517–31.

16. March DR, et al. Potent cyclic antagonists of the complement C5a receptor on human polymorphonuclear leukocytes. Relationships between structures and activity. *Mol Pharmacol* 2004;**65**:868–79.

17. Karsten CM, et al. Monitoring and cell-specific deletion of C5aR1 using a novel floxed GFP-C5aR1 reporter knock-in mouse. *J Immunol* 2015;**194**:1841–55.

18. Dunkelberger J, Zhou L, Miwa T, Song W-C. C5aR expression in a novel GFP reporter gene knockin mouse: implications for the mechanism of action of C5aR signaling in T cell immunity. *J Immunol* 2012;**188**:4032–42.

19. Nataf S, Stahel PF, Davoust N, Barnum SR. Complement anaphylatoxin receptors on neurons: new tricks for old receptors? *Trends Neurosci* 1999;**22**:397–402.

20. O'Barr SA, et al. Neuronal expression of a functional receptor for the C5a complement activation fragment. *J Immunol* 2001;**166**:4154–62.

21. Woodruff TM, Ager RR, Tenner AJ, Noakes PG, Taylor SM. The role of the complement system and the activation fragment C5a in the central nervous system. *Neuromolecular Med* 2010;**12**:179–92.

22. Crane JW, et al. The C5a anaphylatoxin receptor CD88 is expressed in presynaptic terminals of hippocampal mossy fibres. *J Neuroinflammation* 2009;**6**:34.

23. Rahpeymai Y, et al. Complement: a novel factor in basal and ischemia-induced neurogenesis. *EMBO J* 2006;**25**:1364–74.

24. Hawksworth OA, Coulthard LG, Taylor SM, Wolvetang EJ, Woodruff TM. Brief report: complement C5a promotes human embryonic stem cell pluripotency in the absence of FGF2. *Stem Cells* 2014;**32**:3278–84.

25. Hunt JR, Martin CB, Martin BK. Transcriptional regulation of the murine C5a receptor gene: NF-Y is required for basal and LPS induced expression in macrophages and endothelial cells. *Mol Immunol* 2005;**42**:1405–15.

26. Palmer E, Gray LC, Stott M, Bowen DJ, van den Berg CW. Roles of promoter and 3' untranslated motifs in expression of the human C5a receptor. *Mol Immunol* 2012;**52**:88–95.

27. Martin CB, Ingersoll SA, Martin BK. Regulation of the C5a receptor promoter in glial cells: minimal dependence upon the CCAAT element in astrocytes. *Mol Immunol* 2007;**44**:713–21.

28. Martin BK. Transcriptional control of complement receptor gene expression. *Immunol Res* 2007;**39**:146–59.

29. Zheng Y-Y, et al. Association of C5aR1 genetic polymorphisms with coronary artery disease in a Han population in Xinjiang, China. *Diagn Pathol* 2015;**10**:33.

30. Apostolidou E, et al. Genetic analysis of C5a receptors in neutrophils from patients with familial Mediterranean fever. *Mol Biol Rep* 2012;**39**:5503–10.

31. Hopken UE, Lu B, Gerard NP, Gerard C. The C5a chemoattractant receptor mediates mucosal defence to infection. *Nature* September 05, 1996;**383**:86–9. http://dx.doi.org/10.1038/383086a0.

32. Hollmann TJ, Mueller-Ortiz SL, Braun MC, Wetsel RA. Disruption of the C5a receptor gene increases resistance to acute gram-negative bacteremia and endotoxic shock: opposing roles of C3a and C5a. *Mol Immunol* 2008;**45**:1907–15.

33. Höpken UE, Lu B, Gerard NP, Gerard C. The C5a chemoattractant receptor mediates mucosal defence to infection. *Nature* 1996;**383**:86–9.

Chapter 38

C5aR2

Liam G. Coulthard[1,2], Owen A. Hawksworth[3], Trent M. Woodruff[3]

[1]Royal Brisbane and Women's Hospital, Herston, QLD, Australia; [2]University of Queensland, Herston, QLD, Australia; [3]University of Queensland, St. Lucia, QLD, Australia

OTHER NAMES

Complement C5a receptor-2, C5a receptor beta, C5a receptor-like 2, C5L2, G-protein coupled receptor 77, GPR77.

PHYSIOCHEMICAL PROPERTIES

C5aR2 is produced as a single polypeptide chain product of 337 residues. The receptor has a predicted molecular weight of 36.1 kDa and isoelectric point of 8.2. There is a single disulphide bond between the cysteine residues (Cys107, Cys186) of the first and second extracellular loops.[1] C5aR2 also contains an *N*-linked glycosylation site (position 3) and a putative phosphorylation site on the intracellular C-terminus (position 320).

STRUCTURE

C5aR2 is a G-protein coupled receptor with seven transmembrane α-helices and an extracellular N-terminus (Fig. 38.1). It shares close homology (37%), with the other receptor for C5a, C5aR1. The N-terminus and three extracellular loops form a complex that binds C5a.[2] The intracellular C-terminus contains a cluster of serine residues that contribute to signalling and desensitisation.[3] There are distinct differences, however, in the intracellular motifs present between C5aR2 and other members of the complement anaphylatoxin receptor family. C5aR2 lacks both the DRY and NPXXY intracellular motifs necessary for G-protein coupling,[4] which has led to a debate on its role.[5]

FUNCTION

The exact role of C5aR2 in the immune response has been one of controversy, reviewed in Li et al. (2013).[5] Three models of action for the receptor have been described in the literature and are as follows:

The Complement FactsBook. http://dx.doi.org/10.1016/B978-0-12-810420-0.00038-9
407

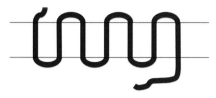

FIGURE 38.1   Domain structure of C5aR2.

1. A nonsignalling decoy receptor.

   This model builds on the lack of G-protein coupling to suggest that C5aR2 acts as a trap for C5a that would otherwise bind and activate C5aR1. However, physical experimental evidence for this model comes from transfected cell types that do not natively express C5aR2.[6] Whether C5aR2 has this action in vivo remains to be demonstrated.

2. A modulator of C5aR1 signalling.

   The intracellular location of C5aR2, as a receptor that binds a water-soluble extracellular ligand, is puzzling. One hypothesis is that the intracellular location of C5aR2 acts as a reservoir for influencing the function of the C5a–C5aR1 interaction. Through β-arrestin coupling, C5aR2 acts to reduce C5aR1 signalling once bound by C5a. This is supported by studies in human cells demonstrating colocalisation, and perhaps heterodimerisation of C5aR1, C5aR2 and β-arrestin on stimulation with C5a.[7,8]

3. G-protein independent signalling.

   Several studies have noted a difference in the inflammatory cytokine profile between C5aR2-deficient and wild-type mice. C5aR2-deficient mice have reduced levels of the proinflammatory mediators, interleukin-6 and tumour necrosis factor-α.[9] This observation is not well explained by the previous models for C5aR2 function. A study in sepsis using C5aR2-deficient animal demonstrated that the presence of C5aR2 was essential for the release of HMGB1 in peritoneal macrophages, and this effect could be attenuated by MAPK pathway inhibition.[10] The signalling likely occurs in a β-arrestin driven manner, as studies have demonstrated modulation of Erk1/2 signalling and β-arrestin recruitment to C5aR2 upon ligand binding.[11,12] This suggests that although C5aR2 lacks the necessary intracellular motifs for G-protein coupling, it continues to act as a receptor and transducer of extracellular signals independent of the actions of C5aR1.

   It is possible that all of these molecular functions for C5aR2 coexist in a mutually exclusive way, dependent on cell type and the availability of intracellular second messengers. In the immune sense, C5aR2 appears to act as both a pro- and antiinflammatory factor, dependent on disease state,[5] supporting the theory that multiple roles for the receptor exist. In addition, C5aR2 has also been implicated in the regulation of metabolism. It is present within the anterior pituitary and is potentially thought to be the receptor for acetylating stimulating protein ($C3a_{desArg}$). There is evidence that loss of C5aR2 signalling results in insulin resistance and increased appetite, contributing to a diabetogenic phenotype.[13,14]

TISSUE DISTRIBUTION

C5aR2 expression is generally found in tissues and cell types that express C5aR1. Cells of the myeloid lineage, which widely express C5aR1, also express C5aR2. C5aR2 is highly expressed on neutrophils, dendritic cells, macrophages and monocytes.[15] However, C5aR2 is expressed widely on cells and tissues that do not have a direct association with the immune response. These include neurons, astrocytes, smooth muscle cells, adipose cells, fibroblasts and hepatocytes.[5,16–20]

There is emerging evidence that C5aR2 is expressed both intracellularly and upon the membrane of CD4+ T cells, where it modulates NLRP3 inflammasome activation.[21] The receptor is also known to participate in modulating the function of these cells through its actions on dendritic cells and other supportive elements of the B- and T cell response.[22]

Although C5aR2 is membrane bound for most cell types, only a small proportion of the expressed protein is in contact with the extracellular environment. The majority of receptors are localised to intracellular vesicles in inflammatory cells.[5] The significance of this subcellular localisation has yet to be adequately explained.

REGULATION OF EXPRESSION

There are few studies examining the function of promoter sequences that lie upstream of C5aR2 and into the regulators of expression. The demonstrated coexpression with C5aR1 suggests that both receptors share components regulating expression. However, several instances document C5aR2 expression distinct from that of C5aR1. For instance, in normal physiology both receptors are found on epithelial cells of the nephron, but they are expressed on distinct and separate parts of the tubule.[23] In pathological circumstances, C5aR2 is downregulated by the presence of LPS on peripheral blood mononuclear cells, in direct contrast to C5aR1.[24]

On several cell types, C5aR2 expression is upregulated by the inflammatory mediators, interferon-γ, sphingosine-1-phosphate and dibutyl-cAMP.[5] Interestingly, given the potential role of C5aR2 as the receptor for acylation-stimulating protein, there are several factors involved in the regulation of metabolism that also contribute to upregulation of C5aR2. Insulin, oestrogen and noradrenaline upregulate C5aR2. In addition, the PPARγ agonist, thiazolidione, used in the treatment of diabetes, also induces C5aR2 upregulation.[25]

HUMAN PROTEIN SEQUENCE

The human peptide sequence for C5aR2 is presented below. Underlined are the glycosylation site (N3) and the phosphoserine of the C-terminus (S320). The cysteine residues (C107, C186) that form the disulphide bond between extracellular loops 1 and 2 are highlighted and underlined.

```
MGNDSVSYEY GDYSDLSDRP VDCLDGACLA IDPLRVAPLP LYAAIFLVGV  50

PGNAMVAWVA GKVARRRVGA TWLLHLAVAD LLCCLSLPIL AVPIARGGHW 100

PYGAVGCRAL PSIILLTMYA SVLLLAALSA DLCFLALGPA WWSTVQRACG 150

VQVACGAAWT LALLLTVPSA IYRRLHQEHF PARLQCVVDY GGSSSTENAV 200

TAIRFLFGFL GPLVAVASCH SALLCWAARR CRPLGTAIVV GFFVCWAPYH 250

LLGLVLTVAA PNSALLARAL RAEPLIVGLA LAHSCLNPML FLYFGRAQLR 300

RSLPAACHWA LRESQGQDES VDSKKSTSHD LVSEMEV              337
```

PROTEIN MODULES

AA1-38	N-Terminus	exons 1/2
AA39-61	Transmembrane domain 1	exon 2
AA62-72	Intracellular loop 1	exon 2
AA73-95	Transmembrane domain 2	exon 2
AA96-114	Extracellular loop 1	exon 2
AA115-137	Transmembrane domain 3	exon 2
AA138-149	Intracellular loop 2	exon 2
AA150-172	Transmembrane domain 4	exon 2
AA173-202	Extracellular loop 2	exon 2
AA203-225	Transmembrane domain 5	exon 2
AA226-237	Intracellular loop 3	exon 2
AA238-260	Transmembrane domain 6	exon 2
AA261-274	Extracellular loop 3	exon 2
AA275-294	Transmembrane domain 7	exon 2
AA295-337	C-Terminus	exon 2

CHROMOSOMAL LOCATION

The genetic locus for C5aR2 is located on chromosome 19q13.3-13.4 in humans and chromosome 7 A1 in the mouse.

HUMAN cDNA SEQUENCE

Human cDNA sequence is presented below. Three variants and one predicted variant exist, which differ only in the length of the 5′ untranslated regions (UTRs). For simplicity only variant 1 has been presented. The first five nucleotides of exon 2 are underlined (position 53–57). Initiation and termination signals are

highlighted and underlined (positions 219–221 and 1230–1232, respectively).
Putative polyadenylation signal is underlined (position 3496–3501).

```
TATAAAGATT CACTGGGACT GGTGAGGTGG CAGTGCTCAG CAGCATCCGA CAGGAGCCCT   60
GGCAAACAGG ACGGATTTCC AGGACTCTAC CAGCTGCCAG ACACGGCAGG GAGAGACCCC  120
AGACCTCCTG GGTCCTGGCT GTGGGCCCGG ATTGGGCTCC CAAGTGGCGT TTGACTCACG  180
TGGGGACACT CTTGGAAGAG ACGACACCAG GAGCCTGAAT GGGGAACGAT TCTGTCAGCT  240
ACGAGTATGG GGATTACAGC GACCTCTCGG ACCGCCCTGT GGACTGCCTG GATGGCGCCT  300
TGGGGGTGCC GGGCAATGCC ATGGTGGCCT GGGTGGCTGG GAAGGTGGCC CGCCGGAGGG  420
TGGGTGCCAC CTGGTTGCTC CACCTGGCCG TGGCGGATTT GCTGTGCTGT TTGTCTCTGC  480
CCATCCTGGC AGTGCCCATT GCCCGTGGAG GCCACTGGCC GTATGGTGCA GTGGGCTGTC  540
GGGCGCTGCC CTCCATCATC CTGCTGACCA TGTATGCCAG CGTCCTGCTC CTGGCAGCTC  600
TCAGTGCCGA CCTCTGCTTC CTGGCTCTCG GGCCTGCCTG GTGGTCTACG GTTCAGCGGG  660
CGTGCGGGGT GCAGGTGGCC TGTGGGGCAG CCTGGACACT GGCCTTGCTG CTCACCGTGC  720
CCTCCGCCAT CTACCGCCGG CTGCACCAGG AGCACTTCCC AGCCCGGCTG CAGTGTGTGG  780
GGACTACGGC GGCTCCTCCA GCACCGAGAA TGCGGTGACT GCCATCCGGT TTCTTTTTGG  840
TCTTCCTGGG GCCCCTGGTG GCCGTGGCCA GCTGCCACAG TGCCCTCCTG TGCTGGGCAG  900
CCCGACGCTG CCGGCCGCTG GGCACAGCCA TTGTGGTGGG GTTTTTTGTC TGCTGGGCAC  960
CCTACCACCT GCTGGGGCTG GTGCTCACTG TGGCGGCCCC GAACTCCGCA CTCCTGGCCA 1020
GGGCCCTGCG GGCTGAACCC CTCATCGTGG GCCTTGCCCT CGCTCACAGC TGCCTCAATC 1080
CCATGCTCTT CCTGTATTTT GGGAGGGCTC AACTCCGCCG GTCACTGCCA GCTGCCTGTC 1140
ACTGGGCCCT GAGGGAGTCC CAGGGCCAGG ACGAAAGTGT GGACAGCAAG AAATCCACCA 1200
GCCATGACCT GGTCTCGGAG ATGGAGGTGT AGGCTGGAGA GACATTGTGG GTGTGTATCT 1260
TCTTATCTCA TTTCACAAGA CTGGCTTCAG GCATAGCTGG ATCCAGGAGC TCAATGATGT 1320
CTTCATTTTA TTCCTTCCTT CATTCAACAG ATATCCATCA TGCACTTGCT ATGTGCAAGG 1380
CCTTTTTAGG CACTAGAGAT ATAGCAGTGA CCAAAACAGA CACAAATCCT GCCC         1434
```

FIGURE 38.2   The structure of the human gene on chromosome 19. Numbers at each end of the locus represent the nucleotide position within the chromosome. The darker grey colour represents the coding sequence.

GENOMIC STRUCTURE

The C5aR2 locus consists of two exons separated by a large (~8 kb) intron (Fig. 38.2). The first exon, the shorter of the two, contains the majority of the 5′ UTR. Exon 2 contains the entire coding sequence for C5aR2 and the 3′ UTR region. C5aR2 shares similarity in genomic structure with the other complement anaphylatoxins (see Chapters 36 and 37). The C5aR2 locus lies just downstream of the C5aR1 locus.

ACCESSION NUMBERS

Homo sapiens	Variant 1	NM_001271749.1
	Variant 2	NM_0018485.2
	Variant 3	NM_001271750.1
	Predicted variant	XM_011526736.2
Mus musculus	Variant 1	NM_176912.4
	Variant 2	NM_001146005.1
	Predicted variant 1	XR_881730.2
	Predicted variant 2	XM_006540057.3
	Predicted variant 3	XM_01125059.1
	Predicted variant 4	XM_006540052.3
	Predicted variant 5	XM_006540053.3
	Predicted variant 6	XM_006540055.3
	Predicted variant 7	XM_006540049.2
	Predicted variant 8	XM_006540056.2
Rattus norvegicus		NM_001003710.2
	Predicted variant 1	XM_006228352.3
	Predicted variant 2	XM_006228353.3
Macaca fascicularis	Predicted variant 1	XM_005589698.2
	Predicted variant 2	XM_005589699.2
	Predicted variant 3	XM_005589700.2
	Predicted variant 2	XM_005589699.2
	Predicted variant 3	XM_005589700.2

DEFICIENCY

There are no reports of C5aR2 deficiency in humans.

POLYMORPHIC VARIANTS

The P233L polymorphism has been demonstrated to have an association with coronary artery disease in the Chinese Han population.[26]

The S323I polymorphism is a rare polymorphism identified in a French Canadian family with predisposition to coronary heart disease. The individuals heterozygous for this variant had higher serum triglycerides and apolipoprotein B.[27] Cell culture studies have demonstrated that this polymorphism prevents the recruitment of β-arrestin after ligand binding.[28]

MUTANT ANIMALS

Knockout mice for C5aR2 are available on the BALB/c and C57BL6/J backgrounds. These were created through targeted knockout of the entire C5aR2 coding sequence with a LacZ/neomycin resistance cassette. The original chimeras were bred into both a B6 and BALB/c background.[18] No gross morphological defects or behavioural abnormalities have been reported for the *C5aR2−/−* mouse.

REFERENCES

1. Monk PN, Scola A-M, Madala P, Fairlie DP. Function, structure and therapeutic potential of complement C5a receptors. *Br J Pharmacol* 2007;**152**:429–48.
2. Klos A, et al. The role of the anaphylatoxins in health and disease. *Mol Immunol* 2009;**46**:2753–66.
3. Giannini E, Brouchon L, Boulay F. Identification of the major phosphorylation sites in human C5a anaphylatoxin receptor in vivo. *J Biol Chem* 1995;**270**:19166–72.
4. Okinaga S, et al. C5L2, a nonsignaling C5A binding protein. *Biochemistry* 2003;**42**:9406–15.
5. Li R, Coulthard LG, Wu MCL, Taylor SM, Woodruff TM. C5L2: a controversial receptor of complement anaphylatoxin, C5a. *FASEB J* 2013;**27**:855–64.
6. Scola A-M, Johswich K-O, Morgan BP, Klos A, Monk PN. The human complement fragment receptor, C5L2, is a recycling decoy receptor. *Mol Immunol* 2009;**46**:1149–62.
7. Bamberg CE, et al. The C5a receptor (C5aR) C5L2 is a modulator of C5aR-mediated signal transduction. *J Biol Chem* 2010;**285**:7633–44.
8. Croker DE, Halai R, Fairlie DP, Cooper MA. C5a, but not C5a-des Arg, induces upregulation of heteromer formation between complement C5a receptors C5aR and C5L2. *Immunol Cell Biol* 2013;**91**:625–33.
9. Gao H, et al. Evidence for a functional role of the second C5a receptor C5L2. *FASEB J* 2005;**19**:1003–5.
10. Rittirsch D, et al. Functional roles for C5a receptors in sepsis. *Nat Med* 2008;**14**:551–7.
11. Croker DE, et al. C5a2 can modulate ERK1/2 signaling in macrophages via heteromer formation with C5a1 and β-arrestin recruitment. *Immunol Cell Biol* 2014;**92**:631–9.
12. Croker DE, et al. Discovery of functionally selective C5aR2 ligands: novel modulators of C5a signalling. *Immunol Cell Biol* 2016. http://dx.doi.org/10.1038/icb.2016.43.
13. Fisette A, et al. C5L2 receptor disruption enhances the development of diet-induced insulin resistance in mice. *Immunobiology* 2012. http://dx.doi.org/10.1016/j.imbio.2012.04.001.
14. Roy C, et al. Acute injection of ASP in the third ventricle inhibits food intake and locomotor activity in rats. *Am J Physiol Endocrinol Metab* 2011;**301**:E232–41.
15. Ohno M, et al. A putative chemoattractant receptor, C5L2, is expressed in granulocyte and immature dendritic cells, but not in mature dendritic cells. *Mol Immunol* 2000;**37**:407–12.
16. Kalant D, et al. C5L2 is a functional receptor for acylation-stimulating protein. *J Biol Chem* 2005;**280**:23936–44.
17. Gavrilyuk V, et al. Identification of complement 5a-like receptor (C5L2) from astrocytes: characterization of anti-inflammatory properties. *J Neurochem* 2005;**92**:1140–9.
18. Chen N-J, et al. C5L2 is critical for the biological activities of the anaphylatoxins C5a and C3a. *Nature* 2007;**446**:203–7.
19. Woodruff TM, et al. The complement factor C5a contributes to pathology in a rat model of amyotrophic lateral sclerosis. *J Immunol* 2008;**181**:8727–34.
20. Lee JD, et al. Dysregulation of the complement cascade in the hSOD1G93A transgenic mouse model of amyotrophic lateral sclerosis. *J Neuroinflammation* 2013;**10**:119.
21. Arbore G, et al. T helper 1 immunity requires complement-driven NLRP3 inflammasome activity in CD4+T cells. *Science* 2016;**352**:aad1210.
22. Zhang X, et al. A critical role for C5L2 in the pathogenesis of experimental allergic asthma. *J Immunol* 2010;**185**:6741–52.
23. van Werkhoven MB, et al. Novel insights in localization and expression levels of C5aR and C5L2 under native and post-transplant conditions in the kidney. *Mol Immunol* 2013;**53**:237–45.

24. Raby A-C, et al. TLR activation enhances C5a-induced pro-inflammatory responses by negatively modulating the second C5a receptor, C5L2. *Eur J Immunol* 2011;**41**:2741–52.

25. MacLaren RM, Kalant DK, Cianflone KC. The ASP receptor C5L2 is regulated by metabolic hormones associated with insulin resistance. *Biochem Cell Biol* 2007. http://dx.doi.org/10.1139/o06-207.

26. Zheng Y-Y, et al. Relationship between a novel polymorphism of the C5L2 gene and coronary artery disease. *PLoS One* 2011;**6**:e20984.

27. Marcil M, et al. Identification of a novel C5L2 variant (S323I) in a French Canadian family with familial combined hyperlipemia. *Arterioscler Thromb Vasc Biol* 2006;**26**:1619–25.

28. Cui W, Simaan M, Laporte S, Lodge R, Cianflone K. C5a- and ASP-mediated C5L2 activation, endocytosis and recycling are lost in S323I-C5L2 mutation. *Mol Immunol* 2009;**46**:3086–98.

Chapter 39

C1q Receptors

Suzanne Bohlson

Des Moines University, Des Moines, IA, United States

INTRODUCTION

Complement component C1q regulates cellular responses independent of its role in activation of the classical complement pathway. There is not a canonical C1q receptor that mediates complement-independent cellular activation, but rather a variety of molecules with diverse structures have been described as putative C1q receptors. This chapter describes the putative C1q receptors including structure, function and expression patterns.

C1q, the recognition component of the classical complement pathway, also regulates numerous cellular processes independent of its role in complement activation. Among these processes is myeloid cell activation leading to enhanced engulfment of apoptotic cells and the regulation of proinflammatory cytokine production which is thought to be important in the prevention of autoimmunity. C1q deficiency results in a lupus-like autoimmune disease in humans, and there has been a significant effort to identify C1q receptors that might be effective targets for the manipulation of C1q-dependent cellular functions such as phagocytosis and cytokine regulation. C1q is a 460-kD hexamer containing a globular head region and collagen-like tail. Receptors for both the collagen-like tail and globular head regions have been proposed and are described below. The described receptors are heterogeneous; two lack cytoplasmic tails and are found both intracellularly and extracellularly (gC1qR and calreticulin), two are Ig superfamily members (LAIR-1 and RAGE), one is a heterodimeric integrin ($\alpha_2\beta_1$), two are scavenger receptors (SCARF1 and Megf10), one is a receptor for multiple complement components (CR1). In addition, two of the receptors have been previously described as receptors for collagen ($\alpha_2\beta_1$ and LAIR-1), consistent with their ability to bind to the collagen-like tail of C1q. C1q is a soluble pattern-recognition receptor with a highly basic charge, it binds to multiple molecules and may regulate cell activation through multiple different receptors and/or receptor complexes. The putative C1q receptors are listed below in order of their identification.

The Complement FactsBook. http://dx.doi.org/10.1016/B978-0-12-810420-0.00039-0

415

CALRETICULIN (cC1q-R, COLLECTIN RECEPTOR, CR)

Structure

Calreticulin has a β-sheet N-terminal globular domain (N-domain), a proline-rich low-affinity calcium-binding P-domain and a C-terminal high-affinity calcium-binding acidic domain. The mature protein is 417 amino acids and 60 kDa.[1]

C1q-Dependent Function

Calreticulin is localised to the endoplasmic reticulum (ER) where it functions as a chaperone, and it also regulates a wide array of cellular processes independent of its role in the ER.[2] Calreticulin contains a C1q-binding site localised to residues 160–283 between N- and P-domains which interacts with the collagen-like tail of C1q (hence cC1q-R).[3] Calreticulin lacks a transmembrane domain and has been proposed to form complexes with membrane spanning receptors such as CD91 and dectin-1 to mediate C1q-dependent signalling.[1] Antibodies directed at calreticulin and gC1qR (see below) block C1q-dependent dendritic cell migration.[4] Cell-associated calreticulin stimulates phagocytosis of apoptotic cells and cancer cells,[5] but the requirement for C1q in this process remains to be defined.

Expression

Ubiquitous; intracellular (ER and all cellular compartments) and extracellular/cell associated.[1]

gC1qR (P33, P32, C1QBP, TAP)

Structure

gC1qR is a 33-kDa, highly charged acidic protein that forms a noncovalently associated tetramer. Similar to cC1qR, it does not contain a transmembrane domain or glycosylphosphatidylinositol-anchor.[6]

C1q-Dependent Function

gC1qR binds to the globular head domain of C1q and may transmit C1q-dependent signals by associating with cell membrane receptors such as β1 integrins [see Section α2β1 (Very Late Antigen-1/VLA-2, GPIa-IIa, ITGA2, CD49B)]. Aggregated C1q-induced platelet aggregation is inhibited by antibodies generated against gC1qR.[7] Antibodies directed at calreticulin (see above) and gC1qR block C1q-dependent dendritic cell migration.[4] Hosszu et al. demonstrated that gC1qR, C1q and dendritic cell-specific ICAM-3-grabbing non-integrin (DC-SIGN/CD209) colocalise on dendritic cells (DC) by confocal microscopy and DC-SIGN bound directly to C1q.[8] Antibodies that recognise DC-SIGN inhibit C1q-induced phosphorylation of Iκκα/β, suggesting that DC-SIGN may serve to transmit C1q-dependent intracellular signals in a gC1qR-dependent process.[8] gC1qR surface ablation/silencing will be important for confirming of the role of gC1qR in C1q-dependent processes.

Expression

Multicompartmental (intracellular and extracellular) and ubiquitous.[1]

CD93 (ORIGINALLY IDENTIFIED AS C1QR$_P$)

Structure

CD93 is a 126 M_r (reduced) type-1 transmembrane protein with an N-terminal C-type lectin-like domain, five epidermal growth factor (EGF)-like repeats, a highly glycosylated mucin-like domain and a short cytoplasmic tail.[9]

C1q-Dependent Functions

Antibodies generated against C1q-binding proteins recognised CD93 and inhibited C1q-dependent enhancement of phagocytic function.[10] While CD93 does not bind directly to C1q,[11] and CD93 is not required for C1q-dependent phagocytosis,[12] it may be part of a signalling complex that mediates C1q-dependent functions.

Expression

Myeloid cells, endothelial cells, platelets[13] and some subsets of lymphocytes and stem cells.[14]

COMPLEMENT RECEPTOR 1 (CR1, CD35)

Structure

See Chapter 28.

C1q-Dependent Function

Klickstein et al. reported that C1q bound specifically to CR1 when CR1 was expressed on K562 cells or expressed as a recombinant protein. C1q binding to recombinant CR1 was inhibited with soluble C1q collagen-like tails.[15] Antibodies against CR1 inhibit human erythrocyte binding to C1q/C1q tails.[16] These studies suggest that CR1 binds to the collagen-like tails of C1q, and the interaction may be involved in the erythrocyte-dependent clearance of complement-laden immune complexes from circulation.[15]

Expression

See Chapter 28.

α2β1 (VERY LATE ANTIGEN-1/VLA-2, GPIA-IIA, ITGA2, CD49B)

Structure

α2β1 is in the integrin family of cell adhesion molecules which are heterodimeric membrane proteins. α2β1 consists of a 129.2 kDa-α-chain and an 88.4 kDa-β-chain.[17]

C1q-Dependent Function

Feng et al. demonstrated that adhesion and spreading of human dermal microvascular endothelial cells (HDMVEC) on C1q was inhibited by anti-β1 integrin antibody. Antibodies directed at calreticulin (cC1qR) or gC1qR also partially inhibited HDMVEC spreading and adhesion on C1q, suggesting that multiple C1q-binding molecules may regulate adhesion and spreading in this system.[18] Edelson et al. demonstrated an α2β1-dependent adhesion of mouse peritoneal mast cells (PMC) to serum opsonized immune complexes that required C1q. Furthermore, direct adhesion to C1q was observed in PMC expressing endogenous α2β1 or transfected cell lines, but not in α2β1-deficient PMC or nonexpressing cell lines. C1q-dependent adhesion of PMC to serum opsonised immune complexes resulted in production of IL-6 that was α2β1 dependent suggesting that the interaction results in both mast cell adhesion and activation.[19]

Expression

Epithelial cells, platelets/megakaryocytes, fibroblasts, mast cells, T cells, endothelial cells.[20]

LDL-RECEPTOR RELATED PROTEIN-1 (LRP-1, CD91 OR α2-MACROGLOBULIN RECEPTOR)

Structure

LRP1/CD91 is a heterodimer of an 85-kDa membrane-bound C-terminal membrane spanning subunit and a noncovalently attached 515-kDa N-terminal subunit.[21] The N-terminal subunit contains cysteine-rich complement-type repeats, EGF repeats, and β-propeller domains. The cytoplasmic tail contains two NPXY motifs.[22]

C1q-Dependent Functions

LRP1/CD91 is an endocytic receptor that was demonstrated to cooperate with calreticulin to mediate C1q-dependent engulfment of apoptotic cells,[23] however macrophages deficient in LRP1/CD91 responded to C1q with enhanced phagocytosis.[24] Duus et al. demonstrated a direct interaction between LRP1/CD91 and C1q; however, its role in regulating C1q-dependent engulfment is yet to be determined.[25]

RECEPTOR FOR ADVANCED GLYCATION END-PRODUCTS (RAGE, AGER)

Structure

RAGE is a 42.8-kDa single-pass type-1 membrane protein and a member of the immunoglobulin superfamily.[26]

C1q-Dependent Function

RAGE has been described as a C1q receptor that mediates C1q-dependent phagocytosis and also triggers activation of the classical complement pathway.[27] Immobilised complement receptor 3 (CR3) precipitated both C1q and RAGE suggesting that these molecules may form a functional complex. In support of this hypothesis, antibodies directed against RAGE or Mac-1 inhibited engulfment of C1q opsonised apoptotic Jurkat cells or C1q opsonised latex beads.[27] Antibodies directed at the globular heads of C1q, but not the collagen-like tails, inhibited C1q binding to immobilised RAGE.[27] These data are consistent with the role for RAGE as a pattern-recognition molecule that binds to molecules associated with inflammation.

Expression

Widely expressed: leucocytes, endothelium, smooth muscle cells, central nervous system.[28]

LEUCOCYTE-ASSOCIATED IMMUNOGLOBULIN-LIKE RECEPTOR-1 (LAIR-1, CD305)

Structure

LAIR-1 is a single-pass type-1 membrane protein of the immunoglobulin superfamily containing two immunoreceptor tyrosine-based inhibitory motifs in the cytoplasmic tail. The predicted molecular weight is 32-kDa, and *N*-linked glycosylation results in a relative migration of 40-kDa.[29]

C1q-Dependent Function

Initially classified as a receptor for collagen,[29] LAIR-1 binds to the collagen-like tail of C1q and inhibits proinflammatory signalling from toll-like receptors in human monocyte-derived DCs and monocytes.[30,31]

Expression

Peripheral mononuclear leucocytes including natural killer cells, T and B lymphocytes, monocytes and DCs.[29]

SCAVENGER RECEPTOR TYPE-FAMILY MEMBER 1 (SCARF1, ORIGINALLY DESCRIBED AS SCAVENGER RECEPTOR EXPRESSED BY ENDOTHELIAL CELLS-1, SREC-1)

Structure

SCARF1 is an 87.3-kDa single-pass type-1 membrane protein of scavenger receptor type F family containing six EGF-like domains in extracellular region.[32]

C1q-Dependent Function

Ramirez-Ortiz identified SCARF1 as a C1q receptor on DCs, macrophages and endothelial cells that mediates engulfment of apoptotic cells.[33] SCARF1 bound to C1q in vitro, and C1q-dependent enhanced engulfment of apoptotic cells required SCARF since DCs, macrophages and endothelial cells from *SCARF1*−/− mice failed to respond to C1q with enhanced engulfment. Mice deficient in SCARF1 accumulated apoptotic cells and had elevated autoantibodies and autoimmunity.[33]

Expression

Endothelial cells, DCs and macrophages.[33]

MULTIPLE EPIDERMAL GROWTH FACTOR-LIKE DOMAINS 10 (MEGF10, SCAVENGER RECEPTOR TYPE F- 3/SR-F3)

Structure

Megf10 is a 122.2-kDa single-pass type-1 membrane protein of scavenger receptor type F family containing 15 EGF-like domains in extracellular region.

C1q-Dependent Function

Recessive mutations in Megf10 result in a rare disease characterised by early-onset myopathy, respiratory distress and and dysphagia (EMARDD). Megf10 bound to C1q in vitro, and uptake of C1q by human embryonic kidney (HEK) cells was enhanced in HEK cells expressing full length Megf10. Cells expressing Megf10 with EMARDD mutations are defective at engulfment of apoptotic cells and internalisation of C1q.[34]

Expression

Astrocytes and myosatellite cells.[34]

REFERENCES

1. Ghebrehiwet B, Peerschke EI. cC1q-R (calreticulin) and gC1q-R/p33: ubiquitously expressed multi-ligand binding cellular proteins involved in inflammation and infection. *Mol Immunol* 2004;**41**(2–3):173–83. http://dx.doi.org/10.1016/j.molimm.2004.03.014.

2. Gold LI, Eggleton P, Sweetwyne MT, et al. Calreticulin: non-endoplasmic reticulum functions in physiology and disease. *FASEB J* 2010;**24**(3):665–83. http://dx.doi.org/10.1096/fj.09-145482.

3. Stuart GR, Lynch NJ, Day AJ, Schwaeble WJ, Sim RB. The C1q and collectin binding site within C1q receptor (cell surface calreticulin). *Immunopharmacology* 1997;**38**(1–2): 73–80.

4. Vegh Z, Kew RR, Gruber BL, Ghebrehiwet B. Chemotaxis of human monocyte-derived dendritic cells to complement component C1q is mediated by the receptors gC1qR and cC1qR. *Mol Immunol* 2006;**43**(9):1402–7. http://dx.doi.org/10.1016/S0161-5890(05)00310-X.

5. Chao MP, Majeti R, Weissman IL. Programmed cell removal: a new obstacle in the road to developing cancer. *Nat Rev Cancer* 2011;**12**(1):58–67. http://dx.doi.org/10.1038/nrc3171.

6. Ghebrehiwet B, Lim BL, Peerschke EI, Willis AC, Reid KB. Isolation, cDNA cloning, and overexpression of a 33-kD cell surface glycoprotein that binds the globular "heads" of C1q. *J Exp Med* 1994;**179**(6):1809–21.

7. Peerschke EI, Reid KB, Ghebrehiwet B. Identification of a novel 33-kDa C1q-binding site on human blood platelets. *J Immunol* 1994;**152**(12):5896–901.

8. Hosszu KK, Valentino A, Vinayagasundaram U, et al. DC-SIGN, C1q, and gC1qR form a trimolecular receptor complex on the surface of monocyte-derived immature dendritic cells. *Blood* 2012;**120**(6):1228–36. http://dx.doi.org/10.1182/blood-2011-07-369728.

9. Nepomuceno RR, Henschen-Edman AH, Burgess WH, Tenner AJ. cDNA cloning and primary structure analysis of C1qR(P), the human C1q/MBL/SPA receptor that mediates enhanced phagocytosis in vitro. *Immunity* 1997;**6**(2):119–29.

10. Guan E, Robinson SL, Goodman EB, Tenner AJ. Cell-surface protein identified on phagocytic cells modulates the C1q-mediated enhancement of phagocytosis. *J Immunol* 1994;**152**(8):4005–16.

11. McGreal EP, Ikewaki N, Akatsu H, Morgan BP, Gasque P. Human C1qRp is identical with CD93 and the mNI-11 antigen but does not bind C1q. *J Immunol* 2002;**168**(10):5222–32.

12. Norsworthy PJ, Fossati-Jimack L, Cortes-Hernandez J, et al. Murine CD93 (C1qRp) contributes to the removal of apoptotic cells in vivo but is not required for C1q-mediated enhancement of phagocytosis. *J Immunol* 2004;**172**(6):3406–14.

13. Nepomuceno RR, Tenner AJ. C1qRP, the C1q receptor that enhances phagocytosis, is detected specifically in human cells of myeloid lineage, endothelial cells, and platelets. *J Immunol* 1998;**160**(4):1929–35.

14. Greenlee-Wacker MC, Galvan MD, Bohlson SS. CD93: recent advances and implications in disease. *Curr Drug Targets* 2012;**13**(3):411–20. pii:BSP/CDT/E-Pub/00369.

15. Klickstein LB, Barbashov SF, Liu T, Jack RM, Nicholson-Weller A. Complement receptor type 1 (CR1, CD35) is a receptor for C1q. *Immunity* 1997;**7**(3):345–55. http://dx.doi.org/10.1016/S1074-7613(00)80356-8.

16. Tas SW, Klickstein LB, Barbashov SF, Nicholson-Weller A. C1q and C4b bind simultaneously to CR1 and additively support erythrocyte adhesion. *J Immunol* 1999;**163**(9):5056–63. pii:ji_v163n9p5056.

17. Takada Y, Hemler ME. The primary structure of the VLA-2/collagen receptor alpha 2 subunit (platelet GPIa): homology to other integrins and the presence of a possible collagen-binding domain. *J Cell Biol* 1989;**109**(1):397–407.

18. Feng X, Tonnesen MG, Peerschke EI, Ghebrehiwet B. Cooperation of C1q receptors and integrins in C1q-mediated endothelial cell adhesion and spreading. *J Immunol* 2002;**168**(5):2441–8.

19. Edelson BT, Stricker TP, Li Z, et al. Novel collectin/C1q receptor mediates mast cell activation and innate immunity. *Blood* 2006;**107**(1):143–50. http://dx.doi.org/10.1182/2005-06-2218.

20. Madamanchi A, Santoro SA, Zutter MM. alpha2beta1 Integrin. *Adv Exp Med Biol* 2014;**819**:41–60. http://dx.doi.org/10.1007/978-94-017-9153-3_3.

21. Strickland DK, Ashcom JD, Williams S, et al. Sequence identity between the alpha 2-macroglobulin receptor and low density lipoprotein receptor-related protein suggests that this molecule is a multifunctional receptor. *J Biol Chem* 1990;**265**(29):17401–4.

22. Lillis AP, Van Duyn LB, Murphy-Ullrich JE, Strickland DK. LDL receptor-related protein 1: unique tissue-specific functions revealed by selective gene knockout studies. *Physiol Rev* 2008;**88**(3):887–918. http://dx.doi.org/10.1152/physrev.00033.2007.

23. Gardai SJ, Xiao YQ, Dickinson M, et al. By binding SIRPalpha or calreticulin/CD91, lung collectins act as dual function surveillance molecules to suppress or enhance inflammation. *Cell* 2003;**115**(1):13–23. http://dx.doi.org/10.106/S009286740300758X.

24. Lillis AP, Greenlee MC, Mikhailenko I, et al. Murine low-density lipoprotein receptor-related protein 1 (LRP) is required for phagocytosis of targets bearing LRP ligands but is not required for C1q-triggered enhancement of phagocytosis. *J Immunol* 2008;**181**(1):364–73.

25. Duus K, Hansen EW, Tacnet P, et al. Direct interaction between CD91 and C1q. *FEBS J* 2010;**277**(17):3526–37. http://dx.doi.org/10.1111/j.1742-4658.2010.07762.x.

26. Neeper M, Schmidt AM, Brett J, et al. Cloning and expression of a cell surface receptor for advanced glycosylation end products of proteins. *J Biol Chem* 1992;**267**(21):14998–5004.

27. Ma W, Rai V, Hudson BI, et al. RAGE binds C1q and enhances C1q-mediated phagocytosis. *Cell Immunol* 2012;**274**(1–2):72–82. http://dx.doi.org/10.1016/j.cellimm.2012.02.001.

28. Kierdorf K, Fritz G. RAGE regulation and signaling in inflammation and beyond. *J Leukoc Biol* 2013;**94**(1):55–68. http://dx.doi.org/10.1189/jlb.1012519.

29. Meyaard L, Adema GJ, Chang C, et al. LAIR-1, a novel inhibitory receptor expressed on human mononuclear leukocytes. *Immunity* 1997;**7**(2):283–90. http://dx.doi.org/10.1016/S1074-7613(00)80530-0.

30. Son M, Santiago-Schwarz F, Al-Abed Y, Diamond B. C1q limits dendritic cell differentiation and activation by engaging LAIR-1. *Proc Natl Acad Sci USA* 2012;**109**(46):E3160–7. http://dx.doi.org/10.1073/pnas.1212753109.

31. Son M, Diamond B. C1q-mediated repression of human monocytes is regulated by leukocyte-associated Ig-like receptor 1 (LAIR-1). *Mol Med* 2015;**20**:559–68. http://dx.doi.org/10.2119/molmed.2014.00185.

32. Adachi H, Tsujimoto M, Arai H, Inoue K. Expression cloning of a novel scavenger receptor from human endothelial cells. *J Biol Chem* 1997;**272**(50):31217–20.

33. Ramirez-Ortiz ZG, Pendergraft 3rd WF, Prasad A, et al. The scavenger receptor SCARF1 mediates the clearance of apoptotic cells and prevents autoimmunity. *Nat Immunol* 2013;**14**(9):917–26. http://dx.doi.org/10.1038/ni.2670; 10.1038/ni.2670.

34. Iram T, Ramirez-Ortiz Z, Byrne MH, et al. Megf10 is a receptor for C1Q that mediates clearance of apoptotic cells by astrocytes. *J Neurosci* 2016;**36**(19):5185–92. http://dx.doi.org/10.1523/JNEUROSCI.3850-15.2016.

Chapter 40

CR2

Susan A. Boackle

University of Colorado School of Medicine, Aurora, CO, United States

OTHER NAMES

Complement C3d receptor, Epstein–Barr virus receptor, CD21.

PHYSICOCHEMICAL PROPERTIES

Immature protein: CR2 is synthesised as two well-characterised precursor molecules of 1092 amino acids [16 complement control protein (CCP) form][1] and 1033 amino acids (15 CCP form),[1–3] and a third 971-amino acid molecule (14 CCP form) that is less well characterised[4] (Ensembl release 86). Each molecule contains a 20 amino acid signal peptide, a 22–24 amino acid transmembrane region and a 34 amino acid intracellular domain.

Mature Protein Isoforms	14 CCP	15 CCP	16 CCP
pI (predicted)	7.52	7.53	7.52
Amino acids	21–659	21–659	21–1092
	719–905	719–1092	
	968–1092		
M_r (K) predicted	106.4	112.9	119.2
Observed	N/A	146	148
N-linked glycosylation sites	10 (121, 127, 294, 372, 492, 623, 741, 859, 882, 970)	11 (same as 14 CCP plus 920)	13 (same as 15 CCP plus 699, 709)

STRUCTURE

CR2 is a glycosylated type I transmembrane protein consisting of a series of 14–16 CCP domains. The 14 CCP domain protein lacks the 11th and 15th CCP domain, and the 15 CCP domain protein lacks the 11th CCP domain. The CCP domains are partially folded back but flexible[5] and binding of C3d does not change the conformation of CR2 (Fig. 40.1).[6]

Crystal structure: The initial crystal structure of the first two CCP domains with the C3 ligand, C3d, showed that CCP1 and 2 formed a tightly compacted

The Complement FactsBook. http://dx.doi.org/10.1016/B978-0-12-810420-0.00040-7

FIGURE 40.1 Diagram of protein domains for CR2.

(A)

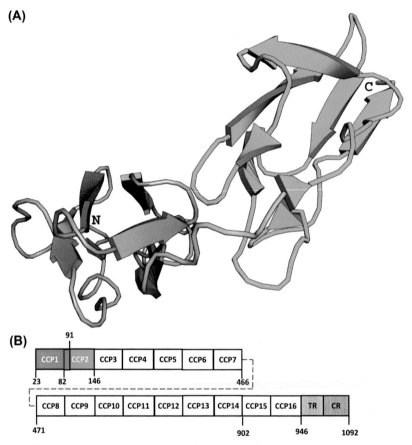

FIGURE 40.2 Diagram of crystal structure for CR2. (A) Ribbon diagram of the human complement receptor type 2 (CR2) CCP1-2 domains crystal structure (PDB ID: 1LY2). (B) Schematic representation of the single-chain CR2 domains.

V-shape, with only CCP2 making contact with C3d.[7] A later report of unbound CCP1 and 2 again showed a compact V-shape but also a high degree of flexibility at CCP1-CCP2,[8] suggesting that with glycosylation of the protein a more extended conformation would result. This was confirmed in a third report, which consistent with biochemical data, showed an extended conformation of CCP1 and CCP2 with contact of both with the acidic pocket on the concave surface of C3d (Fig. 40.2).[9]

FUNCTION

CR2 binds the complement components iC3b, C3dg and C3d[10,11]; the Epstein–Barr virus (EBV) glycoprotein gp350/220[12,13]; the low-affinity IgE receptor CD23;[14,15] interferon-α[16,17] and DNA.[18] C3d, EBV gp350/220, IFN-α and DNA bind to CCPs 1 and 2,[17-20] and CD23 binds to a glycosylation-dependent epitope in CCPs 5–8, as well as a glycosylation-independent epitope in CCPs 1 and 2.[15] CR2 cooperates with the B cell receptor to activate B cells,[21] processes and presents complement-opsonised antigens to T cells,[22] traps and retains immune complexes on follicular dendritic cells in lymphoid tissues to sustain immunologic memory[23] and shapes the natural antibody repertoire.[24]

TISSUE DISTRIBUTION

CR2 is expressed on the surface of mature B lymphocytes,[25] follicular dendritic cells,[26] thymocytes,[27] a subpopulation of CD4+ and CD8+ T lymphocytes,[28,29] basophils,[30] keratinocytes,[31] astrocytes[32] and epithelial cells.[33] It also exists in a soluble form.[34]

REGULATION OF EXPRESSION

Expression of CR2 is regulated in a cell type- and stage-specific developmental pattern using both transcriptional and posttranscriptional mechanisms. It is absent on pre-B cells, appears first on immature B cells and is most highly expressed on marginal zone B cells.[25,35] Within a few days of B cell activation, levels increase, but then rapidly decrease and are undetectable on plasma cells.[36,37] Infection of B cells with EBV increases B cell expression of CR2 through the transcriptional activating effects of EBNA-2[38] as does infection of T cells with HTLV-1.[39] Levels are decreased in systemic lupus erythematosus,[40] rheumatoid arthritis,[41] common variable immunodeficiency disease,[42] human immunodeficiency virus[43] and hepatitis C virus.[44]

Transcription of CR2 proceeds from a narrow ~30 bp region in mature B cells. The core promoter consists of a non-consensus TATA box, a GC box predicted to bind SP1, an initiator element and a downstream promoter element that binds E2A.[45] The CR2 proximal promoter contains four functionally relevant sites at −47, −63, −81 and −120/−93 that bind the transcription factors USF1, E2A proteins, RP58, an AP-2-like transcription factor and SP1 with activating or repressive effects, depending on the cellular context.[46-48] Inducible B cell expression of CR2 occurs after stimulation with IL-4 and anti-CD40 and is mediated by protein kinase A and C signalling pathways, a heterogeneous nuclear ribonucleoprotein and NF-κB via interactions with both proximal and distal promoter elements.[49-51] A silencing region in intron 1 that binds the transcriptional repressor CBF1 regulates cell type and stage-specific expression.[52,53]

Follicular dendritic cells express the 16 short consensus repeat (SCR) iso-form of CR2, whereas B cells express the 15 SCR isoform in which exon 11 (corresponding to CCP 11) has been removed.[54] The regulatory elements medi-ating this cell-type-specific alternative splicing are not known.

HUMAN PROTEIN SEQUENCE

```
MGAAGLLGVF LALVAPGVLG ISCGSPPPIL NGRISYYSTP IAVGTVIRYS   50
CSGTFRLIGE KSLLCITKDK VDGTWDKPAP KCEYFNKYSS CPEPIVPGGY  100
KIRGSTPYRH GDSVTFACKT NFSMNGNKSV WCQANNMWGP TRLPTCVSVF  150
PLECPALPMI HNGHHTSENV GSIAPGLSVT YSCESGYLLV GEKIINCLSS  200
GKWSAVPPTC EEARCKSLGR FPNGKVKEPP ILRVGVTANF FCDEGYRLQG  250
PPSSRCVIAG QGVAWTKMPV CEEIFCPSPP PILNGRHIGN SLANVSYGSI  300
VTYTCDPDPE EGVNFILIGE STLRCTVDSQ KTGTWSGPAP RCELSTSAVQ  350
CPHPQILRGR MVSGQKDRYT YNDTVIFACM FGFTLKGSKQ IRCNAQGTWE  400
PSAPVCEKEC QAPPNILNGQ KEDRHMVRFD PGTSIKYSCN PGYVLVGEES  450
IQCTSEGVWT PPVPQCKVAA CEATGRQLLT KPQHQFVRPD VNSSCGEGYK  500
LSGSVYQECQ GTIPWFMEIR LCKEITCPPP PVIYNGAHTG SSLEDFPYGT  550
TVTYTCNPGP ERGVEFSLIG ESTIRCTSND QERGTWSGPA PLCKLSLLAV  600
QCSHVHIANG YKISGKEAPY FYNDTVTFKC YSGFTLKGSS QIRCKADNTW  650
DPEIPVCEKG CQSPPGLHHG RHTGGNTVFF VSGMTVDYTC DPGYLLVGNK  700
SIHCMPSGNW SPSAPRCEET CQHVRQSLQE LPAGSRVELV NTSCQDGYQL  750
TGHAYQMCQD AENGIWFKKI PLCKVIHCHP PPVIVNGKHT GMMAENFLYG  800
NEVSYECDQG FYLLGEKKLQ CRSDSKGHGS WSGPSPQCLR SPPVTRCPNP  850
EVKHGYKLNK THSAYSHNDI VYVDCNPGFI MNGSRVIRCH TDNTWVPGVP  900
TCIKKAFIGC PPPPKTPNGN HTGGNIARFS PGMSILYSCD QGYLLVGEAL  950
LLCTHEGTWS QPAPHCKEVN CSSPADMDGI QKGLEPRKMY QYGAVVTLEC 1000
EDGYMLEGSP QSQCQSDHQW NPPLAVCRSR SLAPVLCGIA AGLILLTFLI 1050
VITLYVISKH RARNYYTDTS QKEAFHLEAR EVYSVDPYNP AS          1091
```

The leader peptide is underlined, *N*-linked glycosylation sites bolded, and alter-natively spliced sequence italicised.

PROTEIN MODULES

Sequence	Domain	Exon
1–21	Leader sequence	exon 1
21–84	Sushi 1	exon 2
89–148	Sushi 2	exon 2
152–212	Sushi 3	exon 3
213–273	Sushi 4	exons 4/5
274–344	Sushi 5	exon 6
349–408	Sushi 6	exon 6
409–468	Sushi 7	exon 7
469–524	Sushi 8	exons 8/9
525–595	Sushi 9	exon 10
600–659	Sushi 10	exon 10
660–716	Sushi 11	exon 11
717–781	Sushi 12	exons 12/13
786–845	Sushi 13	exon 14
849–909	Sushi 14	exon 14
910–970	Sushi 15	exon 15

CHROMOSOMAL LOCATION

Human chromosome 1: 207,454,230-207,489,895 forward strand; Cyto-Band q32.2.

Telomere … MCP/CD46 … CR1L … CR1/CD35 … CR2/CD21 … DAF/CD55 … C4BPA … C4BPB … Centromere.

Mouse chromosome 1: 195,136,811-195,176,716 reverse strand.

Telomere … Cr2 … Crry … MCP/Cd46 … C4bp … DAF/Cd55 … Centromere.

Rat chromosome 13: 113,890,272-113,927,877 reverse strand.

Telomere … Cr2 … Crry … MCP/CD46 … C4bpa … C4bpb … DAF/Cd55 … Centromere.

HUMAN cDNA SEQUENCE

```
1     CACCGGCGCCGCGTCAGCCCCCAGGCCGCCTGCAGGTGTGCGCTCAGAACTAGCACGTGT    60
61    GCCGGACACTATTTAAGGGCCCGCCTCTCCTGGCTCACAGCTGCTTGCTGCTCCAGCCTT   120
121   GCCCTCCCAGAGCTGCCGGACGCTCGCGGGTCTCGGAACGCATCCCGCCGCGGGGGCTTC   180
181   GGCCGTGGCATGGGCGCCGCGGGCCTGCTCGGGGTTTTCTTGGCTCTCGTCGCACCGGGG   240
241   GTCCTCGGGGATTTCTTGTGGCTCTCCTCCGCCTATCCTAAATGGCCGGATTAGTTATTAT   300
301   TCTACCCCCATTGCTGTTGGTACCGTGATAAGGTACAGTTGTTCAGGTACCTTCCGCCTC   360
361   ATTGGAGAAAAAAGTCTATTATGCATAACTAAAGACAAAGTGGATGGAACCTGGGATAAA   420
421   CCTGCTCCTAAATGTGAATATTTCAATAAATATTCTTCTTGCCCTGAGCCCATAGTACCA   480
481   GGAGGATACAAAATTAGAGGCTCTACACCCTACAGACATGGTGATTCTGTGACATTTGCC   540
541   TGTAAAACCAACTTCTCCATGAACGGAAACAAGTCTGTTTGGTGTCAAGCAAATAATATG   600
601   TGGGGGCCGACACGACTACCAACCTGTGTAAGTGTTTTCCCTCTCGAGTGTCCAGCACTT   660
661   CCTATGATCCACAATGGACATCACACAAGTGAGAATGTTGGCTCCATTGCTCCAGGATTG   720
721   TCTGTGACTTACAGCTGTGAATCTGGTTACTTGCTTGTTGGAGAAAAGATCATTAACTGT   780
781   TTGTCTTCGGGAAAATGGAGTGCTGTCCCCCCCACATGTGAAGAGGCACGCTGTAAATCT   840
841   CTAGGACGATTTCCCAATGGGAAGGTAAAGGAGCCTCCAATTCTCCGGGTTGGTGTAACT   900
901   GCAAACTTTTTCTGTGATGAAGGGTATCGACTGCAAGGCCCACCTTCTAGTCGGTGTGTA   960
961   ATTGCTGGACAGGGAGTTGCTTGGACCAAAATGCCAGTATGTGAAGAAATTTTTTGCCCA  1020
1021  TCACCTCCCCCTATTCTCAATGGAAGACATATAGGCAACTCACTAGCAAATGTCTCATAT  1080
1081  GGAAGCATAGTCACTTACACTTGTGACCCGGACCCAGAGGAAGGAGTGAACTTCATCCTT  1140
1141  ATTGGAGAGAGCACTCTCCGTTGTACAGTTGATAGTCAGAAGACTGGGACCTGGAGTGGC  1200
1201  CCTGCCCCACGCTGTGAACTTTCTACTTCTGCGGTTCAGTGTCCACATCCCCAGATCCTA  1260
1261  AGAGGCCGAATGGTATCTGGGCAGAAAGATCGATATACCTATAACGACACTGTGATATTT  1320
1321  GCTTGCATGTTTGGCTTCACCTTGAAGGGCAGCAAGCAAATCCGATGCAATGCCCAAGGC  1380
1381  ACATGGGAGCCATCTGCACCAGTCTGTGAAAAGGAATGCCAGGCCCCTCCTAACATCCTC  1440
1441  AATGGGCAAAAGGAAGATAGACACATGGTCCGCTTTGACCCTGGAACATCTATAAAATAT  1500
1501  AGCTGTAACCCTGGCTATGTGCTGGTGGGAGAAGAATCCATACAGTGTACCTCTGAGGGG  1560
1561  GTGTGGACACCCCCTGTACCCCAATGCAAAGTGGCAGCGTGTGAAGCTACAGGAAGGCAA  1620
1621  CTCTTGACAAAACCCCAGCACCAATTTGTTAGACCAGATGTCAACTCTTCTTGTGGTGAA  1680
1681  GGGTACAAGTTAAGTGGGAGTGTTTATCAGGAGTGTCAAGGCACAATTCCTTGGTTTATG  1740
1741  GAGATTCGTCTTTGTAAAGAAATCACCTGCCCACCACCCCCTGTTATCTACAATGGGGCA  1800
1801  CACACCGGGAGTTCCTTGGAAGATTTTCCATATGGAACCACGGTCACTTACACATGTAAC  1860
1861  CCTGGGCCAGAAAGAGGAGTGGAATTCAGCCTCATTGGAGAGCACCATCCGTTGTACA  1920
1921  AGCAATGATCAAGAAGAGGCACCTGGAGTGGCCCTGCTCCCCTGTGTAAACTTTCCCTC  1980
1981  CTTGCTGTCCAGTGCTCACATGTCCATATTGCAAATGGATACAAGATATCTGGCAAGGAA  2040
2041  GCCCCATATTCTACAATGACACTGTGACATTCAAGTGTTATAGTGGATTTACTTTGAAG  2100
2101  GGCAGTAGTCAGATTCGTTGCAAAGCTGATAACACCTGGGATCCTGAAATTCCAGTTTGT  2160
2161  GAAAAAGGCTGCCCAGTCACCTCCTGGGCTCCACCATGGTCGTCATACAGGTGGAAATACG  2220
2221  GTCTTCTTTGTCTCTGGGATGACTGTAGACTACACTTGTGACCCTGGCTATTTGCTTGTG  2280
2281  GGAAACAAATCCATTCACTGTATGCCTTCAGGAAATTGGAGTCCTTCTGCCCCACGGTGT  2340
2341  GAAGCAAACATGCCAGCATGTGAGACAGAGTCTTCAAGAACTTCCAGCTGGTTCACGTGTG  2400
2401  GAGCTAGTTAATACGTCCTGCCAAGATGGGTACCAGTTGACTGGACATGCTTATCAGATG  2460
2461  TGTCAAGATGCTGAAAATGGAATTTGGTTCAAAAAGATTCCACTTTGTAAAGTTATTCAC  2520
```

HUMAN cDNA SEQUENCE—*Continued*

```
2521 TGTCACCCTCCACCAGTGATTGTCAATGGGAAGCACACAGGCATGATGGCAGAAAACTTT 2580
2581 CTATATGGAAATGAAGTCTCTTATGAATGTGACCAAGGATTCTATCTCCTGGGAGAGAAA 2640
2641 AAATTGCAGTGCAGAAGTGATTCTAAAGGACATGGATCTTGGAGCGGGCCTTCCCCACAG 2700
2701 TGCTTACGATCTCCTCCTGTGACTCGCTGCCCTAATCCAGAAGTCAAACATGGGTACAAG 2760
2761 CTCAATAAAACACATTCTGCATATTCCCACAATGACATAGTGTATGTTGACTGCAATCCT 2820
2821 GGCTTCATCATGAATGGTAGTCGCGTGATTAGGTGTCATACTGATAACACATGGGTGCCA 2880
2881 GGTGTGCCAACTTGTATCAAAAAAG̲C̲C̲T̲T̲CATAGGGTGTCCACCTCCGCCTAAGACCCCT 2940
2941 AACGGGAACCATACTGGTGGAAACATAGCTCGATTTTCTCCTGGAATGTCAATCCTGTAC 3000
3001 AGCTGTGACCAAGGCTACCTGCTGGTGGGGAGAGGCACTCCTTCTTTGCACACATGAGGGA 3060
3061 ACCTGGAGCCAACCTGCCCCTCATTGTAAAG̲A̲G̲G̲T̲AAACTGTAGCTCACCAGCAGATATG 3120
3121 GATGGAATCCAGAAAGGGCTGGAACCAAGGAAAATGTATCAGTATGGAGCTGTTGTAACT 3180
3181 CTGGAGTGTGAAGATGGGTATATGCTGGAAGGCAGTCCCCAGAGCCAGTGCCAATCGGAT 3240
3241 CACCAATGGAACCCTCCCCTGGCGGTTTGCAGATCCC̲G̲T̲T̲C̲A̲CTTGCTCCTGTCCTTTGT 3300
3301 G̲G̲T̲A̲T̲TGCTGCAGGTTTGATACTTCTTACCTTCTTGATTGTCATTACCTTATACGTGATA 3360
3361 TCAAAACACAGAGCACGC̲A̲A̲T̲T̲ATTATTATACAGATCAAGCCAGAAAGAAGCTTTTCATTTA 3420
3421 GAAGCACGAGAAGTATATTCTGTTGATCCATACAACCCAGCCAGC̲T̲G̲A̲TCAGAAGACAAA 3480
3481 CTGGTG̲T̲G̲T̲G̲CCTCATTGCTTGGAATTCAGCGGAATATTGATTAGAAAGAAACTGCTCTA 3540
3541 ATATCAGCAAGTCTCTTTATATGGCCTCAAGATCAATGAAATGATGTCATAAGCGATCAC 3600
3601 TTCCTATATGCACTTATTCTCAAGAAGAACATCTTTATGGTAAAGATGGGAGCCCAGTTT 3660
3661 CACTGCCATATACTCTTCAAGGACTTTCTGAAGCCTCACTTATGAGATGCCTGAAGCCAG 3720
3721 GCCATGGCTATAAACAATTACATGGCTCTAAAAAGTTTTGCCCTTTTTAAGGAAGGCACT 3780
3781 AAAAAGAGCTGTCCTGGTATCTAGACCCATCTTCTTTTTGAAATCAGCATACTCAATGTT 3840
3841 ACTATCTGCTTTTGGTTATAATGTGTTTTTAATTATCTAAAGTATGAAGCATTTTCTGGG 3900
3901 GTTATGATGGCTTTACCTTTATTAGGAAGTATGGTTTTATTTTGATAGTAGCTTCCTCCT 3960
3961 CTGGTGGTGTTAATCATTTCATTTTTACCCTTACTTGGTTTGAGTTTCTCTCACATTACT 4020
4021 GTATATACTTTGCCTTTCCATAATCACTCAGTGATTGCAATTTGCACAAGTTTTTTTAAA 4080
4081 TTATGGGAATCAAGATTTAATCCTAGAGATTTGGTGTACAATTCAGGCTTTGGATGTTTC 4140
4141 TTTAGCAGTTTTGTGATAAGTTCTAGTTGCTTGTAAAATTTCACTTAATAATGTGTACAT 4200
4201 TAGTCATTC̲A̲A̲T̲A̲A̲A̲TTGTAATTGTAAAGAAAACATACAA                      4240
```

The first five nucleotides in each exon are underlined to mark the intron–exon boundaries. The methionine initiation site (ATG), termination codon (TGA) and probable polyadenylation signal (AATAAA) are indicated by double underlining. This sequence corresponds to NCBI Reference Sequence: NM_001006658.2 and NP_001006659.1.

GENOMIC STRUCTURE

Human *CR2* spans 35.67 kb and includes 20 exons as shown below.[4] Like the membrane-bound complement regulatory proteins MCP, DAF and CR1 and the soluble complement regulatory proteins fH and C4BP, CR2 has evolved from a common structural and functional domain called the CCP, which is an ~60 amino acid sequence with internal disulphide bonds also known as a Sushi domain or SCR (Fig. 40.3).

Exon 1 of *CR2* corresponds to the 5′ untranslated region (UTR) and leader peptide and exon 20 corresponds to the 3′ UTR. Exon 2 encodes CCPs 1 and 2, exon 3 encodes CCP3, and exon 4 and 5 encode CCP4. Exon 6 encodes CCPs 5 and 6, exon 7 encodes CCP7, and exon 8 and 9 encode CCP8. Exon 10

FIGURE 40.3 Schematic of CR2 gene.

encodes CCPs 9 and 10, exon 11 encodes CCP11, and exons 12 and 13 encode CCP12. Exon 14 encodes CCPs 13 and 14, exon 15 encodes CCP15, and exons 16 and 17 encode CCP16. Exons 17 and 18 encode the transmembrane domain, and exon 19 encodes the intracellular domain. Alternative splicing of exons 11 and 15 generates the 14/15/16 CCP isoforms of CR2. The first intron is 12.05 kb and contains the intronic silencer,[52,53] with the CBF1 site located 4.38 kb from the beginning of the intron. The 5′ region of intron 1 also contains B cell-specific enhancer elements based on histone marks and DNAse hypersensitivity [ENcyclopedia Of DNA Elements (ENCODE project)[55,56]].

The genome structure for *CR2* is similar in humans and nonhuman primates, but in mice, as in other non-primate species, the 25-exon *Cr2* gene also encodes CR1 by alternative splicing of exons 2–8 to encode an additional 6 N-terminal CCPs.[57,58] The exons encoding these CCPs remain in human *CR2* but are not incorporated into functional transcripts.

ACCESSION NUMBERS

Human CR2	NM_001006658
	NM_001877
Mouse Cr2	NM_007758

DEFICIENCY

CR2 deficiency has been described in a 28-year-old man with common variable immunodeficiency syndrome who had a compound heterozygous mutation in *CR2* and reduced serum immunoglobulins and class-switched memory B cells, abolition of C3b costimulatory activity and impaired responses to polysaccharide antigens.[59]

POLYMORPHIC VARIANTS

There are 2416 known variants in *CR2* (NCBI Variation Viewer), including 2197 single-nucleotide variants, 38 copy number variants, 107 deletions, 70 insertions, four indels, and two inversions. 1738 are intronic, 568 are in coding domains (407 nonsynonymous, 161 synonymous), 116 are in the 5′ or 3′ UTR, 86 are in the 2 kb upstream region, and 34 have other potential molecular consequences (21 frame-shift, 11 splice donor, 1 splice acceptor, 1 inframe). Many of these are rare, occurring in less than 0.5% of the population according to the 1000 Genomes Project.

The following are functional *CR2* variants.

Variant ID	Type	Functional Effect	Disease Association
rs182309299	5′ UTR	Increased CR2 transcription	–
rs3813946	5′ UTR	Decreased CR2 promoter activity	SLE
rs1876453	Intron 1	Increased transcription of *CR1* gene	SLE
rs1048971	Synonymous, exon 10	Decreased splicing effect of exon 11	SLE
rs17615	Missense exon 10		
rs4308977	Missense exon 11		

MUTANT ANIMALS

Cr2-knockout mice have been generated by targeted gene disruption and demonstrate hypogammaglobulinemia, impaired humoral immune responses and altered development of autoimmune disease.[60–68] One line is available through Jackson Labs.[60] Mice with natural mutations in *Cr2* that alter receptor function have also been described.[69] A new mouse anti-mouse antibody to CR1/CR2 has been developed that blocks C3dg binding and enables the examination of the in vivo effects of receptor blockage on immune responses and disease.[70]

REFERENCES

1. Moore MD, Cooper NR, Tack BF, Nemerow GR. Molecular cloning of the cDNA encoding the Epstein–Barr virus/C3d receptor (complement receptor type 2) of human B lymphocytes. *Proc Natl Acad Sci USA* 1987;**84**:9194–8.
2. Fujisaku A, Harley JB, Frank MB, Gruner BA, Frazier B, Holers VM. Genomic organization and polymorphisms of the human C3d/Epstein–Barr virus receptor. *J Biol Chem* 1989;**264**:2118–25.
3. Weis JJ, Toothaker LE, Smith JA, Weis JH, Fearon DT. Structure of the human B lymphocyte receptor for C3d and the Epstein–Barr virus and relatedness to other members of the family of C3/C4 proteins. *J Exp Med* 1988;**167**:1047–66.
4. Aken BL, Ayling S, Barrell D, et al. The Ensembl gene annotation system. *Database J Biol Databases Curation* 2016;**2016**.
5. Gilbert HE, Asokan R, Holers VM, Perkins SJ. The 15 SCR flexible extracellular domains of human complement receptor type 2 can mediate multiple ligand and antigen interactions. *J Mol Biol* 2006;**362**:1132–47.
6. Li K, Okemefuna AI, Gor J, et al. Solution structure of the complex formed between human complement C3d and full-length complement receptor type 2. *J Mol Biol* 2008;**384**:137–50.
7. Szakonyi G, Guthridge JM, Li D, Young K, Holers VM, Chen XS. Structure of complement receptor 2 in complex with its C3d ligand. *Science* 2001;**292**:1725–8.
8. Prota AE, Sage DR, Stehle T, Fingeroth JD. The crystal structure of human CD21: implications for Epstein–Barr virus and C3d binding. *Proc Natl Acad Sci USA* 2002;**99**:10641–6.
9. van den Elsen JM, Isenman DE. A crystal structure of the complex between human complement receptor 2 and its ligand C3d. *Science* 2011;**332**:608–11.

10. Iida K, Nadler L, Nussenzweig V. Identification of the membrane receptor for the complement fragment C3d by means of a monoclonal antibody. *J Exp Med* 1983;**158**:1021–33.

11. Weis JJ, Tedder TF, Fearon DT. Identification of a 145,000 Mr membrane protein as the C3d receptor (CR2) of human B lymphocytes. *Proc Natl Acad Sci USA* 1984;**81**:881–5.

12. Fingeroth JD, Weis JJ, Tedder TF, Strominger JL, Biro PA, Fearon DT. Epstein–Barr virus receptor of human B lymphocytes is the C3d receptor CR2. *Proc Natl Acad Sci USA* 1984;**81**:4510–4.

13. Nemerow GR, Wolfert R, McNaughton ME, Cooper NR. Identification and characterization of the Epstein–Barr virus receptor on human B lymphocytes and its relationship to the C3d complement receptor (CR2). *J Virol* 1985;**55**:347–51.

14. Aubry J-P, Pochon S, Graber P, Jansen KU, Bonnefoy J-Y. CD21 is a ligand for CD23 and regulates IgE production. *Nature* 1992;**358**:505–7.

15. Aubry J-P, Pochon S, Gauchat J-F, et al. CD23 interacts with a new functional extracytoplasmic domain involving N-linked oligosaccharides on CD21. *J Immunol* 1994;**152**:5806–13.

16. Delcayre AX, Salas F, Mathur S, Kovats K, Lotz M, Lernhardt W. Epstein Barr virus/complement C3d receptor is an interferon alpha receptor. *EMBO J* 1991;**10**:919–26.

17. Asokan R, Hua J, Young KA, et al. Characterization of human complement receptor type 2 (CR2/CD21) as a receptor for IFN-α: a potential role in systemic lupus erythematosus. *J Immunol* 2006;**177**:383–94.

18. Asokan R, Banda NK, Szakonyi G, Chen XS, Holers VM. Human complement receptor 2 (CR2/CD21) as a receptor for DNA: implications for its roles in the immune response and the pathogenesis of systemic lupus erythematosus (SLE). *Mol Immunol* 2013;**53**:99–110.

19. Carel J-C, Myones BL, Frazier B, Holers VM. Structural requirements for C3d,g/Epstein–Barr virus receptor (CR2/CD21) ligand binding, internalization, and viral infection. *J Biol Chem* 1990;**265**:12293–9.

20. Lowell CA, Klickstein LB, Carter RH, Mitchell JA, Fearon DT, Ahearn JM. Mapping of the Epstein–Barr virus and C3dg binding sites to a common domain on complement receptor type 2. *J Exp Med* 1989;**170**:1931–46.

21. Carter RH, Spycher MO, Ng YC, Hoffman R, Fearon DT. Synergistic interaction between complement receptor type 2 and membrane IgM on B lymphocytes. *J Immunol* 1988;**141**:457–63.

22. Boackle SA, Morris MA, Holers VM, Karp DR. Complement opsonization is required for the presentation of immune complexes by resting peripheral blood B cells. *J Immunol* 1998;**161**:6537–43.

23. Heesters BA, Myers RC, Carroll MC. Follicular dendritic cells: dynamic antigen libraries. *Nat Rev Immunol* 2014;**14**:495–504.

24. Holers VM. Complement receptors and the shaping of the natural antibody repertoire. *Springer Semin Immun* 2005;**26**:405–23.

25. Tedder TF, Clement LT, Cooper MD. Expression of C3d receptors during human B cell differentiation: immunofluorescence analysis with the HB-5 monoclonal antibody. *J Immunol* 1984;**133**:678–83.

26. Reynes M, Aubert JP, Cohen JHM, et al. Human follicular dendritic cells express CR1, CR2, and CR3 complement receptor antigens. *J Immunol* 1985;**135**:2687–94.

27. Watry D, Hedrick JA, Siervo S, et al. Infection of human thymocytes by Epstein–Barr virus. *J Exp Med* 1991;**173**:971–80.

28. Fischer E, Delibrias C, Kazatchkine MD. Expression of CR2 (the C3dg/EBV receptor, CD21) on normal human peripheral blood T lymphocytes. *J Immunol* 1991;**146**:865–9.

29. Levy E, Ambrus J, Kahl L, Molina H, Tung K, Holers VM. T lymphocyte expression of complement receptor 2 (CR2/CD21): a role in adhesive cell-cell interactions and dysregulation in a patient with systemic lupus erythematosus (SLE). *Clin Exp Immunol* 1992;**90**:235–44.

30. Bacon K, Gauchat JF, Aubry JP, et al. CD21 expressed on basophilic cells is involved in histamine release triggered by CD23 and anti-CD21 antibodies. *EurJImmunol* 1993;**23**:2721–4.

31. Hunyadi J, Simon M, Kenderessy AS, Dobozy A. Expression of complement receptor 2 (CD21) on human subcorneal keratinocytes in normal and diseased skin. *Dermatologica* 1991;**183**: 184–6.

32. Gasque P, Chan P, Mauger C, et al. Identification and characterization of complement C3 receptors on human astrocytes. *J Immunol* 1996;**156**:2247–55.

33. Levine J, Pflugfelder SC, Yen M, Crouse CA, Atherton SS. Detection of the complement (CD21)/Epstein–Barr virus receptor in human lacrimal gland and ocular surface epithelia. *Reg Immunol* 1990;**3**:164–70.

34. Huemer HP, Larcher C, Prodinger WM, Petzer AL, Mitterer M, Falser N. Determination of soluble CD21 as a parameter of B cell activation. *Clin Exp Immunol* 1995;**93**:195–9.

35. Timens W, Boes A, Poppema S. Human marginal zone B cells are not an activated B cell subset: strong expression of CD21 as a putative mediator for rapid B cell activation. *Eur J Immunol* 1989;**19**:2163–6.

36. Boyd AW, Anderson KC, Freedman AS, et al. Studies of in vitro activation and differentiation of human B lymphocytes: phenotypic and functional characterization of the B cell population responding to anti-Ig antibody. *J Immunol* 1985;**134**:1516–23.

37. Stashenko P, Nadler LM, Hardy R, Schlossman SF. Expression of cell surface markers after human B lymphocyte activation. *Proc Natl Acad Sci* 1981;**78**:3848–52.

38. Cordier-Bussat M, Billaud M, Calender A, Lenoir GM. Epstein–Barr virus (EBV) nuclear-antigen-2-induced up-regulation of CD21 and CD23 molecules is dependent on a permissive cellular context. *Int J Cancer* 1993;**53**:153–60.

39. McNearney TA, Ebenbichler CF, Totsch M, Dierich MP. Expression of complement receptor 2 (CR2) on HTLV-1-infected lymphocytes and transformed cell lines. *Eur J Immunol* 1993;**23**:1266–70.

40. Wilson JG, Ratnoff WD, Schur PH, Fearon DT. Decreased expression of the C3b/C4b receptor (CR1) and the C3d receptor (CR2) on B lymphocytes and of CR1 on neutrophils of patients with systemic lupus erythematosus. *Arthritis Rheum* 1986;**29**:739–47.

41. Isnardi I, Ng YS, Menard L, et al. Complement receptor 2/CD21- human naive B cells contain mostly autoreactive unresponsive clones. *Blood* 2010;**115**:5026–36.

42. Warnatz K, Wehr C, Drager R, et al. Expansion of CD19(hi)CD21(lo/neg) B cells in common variable immunodeficiency (CVID) patients with autoimmune cytopenia. *Immunobiology* 2002;**206**:502–13.

43. Moir S, Ho J, Malaspina A, et al. Evidence for HIV-associated B cell exhaustion in a dysfunctional memory B cell compartment in HIV-infected viremic individuals. *J Exp Med* 2008;**205**:1797–805.

44. Charles ED, Brunetti C, Marukian S, et al. Clonal B cells in patients with hepatitis C virus-associated mixed cryoglobulinemia contain an expanded anergic CD21low B-cell subset. *Blood* 2011;**117**:5425–37.

45. Taylor RL, Cruickshank MN, Karimi M, et al. Focused transcription from the human CR2/CD21 core promoter is regulated by synergistic activity of TATA and initiator elements in mature B cells. *Cell Mol Immunol* 2015. http://dx.doi.org/10.1038/cmi.2014.138.

46. Ulgiati D, Holers VM. CR2/CD21 proximal promoter activity is critically dependent on a cell type-specific repressor. *J Immunol* 2001;**167**:6912–9.

47. Ulgiati D, Pham C, Holers VM. Functional analysis of the human complement receptor 2 (CR2/CD21) promoter: characterization of basal transcriptional mechanisms. *J Immunol* 2002;**168**:6279–85.

48. Cruickshank MN, Dods J, Taylor RL, et al. Analysis of tandem E-box motifs within human complement receptor 2 (CR2/CD21) promoter reveals cell specific roles for RP58, E2A, USF and localized chromatin accessibility. *Int J Biochem Cell Biol* 2015;**64**:107–19.

49. Vereshchagina LA, Tolnay M, Tsokos GC. Multiple transcriptional factors regulate the inducible expression of the human complement receptor 2 promoter. *J Immunol* 2001;**166**: 6156–63.

50. Tolnay M, Vereshchagina LA, Tsokos GC. NF-kappaB regulates the expression of the human complement receptor 2 gene. *J Immunol* 2002;**169**:6236–43.

51. Tolnay M, Lambris JD, Tsokos GC. Transcriptional regulation of the complement receptor 2 gene: role of heterogeneous nuclear ribonucleoprotein. *J Immunol* 1998;**159**:5492–501.

52. Makar KW, Pham CTN, Dehoff MH, O'Connor SM, Jacobi SM, Holers VM. An intronic silencer regulates B lymphocyte cell- and stage-specific expression of the human complement receptor type 2 (CR2, CD21) gene. *J Immunol* 1998;**160**:1268–78.

53. Makar KW, Ulgiati D, Hagman J, Holers VM. A site in the complement receptor 2 (CR2/CD21) silencer is necessary for lineage specific transcriptional regulation. *Int Immunol* 2001;**13**:657–64.

54. Liu Y-J, Xu J, de Bouteiller O, et al. Follicular dendritic cells specifically express the long CR2/CD21 isoform. *J Exp Med* 1997;**185**:165–70.

55. Rosenbloom KR, Armstrong J, Barber GP, et al. The UCSC Genome Browser database: 2015 update. *Nucleic Acids Res* 2015;**43**:D670–81.

56. Rosenbloom KR, Sloan CA, Malladi VS, et al. ENCODE data in the UCSC Genome Browser: year 5 update. *Nucleic Acids Res* 2013;**41**:D56–63.

57. Kurtz CB, O'Toole E, Christensen SM, Weis JH. The murine complement receptor gene family. IV. Alternative splicing of Cr2 gene transcripts predicts two distinct gene products that share homologous domains with both human CR2 and CR1. *J Immunol* 1990;**144**:3581–91.

58. Fingeroth JD. Comparative structure and evolution of murine CR2: the homolog of the human C3d/EBV receptor (CD21). *J Immunol* 1990;**144**:3458–67.

59. Thiel J, Kimmig L, Salzer U, et al. Genetic CD21 deficiency is associated with hypogamma-globulinemia. *J Allergy Clin Immunol* 2012;**129**:801–10.

60. Molina H, Holers VM, Li B, et al. Markedly impaired humoral immune response in mice deficient in complement receptors 1 and 2. *Proc Natl Acad Sci USA* 1996;**93**:3357–61.

61. Croix DA, Ahearn JM, Rosengard AM, et al. Antibody response to a T-dependent antigen requires B cell expression of complement receptors. *J Exp Med* 1996;**183**:1857–64.

62. Ahearn JM, Fischer MB, Croix D, et al. Disruption of the Cr2 locus results in a reduction in B-1a cells and in an impaired B cell response to T-dependent antigen. *Immunity* 1996;**4**:251–62.

63. Haas KM, Hasegawa M, Steeber DA, et al. Complement receptors CD21/35 link innate and protective immunity during *Streptococcus pneumoniae* infection by regulating IgG3 antibody responses. *Immunity* 2002;**17**:713–23.

64. Boackle SA, Culhane KK, Brown JM, et al. CR1/CR2 deficiency alters IgG3 autoantibody production and IgA glomerular deposition in the MRL/lpr model of SLE. *Autoimmunity* 2004;**37**:111–23.

65. Wu X, Jiang N, Deppong C, et al. A role for the Cr2 gene in modifying autoantibody production in systemic lupus erythematosus. *J Immunol* 2002;**169**:1587–92.

66. Prodeus AP, Georg S, Shen L-M, et al. A critical role for complement in the maintenance of self-tolerance. *Immunity* 1998;**9**:721–31.

67. Kaya Z, Afanasyeva M, Wang Y, et al. Contribution of the innate immune system to autoimmune myocarditis: a role for complement. *Nat Immunol* 2001;**2**:739–45.

68. Del Nagro CJ, Kolla RV, Rickert RC. A critical role for complement C3d and the B cell coreceptor (CD19/CD21) complex in the initiation of inflammatory arthritis. *J Immunol* 2005;**175**:5379–89.

69. Boackle SA, Holers VM, Chen X, et al. Cr2, a candidate gene in the murine Sle1c lupus susceptibility locus, encodes a dysfunctional protein. *Immunity* 2001;**15**:775–85.

70. Kulik L, Hewitt FB, Willis VC, Rodriguez R, Tomlinson S, Holers VM. A new mouse anti-mouse complement receptor type 2 and 1 (CR2/CR1) monoclonal antibody as a tool to study receptor involvement in chronic models of immune responses and disease. *Mol Immunol* 2015;**63**:479–88.

Chapter 41

CR3

Daniel C. Bullard

University of Alabama at Birmingham, Birmingham, AL, United States

OTHER NAMES

Mac-1, CD11b/CD18, α_M/β_2, *ITGAM/ITGB2*.

PHYSICOCHEMICAL PROPERTIES

	CD11b (*ITGAM*)	CD18 (*ITGB2*)
Amino acids	1152	769
Signal peptide	1–16	1–22
Isoelectric point	7.21	6.95
Predicted MW (K)	127.2	87.8
Actual MW (K)	160–170	95
N-linked glycosylation sites	86, 240, 391, 469 692, 696, 734, 801 880, 900, 911, 940 946, 978, 993, 1021 1044, 1050, 1075	50, 116, 212, 213, 215, 254 501, 642
Interchain disulphide bonds	66–73, 105–123, 654–711 770–776, 847–864 998–1022, 1027–1032	28 predicted

STRUCTURE

CR3 shares a similar structure to all four members of β_2 integrins.[1] The α-chain for this complement receptor is encoded by the *CD11b* (α_m or *ITGAM*) gene, which noncovalently associates with the β_2-subunit (encoded by the CD18 or *ITGB2* gene). The extracellular portion of CR3 consists of several different structures important for binding to many of the published ligands for this integrin such as iC3b, ICAM-1 and fibrinogen. These include the inserted αI- and β-propeller domains (α-chain) and the βI-like domain in the β_2-chain that are present at the top of the integrin headpiece[2–5] (Fig. 41.1). The αI- and βI-like domains bind Mg^{+2} at a site termed the metal ion-dependent adhesion site or MIDAS, which is a critical event for promoting ligand interactions (Fig. 41.2).[6] The α- and β-subunits also contain additional domains proximal

The Complement FactsBook. http://dx.doi.org/10.1016/B978-0-12-810420-0.00041-9

435

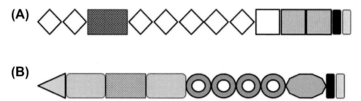

FIGURE 41.1 Diagram of protein domains for (A) CD11b and (B) CD18.

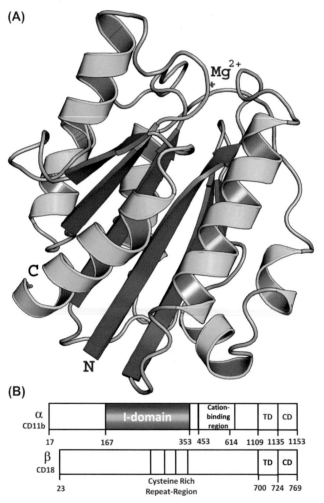

FIGURE 41.2 Diagram of crystal structure of CR3. (A) Ribbon diagram of α-chain I-domain of human complement receptor type 3 (CR3) crystal structure (PDB ID: 1NA5). Mg²⁺ binding site is labelled. (B) Schematic representation of α- and β-chains of CR3 and its domains.

to the membrane spanning region, including the thigh, genu and calf 1 and 2 regions (α- chain) and hybrid domain, plexin-semaphorin-integrin domain, integrin-EGF-like (I-EGF) repeats and a membrane proximal β-tail region (β-chain).[2–4] The short $α_m$ cytoplasmic tail contains a serine at position 1142 that has been shown to be constitutively phosphorylated and involved in regulating binding to certain CR3 ligands.[7] The cytoplasmic tail of the β-subunit is involved in linking CR3 to the cytoskeleton through interactions with different proteins such as talin.[8]

FUNCTION

CR3 is expressed predominately by myeloid cells and has been reported to bind to over 50 ligands, including iC3b, ICAM-1, ICAM-2, fibrinogen, the receptor for advanced glycation end products and clotting factor X.[9–11] CR3 serves key functions in many different leucocyte processes including adhesion and recruitment, iC3b-mediated phagocytosis, neutrophil activation, respiratory burst, degranulation, netosis, apoptosis and regulating cytokine production in macrophages and some dendritic cell subtypes.[9–21] Ligand binding to CR3 can trigger an 'outside-in' cell activation programmed leading to cytokine release and recruitment of additional Mac-1 receptors (clustering) that bolster the activation response.[22,23] Alternatively, when Toll-like receptors, Fc receptors, chemokines and other mediators are engaged by their own cognate ligands, their signalling machinery can drive 'inside-out' activation that increases CR3 affinity/avidity (leading to conformational changes of this integrin[24,25]). This bidirectional signalling ability allows CR3 to control its immune and inflammatory functions.[1]

CR3 plays key roles in host defence and in loss or inhibition of CR3 in several infectious model systems to lead to increased bacterial load, organ damage and in some cases decreased survival.[26–28] In contrast, CR3 is also used by several organisms to gain entry into myeloid cells and suppress proinflammatory responses through triggering of outside-in signalling.[29–31] Early studies of CR3, which primarily used inhibitory antibodies to block this receptor, largely suggested that this integrin served to promote inflammatory processes. However, it has become clear from more studies using CD11b mutant mice that CR3 has additional roles in inhibiting immune and inflammatory responses.[3,32–36] For example, Mac-1-mediated uptake of iC3b-decorated antigen can induce a tolerogenic state in dendritic cells,[19,37,38] while ligand-activated Mac-1 can lead to decreased inflammatory cytokine expression and increased antiinflammatory cytokine production by antigen-presenting cells,[15–18] as well as reduce their ability to activate T cells.[18,39] The involvement of Mac-1 in lymphocyte functions and activities is less clear, although several publications suggest that Mac-1 is involved in B and T cell activation processes.[40–42]

TISSUE DISTRIBUTION

CR3 is expressed mainly on neutrophils, monocytes/macrophages, certain dendritic cell populations and also on natural killer cells.[3] Only a small percentage of T or B lymphocytes have been shown to express CR3.[43,44]

REGULATION OF EXPRESSION

CR3 expression is controlled by several mechanisms, including at the transcriptional level, through release from intracellular granules following cellular activation with many different mediators, and by cleavage from the cell surface.[45–47] Transcription of CD11b has been shown to be regulated by a number of different factors, including Sp1, Sp3, PU.1 and NF-κB.[48–51] Previous analyses of the CD11b promoter have identified a 1.7 kb region upstream of the transcription initiation site that contains many of the regulatory sequences that control expression.[45] A number of different published studies have used this 1.7 kb fragment to drive expression in myeloid cells in different model systems including transgenic mice.[52]

HUMAN PROTEIN SEQUENCE

Alpha-Subunit

```
MALRVLLLTALTLCHGFNLDTENAMTFQENARGFGQSVVQLQGSRVVVGA    50

PQEIVAANQRGSLYQCDYSTGSCEPIRLQVPVEAVNMSLGLSLAATTSPP   100

QLLACGPTVHQTCSENTYVKGLCFLFGSNLRQQPQKFPEALRGCPQEDSD   150

IAFLIDGSGSIIPHDFRRMKEFVSTVMEQLKKSKTLFSLMQYSEEFRIHF   200

TFKEFQNNPNPRSLVKPITQLLGRTHTATGIRKVVRELFNITNGARKNAF   250

KILVVITDGEKFGDPLGYEDVIPEADREGVIRYVIGVGDAFRSEKSRQEL   300

NTIASKPPRDHVFQVNNFEALKTIQNQLREKIFAIEGTQTGSSSSFEHEM   350

SQEGFSAAITSNGPLLSTVGSYDWAGGVFLYTSKEKSTFINMTRVDSDMN   400

DAYLGYAAAIILRNRVQSLVLGAPRYQHIGLVAMFRQNTGMWESNANVKG   450

TQIGAYFGASLCSVDVDSNGSTDLVLIGAPHYYEQTRGGQVSVCPLPRGR   500

ARWQCDAVLYGEQGQPWGRFGAALTVLGDVNGDKLTDVAIGAPGEEDNRG   550

AVYLFHGTSGSGISPSHSQRIAGSKLSPRLQYFGQSLSGGQDLTMDGLVD   600

LTVGAQGHVLLLRSQPVLRVKAIMEFNPREVARNVFECNDQVVKGKEAGE   650

VRVCLHVQKSTRDRLREGQIQSVVTYDLALDSGRPHSRAVFNETKNSTRR   700

QTQVLGLTQTCETLKLQLPNCIEDPVSPIVLRLNFSLVGTPLSAFGNLRP   750
```

```
VLAEDAQRLFTALFPFEKNCGNDNICQDDLSITFSFMSLDCLVVGGPREF    800

NVTVTVRNDGEDSYRTQVTFFFPLDLSYRKVSTLQNQRSQRSWRLACESA    850

SSTEVSGALKSTSCSINHPIFPENSEVTFNITFDVDSKASLGNKLLLKAN    900

VTSENNMPRTNKTEFQLELPVKYAVYMVVTSHGVSTKYLNFTASENTSRV    950

MQHQYQVSNLGQRSLPISLVFLVPVRLNQTVIWDRPQVTFSENLSSTCHT   1000

KERLPSHSDFLAELRKAPVVNCSIAVCQRIQCDIPFFGIQEEFNATLKGN   1050

LSFDWYIKTSHNHLLIVSTAEILFNDSVFTLLPGQGAFVRSQTETKVEPF   1100

EVPNPLPLIVGSSVGGLLLLALITAALYKLGFFKRQYKDMMSEGGPPGAE   1150

PQ                                                   1152
```

Beta-Subunit

```
MLGLRPPLLALVGLLSLGCVLSQECTKFKVSSCRECIESGPGCTWCQKLN     50

FTGPGDPDSIRCDTRPQLLMRGCAADDIMDPTSLAETQEDHNGGQKQLSP    100

QKVTLYLRPGQAAAFNVTFRRAKGYPIDLYYLMDLSYSMLDDLRNVKKLG    150

GDLLRALNEITESGRIGFGSFVDKTVLPFVNTHPDKLRNPCPNKEKECQP    200

PFAFRHVLKLTNNSNQFQTEVGKQLISGNLDAPEGGLDAMMQVAACPEEI    250

GWRNVTRLLVFATDDGFHFAGDGKLGAILTPNDGRCHLEDNLYKRSNEFD    300

YPSVGQLAHKLAENNIQPIFAVTSRMVKTYEKLTEIIPKSAVGELSEDSS    350

NVVHLIKNAYNKLSSRVFLDHNALPDTLKVTYDSFCSNGVTHRNQPRGDC    400

DGVQINVPITFQVKVTATECIQEQSFVIRALGFTDIVTVQVLPQCECRCR    450

DQSRDRSLCHGKGFLECGICRCDTGYIGKNCECQTQGRSSQELEGSCRKD    500

NNSIICSGLGDCVCGQCLCHTSDVPGKLIYGQYCECDTINCERYNGQVCG    550

GPGRGLCFCGKCRCHPGFEGSACQCERTTEGCLNPRRVECSGRGRCRCNV    600

CECHSGYQLPLCQECPGCPSPCGKYISCAECLKFEKGPFGKNCSAACPGL    650

QLSNNPVKGRTCKERDSEGCWVAYTLEQQDGMDRYLIYVDESRECVAGPN    700

IAAIVGGTVAGIVLIGILLLVIWKALIHLSDLREYRRFEKEKLKSQWNND    750

NPLFKSATTTVMNPKFAES                                  769
```

The leader peptide is underlined. *N*-linked glycosylation sites (**N**) were determined using the UniProt database (http://www.uniprot.org/).

PROTEIN MODULES

Alpha-Subunit*

1–16	Leader peptide	exons 1 and 2
150–328	I-domain	exons 6–9
1105–1128	Transmembrane domain	exon 29
1129–1152	Cytoplasmic domain	exon 29 and 30

Beta-Subunit*

1–22	Leader peptide	exon 2 and 3
124–363	βI-like domain	exons 5–10
701–723	Transmembrane domain	exon 15
724–769	Cytoplasmic domain	exons 15 and 16

*Determined using the UniProt database

CHROMOSOMAL LOCATION

Alpha-Subunit

Human	Chromosome 16: 31,259,990-31,331,768
Mouse	Chromosome 7: 128,062,640-128,118,491
Rat	Chromosome 1, exact position still pending

All four β_2 integrins are closely linked on each of the chromosomes listed above.

Beta-Subunit

Human	Chromosome 21: 44,885,956-44,921,050
Mouse	Chromosome 10: 77,530,252-77,565,708
Rat	Chromosome 20: 11,777,783-11,815,647

HUMAN cDNA SEQUENCE

Alpha-Subunit

```
TGGCTTCCTTGTGGTTCCTCAGTGGTGCCTGCAACCCCTGGTTCACCTCCTTCCAGGTTC   60

TGGCTCCTTCCAGCCATGGCTCTCAGAGTCCTTCTGTTAACAGCCTTGACCTTATGTCAT   120

GGGTTCAACTTGGACACTGAAAACGCAATGACCTTCCAAGAGAACGCAAGGGGCTTCGGG   180

CAGAGCGTGGTCCAGCTTCAGGGATCCAGGGTGGTGGTTGGAGCCCCCCAGGAGATAGTG   240

GCTGCCAACCAAAGGGGCAGCCTCTACCAGTGCGACTACAGCACAGGCTCATGCGAGCCC   300

ATCCGCCTGCAGGTCCCCGTGGAGGCCGTGAACATGTCCCTGGGCCTGTCCCTGGCAGCC   360

ACCACCAGCCCCCCTCAGCTGCTGGCCTGTGGTCCCACCGTGCACCAGACTTGCAGTGAG   420

AACACGTATGTGAAAGGGCTCTGCTTCCTGTTTGGATCCAACCTACGGCAGCAGCCCCAG   480
```

```
AAGTTCCCAGAGGCCCTCCGAGGGTGTCCTCAAGAGGATAGTGACATTGCCTTCTTGATT   540

GATGGCTCTGGTAGCATCATCCCACAGATCTTTCGGCGGATGAAGGAGTTTGTCTCAACT   600

GTGATGGAGCAATTAAAAAAGTCCAAAACCTTGTTCTCTTTGATGCAGTACTCTGAAGAA   660

TTCCGGATTCACTTTACCTTCAAAGAGTTCCAGAACAACCCTAACCCAAGATCACTGGTG   720

AAGCCAATAACGCAGCTGCTTGGGCGGACACACACGGCCACGGGCATCCGCAAAGTGGTA   780

CGAGAGCTGTTTAACATCACCAACGGAGCCCGAAAGAATGCCTTTAAGATCCTAGTTGTC   840

ATCACGGATGGAGAAAAGTTTGGCGATCCCTTGGGATATGAGGATGTCATCCCTGAGGCA   900

GACAGAGAGGGAGTCATTCGCTACGTCATTGGGGTGGGAGATGCCTTCCGCAGTGAGAAA   960

TCCCGCCAAGAGCTTAATACCATCGCATCCAAGCCGCCTCGTGATCACGTGTTCCAGGTG   1020

AATAACTTTGAGGCTCTGAAGACCATTCAGAACCAGCTTCGGGAGAAGATCTTTGCGATC   1080

GAGGGTACTCAGACAGGAAGTAGCAGCTCCTTTGAGCATGAGATGTCTCAGGAAGGCTTC   1140

AGCGCTGCCATCACCTCTAATGGCCCCTTGCTGAGCACTGTGGGGAGCTATGACTGGGCT   1200

GGTGGAGTCTTTCTATATACATCAAAGGAGAAAAGCACCTTCATCAACATGACCAGAGTG   1260

GATTCAGACATGAATGATGCTTACTTGGGTTATGCTGCCGCCATCATCTTACGGAACCGG   1320

GTGCAAAGCCTGGTTCTGGGGGCACCTCGATATCAGCACATCGGCCTGGTAGCGATGTTC   1380

AGGCAGAACACTGGCATGTGGGAGTCCAACGCTAATGTCAAGGGCACCCAGATCGGCGCC   1440

TACTTCGGGGCCTCCCTCTGCTCCGTGGACGTGGACAGCAACGGCAGCACCGACCTGGTC   1500

CTCATCGGGGCCCCCCATTACTACGAGCAGACCCGAGGGGGCCAGGTGTCCGTGTGCCCC   1560

TTGCCCAGGGGGAGGGCTCGGTGGCAGTGTGATGCTGTTCTCTACGGGGAGCAGGGCCAA   1620

CCCTGGGGCCGCTTTGGGGCAGCCCTAACAGTGCTGGGGGACGTAAATGGGGACAAGCTG   1680

ACGGACGTGGCCATTGGGGCCCCAGGAGAGGAGGACAACCGGGGTGCTGTTTACCTGTTT   1740

CACGGAACCTCAGGATCTGGCATCAGCCCCTCCCATAGCCAGCGGATAGCAGGCTCCAAG   1800

CTCTCTCCCAGGCTCCAGTATTTTGGTCAGTCACTGAGTGGGGGCCAGGACCTCACAATG   1860

GATGGACTGGTAGACCTGACTGTAGGAGCCCAGGGGCACGTGCTGCTGCTCAGGTCCCAG   1920

CCAGTACTGAGAGTCAAGGCAATCATGGAGTTCAATCCCAGGGAAGTGGCAAGGAATGTA   1980

TTTGAGTGTAATGATCAGGTGGTGAAAGGCAAGGAAGCCGGAGAGGTCAGAGTCTGCCTC   2040

CATGTCCAGAAGAGCACACGGGATCGGCTAAGAGAAGGACAGATCCAGAGTGTTGTGACT   2100

TATGACCTGGCTCTGGACTCCGGCCGCCCACATTCCCGCGCCGTCTTCAATGAGACAAAG   2160

AACAGCACACGCAGACAGACACAGGTCTTGGGGCTGACCCAGACTTGTGAGACCCTGAAA   2220

CTACAGTTGCCGAATTGCATCGAGGACCCAGTGAGCCCCATTGTGCTGCGCCTGAACTTC   2280

TCTCTGGTGGGAACGCCATTGTCTGCTTTCGGGAACCTCCGGCCAGTGCTGGCGGAGGAT   2340

GCTCAGAGACTCTTCACAGCCTTGTTTCCCTTTGAGAAGAATTGTGGCAATGACAACATC   2400
```

TGCCAGGATGACCTCAGCATCACCTTCAGTTTCATGAG<u>CCTGG</u>ACTGCCTCGTGGTGGGT 2460

GGGCCCCGGGAGTTCAACGTGACAGTGACTGTGAGAAATGATGGTGAGGACTCCTACAGG 2520

ACACAGGTCACCTTCTTCTTCCCGCTTGACCTGTCCTACCGGAAGGTGTCCACGCTCCAG 2580

<u>AACCA</u>GCGCTCACAGCGATCCTGGCGCCTGGCCTGTGAGTCTGCCTCCTCCACCGAAGTG 2640

TCTGGGGCCTTGAAGAGCACCAGCTGCAGCATAAACCACCCCATCTTCCCGGAAAACTCA 2700

GAG<u>GTCAC</u>CTTTAATATCACGTTTGATGTAGACTCTAAGGCTTCCCTTGGAAACAAACTG 2760

CTCCTCAAGGCCAATGTGACCAG<u>TGAGA</u>ACAACATGCCCAGAACCAACAAAACCGAATTC 2820

CAACTGGAGCTGCCGGTGAAATATGCTGTCTACATGGTGGTCACCAG<u>CCATG</u>GGGTCTCC 2880

ACTAAATATCTCAACTTCACGGCCTCAGAGAATACCAGTCGGGTCATGCAGCATCAATAT 2940

CAG<u>GTCAG</u>CAACCTGGGGCAGAGGAGCCTCCCCATCAGCCTGGTGTTCTTGGTGCCCGTC 3000

CGGCTGAACCAGACTGTCATATGGGACCGCCCCCAGGTCACCTTCTCCGAG<u>AACCT</u>CTCG 3060

AGTACGTGCCACACCAAGGAGCGCTTGCCCTCTCACTCCGACTTTCTGGCTGAGCTTCGG 3120

AAGGCCCCCGTGGTG<u>AACTG</u>CTCCATCGCTGTCTGCCAGAGAATCCAGTGTGACATCCCG 3180

TTCTTTGGCATCCAGGAAGAATTCAATGCTACCCTCAAAGGCAACCTCTCGTTTGACTGG 3240

TACATCAAG<u>ACCTC</u>GCATAACCACCTCCTGATCGTGAGCACAGCTGAGATCTTGTTTAAC 3300

GATTCCGTGTTCACCCTGCTGCCGGGACAGGGGGCGTTTGTGAGGTCCCAG<u>ACGGA</u>GACC 3360

AAAGTGGAGCCGTTCGAGGTCCCCAACCCCCTGCCGCTCATCGTGGGCAGCTCTGTCGGG 3420

GGACTGCTGCTCCTGGCCCTCATCACCGCCGCGCTGTACAAG<u>CTCGG</u>CTTCTTCAAGCGG 3480

CAATACAAGGACATGATGAGTGAAGGGGGTCCCCCGGGGGCCGAACCCCAG<u>TAG</u>CGGCTC 3540

CTTCCCGACAGAGCTGCCTCTCGGTGGCCAGCAGGACTCTGCCCAGACCACACGTAGCCC 3600

CCAGGCTGCTGGACACGTCGGACAGCGAAGTATCCCCGACAGGACGGGCTTGGGCTTCCA 3660

TTTGTGTGTGTGCAAGTGTGTATGTGCGTGTGTGCAAGTGTCTGTGTGCAAGTGTGTGCA 3720

CATGTGTGCGTGTGCGTGCATGTGCACTTGCACGCCCATGTGTGAGTGTGTGCAAGTATG 3780

TGAGTGTGTCCAAGTGTGTGTGCGTGTGTCCATGTGTGTGCAAGTGTGTGCATGTGTGCG 3840

AGTGTGTGCATGTGTGTGCTCAGGGGCGTGTGGCTCACGTGTGTGACTCAGATGTCTCTG 3900

GCGTGTGGGTAGGTGACGGCAGCGTAGCCTCTCCGGCAGAAGGGAACTGCCTGGGCTCCC 3960

TTGTGCGTGGGTGAAGCCGCTGCTGGGTTTTCCTCCGGGAGAGGGGACGGTCAATCCTGT 4020

GGGTGAAGACAGAGGGAAACACAGCAGCTTCTCTCCACTGAAAGAAGTGGGACTTCCCGT 4080

CGCCTGCGAGCCTGCGGCCTGCTGGAGCCTGCGCAGCTTGGATGGAGACTCCATGAGAAG 4140

CCGTGGGTGGAACCAGGAACCTCCTCCACACCAGCGCTGATGCCC<u>AATAAA</u>GATGCCCAC 4200

TGAGGAATGATGAAGCTTCCTTTCTGGATTCATTTATTATTTCAATGTGACTTTAATTTT 4260

TTGGATGGATAAGCTTGTCTATGGTACAAAAATCACAAGGCATTCAAGTGTACAGTGAAA 4320

AGTCTCCCTTTCCAGATATTCAAGTCACCTCCTTAAAGGTAGTCAAGATTGTGTTTTGAG 4380

GTTTCCTTCAGACAGATTCCAGGCGATGTGCAAGTGTATGCACGTGTGCACACACACCAC 4440

ACATACACACACACAAGCTTTTTTACACAAATGGTAGCATACTTTATATTGGTCTGTATC 4500

TTGCTTTTTTTCACCAATATTTCTCAGACATCGGTTCATATTAAGACATAAATTACTTTT 4560

TCATTCTTTTATACCGCTGCATAGTATTCCATTGTGTGAGTGTACCATAATGTATTTAAC 4620

CAGTCTTCTTTTGATATACTATTTTCATTCTCTTGTTATTGCATCAATGCTGAGTTAATA 4680

AATCAAATATATGTCATTTTTGCATATATGTAAGGATAA 4719

Beta-Subunit

GGGCCGCTCTCTGACATCAGAGCTGCTGTAGAGCGGAGAGGGGCAGGGGTGAAGGGCCAC 60

GGTGGTGCAACCCACCACTTCCTCCAAGGAGGAGCTGAGAGGAACAGGAAGTGTCAGGAC 120

TTTACGACCCGCGCCTCCAGCTGAGGTTTCTAGACGTGACCCAGGGCAGACTGGTAGCAA 180

AGCCCCCACGCCCAGCCAGGAGCACCGCCGAGGACTCCAGCACACCGAGG<u>GACATG</u>CTGG 240

GCCTGCGCCCCCCACTGCTCGCCCTGGTGGGGCTGCTCTCCCTCGGGTGCG<u>TCCTC</u>TCTC 300

AGGAGTGCACGAAGTTCAAGGTCAGCAGCTGCCGGGAATGCATCGAGTCGGGGCCCGGCT 360

GCACCTGGTGCCAGAAGCTG<u>AACTT</u>CACAGGGCCGGGGGATCCTGACTCCATTCGCTGCG 420

ACACCCGGCCACAGCTGCTCATGAGGGGCTGTGCGGCTGACGACATCATGGACCCCACAA 480

GCCTCGCTGAAACCCAGGAAGACCACAATGGGGGCCAGAAGCAGCTGTCCCCACAAAAAG 540

TGACGCTTTACCTGCGACCAG<u>GCCAG</u>GCAGCAGCGTTCAACGTGACCTTCCGGCGGGCCA 600

AGGGCTACCCCATCGACCTGTACTATCTGATGGACCTCTCCTACTCCATGCTTGATGACC 660

TCAGGAATGTCAAGAAGCTAGGTGGCGACCTGCTCCGGGCCCTCAACGAGATCACCGAGT 720

CCGGCCGCATTG<u>GCTTC</u>GGGTCCTTCGTGGACAAGACCGTGCTGCCGTTCGTGAACACGC 780

ACCCTGATAAGCTGCGAAACCCATGCCCCAACAAGGAGAAAGAGTGCCAGCCCCCGTTTG 840

CCTTCAGGCACGTGCTGAAGCTGACCAACAACTCCAACCAGTTTCAGACCGAGGTCGGGA 900

AGCAGCTGATTTCCGGAAACCTGGATGCACCCGAGGGTGGGCTGGACGCCATGATGCAGG 960

TCGCCGCCTGCCCG<u>GAGGA</u>AATCGGCTGGCGCAACGTCACGCGGCTGCTGGTGTTTGCCA 1020

CTGATGACGGCTTCCATTTCGCGGGCGACGGGAAGCTGGGCGCCATCCTGACCCCCAACG 1080

ACGGCCGCTGTCACCTGGAGGACAACTTGTACAAGAGGAGCAACGAATTC<u>GACTA</u>CCCAT 1140

CGGTGGGCCAGCTGGCGCACAAGCTGGCTGAAAACAACATCCAGCCCATCTTCGCGGTGA 1200

```
CCAGTAGGATGGTGAAGACCTACGAGAAACTCACCGAGATCATCCCCAAGTCAGCCGTGG   1260

GGGAGCTGTCTGAGGACTCCAGCAATGTGGTCCATCTCATTAAGAATGCTTACAATAAAC   1320

TCTCCTCCAGGGTCTTCCTGGATCACAACGCCCTCCCCGACACCCTGAAAGTCACCTACG   1380

ACTCCTTCTGCAGCAATGGAGTGACGCACAGGAACCAGCCCAGAGGTGACTGTGATGGCG   1440

TGCAGATCAATGTCCCGATCACCTTCCAGGTGAAGGTCACGGCCACAGAGTGCATCCAGG   1500

AGCAGTCGTTTGTCATCCGGGCGCTGGGCTTCACGGACATAGTGACCGTGCAGGTTCTTC   1560

CCCAGTGTGAGTGCCGGTGCCGGGACCAGAGCAGAGACCGCAGCCTCTGCCATGGCAAGG   1620

GCTTCTTGGAGTGCGGCATCTGCAGGTGTGACACTGGCTACATTGGGAAAAACTGTGAGT   1680

GCCAGACACAGGGCCGGAGCAGCCAGGAGCTGGAAGGAAGCTGCCGGAAGGACAACAACT   1740

CCATCATCTGCTCAGGGCTGGGGGACTGTGTCTGCGGGCAGTGCCTGTGCCACACCAGCG   1800

ACGTCCCCGGCAAGCTGATATACGGGCAGTACTGCGAGTGTGACACCATCAACTGTGAGC   1860

GCTACAACGGCCAGGTCTGCGGCGGCCCGGGGAGGGGGCTCTGCTTCTGCGGGAAGTGCC   1920

GCTGCCACCCGGGCTTTGAGGGCTCAGCGTGCCAGTGCGAGAGGACCACTGAGGGCTGCC   1980

TGAACCCGCGGCGTGTTGAGTGTAGTGGTCGTGGCCGGTGCCGCTGCAACGTATGCGAGT   2040

GCCATTCAGGCTACCAGCTGCCTCTGTGCCAGGAGTGCCCCGGCTGCCCCTCACCCTGTG   2100

GCAAGTACATCTCCTGCGCCGAGTGCCTGAAGTTCGAAAAGGGCCCCTTTGGGAAGAACT   2160

GCAGCGCGGCGTGTCCGGGCCTGCAGCTGTCGAACAACCCCGTGAAGGGCAGGACCTGCA   2220

AGGAGAGGGACTCAGAGGGCTGCTGGGTGGCCTACACGCTGGAGCAGCAGGACGGGATGG   2280

ACCGCTACCTCATCTATGTGGATGAGAGCCGAGAGTGTGTGGCAGGCCCCAACATCGCCG   2340

CCATCGTCGGGGGCACCGTGGCAGGCATCGTGCTGATCGGCATTCTCCTGCTGGTCATCT   2400

GGAAGGCTCTGATCCACCTGAGCGACCTCCGGGAGTACAGGCGCTTTGAGAAGGAGAAGC   2460

TCAAGTCCCAGTGGAACAATGATAATCCCCTTTTCAAGAGCGCCACCACGCACGGTCATGA   2520

ACCCCAAGTTTGCTGAGAGTTAGGAGCACTTGGTGAAGACAAGGCCGTCAGGACCCACCA   2580

TGTCTGCCCCATCACGCGGCCGAGACATGGCTTGCCACAGCTCTTGAGGATGTCACCAAT   2640

TAACCAGAAATCCAGTTATTTTCCGCCCTCAAAATGACAGCCATGGCCGGCCGGGTGCTT   2700

CTGGGGGCTCGTCGGGGGACAGCTCCACTCTGACTGGCACAGTCTTTGCATGGAGACTT   2760

GAGGAGGGAGGGCTTGAGGTTGGTGAGGTTAGGTGCGTGTTTCCTGTGCAAGTCAGGACA   2820

TCAGTCTGATTAAAGGTGGTGCCAATTTATTTACATTTAAACTTGTCAGGGTATAAAATG   2880

ACATCCCATTAATTATATTGTTAATCAATCACGTGTATAGAAAAAAAATAAAACTTCAAT   2940

ACAGGCTGTCCATGG                                               2955
```

The methionine start codon (ATG), termination codon (TGA) and polyadenylation signal are shown (AATAAA). The first five nucleotides of each exon are underlined.

CD11b - 71.78 kb

FIGURE 41.3 Genomic organisation of the α-chain (CD11b) of CR3.

CD18 - 35.09 kb

FIGURE 41.4 Genomic organisation of the β-chain (CD18) of CR3.

GENOMIC STRUCTURE

The CD11b gene spans 71.78 kb and contains 30 exons (Fig. 41.3).
The CD18 gene spans 35.09 kb and has 16 exons (Fig. 41.4).

ACCESSION NUMBERS

Human α-subunit	NM000632, NP000632
Human β-subunit	NM000211, NP000202
Mouse α-subunit	NM008401, NP032427
Mouse β-subunit	NM008404, NP032430

DEFICIENCY

There have been no reports to date of humans with complete deficiency of CR3, except for patients with leucocyte adhesion deficiency type I (LAD I) that results from mutations in the CD18 gene and thus affects expression or function of all four β$_2$ integrins.[53] LAD I is a life-threatening disorder and is characterised by increased granulocyte counts; impaired leucocyte adhesion and transendothelial migration; and severe susceptibility to bacterial infections. Both moderate and severe forms of this disease have been described and are related to the levels of CD18 expression.[53] Several of these phenotypes are consistent with loss of CR3 functions, but cannot solely be contributed to deficiency of this integrin.

POLYMORPHIC VARIANTS

Many different single nucleotide polymorphisms (SNPs) have been identified in the coding and noncoding regions of the CD11b/*ITGAM* gene. Several of these SNPs have been strongly implicated in the pathogenesis of systemic lupus erythematosus (SLE), systemic sclerosis, melanoma and other diseases through genome-wide association studies.[54–57] Functional genomic analyses to date have focused on two different SLE-associated CD11b/*ITGAM* SNP variants, rs1143679, encoding an [77]Arg→[77]His change in the extracellular domain of

CD11b (exon 3) and rs1143678, encoding a ^{1146}Pro→^{1146}Ser change in its cytoplasmic tail (exon 30). Both variants have been shown to inhibit or alter CR3-mediated leucocyte adhesion, phagocytosis and cytokine production.[4,36,58–63]

MUTANT ANIMALS

Two different lines of gene-targeted CD11b null mutant mice (CR3 or Mac-1-deficient mice) were developed in the mid- to late 1990s.[64,65] Both of these lines were generated in 129/Sv-derived ES cell lines and used a replacement strategy to delete several exons in the 5′ region of the gene. Early reported studies predominately used CR3-deficient mice on a mixed 129/Sv× C57BL/6 strain background, while more recent analyses have used lines where the CD11b mutation has been extensively backcrossed onto C57BL/6. Complete loss of CR3 in mice did not lead to increased infectious susceptibility or other obvious phenotypic abnormalities in unmanipulated CD11b mutant mice.[64,65] These mice have been examined in a number of different infection models, which have further illustrated the important role of CR3 in promoting host defence responses.[26–28] Investigations of CD11b mutant mice in inflammatory disease models such as experimental autoimmune encephalomyelitis have elucidated important roles for CR3 in promoting inflammation,[42] while analyses in other murine systems, such as those involving immune complex glomerulonephritis have shown a protective role for this complement receptor in restricting tissue injury.[32,36,66,67] A single line of CD11b mutant mice (mutation backcrossed onto the C57BL/6 inbred background) is currently available from The Jackson Laboratory. Information on other lines of CD11b mutant mice can be found at http://www.informatics.jax.org/.

REFERENCES

1. Tan SM. The leucocyte beta2 (CD18) integrins: the structure, functional regulation and signalling properties. *Biosci Rep* 2012;**32**:241–69.
2. Hogg N, Patzak I, Willenbrock F. The insider's guide to leukocyte integrin signalling and function. *Nat Rev Immunol* 2011;**11**:416–26.
3. Rosetti F, Mayadas TN. The many faces of Mac-1 in autoimmune disease. *Immunol Rev* 2016;**269**:175–93.
4. Rosetti F, Chen Y, Sen M, Thayer E, Azcutia V, Herter JM, et al. A lupus-associated Mac-1 variant has defects in integrin allostery and interaction with ligands under force. *Cell Rep* 2015. http://dx.doi.org/10.1016/j.celrep.2015.02.037.
5. Bajic G, Yatime L, Sim RB, Vorup-Jensen T, Andersen GR. Structural insight on the recognition of surface-bound opsonins by the integrin I domain of complement receptor 3. *Proc Natl Acad Sci USA* 2013;**110**:16426–31.
6. Lee JO, Rieu P, Arnaout MA, Liddington R. Crystal structure of the A domain from the alpha subunit of integrin CR3 (CD11b/CD18). *Cell* 1995;**80**:631–8.
7. Fagerholm SC, Varis M, Stefanidakis M, Hilden TJ, Gahmberg CG. alpha-Chain phosphorylation of the human leukocyte CD11b/CD18 (Mac-1) integrin is pivotal for integrin activation to bind ICAMs and leukocyte extravasation. *Blood* 2006;**108**:3379–86.

8. Lim J, Wiedemann A, Tzircotis G, Monkley SJ, Critchley DR, Caron E. An essential role for talin during alpha(M)beta(2)-mediated phagocytosis. *Mol Biol Cell* 2007;**18**:976–85.

9. Springer TA. Traffic signals for lymphocyte recirculation and leukocyte emigration: the multi-step paradigm. *Cell* 1994;**76**:301–14.

10. Flick MJ, Du X, Witte DP, Jirouskova M, Soloviev DA, Busuttil SJ, et al. Leukocyte engagement of fibrin(ogen) via the integrin receptor alphaMbeta2/Mac-1 is critical for host inflammatory response in vivo. *J Clin Invest* 2004;**113**:1596–606.

11. Herold K, Moser B, Chen Y, Zeng S, Yan SF, Ramasamy R, et al. Receptor for advanced glycation end products (RAGE) in a dash to the rescue: inflammatory signals gone awry in the primal response to stress. *J Leukoc Biol* 2007;**82**:204–12.

12. Zhang B, Hirahashi J, Cullere X, Mayadas TN. Elucidation of molecular events leading to neutrophil apoptosis following phagocytosis: cross-talk between caspase 8, reactive oxygen species, and MAPK/ERK activation. *J Biol Chem* 2003;**278**:28443–54.

13. Mayadas TN, Cullere X. Neutrophil beta2 integrins: moderators of life or death decisions. *Trends Immunol* 2005;**26**:388–95.

14. Mevorach D, Mascarenhas JO, Gershov D, Elkon KB. Complement-dependent clearance of apoptotic cells by human macrophages. *J Exp Med* 1998;**188**:2313–20.

15. Morelli AE, Larregina AT, Shufesky WJ, Zahorchak AF, Logar AJ, Papworth GD, et al. Internalization of circulating apoptotic cells by splenic marginal zone dendritic cells: dependence on complement receptors and effect on cytokine production. *Blood* 2003;**101**:611–20.

16. Marth T, Kelsall BL. Regulation of interleukin-12 by complement receptor 3 signaling. *J Exp Med* 1997;**185**:1987–95.

17. Sohn JH, Bora PS, Suk HJ, Molina H, Kaplan HJ, Bora NS. Tolerance is dependent on complement C3 fragment iC3b binding to antigen-presenting cells. *Nat Med* 2003;**9**:206–12.

18. Skoberne M, Somersan S, Almodovar W, Truong T, Petrova K, Henson PM, et al. The apoptotic-cell receptor CR3, but not alphavbeta5, is a regulator of human dendritic-cell immunostimulatory function. *Blood* 2006;**108**:947–55.

19. Behrens EM, Sriram U, Shivers DK, Gallucci M, Ma Z, Finkel TH, et al. Complement receptor 3 ligation of dendritic cells suppresses their stimulatory capacity. *J Immunol* 2007;**178**:6268–79.

20. Byrd AS, O'Brien XM, Johnson CM, Lavigne LM, Reichner JS. An extracellular matrix-based mechanism of rapid neutrophil extracellular trap formation in response to *Candida albicans*. *J Immunol* 2013;**190**:4136–48.

21. Chen K, Nishi H, Travers R, Tsuboi N, Martinod K, Wagner DD, et al. Endocytosis of soluble immune complexes leads to their clearance by FcgammaRIIIB but induces neutrophil extracellular traps via FcgammaRIIA in vivo. *Blood* 2012;**120**:4421–31.

22. Lefort CT, Hyun YM, Schultz JB, Law FY, Waugh RE, Knauf PA, et al. Outside-in signal transmission by conformational changes in integrin Mac-1. *J Immunol* 2009;**183**:6460–8.

23. Legate KR, Wickstrom SA, Fassler R. Genetic and cell biological analysis of integrin outside-in signaling. *Genes Dev* 2009;**23**:397–418.

24. Evans R, Patzak I, Svensson L, De Filippo K, Jones K, McDowall A, et al. Integrins in immunity. *J Cell Sci* 2009;**122**:215–25.

25. Abram CL, Lowell CA. The ins and outs of leukocyte integrin signaling. *Annu Rev Immunol* 2009;**27**:339–62.

26. Ren B, McCrory MA, Pass C, Bullard DC, Ballantyne CM, Xu Y, et al. The virulence function of *Streptococcus pneumoniae* surface protein A involves inhibition of complement activation and impairment of complement receptor-mediated protection. *J Immunol* 2004;**173**:7506–12.

27. Pilione MR, Agosto LM, Kennett MJ, Harvill ET. CD11b is required for the resolution of inflammation induced by *Bordetella bronchiseptica* respiratory infection. *Cell Microbiol* 2006;**8**:758–68.

28. Soloviev DA, Jawhara S, Fonzi WA. Regulation of innate immune response to *Candida albicans* infections by alphaMbeta2-Pra1p interaction. *Infect Immun* 2011;**79**:1546–58.

29. Oliva CR, Swiecki MK, Griguer CE, Lisanby MW, Bullard DC, Turnbough Jr CL, et al. The integrin Mac-1 (CR3) mediates internalization and directs *Bacillus anthracis* spores into professional phagocytes. *Proc Natl Acad Sci USA* 2008;**105**:1261–6.

30. Carter CR, Whitcomb JP, Campbell JA, Mukbel RM, McDowell MA. Complement receptor 3 deficiency influences lesion progression during *Leishmania* major infection in BALB/c mice. *Infect Immun* 2009;**77**:5668–75.

31. Hajishengallis G, Shakhatreh MA, Wang M, Liang S. Complement receptor 3 blockade promotes IL-12-mediated clearance of *Porphyromonas gingivalis* and negates its virulence in vivo. *J Immunol* 2007;**179**:2359–67.

32. Kevil CG, Hicks MJ, He X, Zhang J, Ballantyne CM, Raman C, et al. Loss of LFA-1, but not Mac-1, protects MRL/MpJ-Fas(lpr) mice from autoimmune disease. *Am J Pathol* 2004;**165**:609–16.

33. Kanwar S, Smith CW, Shardonofsky FR, Burns AR. The role of Mac-1 (CD11b/CD18) in antigen-induced airway eosinophilia in mice. *Am J Respir Cell Mol Biol* 2001;**25**:170–7.

34. Ehirchiou D, Xiong Y, Xu G, Chen W, Shi Y, Zhang L. CD11b facilitates the development of peripheral tolerance by suppressing Th17 differentiation. *J Exp Med* 2007;**204**:1519–24.

35. Han C, Jin J, Xu S, Liu H, Li N, Cao X. Integrin CD11b negatively regulates TLR-triggered inflammatory responses by activating Syk and promoting degradation of MyD88 and TRIF via Cbl-b. *Nat Immunol* 2010;**11**:734–42.

36. Rosetti F, Tsuboi N, Chen K, Nishi H, Ernandez T, Sethi S, et al. Human lupus serum induces neutrophil-mediated organ damage in mice that is enabled by Mac-1 deficiency. *J Immunol* 2012;**189**:3714–23.

37. Monrad S, Kaplan MJ. Dendritic cells and the immunopathogenesis of systemic lupus erythematosus. *Immunol Res* 2007;**37**:135–45.

38. Schmidt J, Klempp C, Buchler MW, Marten A. Release of iC3b from apoptotic tumor cells induces tolerance by binding to immature dendritic cells in vitro and in vivo. *Cancer Immunol Immunother* 2006;**55**:31–8.

39. Varga G, Balkow S, Wild MK, Stadtbaeumer A, Krummen M, Rothoeft T, et al. Active MAC-1 (CD11b/CD18) on DCs inhibits full T-cell activation. *Blood* 2007;**109**:661–9.

40. Ding C, Ma Y, Chen X, Liu M, Cai Y, Hu X, et al. Integrin CD11b negatively regulates BCR signalling to maintain autoreactive B cell tolerance. *Nat Commun* 2013;**4**:2813.

41. Wu H, Rodgers JR, Perrard XY, Perrard JL, Prince JE, Abe Y, et al. Deficiency of CD11b or CD11d results in reduced staphylococcal enterotoxin-induced T cell response and T cell phenotypic changes. *J Immunol* 2004;**173**:297–306.

42. Bullard DC, Hu X, Schoeb TR, Axtell RC, Raman C, Barnum SR. Critical requirement of CD11b (Mac-1) on T cells and accessory cells for development of experimental autoimmune encephalomyelitis. *J Immunol* 2005;**175**:6327–33.

43. Griffin DO, Rothstein TL. A small CD11b(+) human B1 cell subpopulation stimulates T cells and is expanded in lupus. *J Exp Med* 2011;**208**:2591–8.

44. Wagner C, Hansch GM, Stegmaier S, Denefleh B, Hug F, Schoels M. The complement receptor 3, CR3 (CD11b/CD18), on T lymphocytes: activation-dependent up-regulation and regulatory function. *Eur J Immunol* 2001;**31**:1173–80.

45. Pahl HL, Rosmarin AG, Tenen DG. Characterization of the myeloid-specific CD11b promoter. *Blood* 1992;**79**:865–70.

46. Lacal P, Pulido R, Sanchez-Madrid F, Cabanas C, Mollinedo F. Intracellular localization of a leukocyte adhesion glycoprotein family in the tertiary granules of human neutrophils. *Biochem Biophys Res Commun* 1988;**154**:641–7.

47. Tsubota Y, Frey JM, Tai PW, Welikson RE, Raines EW. Monocyte ADAM17 promotes diapedesis during transendothelial migration: identification of steps and substrates targeted by metalloproteinases. *J Immunol* 2013;**190**:4236–44.

48. Chen HM, Pahl HL, Scheibe RJ, Zhang DE, Tenen DG. The Sp1 transcription factor binds the CD11b promoter specifically in myeloid cells in vivo and is essential for myeloid-specific promoter activity. *J Biol Chem* 1993;**268**:8230–9.

49. Noti JD. Sp3 mediates transcriptional activation of the leukocyte integrin genes CD11C and CD11B and cooperates with c-Jun to activate CD11C. *J Biol Chem* 1997;**272**:24038–45.

50. Pahl HL, Scheibe RJ, Zhang DE, Chen HM, Galson DL, Maki RA, et al. The proto-oncogene PU.1 regulates expression of the myeloid-specific CD11b promoter. *J Biol Chem* 1993;**268**:5014–20.

51. Zhou X, Gao XP, Fan J, Liu Q, Anwar KN, Frey RS, et al. LPS activation of Toll-like receptor 4 signals CD11b/CD18 expression in neutrophils. *Am J Physiol Lung Cell Mol Physiol* 2005;**288**:L655–62.

52. Dziennis S, Van Etten RA, Pahl HL, Morris DL, Rothstein TL, Blosch CM, et al. The CD11b promoter directs high-level expression of reporter genes in macrophages in transgenic mice. *Blood* 1995;**85**:319–29.

53. Anderson DC, Springer TA. Leukocyte adhesion deficiency: an inherited defect in the Mac-1, LFA-1 and p150,95 glycoproteins. *Annu Rev Med* 1987;**38**:175–94.

54. Hom G, Graham RR, Modrek B, Taylor KE, Ortmann W, Garnier S, et al. Association of systemic lupus erythematosus with C8orf13-BLK and ITGAM-ITGAX. *N Engl J Med* 2008;**358**:900–9.

55. Nath SK, Han S, Kim-Howard X, Kelly JA, Viswanathan P, Gilkeson GS, et al. A nonsynonymous functional variant in integrin-alpha(M) (encoded by ITGAM) is associated with systemic lupus erythematosus. *Nat Genet* 2008;**40**:152–4.

56. Anaya JM, Kim-Howard X, Prahalad S, Chernavsky A, Canas C, Rojas-Villarraga A, et al. Evaluation of genetic association between an ITGAM non-synonymous SNP (rs1143679) and multiple autoimmune diseases. *Autoimmun Rev* 2012;**11**:276–80.

57. Lenci RE, Rachakonda PS, Kubarenko AV, Weber AN, Brandt A, Gast A, et al. Integrin genes and susceptibility to human melanoma. *Mutagenesis* 2012;**27**:367–73.

58. MacPherson M, Lek HS, Prescott A, Fagerholm SC. A systemic lupus erythematosus-associated R77H substitution in the CD11b chain of the Mac-1 integrin compromises leukocyte adhesion and phagocytosis. *J Biol Chem* 2011;**286**:17303–10.

59. Rhodes B, Furnrohr BG, Roberts AL, Tzircotis G, Schett G, Spector TD, et al. The rs1143679 (R77H) lupus associated variant of ITGAM (CD11b) impairs complement receptor 3 mediated functions in human monocytes. *Ann Rheum Dis* 2012;**71**:2028–34.

60. Reed JH, Jain M, Lee K, Kandimalla ER, Faridi MH, Buyon JP, et al. Complement receptor 3 influences toll-like receptor 7/8-dependent inflammation: implications for autoimmune diseases characterized by antibody reactivity to ribonucleoproteins. *J Biol Chem* 2013;**288**:9077–83.

61. Zhou Y, Wu J, Kucik DF, White NB, Redden DT, Szalai AJ, et al. Multiple lupus-associated ITGAM variants alter Mac-1 functions on neutrophils. *Arthritis Rheum* 2013;**65**:2907–16.

62. Fossati-Jimack L, Ling GS, Cortini A, Szajna M, Malik TH, McDonald JU, et al. Phagocytosis is the main CR3-mediated function affected by the lupus-associated variant of CD11b in human myeloid cells. *PLoS One* 2013;**8**:e57082.

63. Maiti AK, Kim-Howard X, Motghare P, Pradhan V, Chua KH, Sun C, et al. Combined protein- and nucleic acid-level effects of rs1143679 (R77H), a lupus-predisposing variant within ITGAM. *Hum Mol Genet* 2014;**23**:4161–76.

64. Coxon A, Rieu P, Barkalow FJ, Askari S, Sharpe AH, von Andrian UH, et al. A novel role for the beta 2 integrin CD11b/CD18 in neutrophil apoptosis: a homeostatic mechanism in inflammation. *Immunity* 1996;**5**:653–66.

65. Lu H, Smith CW, Perrard J, Bullard DC, Tang L, Beaudet AL, et al. LFA-1 is sufficient in mediating neutrophil transmigration in Mac-1 deficient mice. *J Clin Invest* 1997;**99**:1340–50.

66. Alexander JJ, Chaves LD, Chang A, Jacob A, Ritchie M, Quigg RJ. CD11b is protective in complement-mediated immune complex glomerulonephritis. *Kidney Int* 2015;**87**:930–9.

67. Chaves LD, Bao L, Wang Y, Chang A, Haas M, Quigg RJ. Loss of CD11b exacerbates murine complement-mediated tubulointerstitial nephritis. *PLoS One* 2014;**9**:e92051.

Chapter 42

CR4

Daniel C. Bullard

University of Alabama at Birmingham, Birmingham, AL, United States

OTHER NAMES

p150/95, CD11c/CD18, α_x/β_2, *ITGAX/ITGB2*.

PHYSICOCHEMICAL PROPERTIES

	CD11c (*ITGAX*)	CD18 (*ITGB2*)
Amino acids	1163	769
Signal peptide	1–19	1–22
Isoelectric point	6.61	6.95
Predicted MW (K)	127.8	87.8
Actual MW (K)	150	95
N-linked glycosylation sites	61, 89, 392, 697, 735, 899,	50, 116, 212, 213, 215, 254,
	939, 1050	501, 642
Interchain disulphide bonds	69–76, 108–126, 116–145, 495–506, 639–722, 655–712, 771–777, 848–863, 998–1022, 1027–1032	28 predicted

STRUCTURE

CR4 shares a similar structure with all four members of β_2 integrins that is described in the previous chapter on CR3.[1,2] The alpha chain for this complement receptor is encoded by the CD11c (α_x or *ITGAX*) gene, which noncovalently associates with the β_2 subunit (encoded by the CD18 or *ITGB2* gene). The cytoplasmic domain of the α_x subunit contains a serine at position 1158 that has been shown to be phosphorylated and regulates adhesion to iC3b and phagocytosis[3] (Fig. 42.1). The crystal structure of the I domain of CR4 is shown in Fig. 42.2.

The Complement FactsBook. http://dx.doi.org/10.1016/B978-0-12-810420-0.00042-0

FIGURE 42.1 Diagram of protein domains for (A) CD11c and (B) CD18.

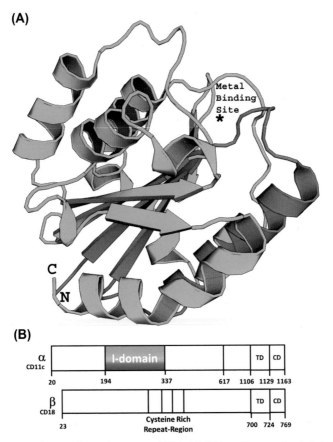

FIGURE 42.2 Diagram of crystal structure of CR4. (A) Ribbon diagram of α-chain I-domain of human complement receptor type 4 (CR4) crystal structure (PDB ID: 1N3Y). Metal binding site labelled. (B) Schematic representation of α- and β chains of the CR4 and its domains.

FUNCTION

CR4 is expressed by dendritic cells (DCs), neutrophils, different monocyte/macrophage populations, NK cells and certain T and B lymphocyte subsets.[4–7] This complement receptor binds to a number of different ligands including iC3b, fibrinogen, ICAM-1, ICAM-2 and VCAM-1.[8–11] Similar to CR3, ligand binding to CR4 can trigger an 'outside-in' cell activation programme leading to cytokine release and clustering of additional molecules.[12,13] CR4 affinity/avidity changes resulting in conformational changes also occur following 'inside-out' activation with various mediators.[14,15] Previous investigations of CR4 functions have shown that it is important in the phagocytic clearance of bacteria and apoptotic cells through interactions with iC3b, promotes cytokine production from monocytes, triggers neutrophil spreading and the respiratory burst and plays a role in monocyte and neutrophil adhesion.[13,16–22] CR4 expression increases on activated B and T cells, particularly cytotoxic T cells, suggesting a role in adhesive events.[5,6,23] CD11c expression is commonly used to identify DCs, although the functions of CR4 on these different DC populations have been poorly defined.

TISSUE DISTRIBUTION

As described above, CR4 is expressed by neutrophils, DCs, certain monocyte/macrophage populations, NK cells and some T and B lymphocyte subsets.[4–7]

REGULATION OF EXPRESSION

CR4 expression increases on treatment with a variety of chemoattractants, cytokines, phorbol esters or on antigen-mediated activation.[4,5] During hypertriglyceridemia, monocytes were shown to internalise lipids and upregulate CR4 expression, which leads to increased adhesion to VCAM-1.[24] Similar to CR3, CR4 expression is stored in cytoplasmic granules and can be rapidly mobilised to the cell surface following activation.[25] Transcription of CD11c has been shown to be regulated by a number of different transcription factors, including AP-1 family members, SP1, SP3, C/EBP.[26–28] Different 5′ regions of the CD11c gene have been used in a number of different published studies to express proteins specifically in DCs,[29] although caution should be exercised when using this strategy since CD11c/CR4 is also expressed on other leucocyte populations as described above.

HUMAN PROTEIN SEQUENCE
Alpha Subunit

```
MTRTRAALLL FTALATSLGF NLDTEELTAF RVDSAGFGDS VVQYANSWVV   50
VGAPQKITAA NQTGGLYQCG YSTGACEPIG LQVPPEAVNM SLGLSLASTT  100
SPSQLLACGP TVHHECGRNM YLTGLCFLLG PTQLTQRLPV SRQECPRQEQ  150
DIVFLIDGSG SISSRNFATM MNFVRAVISQ FQRPSTQFSL MQFSNKFQTH  200
FTFEEFRRSS NPLSLLASVH QLQGFTYTAT AIQNVVHRLF HASYGARRDA  250
AKILIVITDG KKEGDSLDYK DVIPMADAAG IIRYAIGVGL AFQNRNSWKE  300
LNDIASKPSQ EHIFKVEDFD ALKDIQNQLK EKIFAIEGTE TTSSSSFELE  350
MAQEGFSAVF TPDGPVLGAV GSFTWSGGAF LYPPNMSPTF INMSQENVDM  400
RDSYLGYSTE LALWKGVQSL VLGAPRYQHT GKAVIFTQVS RQWRMKAEVT  450
GTQIGSYFGA SLCSVDVDSD GSTDLVLIGA PHYYEQTRGG QVSVCPLPRG  500
WRRWWCDAVL YGEQGHPWGR FGAALTVLGD VNGDKLTDVV IGAPGEEENR  550
GAVYLFHGVL GPSISPSHSQ RIAGSQLSSR LQYFGQALSG GQDLTQDGLV  600
DLAVGARGQV LLLRTRPVLW VGVSMQFIPA EIPRSAFECR EQVVSEQTLV  650
QSNICLYIDK RSKNLLGSRD LQSSVTLDLA LDPGRLSPRA TFQETKNRSL  700
SRVRVLGLKA HCENFNLLLP SCVEDSVTPI TLRLNFTLVG KPLLAFRNLR  750
PMLAADAQRY FTASLPFEKN CGADHICQDN LGISFSFPGL KSLLVGSNLE  800
LNAEVMVWND GEDSYGTTIT FSHPAGLSYR YVAEGQKQGQ LRSLHLTCDS  850
APVGSQGTWS TSCRINHLIF RGGAQITFLA TFDVSPKAVL GDRLLLTANV  900
SSENNTPRTS KTTFQLELPV KYAVYTVVSS HEQFTKYLNF SESEEKESHV  950
AMHRYQVNNL GQRDLPVSIN FWVPVELNQE AVWMDVEVSH PQNPSLRCSS 1000
EKIAPPASDF LAHIQKNPVL DCSIAGCLRF RCDVPSFSVQ EELDFTLKGN 1050
LSFGWVRQIL QKKVSVVSVA EITFDTSVYS QLPGQEAFMR AQTTTVLEKY 1100
KVHNPTPLIV GSSIGGLLLL ALITAVLYKV GFFKRQYKEM MEEANGQIAP 1150
ENGTQTPSPP SEK                                        1163
```

The leader peptide is underlined. *N*-linked glycosylation sites (**N**) were determined using the UniProt database (http://www.uniprot.org/).

Beta Subunit

See Chapter 41.

PROTEIN MODULES
Alpha Subunit[a]

1–19	Leader peptide	exons 1/2
165–339	I-domain	exons 6–10
1108–1128	Transmembrane domain	exon 29
1129–1163	Cytoplasmic domain	exons 29/30

[a]Determined using the UniProt database.

Beta Subunit

See Chapter 41.

CHROMOSOMAL LOCATION

Alpha Subunit

Human: Chromosome 16: 31,355,134-31,382,997.
Mouse: Chromosome 7: 128,129,547-128,150,657.
Rat: Chromosome 1: 199,555,722-199,576,932.
    All four β₂ integrins are closely linked to each of the chromosomes listed above.

Beta Subunit

See Chapter 41.

HUMAN cDNA SEQUENCE

Alpha Subunit

```
ATGCTGACAA TCTTCTTCCT TCCCCTGGCC ACCTCTCTGC CCACTTGCTT CCTCAGTACC    60
TTGGTCCAGC TCTTCCTGCA ACGGCCCAGG AGCTCAGAGC TCCACATCTG ACCTTCTAGT   120
CATGACCAGG ACCAGGGCAG CACTCCTCCT GTTCACAGCC TTAGCAACTT CTCTAGGTTT   180
CAACTTGGAC ACAGAGGAGC TGACAGCCTT CCGTGTGGAC AGCGCTGGGT TGGAGACAG    240
CGTGGTCCAG TATGCCAACT CCTGGGTGGT GGTTGGAGCC CCCCAAAAGA TAACAGCTGC   300
CAACCAAACG GGTGGCCTCT ACCAGTGTGG CTACAGCACT GGTGCCTGTG AGCCCATCGG   360
CCTGCAGGTG CCCCCGGAGG CCGTGAACAT GTCCCTGGGC CTGTCCCTGG CGTCTACCAC   420
CAGCCCTTCC CAGCTGCTGG CCTGCGGGCC CACCGTGCAC CACGAGTGCG GGAGGAACAT   480
GTACCTCACC GGACTCTGCT TCCTCCTGGG CCCCACCCAG CTCACCCAGA GGCTCCCGGT   540
GTCCAGGCAG GAGTGCCCAA GACAGGAGCA GGACATTGTG TTCCTGATCG ATGGCTCAGG   600
CAGCATCTCC TCCCGCAACT TTGCCACGAT GATGAACTTC GTGAGAGCTG TGATAAGCCA   660
GTTCCAGAGA CCCAGCACCC AGTTTTCCCT GATGCAGTTC TCCAACAAAT TCCAAACACA   720
CTTCACTTTC GAGGAATTCA GGCGCAGCTC AAACCCCCTC AGCCTGTTGG CTTCTGTTCA   780
CCAGCTGCAA GGGTTTACAT ACACGGCCAC CGCCATCCAA AATGTCGTGC ACCGATTGTT   840
CCATGCCTCA TATGGGGCCC GTAGGGATGC CGCCAAAATT CTCATTGTCA TCACTGATGG   900
GAAGAAAGAA GGCGACAGCC TGGATTATAA GGATGTCATC CCCATGGCTG ATGCAGCAGG   960
CATCATCCGC TATGCAATTG GGGTTGGATT AGCTTTTCAA AACAGAAATT CTTGGAAAGA  1020
ATTAAATGAC ATTGCATCGA AGCCCTCCCA GGAACACATA TTTAAAGTGG AGGACTTTGA  1080
TGCTCTGAAA GATATTCAAA ACCAACTGAA GGAGAAGATC TTTGCCATTG AGGGTACGGA  1140
GACCACAAGC AGTAGCTCCT TCGAATTGGA GATGGCACAG GAGGGCTTCA GCGCTGTGTT  1200
```

Continued

HUMAN cDNA SEQUENCE—*Continued*

```
CACACCTGAT GGCCCCGTTC TGGGGGCTGT GGGGAGCTTC ACCTGGTCTG GAGGTGCCTT 1260
CCTGTACCCC CCAAATATGA GCCCTACCTT CATCAACATG TCTCAGGAGA ATGTGGACAT 1320
GAGGGACTCT TACCTGGGTT ACTCCACCGA GCTGGCCCTC TGGAAAGGGG TGCAGAGCCT 1380
TGCTCTGAAA GATATTCAAA ACCAACTGAA GGAGAAGATC TTTGCCATTG AGGGTACGGA 1140
GGTCCTGGGG GCCCCCCGCT ACCAGCACAC CGGGAAGGCT GTCATCTTCA CCCAGGTGTC 1440
CAGGCAATGG AGGATGAAGG CCGAAGTCAC GGGGACTCAG ATCGGCTCCT ACTTCGGGGC 1500
CTCCCTCTGC TCCGTGGACG TAGACAGCGA CGGCAGCACC GACCTGGTCC TCATCGGGGC 1560
CCCCCATTAC TACGAGCAGA CCCGAGGGGG CCAGGTGTCT GTGTGTCCCT TGCCCAGGGG 1620
GTGGAGAAGG TGGTGGTGTG ATGCTGTTCT CTACGGGGAG CAGGGCCACC CCTGGGGTCG 1680
CTTTGGGGCG GCTCTGACAG TGCTGGGGGA TGTGAATGGG GACAAGCTGA CAGACGTGGT 1740
CATCGGGGCC CCAGGAGAGG AGGAGAACCG GGGTGCTGTC TACCTGTTTC ACGGAGTCTT 1800
GGGACCCAGC ATCAGCCCCT CCCACAGCCA GCGGATCGCG GGCTCCCAGC TCTCCTCCAG 1860
GCTGCAGTAT TTTGGGCAGG CACTGAGCGG GGGTCAAGAC CTCACCCAGG ATGGACTGGT 1920
GGACCTGGCT GTGGGGGCCC GGGGCCAGGT GCTCCTGCTC AGGACCAGAC CTGTGCTCTG 1980
GGTGGGGGTG AGCATGCAGT TCATACCTGC CGAGATCCCC AGGTCTGCGT TTGAGTGTCG 2040
GGAGCAGGTG GTCTCTGAGC AGACCCTGGT ACAGTCCAAC ATCTGCCTTT ACATTGACAA 2100
ACGTTCTAAG AACCTGCTTG GGAGCCGTGA CCTCCAAAGC TCTGTGACCT TGGACCTGGC 2160
CCTCGACCCT GGCCGCCTGA GTCCCCGTGC CACCTTCCAG GAAACAAAGA ACCGGAGTCT 2220
GAGCCGAGTC CGAGTCCTCG GGCTGAAGGC ACACTGTGAA AACTTCAACC TGCTGCTCCC 2280
GAGCTGCGTG GAGGACTCTG TGACCCCCAT TACCTTGCGT CTGAACTTCA CGCTGGTGGG 2340
CAAGCCCCTC CTTGCCTTCA GAAACCTGCG GCCTATGCTG GCCGCCGATG CTCAGAGATA 2400
CTTCACGGCC TCCCTACCCT TTGAGAAGAA CTGTGGAGCC GACCATATCT GCCAGGACAA 2460
TCTCGGCATC TCCTTCAGCT TCCCAGGCTT GAAGTCCCTG CTGGTGGGGA GTAACCTGGA 2520
GCTGAACGCA GAAGTGATGG TGTGGAATGA CGGGGAAGAC TCCTACGGAA CCACCATCAC 2580
CTTCTCCCAC CCCGCAGGAC TGTCCTACCG CTACGTGGCA GAGGGCCAGA AACAAGGGCA 2640
GCTGCGTTCC CTGCACCTGA CATGTGACAG CGCCCCAGTT GGGAGCCAGG GCACCTGGAG 2700
CACCAGCTGC AGAATCAACC ACCTCATCTT CCGTGGCGGC GCCCAGATCA CCTTCTTGGC 2760
TACCTTTGAC GTCTCCCCCA AGGCTGTCCT GGGAGACCGG CTGCTTCTGA CAGCCAATGT 2820
GAGCAGTGAG AACAACACTC CCAGGACCAG CAAGACCACC TTCCAGCTGG AGCTCCCGGT 2880
GAAGTATGCT GTCTACACTG TGGTTAGCAG CCACGAACAA TTCACCAAAT ACCTCAACTT 2940
CTCAGAGTCT GAGGAGAAGG AAAGCCATGT GGCCATGCAC AGATACCAGG TCAATAACCT 3000
GGGACAGAGG GACCTGCCTG TCAGCATCAA CTTCTGGGTG CCTGTGGGAGC TGAACCAGGA 3060
GGCTGTGTGG ATGGATGTGG AGGTCTCCCA CCCCCAGAAC CCATCCCTTC GGTGCTCCTC 3120
AGAGAAAATC GCACCCCCAG CATCTGACTT CCTGGCGCAC ATTCAGAAGA ATCCCGTGCT 3180
GGACTGCTCC ATTGCTGGCT GCCTGCGGTT CCGCTGTGAC GTCCCCTCCT TCAGCGTCCA 3240
GGAGGAGCTG GATTTCACCC TGAAGGGCAA CCTCAGCTTT GGCTGGGTCC GCCAGATATT 3300
GCAGAAGAAG GTGTCGGTCG TGAGTGTGGC TGAAATTACG TTCGACACAT CCGTGTACTC 3360
CCAGCTTCCA GGACAGGAGG CATTTATGAG AGCTCAGACG ACAACGGTGC TGGAGAAGTA 3420
CAAGGTCCAC AACCCCACCC CCCTCATCGT AGGCAGCTCC ATTGGGGGTC TGTTGCTGCT 3480
GGCACTCATC ACAGCGGTAC TGTACAAAGT TGGCTTCTTC AAGCGTCAGT ACAAGGAAAT 3540
```

HUMAN cDNA SEQUENCE—*Continued*

```
GATGGAGGAG GCAAATGGAC AAAATTGCCCC AGAAAACGGG ACACAGACCC CCAGCCCGCC 3600
CAGTGAGAAA TGATCCCCTC TTTGCCTTGG ACTTCTTCTC CCCCGCGAGT TTTCCCCACT 3660
TACTTACCCT CACCTGTCAG GCCTGACGGG GAGGAACCAC TGCACCACCG AGAGAGGCTG 3720
GGATGGGCCT GCTTCCTGTC TTTGGGAGAA AACGTCTTGC TTGGGAAGGG GCCTTTGTCT 3780
TGTCAAGGTT CCAACTGGAA ACCCTTAGGA CAGGGTCCCT GCTGTGTTCC CCAAAGGACT 3840
TGACTTGCAA TTTCTACCTA GAAATACATG GACAATACCC CCAGGCCTCA GTCTCCCTTC 3900
TCCCATGAGG CACGAATGAT CTTTCTTTCC TTTCTTTTTT TTTTTTTTTC TTTTCTTTTT 3960
TTTTTTTTTG AGACGGAGTC TCGCTCTGTC ACCCAGGCTG GAGTGCAATG GCGTGATCTC 4020
GGCTCACTGC AACCTCCGCC TCCCGGGTTC AAGTAATTCT GCTGTCTCAG CCTCCTGAGT 4080
AGCTGGGACT ACAGGCACAC GCCACCTCGC CCGGCCCGAT CTTTCTAAAA TACAGTTCTG 4140
AATATGCTGC TCATCCCCAC CTGTCTTCAA CAGCTCCCCA TTACCCTCAG GACAATGTCT 4200
GAACTCTCCA GCTTCGCGTG AGAAGTCCCC TTCCATCCCA GAGGGTGGGC TTCAGGGCGC 4260
ACAGCATGAG AGGCTCTGTG CCCCCATCAC CCTCGTTTCC AGTGAATTAG TGTCATGTCA 4320
GCATCAGCTC AGGGCTTCAT CGTGGGGCTC TCAGTTCCGA TTTCCCAGGC TGAATTGGGA 4380
GTGAGATGCC TGCATGCTGG GTTCTGCACA GCTGGCCTCC CGCGTTGGGC AACATTGCTG 4440
GCTGGAAGGG AGGAGCGCCC TCTAGGGAGG GACATGGCCC CGGTGCGGCT GCAGCTCACC 4500
CAGCCCCAGG GGCAGAAGAG ACCCAACCAC TTCTATTTTT TGAGGCTATG AATATAGTAC 4560
CTGAAAAAAT GCCAAGACAT GATTATTTTT TTAAAAAGCG TACTTTAAAT GTTTGTGTTA 4620
CTGAAAAAAT GCCAAGACAT GATTATTTTT TTAAAAAGCG TACTTTAAAT GTTTGTGTTA 4620
ATAAATTAAA ACATGCACAA AAAGATGCAT CTACCGCTCT TGGGAAATAT GTCAAAGGTC 4680
TAAAAATAAA AAAGCCTTCT GTG                                      4703
```

The methionine start codon (<u>ATG</u>), termination codon (<u>TGA</u>) and polyadenylation signal are shown (<u>AATAAA</u>). The first five nucleotides of each exon are underlined.

Beta Subunit

See Chapter 41.

GENOMIC STRUCTURE

The alpha chain spans 27.86 kb with 30 exons (Figs 42.2 and 42.3). The CD18 gene spans 42.92 kb and has 17 exons (see Chapter 41).

CD11c - 27.86 kb

FIGURE 42.3   Genomic organisation of the α chain (CD11c) of CR4.

ACCESSION NUMBERS

Human alpha subunit: NM_000887, NP_000878.
Human beta subunit: See Chapter 41.
Mouse alpha subunit: NM_021334, NP_067309.
Mouse beta subunit: See Chapter 41.

DEFICIENCY

Similar to CR3, there have been no reports to date of human mutations in the CD11c (*ITGAX*) gene that lead to loss of expression. Mutations in CD18 (*ITGB2*) have been reported and results in complete or partial loss of function of all the four β_2 integrins, including CR4 (termed leucocyte adhesion disorder, or LAD I).[30] LAD I is characterised by increased granulocyte counts, impaired leucocyte firm adhesion and transendothelial migration and severe susceptibility to bacterial infections, and it is likely that loss of CR4 directly impacts one or more of these phenotypes.

POLYMORPHIC VARIANTS

A number of single nucleotide polymorphisms (SNPs) have been identified in the coding and noncoding regions of the CD11c/*ITGAX* gene. Two intronic SNPs in the CD11c/*ITGAX* gene, rs7190997 and rs11574637, have been implicated through genome-wide association studies (GWAS) in the pathogenesis of IgA nephropathy.[31,32] In separate GWAS, rs11574637 has also been shown to be genetically associated with systemic lupus erythematosus.[33]

MUTANT ANIMALS

Only a single line of CR4-deficient (CD11c null mutant) mice has been described to date.[16,34,35] These mice were generated by gene targeting using 129/Sv ES cell line. The replacement construct was designed to delete the coding region of exon 1 and all of exons 2 and 3. The mutation was then backcrossed onto the C57BL/6 inbred strain background. Similar to CR3-deficient mice, complete loss of CR4 did not lead to increased infectious susceptibility or other obvious phenotypic abnormalities in unmanipulated mice.[16,34,35] Published analyses of CD11c mutant mice suggest that CR4 plays an important role in the development of experimental autoimmune encephalomyelitis and atherosclerosis.[35,36] In contrast to these studies, loss of CR4 expression led to a more severe form of carditis and increased production of CCL2 from bone marrow-derived and cultured dendritic cells in a murine Lyme disease model.[37] In addition, CR4-deficient mice were shown to be more susceptible to candida, as well as in mediating host defence responses to *Streptococcus pneumoniae* infection.[16,38] Currently, there are no lines of CR4-deficient (CD11c mutant) mice commercially available.

REFERENCES

1. Xie C, Zhu J, Chen X, Mi L, Nishida N, Springer TA. Structure of an integrin with an alphaI domain, complement receptor type 4. *EMBO J* 2010;**29**:666–79.
2. Sen M, Yuki K, Springer TA. An internal ligand-bound, metastable state of a leukocyte integrin, alphaXbeta2. *J Cell Biol* 2013;**203**:629–42.
3. Uotila LM, Aatonen M, Gahmberg CG. Integrin CD11c/CD18 alpha-chain phosphorylation is functionally important. *J Biol Chem* 2013;**288**:33494–9.
4. Tan SM. The leucocyte beta2 (CD18) integrins: the structure, functional regulation and signalling properties. *Biosci Rep* 2012;**32**:241–69.
5. Postigo AA, Corbi AL, Sanchez-Madrid F, de Landazuri MO. Regulated expression and function of CD11c/CD18 integrin on human B lymphocytes. Relation between attachment to fibrinogen and triggering of proliferation through CD11c/CD18. *J Exp Med* 1991;**174**:1313–22.
6. Huleatt JW, Lefrancois L. Antigen-driven induction of CD11c on intestinal intraepithelial lymphocytes and CD8⁺ T cells in vivo. *J Immunol* 1995;**154**:5684–93.
7. Aranami T, Miyake S, Yamamura T. Differential expression of CD11c by peripheral blood NK cells reflects temporal activity of multiple sclerosis. *J Immunol* 2006;**177**:5659–67.
8. Myones BL, Dalzell JG, Hogg N, Ross GD. Neutrophil and monocyte cell surface p150,95 has iC3b-receptor (CR4) activity resembling CR3. *J Clin Invest* 1988;**82**:640–51.
9. Loike JD, Sodeik B, Cao L, Leucona S, Weitz JI, Detmers PA, et al. CD11c/CD18 on neutrophils recognizes a domain at the N terminus of the A alpha chain of fibrinogen. *Proc Natl Acad Sci USA* 1991;**88**:1044–8.
10. Frick C, Odermatt A, Zen K, Mandell KJ, Edens H, Portmann R, et al. Interaction of ICAM-1 with beta 2-integrin CD11c/CD18: characterization of a peptide ligand that mimics a putative binding site on domain D4 of ICAM-1. *Eur J Immunol* 2005;**35**:3610–21.
11. Sadhu C, Ting HJ, Lipsky B, Hensley K, Garcia-Martinez LF, Simon SI, et al. CD11c/CD18: novel ligands and a role in delayed-type hypersensitivity. *J Leukoc Biol* 2007;**81**:1395–403.
12. Menegazzi R, Busetto S, Decleva E, Cramer R, Dri P, Patriarca P. Triggering of chloride ion efflux from human neutrophils as a novel function of leukocyte beta 2 integrins: relationship with spreading and activation of the respiratory burst. *J Immunol* 1999;**162**:423–34.
13. Georgakopoulos T, Moss ST, Kanagasundaram V. Integrin CD11c contributes to monocyte adhesion with CD11b in a differential manner and requires Src family kinase activity. *Mol Immunol* 2008;**45**:3671–81.
14. Nishida N, Xie C, Shimaoka M, Cheng Y, Walz T, Springer TA. Activation of leukocyte beta2 integrins by conversion from bent to extended conformations. *Immunity* 2006;**25**:583–94.
15. Abram CL, Lowell CA. The ins and outs of leukocyte integrin signaling. *Annu Rev Immunol* 2009;**27**:339–62.
16. Ren B, McCrory MA, Pass C, Bullard DC, Ballantyne CM, Xu Y, et al. The virulence function of *Streptococcus pneumoniae* surface protein A involves inhibition of complement activation and impairment of complement receptor-mediated protection. *J Immunol* 2004;**173**:7506–12.
17. Keizer GD, Te Velde AA, Schwarting R, Figdor CG, De Vries JE. Role of p150,95 in adhesion, migration, chemotaxis and phagocytosis of human monocytes. *Eur J Immunol* 1987;**17**:1317–22.
18. Stacker SA, Springer TA. Leukocyte integrin P150,95 (CD11c/CD18) functions as an adhesion molecule binding to a counter-receptor on stimulated endothelium. *J Immunol* 1991;**146**:648–55.

19. Skoberne M, Somersan S, Almodovar W, Truong T, Petrova K, Henson PM, et al. The apoptotic-cell receptor CR3, but not alphavbeta5, is a regulator of human dendritic-cell immunostimulatory function. *Blood* 2006;**108**:947–55.

20. Dalgaard J, Beckstrom KJ, Jahnsen FL, Brinchmann JE. Differential capability for phagocytosis of apoptotic and necrotic leukemia cells by human peripheral blood dendritic cell subsets. *J Leukoc Biol* 2005;**77**:689–98.

21. Ruf A, Patscheke H. Platelet-induced neutrophil activation: platelet-expressed fibrinogen induces the oxidative burst in neutrophils by an interaction with CD11C/CD18. *Br J Haematol* 1995;**90**:791–6.

22. Yan SR, Berton G. Antibody-induced engagement of beta2 integrins in human neutrophils causes a rapid redistribution of cytoskeletal proteins, Src-family tyrosine kinases, and p72syk that precedes de novo actin polymerization. *J Leukoc Biol* 1998;**64**:401–8.

23. Lin Y, Roberts TJ, Sriram V, Cho S, Brutkiewicz RR. Myeloid marker expression on antiviral CD8[+] T cells following an acute virus infection. *Eur J Immunol* 2003;**33**:2736–43.

24. Gower RM, Wu H, Foster GA, Devaraj S, Jialal I, Ballantyne CM, et al. CD11c/CD18 expression is upregulated on blood monocytes during hypertriglyceridemia and enhances adhesion to vascular cell adhesion molecule-1. *Arterioscler Thromb Vasc Biol* 2011;**31**:160–6.

25. Lacal P, Pulido R, Sanchez-Madrid F, Cabanas C, Mollinedo F. Intracellular localization of a leukocyte adhesion glycoprotein family in the tertiary granules of human neutrophils. *Biochem Biophys Res Commun* 1988;**154**:641–7.

26. Hara M, Yokoyama H, Fukuyama K, Kitamura N, Shimokawa N, Maeda K, et al. Transcriptional regulation of the mouse CD11c promoter by AP-1 complex with JunD and Fra2 in dendritic cells. *Mol Immunol* 2013;**53**:295–301.

27. Noti JD. Sp3 mediates transcriptional activation of the leukocyte integrin genes CD11C and CD11B and cooperates with c-Jun to activate CD11C. *J Biol Chem* 1997;**272**:24038–45.

28. Lopez-Rodriguez C, Botella L, Corbi AL. CCAAT-enhancer-binding proteins (C/EBP) regulate the tissue specific activity of the CD11c integrin gene promoter through functional interactions with Sp1 proteins. *J Biol Chem* 1997;**272**:29120–6.

29. Jung S, Unutmaz D, Wong P, Sano G, De los Santos K, Sparwasser T, et al. In vivo depletion of CD11c(+) dendritic cells abrogates priming of CD8(+) T cells by exogenous cell-associated antigens. *Immunity* 2002;**17**:211–20.

30. Anderson DC, Springer TA. Leukocyte adhesion deficiency: an inherited defect in the Mac-1, LFA-1 and p150,95 glycoproteins. *Annu Rev Med* 1987;**38**:175–94.

31. Li M, Foo JN, Wang JQ, Low HQ, Tang XQ, Toh KY, et al. Identification of new susceptibility loci for IgA nephropathy in Han Chinese. *Nat Commun* 2015;**6**:7270.

32. Kiryluk K, Li Y, Scolari F, Sanna-Cherchi S, Choi M, Verbitsky M, et al. Discovery of new risk loci for IgA nephropathy implicates genes involved in immunity against intestinal pathogens. *Nat Genet* 2014;**46**:1187–96.

33. Hom G, Graham RR, Modrek B, Taylor KE, Ortmann W, Garnier S, et al. Association of systemic lupus erythematosus with C8orf13-BLK and ITGAM-ITGAX. *N Engl J Med* 2008;**358**:900–9.

34. Wu H, Rodgers JR, Perrard XY, Perrard JL, Prince JE, Abe Y, et al. Deficiency of CD11b or CD11d results in reduced staphylococcal enterotoxin-induced T cell response and T cell phenotypic changes. *J Immunol* 2004;**173**:297–306.

35. Wu H, Gower RM, Wang H, Perrard XY, Ma R, Bullard DC, et al. Functional role of CD11c[+] monocytes in atherogenesis associated with hypercholesterolemia. *Circulation* 2009;**119**:2708–17.

36. Bullard DC, Hu X, Adams JE, Schoeb TR, Barnum SR. p150/95 (CD11c/CD18) expression is required for the development of experimental autoimmune encephalomyelitis. *Am J Pathol* 2007;**170**:2001–8.

37. Guerau-de-Arellano M, Alroy J, Bullard D, Huber BT. Aggravated Lyme carditis in CD11a$^{-/-}$ and CD11c$^{-/-}$ mice. *Infect Immun* 2005;**73**:7637–43.

38. Jawhara S, Pluskota E, Verbovetskiy D, Skomorovska-Prokvolit O, Plow EF, Soloviev DA. Integrin alphaXbeta(2) is a leukocyte receptor for *Candida albicans* and is essential for protection against fungal infections. *J Immunol* 2012;**189**:2468–77.

Appendix A

Complement Nomenclature 2014[☆]

Claudia Kemper[1], Michael K. Pangburn[2], Zvi Fishelson[3]

[1]King's College London, London, United Kingdom; [2]The University of Texas Health Science Center at Tyler, Tyler, TX, United States; [3]Tel Aviv University, Tel Aviv, Israel

INTRODUCTION

In 2009 the International Complement Society (ICS)[1] and the European Complement Network (ECN)[1] began a joint re-evaluation of the nomenclature for the complement system. Effective communication of scientific results requires terminology that is as simple, as clear and as unambiguous as possible. The last review was published in 1981[1] and much has changed since then. Complement is now known to be initiated by three independent pathways - the classical, alternative and lectin pathways - and many new proteins and receptors have been discovered.

☆ XXV ICW Rio 2014.

1. This complement nomenclature list was initially constructed by the Complement Nomenclature Committee (appointed by the ICS Board in 2009) and reviewed and edited by the board members of the ICS and ECN, listed below. In total, 90 members of the ICS and the ECN have reviewed and accepted this list.

> *Complement Nomenclature Committee* (CNC): Claudia Kemper (Head), Michael Pang-burn (ICS President 2008–2010), Dennis Hourcade, David Isenman, SakariJokiranta, Dan Ricklin, Richard Smith, Andrea Tenner.
>
> *International Complement Society* (ICS) Board: Zvi Fishelson (President), Andrea Tenner, Paul Morgan, Claudia Kemper, Wenchao Song, Peter Garred, Michael Holers, Tom-Eirik Mollnes, Bo Nilsson, Matthew Pickering, Santiago Rodriguez de Cordoba, Denise Tambourgi, Trent M. Woodruff.
>
> *European Complement Network* (ECN) Board: Tom-Eirik Mollnes (President), Anna Blom, Matthew Pickering, Zvi Fishelson, Veronique Fremeaux-Bacchi, Peter Gal, Jörg Köhl, Seppo Meri, Marina Noris, Robert Rieben, Pilar Sanchez-Corral, Heribert Stoiber, Steffen Thiel, Leendert Trouw.

At a meeting in 1963 at the National Institute of Health[2] it was agreed that the components of complement were to be designated by the letter "C," the prime symbol, and numbers with activated products designated by the letter "a" added to the symbols (as in C'1a for activated C'1 and C'2a for activated C'2). At the time known components consisted of only nine proteins C'1, C'2, C'3, C'4, C'5, C'6 and C'7 with C'1 known to be a complex between C'1q, C'1r and C'1s. The letters q, r and s designated their order of elution from ion exchange chromatography on DEAE-cellulose. In 1968 complement nomenclature was revisited.[3,4] Among the many changes recommended, was the removal of the prime symbol (C3 instead of C'3) and the extension of the number of proteins to C9. In addition, in recognition of the importance of proteolytic fragmentation it was agreed to use the letters a, b, c, etc. to designate the series of activation fragments deriving from a native component (e.g., C3a and C3b generated from C3). Unfortunately this necessitated a new designation for activated components and it was agreed that a bar over the activated component name would be used to indicate that it was in an activated form ($\bar{C}1$ for C'1a and EAC14$\overline{23}$ for activated complexes on cells).

By 1981 an entirely new pathway had been characterized and new proteins and complexes had been described. A subcommittee of the Nomenclature Committee of the International Union of Immunological Societies formalized the nomenclature for this "alternative" pathway of complement activation.[1] Rather than continue numbering the components beyond C9 it was agreed that the components of this new pathway would be designated by letter symbols except for the C3 protein which was shared with the original activation pathway. The originally described pathway has since been referred to as the "classical" pathway of complement activation. Thus, constituents of the alternative pathway were designated factors B, D, H, I and P plus C3 and enzymatically active complexes were given designations such as C3b,Bb and C3b,Bb,C3b,P.

With the discovery of the lectin pathway and the characterization of many cellular receptors for complement activation fragments an entirely new set of components are now known to be involved in the activation of and the biological responses to complement. As is characteristic of any discovery process the nomenclature used for these new components is not uniform and is often redundant. Thus, the committee[1] was charged with conducting the first formal evaluation of complement nomenclature since 1981 and the first set of recommendations are presented in Table A.1. As with previous considerations of nomenclature the committee attempted to harmonize and simplify the language of complement while making minimal changes to long established conventions.

NOMENCLATURE FOR COMPLEMENT

Table A.1 presents the nomenclature for complement proposed by the Complement Nomenclature Committee[1] and the boards of directors of both the International Complement Society and the European Complement Network. A

TABLE A.1 Nomenclature of Complement Components (Old Names are Indicated in Parenthesis)

Recommended Name	Comments	Recommended Name	Comments
Pathways		**Proteins (cont.)**	
CP	Classical pathway	MBL	Mannose binding lectin
AP	Alternative pathway	Ficolin-1	(M-Ficolin)
LP	Lectin pathway	Ficolin-2	(L- Ficolin)
TP	Terminal pathway	Ficolin-3	(H-Ficolin)
Proteins		MASP-1	MBL-associated serine protease 1
C1	Complex of C1q, 2C1r, 2C1s	MASP-2	MBL-associated serine protease 2
C1q		MASP-3	MBL-associated serine protease 3
C1r		FHL-1	Factor H-like protein 1
C1s		FHR-1	Factor H-related protein 1
C1-INH	C1 inhibitor (C1 esterase inhibitor)	FHR-2	Factor H-related protein 2
C2		FHR-3	Factor H-related protein 3
C3		FHR-4	Factor H-related protein 4
$C3(H_2O)$	Thioester-hydrolyzed form of C3	FHR-5	Factor H-related protein 5
C3a	Anaphylatoxin from C3	CD59	(Protectin, homologous restriction factor)
C3a-desArg	C3a without C-terminal Arginine	Cn	Clusterin (Apolipoprotein J, SP-40,40)

Continued

TABLE A.1 Nomenclature of Complement Components (Old Names are Indicated in Parenthesis)—cont'd

Recommended Name	Comments	Recommended Name	Comments
C3b		**Protein complexes**	
iC3b	Inactivated C3b	C5b6	Terminal pathway complex of C5b + C6
C3dg		C5b-7	Terminal pathway complex of C5b6 + C7
C3d		C5b-8	Terminal pathway complex of C5b-7 +C8
C4		C5b-9	Terminal pathway complete complex
C4a		sC5b-9	Soluble C5b-9 with Vn or Cn bound
C4a-desArg	C4a without C-terminal Arginine	C3bBb	AP C3 convertase
C4b		C3bBbP	AP C3 convertase with properdin
C4d		C3bBbC3b	AP C3/C5 convertase
C4BP	C4b binding protein	C4BP-protein S	C4BP bound to protein S
C5		**Receptors**	
C5a	Anaphylatoxin from C5	CR1	CD35 (C3b/C4b Receptor)
C5a-desArg	C5a without C-terminal Arginine	CR2	CD21 (C3d receptor)
C5b		CR3	CD11b/CD18 complex
C6		CR4	CD11c/CD18 complex
C7		C3aR	Requesting CD number

TABLE A.1 Nomenclature of Complement Components (Old Names are Indicated in Parenthesis) — cont'd

Recommended Name	Comments	Recommended Name	Comments
C8		C5aR1	CD88 (C5aR)
C9		C5aR2	(C5L2) Requesting CD number
Vn	Vitronectin (S protein, S40)	CRIg	Complement receptor of the Ig family
FB	Factor B	C1qR	C1q receptor
FD	Factor D	gC1qR	Recognizes globular C1q domains
FH	Factor H	cC1qR	Calreticulin, Recognizes collagen domain
FI	Factor I	LHR	Long homologous repeat [in CR1]

consensus on a number of names has still not been reached and these will be the subject of continuing discussion.

Some of the issues resolved include using two letter names for the four pathways and using a hyphen in names for C1-INH, C3a-desArg, C4a-desArg, and C5a-desArg. In the CP the designation for C4b binding protein was decided to be C4BP. The form of C4BP that circulates in complex with Protein S is to be designated C4BP-Protein S. In the AP it was agreed that the factors shall be abbreviated with the capital letter "F" in front of the letter designating the component (FB, FD, FH, and FI). A consensus was not reached for the designation of properdin (where it was to be decided whether to retain the designation properdin or to change the name to factor P (FP)). The C3b-like form of C3 in which the thioester has been hydrolysed is designated $C3(H_2O)$ and the form of C3b inactivated by the action of FH and FI is designated iC3b. The nomenclature for AP complexes was shortened and simplified by removing the commas (C3bBb, C3bBbP, and C3bBbC3b). The family of small FH-like and FH-related proteins should be designated by capitalized letters, a hyphen and the number of the protein (FHL-1, FHR-1, FHR-2, FHR-3, FHR-4 and FHR-5).

In the LP the ficolins are to be referred to by their numerical designations (Ficolin-1, Ficolin-2 and Ficolin-3) and MBL is to be used for mannose binding lectin. It was decided that the proteolytic subunits associated with MBL, the ficolins and the other lectins of the pathway should be designated as MASP-1, MASP-2 and MASP-3.

Regarding the TP it was decided to harmonize the many designations for the intermediates by calling the activated complex between C5b and C6, C5b6 and to use a hyphen in the other complexes (C5b-7, C5b-8 and C5b-9). The soluble form of the complex C5b-9 which has been inhibited from entering a membrane by binding vitronectin (now abbreviated Vn) shall be designated sC5b-9. The lower case "s" signifies "soluble" and it is understood that the complex contains one or more Vn molecules. Another inhibitor of C5b-9 membrane insertion is clusterin which is now to be abbreviated Cn. The membrane-bound protein that inhibits the lytic action of C5b-9 shall be referred to only by its CD59 designation.

Since the last nomenclature review many receptors have been identified. The C3a receptor (C3aR) does not yet have a CD number. Two C5a receptors have been found and it has been decided to designate these C5aR1 (previously C5aR) and C5aR2 (previously C5L2). None of these receptors have CD numbers, but applications for these designations will be made. The complement receptor for C3b and iC3b with structural similarity to the immunoglobulin family is to be designated by CRIg. Again, this protein does not yet have a CD number. The family of C1q receptors should be referred to by their C1qR, gC1qR and cC1qR names (none have CD numbers yet). For the CR3 and CR4 complement receptors it is recommended that these simpler names be used, but that their respective CD designations (CD11b/CD18 and CD11c/CD18) be cited at least once in each publication. In the receptor CR1 there is a long homologous repeating domain structure which should be referred to as the LHR.

Due to the nature of scientific research the work on complement nomenclature must be ongoing. Indeed, there are numerous issues where consensus has not yet been reached. The lectin pathway is now known to include many MBL-like recognition proteins which have been given many different names depending on the origins of the discoveries. There is also no simple designation for the complexes between MBL and MASPs of the LP such as C1 in the CP nor is there any system for distinguishing zymogen complexes from activated complexes in the LP. The issue of the nomenclature for the activation fragments of C2 remains unresolved. Originally,[2] C2a indicated the activated C2 fragment in the C3/C5 convertase of the CP. Since this fragment turned out to be the larger fragment of C2 this nomenclature was no longer consistent with the convention of naming the smaller fragments with earlier letters. An attempt to reverse these "a" and "b" designations was not universally accepted by experts in the complement field leading to confusion in textbooks and the scientific literature. With the removal of the superscript "bar" above activated molecules our field lacks a definitive designation for distinguishing activated from native or zymogen complexes. On these and many other issues discussions will continue with scientists in the complement field.

CONCLUDING COMMENTS

While there are many issues remaining to be resolved the active complement committees[1] urge authors to use the recommended abbreviations given in Table A.1 and in the paragraphs above in their published and presented works. It is critical that reviewers and editors similarly monitor and enforce conformity in the field. This is in the best interest of the clear communication of science both among complement scientists and between this community and related fields of immunology. With the introduction of complement-targeted drugs and growing public awareness of complement it has become even more important for scientists in the complement field to keep their nomenclature as simple, uniform, and unambiguous as possible.

REFERENCES

1. Alper CA, Austen KF, Cooper NR, Fearon DT, Gigli I, Hadding U, Lachmann PJ, Lambert PH, Lepow IH, Mayer MM, Muller-Eberhard HJ, Nishioka K, Pondman K, Rosen FS, Stroud RM. Nomenclature of the alternative activating pathway of complement. *J Immunol* 1981;**127**:1261–2 (Reprinted from Bull. WHO 1981,59: 489).
2. Rapp HJ, Borsos T. Complement and hemolysis. *Science* 1963;**141**:738–40.
3. Austen KF, Becker EL, Biro CE, Borsos T, Dalmasso AP, Dias Da Silva W, Isliker H, Klein P, Lachmann PJ, Leon MA, Lepow IH, Mayer MM, MullerEberhard HJ, Nelson RA, Nilsson U, Nishioka I, Rapp HJ, Rosen FS, Trnka Z, Ward PA, Wardlaw AC. Nomenclature of complement. *Bull World Health Organ* 1968;**39**:935–8 (Immunochemistry 1970, 7:137-142).
4. Karlson P, Bielka H, Liebecq C, Sharon N, Velick SF, Vliegenthart JFG, Dixon HBF, Webb EC. Enzyme nomenclature. Recommendations 1978, Suppl. 2, corrections and addition. *Eur J Biochem* 1981;**116**:423–35.

Index

H

Printed in the United States
By Bookmasters